BROADBAND TELECOMMUNICATIONS HANDBOOK

MCGRAW-HILL TELECOMMUNICATIONS

Bass	Fiber Optics Handbook
Bates	Broadband Telecommunications Handbook
Bates	GPRS
Bates	Optical Switching and Networking Handbook
Bates	Wireless Broadband Handbook
Bedell	Wireless Crash Course
Benner	Fibre Channel for SANs
Camarillo	SIP Demystified
Chernock	Data Broadcasting
Clayton	McGraw-Hill Illustrated Telecom Dictionary 3/e
Collins	Carrier Class Voice Over IP
Faigen	Wireless Data for the Enterprise
Guthery	Mobile Application Development
Harte	3G Wireless Demystified
Harte	Delivering xDSL
Held	Deploying Optical Networking Components
Kobb	Wireless Spectrum Finder
Lee	Mobile Cellular Telecommunications 2/e
Lee	Lee's Essentials of Wireless
Louis	Broadband Crash Course
Louis	M-Commerce Crash Course
Louis	Telecommunications Internetworking
Muller	Bluetooth Demystified
Muller	Desktop Encyclopedia of Telecommunications 3/e
Nichols	Wireless Security
Pecar	Telecommunications Factbook 2/e
Radcom	Telecom Protocol Finder
Richard	Service Discovery: Protocols and Programming
Roddy	Satellite Communications 3/e
Rohde/Whitaker	Communications Receivers 3/e
Russell	Signaling System #7 3/e
Russell	Telecommunications Protocols 2/e
Russell	Telecommunications Pocket Reference
Sayre	Complete Wireless Design
Shepard	Telecommunications Convergence
Shepard	Telecom Crash Course
Shepard	Optical Networking Demystified
Shepard	SONET/SDH Demystified
Simon	Spread Spectrum Communications Handbook
Smith	Cellular System Design and Optimization
Smith	Wireless Telecom FAQs
Snyder	Wireless Telecommunications Networking with ANSI-41 2/e
Sulkin	Implementing the IP-PBX
Vacca	I-Mode Crash Course
Wetteroth	OSI Reference Model for Telecommunications
Whitaker	Interactive Television Demystified

Broadband Telecommunications Handbook

Regis J. (Bud) Bates

Second Edition

McGraw-Hill
New York Chicago San Francisco Lisbon
London Madrid Mexico City Milan New Delhi
San Juan Seoul Singapore Sydney Toronto

Library of Congress Cataloging-in-Publication Data

Bates, Regis J.
Broadband telecommunications handbook / Regis J. "Bud" Bates.—2nd ed.
p. cm.—(McGraw-Hill telecommunications)
ISBN 0-07-139851-1 (alk. paper)
1. Broadband communication systems—handbooks, manuals, etc.
2. Telecommunication systems—Handbooks, manuals, etc. I. Title II. Series.
TK5103.4.B38 2002
384—dc21 2002021281

McGraw-Hill

A Division of The ***McGraw-Hill*** *Companies*

1 2 3 4 5 6 7 8 9 0 DOC/DOC 9 8 7 6 5 4 3 2

ISBN 0-07-139851-1

The sponsoring editor for this book was Steve Chapman and the production supervisor was Pamela Pelton. It was set in Century Schoolbook by MacAllister Publishing Services, LLC.

Printed and bound by R. R. Donnelley and Sons.

This book is printed on recycled, acid-free paper containing a minimum of 50 percent recycled de-inked fiber.

ABOUT THE AUTHOR

Regis J. (Bud) Bates Jr.
President
TC International Consulting, Inc.
PO Box 51108
Phoenix, AZ 85076-1108
Tel. (800) 322-2202
Fax (800) 260-6440
http://www.tcic.com

Mr. Bates has more than 36 years of experience in telecommunications and information systems. He oversees the overall operation of TC International Consulting, Inc. (TCIC) of Phoenix, Arizona. TCIC is a full-service management consulting organization that specializes in designing and integrating information technologies. TC International Consulting leads the pack in strategic development and implementation of new technologies for carriers and corporations alike.

Bud's experience served in major network designs from Local Area Networks (LANs) to Wide Area Networks (WANs) using high-quality, all-digital transmission services: T1, T3, and SONET/SDH. His studies and recommendations resulted in significant financial savings. One project included the design and implementation of a Frame Relay network that spanned over 14 countries and 80 locations. This project resulted in huge monthly savings while preserving subsecond response times across the network.

His articles have been published in *Network World*, *Information Week*, *International Journal of Information Management*, and others. He has authored numerous books published by McGraw-Hill and Artech House. His recent published books *Voice and Data Communications, Fourth Edition, GPRS*, and *Optical Switching and Networking* continue to fall on McGraw-Hill's best-seller list. Bud also develops and conducts various public seminars throughout the world, ranging from a managerial overview to very technical instruction on voice, data, and LAN communications. He spends much of his time working with the major telecommunications manufacturers in training their staff members on the innovations of technology and the convergence of voice and data networks for the future. Many of his materials are used throughout the higher education institutions in certification and graduate-level classes in telecommunications management.

Mr. Bates holds a degree in Business Management from Stonehill College, Easton, MA. He has completed graduate-level courses at Lehigh University and Saint Joseph's University, specifically in Financial Management and Advanced Mathematics.

MEMBER

NATIONAL
SPEAKERS
ASSOCIATION

CONTENTS

ACKNOWLEDGMENTS

I would like to take the opportunity to recognize several people who had a considerable influence on my ability to complete this project. One cannot produce a book or write a manuscript in a vacuum. Therefore, without the people who aided me, this book might not exist.

First, I have to readily acknowledge and thank all the folks at McGraw-Hill for their continued support of this author and their exceptional patience. This holds especially true for my Senior Editor Steve Chapman. Steve has become a friend and editor all rolled up in one. He knows when to push and when to back off when following up on a manuscript. Somewhere is an unwritten rule that an author is supposed to have unlimited time available and unmitigated commitment to completing the book early. Well, in my case, it is not true! Too many challenges and changes crept into our lives and postponed the inevitable completion of this project. As the radical changes and slowdowns in the industry cause major changes in the providers, the protocols, and the acceptance of any specific product, we had to juggle all the schedules to try to get to a completion of this second edition. I put the McGraw-Hill people through the paces, promising to get the manuscript to them and missing just about every date.

I thank Steve Chapman for his patience and his periodic prods to remind me to stick with it. I also appreciate the efforts of all the folks I never saw or talked with who remain in the background. These unsung heroes of the production department never get their credit, but we all should be grateful to them for their dedication and stick-to-it attitudes.

Beyond the folks at McGraw-Hill, there is a special person who has held the entire project together many times now. Her ability to keep after me to complete the project without creating a lot of friction was outstanding. Gabriele Bates has been the anchor in all these books, keeping track of what is done, what is in motion, and what needs attention. Her dedication to the overall success of my books never gets the credit she truly deserves. Gabriele provides the gentle push I need from time to time, keeping me focused and working at it. Even when she knows I am behind, there is no panic—just constant reinforcement and encouragement.

Several vendors and friends were supportive and helpful in garnering information for the development of this manuscript. I thank all of them, who are too numerous to mention each of them individually. However, they know who they are and can take silent comfort in knowing they got us here.

This book is also dedicated to you who buy the books we develop for your understanding. It is you, the reader of this material, who should also be praised for the demand for more information. In many cases, the ideas of broadband communications are still emerging for some of the areas discussed herein. However, we hope we were able to capture the spirit and the letter of the concept even before it truly develops. Enjoy this book as you would a version of a 201 series after the *Voice and Data Communications Handbook, Fourth Edition.* Convergence is the name of our industry today, yet we must continue to seek new ways of providing the information and using the technology. As long as you, the reader, continue to demand high-speed services, reliability, and mobility, I will have a job. That job will be to seek the ways of describing and applying the technologies so that you can use them.

I personally appreciate talking with readers who have bought a book and call (or e-mail) me with a question. As long as I can continue to get your feedback, I will continue to try to explain things in a way that hopefully makes sense. I thoroughly enjoy it when a reader calls (or e-mails) to tell me that he or she understood the materials better having read the book. Moreover, I hope that I can continue to offer one-on-one explanations to those of you who have a difficult time understanding a point I make in this book. Once again, I appreciate your support!

Good luck and happy reading!

CHAPTER 1

Introduction to Telecommunications Concepts

Welcome to the world of Broadband Telecommunications again in this second edition! In this book, we attempt to deliver a series of different approaches to the use and application of telecommunications' principles, concepts, and guidelines and offer new approaches to the use of voice and data communications.

Last year, I wrote *The Voice and Data Communications Handbook, Fourth Edition*, as a means of introducing several new ways of looking at the telecommunications industry. The Voice and Data Handbook is so successful that it begs for a sequel with a more in-depth approach to the more technical aspect of the use of telecommunications. Therefore, my goal is to delve into the topics of broadband communications. For those who have not read other books on this topic, I will attempt to simplify the concepts discussed. For those who had a chance to read the first book (or others on this topic), I will attempt to pick up where we left off during the first volume. This book is structured by groupings of topics. For example, the first few chapters work with the convergence of voice and data networks as we see the virtual private networks, intelligent networks, and the portability of our systems for today and the future. Using a combined wired and wireless networking approach, we shall take one component at a time to determine what it is, what it does, and what it typically costs (not so much in actual cost as in opportunity costs).

After the first grouping of chapters, we step into a discussion of signaling systems that make wonderful things happen in the convergence world—coupled with that discussion is the idea of computer and telephony integration. (What better way to describe convergence!) We also look at the concept of *Integrated Services Digital Network* (ISDN), which is not as popular in the North American countries as in many international markets. However, there is still a need to understand what it is and how it works.

After a few ideas have sunk in, we move on to a higher-speed data networking strategy, with the use of Frame Relay. After Frame Relay, we discuss the use of *Asynchronous Transfer Mode* (ATM) for its merits and benefits. Next, we take the convergence a step farther and delve into the Frame and ATM internetworking applications—still a great way to carry our voice and data no matter how we slice and dice it. We will also look at the IP-enabled Frame Relay services and Frame over xDSL.

Just when we thought it was safe to use these high-speed services across the *Wide Area Network* (WAN), we realized that local access is a problem. Entering into the discussion is the high-speed convergence in the local loop arena with the use of CATV and cable modems to access the Internet at *Local Area Network* (LAN) speeds. Mix in a little xDSL, and we start the fires burning on the local wires. The use of copper wires or cable TV is the hot issue in data access.

From the discussion of the local loop, we then see the comparisons of a wireless local loop with *Local Multipoint Distribution Service* (LMDS) and *Multichannel Multipoint Distribution Service* (MMDS). These techniques are all based on a form of Microwave, so the comparison of microwave radio techniques is shown.

Wireless portability is another hot area in the marketplace. Therefore, we compare and contrast the use of the *Global System for Mobile Communications* (GSM), cellular, and personal communications' services and capacities. Convergence is only as good as one's ability to place the voice and data on the same links. We will look at the choices available in the market for *Time Division Multiple Access* (TDMA) and *Code Division Multiple Access* (CDMA) options at the radio level.

We then look at the wireless data operation such as wireless IP and the "always on" services of the Internet at the handset level using *General Packet Radio Services* (GPRS). We will even dip into the future to see where the 3G wireless applications are developing and where they may have a use for the future of our communications architectures. This will include the Wideband CDMA approaches for the future and the *Universal Mobile Telecommunication System* (UMTS) application.

Leaving the low-end wireless services behind, we then enter into a discussion of the sky wave and satellite transmission for voice and data. No satellite transmission discussion would be worth anything without paying homage to the *Transmission Control Protocol* (TCP) and *Internet Protocol* (IP) on the satellite networks. Yet, the satellite services are now facing direct competition where the *low-Earth-orbit* (LEO) satellite strategies are becoming ever popular. The use of Teledesic, Iridium, or Globalstar systems is merely a transport system. These pull the pieces together and will offer voice and data transmission for years to come.

One could not go too far with the wireless-only world, so we back up and begin to contrast the use of the wired world again. This time, we look at T1, T2, and T3 on copper or coax cable, which is a journey down memory lane for some. We also contrast the international market opportunities with E1, E2, and E3.

However, by adding a little fiber to the diet, we provide these digital architectures on *Synchronous Optical Network* (SONET) or *Synchronous Digital Hierarchy* (SDH) services. SONET makes the T1 and T3 look like fun! Topics include the ability to carry Frame Relay and ATM as the networks are now beginning to meld together. SONET is good, but if we use an older form of multiplexing (wavelength), we can get more yet from the fibers. So, we look at the benefits of *dense wavelength division multiplexing* (DWDM) on the fiber to carry more SONET and more data. SDH is compared to the SONET architecture to see what the main differences are between the two services.

With the infrastructure kicked around, the logical step is to complete this tour of the telecommunications arena with the introduction of the Internet, intranets, and extranets. Wow, this stuff really does come together! Using the Internet or the other two forms of nets, we can then carry our data transparently. What would convergence be without the voice? Therefore, the next step is to look at the use of *Voice over Internet Protocol* (VoIP). A good deal of activity has been placed on the development of *Multiprotocol Label Switching* (MPLS), so we have to analyze what and where the application of the multiprotocol label service fits in the overall networking structure. Lastly, we have to come up with a management system to control all the pieces that we have grouped and bonded together. This is in the form of a *Simple Network Management Protocol* (SNMP) as the network management tool of choice. If all the converged pieces work, there is no issue. However, with all the variants discussed in this book, we must believe that Murphy is alive and well! Thus, all the pieces are blended together by groups, to form a homogenous network of internets.

Basic Telecommunications Systems

When the *Federal Communications Commission* (FCC) began removing regulatory barriers for the long distance and *customer premises equipment* (CPE) markets, its goal was to increase competition through the number of suppliers in these markets. Recently, consumers have begun to enjoy lower prices and new bundled service offerings. The local and long distance markets are examples of the new direction taken by the FCC in the 1980s to eliminate and mitigate the traditional telephone monopoly into a set of competitive markets. Although these two components of the monopoly have been stripped away, barriers still exist at the local access network—the portion of the public network that extends between the *interexchange carrier* (IEC) network and the end user. The local loop and the basic telecommunications infrastructure are not as readily available as one would like to think.

The growth of private network alternatives improves with facilities-based competition in the transport of communications services. The industry realizes that more than 500 *competitive local exchange carriers* (CLECs) have grown out of the deregulation of the monopolies. These CLECs include cable television networks, wireless telephone networks, LANs, and metropolitan area networks. *Incumbent local exchange carriers* (ILECs) indicate that their networks are continually evolving into a multimedia platform capable of delivering a rich variety of text, imaging, and messaging services as a direct response to the competition. Many suggest that their networks are wide open, for all competitors. Imagine an open network—a network with well-defined interfaces accessible to all—allowing an unlimited number of entrants a means to offer competitive services limited only by their imagination and the capabilities of the local loop network facilities.

If natural monopolies are still in the local exchange network, open access to these network resources must be fostered to promote a competitive market in spite of the monopolistic nature of the ILECs. The FCC continues to wrestle with how far it has to go and what requirements are necessary for open and equal access to the network.

Network unbundling, the process of breaking the network into separate functional elements, opens the local access to competition. The CLECs that managed to survive the great fallout of 2001, select the unbundled components they need to provide their own service. If the unbundled price is still too expensive, the service provider will build its own private resources. This is the facilities-based provider. All too often, we hear about new suppliers who offer high-speed services, better than the incumbent. Yet, these suppliers are typically using the Bell System's wires to get to the consumer's door. The only change that occurs is the person to whom we send the bill, hardly a competitive local networking strategy. As a result, the new providers (CATV, wireless local loop, IEC, and facilities-based CLEC) are now in the mode to provide their own facilities.

Components of the Telecommunications Networks

Telecommunications network components fall into logical or physical elements. A logical element is a *Software-Defined Network* (SDN) or voice *Virtual Private Network* (VPN) feature or capability. This SDN or VPN feature can be as simple as the number translations performed in a switch to establish a call. Switching systems have evolved into the use of external signaling systems to set up and tear down the call. These external physical and logical components formulate the basis of a network element. Moreover,

Intelligent Networks (and Advanced Intelligent Networks) have surpassed the wildest expectations of the service provider. These logical extensions of the network bear higher revenue while opening the network to a myriad of new services. Number portability can also be categorized with the logical elements because the number switching and logic are no longer bound to a specific system. A physical element is the actual switching element, such as the link or the matrices used internally. A network is made up of a unique sequence of logical elements implemented by physical elements.

Given the local exchange network and local transport markets, open mandates had to be considered because the LEC has the power to stall competition. In many documented cases the LECs have purposefully dragged their feet to stall the competition and to discredit the new provider in the eyes of the customer. This is a matter of survival of the fittest. The ILECs have the edge over the network components because their networks were built over the past 120 years. This is the basis for the deregulatory efforts in the networks because the LECs are fighting to survive the onslaught of new providers who are in the cream-skimming mode. If access mandates are necessary, to what degree? These and other issues are driving the technological innovation, competition at the local loop, and the development of higher-capacity services in a very competitive manner.

Communications Network Architectures

In any communications network there is architecture planned to make the interconnection work and to add the necessary features and functions. The *Public Switched Telephone Network* (PSTN) evolved using a five-level hierarchy to switch calls across the country or the globe. However, as with any network architecture, there are rules for how the network adapted to the user need. Later in the evolution of the network, we saw the use of a *dynamic nonhierarchical routing protocol* (DNHR) that was instituted to reduce the rigidity of the network protocols to something more on a peer-to-peer arrangement. The DNHR protocols and implementations were transparent to the user, but the operator certainly had to manage the operation and maintenance of the PSTN. The operators did gain a sufficient amount of flexibility using the newer architectural models.

Conversely, in a data model, we saw several protocol stacks that emerged as proprietary architectures consistent with the computing manufacturers. The data network architecture had as many flavors as many ice cream companies. We saw the emergence of communications architecture that satisfied specific vendor products (like IBM's *Systems Network Architecture* [SNA], or DEC's *Digital Network Architecture* [DNA], and so on). These models and architectures used a hierarchy that added some value in the connection and transmission of data between and among computing systems. Openness was a bad word in the data communications industry. Yet, users all screamed for some form of standardization to solve the incompatibility problems at the time.

To solve the problem, we saw the emergence of the *Open Systems Interconnect* (OSI) model that if implemented, would create open communications architecture. Unfortunately this is too expensive and offers little *return on investment* (ROI) to the manufacturing community. An alternative to the open architecture was an open de facto standard such as the TCP/IP architecture. This was one that met with optimism in the early stages of the networking development, and then with pessimism because the openness was too much for many managers to handle. More operations are geared toward a structure rather than a fluid opportunity. Finally, as the old saying goes, what goes around comes around—the TCP/IP model has become one of the most widely implemented standards; albeit a de facto standard, in the world.

Ultimately we have seen the role of packet-switching-based network architecture emerge to be the choice of many providers and users alike. The packet-based technological model assumes that all data traffic is the same and can be dealt with equally. As a data model works, this is fine. However, the emergence of this packet-based architecture changes when we add real-time applications such as voice, video, and audio needs. These applications demand that certain precedence is placed on the real-time application and a lower priority model is applied for strictly a data application. Enter the discussions of *quality of service* (QoS) and the demands for flexibility in handling the data and voice applications on the same links. Through newer technological models we see the overall structure of a layer 2 circuit switching architecture underlying layer 3 packet switching protocols in the form of MPLS. This is all very confusing to the average human, and gets the architecture wizards excited at the same time.

The Local Loop

So much attention has been parlayed on the local loop. Nevertheless, is it a realistic expectation to use the network facilities for future high-speed services? Would the newer providers, such as the CATV companies, have an edge over the ILECs? These issues are the foundation of the network of the new millennium. The new providers will use whatever technology is available to attack the competition, including

- CATV
- Fiber-based architectures (*fiber to the curb* [FTTC], *fiber to the home* [FTTH], *Hybrid Fiber Coax* [HFC])
- Wireless microwave systems
- Wireless third-generation cellular systems
- Infrared and laser-based wireless architectures
- Satellite and *Digital Signature Standard* (DSS) type services

Regardless of the technology used, the demand never seems to be satisfied. Therefore, the field of competitors will continue to metamorphose as the demand dictates and as the revenues continue to attract new business.

The Movement Toward Fiberoptic Networks

A transmission link transports information from one location to another in a usable and understandable format. The three functional attributes of this link are

- Capacity
- Condition
- Quality of Service

The deregulation of the local exchange networks has led to significant improvements in the following criteria:

- Access to network capacity
- Access to intermediate points along the transmission path

The transmission path may include pieces of the existing copper or newer fiber-based network architectures. The current copper-based loop limits opportunities.

- The transmission distances associated with the subscriber loop limit the amount of bandwidth available over twisted wire pair roughly to the DS1 rate of 1.5 Mbps. As broadband services become increasingly popular, the copper network severely constrains the broadband services.

- The current switched-star architecture runs at least one dedicated twisted pair from the central office to each customer's door without any intermediate locations available to unbundle the transport segment. This precludes a lot of the innovation desired by the end user.

Although the current copper-based network is unattractive when unbundling the physical transmission components, fiber-based networks offer many more opportunities. Telephone companies can improve the local access network by deploying fiber in the future. The central office, nodes at remote sites, and the curbside pedestal can all be improved with fiber-based architectures. These nodes serve as flexibility points where signals can be switched or multiplexed to the appropriate destination. A small percentage of lines are served by *digital loop carrier* (DLC) systems that incorporate a second flexibility point into the architecture at the remote node. The third flexibility point at the pedestal has been proposed for FTTC systems in the future.

The bandwidth limitations of a fiber system are not due to the intrinsic properties of the fiber, but the limitations of the switching, multiplexing, and transmission equipment connected to the fiber. This opens the world up for a myriad of new service offerings when fiber makes it to the consumer's door. Third parties like Qwest Communications, Global Crossing,[1] and Level 3 are becoming the carrier's carrier. They will install the fiber to the pedestal, the door, or to the backbone and sell the capacity to the Enterprise (end user), the ILEC, or the CLEC. This produces many attractive alternatives to the broadband networks for the future. No longer will bandwidth be the constraining factor; the application or the computer will be the bottleneck.

Because of the tremendous bandwidth available with fiberoptic cable and the technological improvements in SONET and DWDM, virtually unlimited bandwidth will be available. This statement of course is contingent on the following caveats:

- The overabundance of bandwidth is not likely to appear for some time.
- This bandwidth is available only over the fiber links. Yet, installation of new technology is a slow process. Fiber will be deployed in hybrid network architectures, which continue to utilize existing portions of the copper network.

---

[1]Global Crossing may not be as viable a player in this market. Global Crossing filed for Chapter 11 protection in February 2002. The outcome is anyone's guess right now.

Several times during 2000 and 2001, published reports were released decrying the overabundance of bandwidth in the local and long haul networks. The reports espoused that there is more fiber in the ground than we can ever use, and that the overabundance (estimates are that only 10–20 percent of the fiber is actually lit in use) will drive the prices down to unbelievable deals for the end user and to the chagrin of the carriers. Unfortunately these reports are both correct and wrong at the same time. True, there is a lot of fiber in the ground and much of it is dark. Unfortunately, many people ignore the fact that the fiber is old technology (having been displaced by the newer forms of glass and electronics) and therefore it is not economical to attempt using it. This means that much of the glut that is being discussed really doesn't exist; it means that it is too expensive to remove the glass in its current condition.

Also true is the fact that the emerging data networking standards and demands cooled off during 2000 and 2001 while the bottom fell out of the telecommunications market as well as the Internet suppliers. What everyone fails to see is that this was a cyclic correction of the market and that the true bandwidth demands for real-time packet-switched networks, real-time voice applications, and high-speed multimedia applications are all still developing. The next set of explosive demand will start rolling again when we see the true value of the real-time QoS-oriented and multimedia demands of our networks. Moreover, when the Internet finally starts carrying the time-sensitive data demands of the mission-critical services in an enterprise, the demand for faster, better, and cheaper will roll again.

Consequently, until fiber is deployed all the way to the customer premises, portions of the network will continue to present the same speed and throughput limitations hindering the rollout of the true time-sensitive applications. A caveat here is that the vendors will continue to develop "band-aid" approaches to using copper and coax services until and when fiber reaches the door. These band-aid approaches help to keep the network one step ahead of the demand curve, but they will ultimately become the bottleneck that will force the changeover from copper-based architecture to fiber and broadband wireless solutions to the door.

Digital Transfer Systems

The switching and multiplexing techniques characteristic of the transmission systems within the network are all digital. Currently, the network employs a *Synchronous Transfer Mode* (STM) technique for switching and multiplexing these digital signals. The broadband networks of the future will continue to utilize a synchronous transmission hierarchy using the SONET standards defined by the *International Telecommunications Union* (ITU). SONET describes a family of broadband digital transport signals operating in 50 Mbps increments. As a result, wherever SONET equipment is used, the standard interfaces at the central office, remote nodes, or subscriber premises will be multiples of these rates.

Above the physical layer, however, changes are now underway that move away from the synchronous communications modes. The ATM is the preferred method of transporting at the data link layer. ATM uses the best of packet-switching and routing techniques to carry information signals, regardless of the desired bandwidth, over one high-speed switching fabric. Using fixed-length cells, the information is processed at higher speeds, reducing some of the original latency in the network. These cells then combine with the cells of other signals across a single high-speed channel like a SONET OC-48. In *time-division multiplexing* (TDM), timing is crucial. In ATM, *statistical time-division multiplexing* (STDM) timing is used, so the timing is less crucial at the data link layer. The cells fit into the payload of the SONET frame structure for transmission where the timing is again used by the physical layer devices. ATM will use a combined switching and multiplexing service at the cell level. Continued use of SONET multiplexers will combine and separate SONET signals carrying ATM cells.

What distinguishes ATM from a synchronous approach is that subscribers have the ability to customize their use of the bandwidth without being constrained to the channel data rates.

When the intelligent networks are fully implemented, the logical network components will be separated from the physical switching element—where the physical component of a current digital switch consists of 64 Kbps (DS0) access to the network switch.

ATM should improve the capability to separate the physical switching elements of the network. The key attribute of the ATM switch, which could facilitate more modularity, is the bandwidth flexibility. Because each information signal is segmented into cells, switching is performed in much smaller increments. Current digital switching elements switch a DS0 signal whether the full bandwidth is needed or not. With ATM, the switching element resources can be much more efficiently matched to the bandwidth requirements of the user. Access to the ATM switch will be specified according to the maximum data rate forecasted for the particular access arrangement, instead of specifying the number of DS0 circuits required, as is the case today with digital switches.

The Intelligent Networks of Tomorrow

The ILECs developed the *Advanced Intelligent Network* (AIN) to provide new services or to customize current services based on the user demand. The central office switches contain the necessary software to facilitate these enhanced features. The manufacturers of the systems have fully embodied their application software with the operating system's software within the switch to create a simple interface for the carriers. When new features are added, the integrated software must be fully tested by the switch manufacturer.

The limitations of a centralized architecture caused the vendors and manufacturers concern. Now, as intelligent services are deployed, the movement is to a distributed architecture and intelligent peripheral devices on the network. The LECs use a network architecture, which enables efficient and rapid network deployment.

The single most important feature of AIN is its flexibility to configure the network according to the characteristics of the service. The modular architecture allows the addition of adjunct processors, such as voice processing equipment, data communication gateways, video services, and directory look-up features to the network without major modifications. These peripheral devices (servers) provide local customer database information and act like the intelligent centralized architectures of old.

The basic architecture of the AIN takes these application functions and breaks them into a collection of functionally specific components. Ultimately, AIN allows modifications to application software without having to alter the operating system of the switch.

Summary

The telecommunications systems include the variations of the local loop and the changes taking place within that first (or last) mile. As the migration moves away from the local copper-based cable plant (a slow evolution for sure), the movement will be to other forms of communications subsystems to include the use of

- Fiber optics
- Coax cable

- Radio-based systems
- Light-based systems
- Hybrids of the preceding

These changes will take users and carriers alike into the new millennium. Using the CATV modem technologies on coax, the fiber-based SONET architectures in the backbone (and ultimately in the local loop), and copper wires in the xDSL technologies all combine to bring higher-speed access.

After access is accomplished, the use of the SONET-based protocols and multiplexing systems creates an environment for the orchestration of newer services and features that will be bandwidth intensive. The SONET systems will be used to step up to the challenges of the new millenium. ATM will add a new dimension to the access methods and the transport of the broadband information through the use of STDM and cell-based transmission. No longer will the network suppliers have to commit specific fixed bandwidth to an application that only rarely uses the service. Instead, the services will merely use the cells as necessary to perform the functionality needed.

Wireless local loop services are relatively new in the broadband arena, but will play a significant role in the future. The untethered ability to access the network no matter where you are will be attractive to a large new population of users. Access to low-speed voice and data services are achievable today. However, the demand for real-time voice, data, video, and multimedia applications from a portable device is what the new generation of networks must accommodate. The broadband convergence will set the stage for all future development.

Today speeds are set in the kilobits to megabits per-second range. The broadband networks of the future will have to deal with demands for multi-megabit up to multi-gigabit per-second speeds.

Through each interface, the carriers must be able to preserve as much of their infrastructure as possible so that forklift technological changes are not forced upon them. The business case for the evolution of the broadband convergence is one that mimics a classical business model. Using a 7–15 year return-on-investment model, the carriers must see the benefit of profitability before they install the architectural changes demanded today.

CHAPTER 2

Telecommunications Systems

Before going into the overall technologies of this book, now is a good time to review the goal of the book. First, we plan to discuss technologies that are based on the current world of voice and data convergence. This convergence is one that has been sliding along for two decades, yet seems to have caught everyone by surprise. Second, we will be talking about applications and some cost issues throughout the book. Regardless of which discipline you come from, you cannot escape the ultimate strategy management expects: increase productivity yet hold the line on costs. Lose either one of these in the equation and you will be sitting there trying to figure out why management never buys into any of your great ideas. The answer comes to us in the form of packaging. No matter how great your ideas are, if you cannot sell them, you cannot implement them.

So as we proceed through this material, try not to get frustrated with the constant mix of services, technological discussions, and costing issues. From time to time, we may also introduce some extra technical notes that are for the more technically astute but can be ignored by the novice trying to progress through the industry. As you read about a topic, do so with a focus on systems, rather than individual technologies. We have tried to make these somewhat stand-alone chapters, yet we have also tied them together in bundles of three or four chapters to formulate a final telecommunications system. Do what you must to understand the information, but do not force it as you read. The pieces will all come together throughout the grouping of topics.

What Constitutes a Telecommunications System

A network is a series of interconnections that form a cohesive and ubiquitous connectivity arrangement when all tied together. That sounds rather vague, so let's look at the components of what constitutes the telecommunications network. The telecommunications network referred to here is the one that was built around voice communications but has been undergoing a metamorphosis for the past two decades. The convergence of voice and data is nothing new; we have been trying to run data over a voice network since the 1970s. However, to run data over the voice network, we had to make the data look like voice. This caused significant problems for the data because the voice network was noisy and error-prone. Reliability was a dream and integrity was unattainable, no matter what the price.

Generally speaking, a network is a series of interconnection points. The telephone companies over the years have been developing the connections throughout the world so that a level of cost-effective services can be achieved and their *return on investment* (ROI) can be met. As a matter of due course, whenever a customer wants a particular form of service, the traditional carriers offer two answers:

- It cannot be done technically.
- The tariff will not allow us to do that!

Regardless what the question happened to be, the telephone carriers were constantly the delay and the limiting factor in meeting the needs and demands for data and voice communications.

In order to facilitate our interconnections, the telephone companies installed wires to the customer's door. The wiring was selected as the most economical way to satisfy the need and the ROI equation. Consequently, the telephone companies installed the least expensive wiring possible. Because they were primarily satisfying the demand for voice communications, they installed a thin wire (26-gauge) to most customers whose locations were within a mile or two from the central office. At the demarcation point, they installed the least expensive termination device (RJ-11), satisfying the standard two-wire unshielded twisted pair communications infrastructure. The position of the demarcation point depended on the legal issues involved. In the early days of the telephone network, the telephone companies owned everything, so they ran the wires to an interface point and then connected their telephone equipment to the wires at the customer's end. The point here is that the telephone sets were essentially commodity-priced items requiring little special effect or treatment.

When the data communications industry began during the late 1950s, the telephone companies began to charge an inordinate amount of money to accommodate this different service. Functionally, they were in the voice business and not the data business. As a matter of fact, to this day, most telephone companies do not know how to spell the word *data*! They profess that they understand this technology, but when faced with tough decisions or generic questions, few of their people can even talk about the services. How sad, they will be left behind if they do not change quickly.

New regulations in the United States, in effect since the divestiture agreement, changed this demarcation point to the entrance of the customer's building. From there, the customer hooked up whatever equipment was desired. Few people remember that in early 1980, a 2400 bps modem cost $10,000. The items that customers purchase from myriad other sources include all the pieces we see during the convergence process.

In the rest of the world today, where full divestiture or privatization has not yet taken place, the telephone companies (or *Post, Telephone, and Telegraph* [PTTs]) still own the equipment. Other areas of the world have a hybrid system under which customers might or might not own their equipment. The combinations of this arrangement are almost limitless, depending on the degree of privatization and deregulation. However, the one characteristic that is common in most of the world to date is that the local provider owns the wires from the outside world to the entrance of the customer's building. This local loop is now under constant attack from the wireless providers offering satellite service, *local multipoint distribution services* (LMDS), and *multichannel multipoint distribution services* (MMDS). Moreover, the CATV companies have installed coaxial cable or fiber, if new wiring has been installed, and they offer the interconnection to business and residential consumers alike.

The *Competitive Local Exchange Carriers* (CLECs) who survived the bloodbath and fallout of 2000 and 2001 still remain as formidable foes to the local providers. They are installing fiber to many corporate clients (or buildings) with less expense and long-term write-off issues. The CLECs are literally walking away from the telephone companies' local loop and using their own infrastructure. Add the *x-Type Digital Subscriber Line* (xDSL) family of products to this equation and the telephone companies are running out of options. The *Community Antenna Television* (CATV) companies are still outpacing the installation of Internet cable modems compared to the use of DSL services by the *Regional Bell Operating Company* (RBOC) and the CLECs. The numbers will probably change over time, but the current rate of installation is in the favor of the cable companies. This is where the CATV companies see the convergence occurring.

A Topology of Connections Is Used

In the local loop, the topological layout of the wires has traditionally been a single-wire pair or multiple pairs of wires strung to the customer's location. Just how many pairs of wires are needed for the connection of a single line set to a telecommunications system and network? The answer (one pair) is obvious. However, other types of services, such as digital circuits and connections, require two pairs. The use of a single or dual pair of wires has been the norm. More recently, the local providers have been installing a four-pair (eight wires) connection to the customer location. The end user is now using separate voice lines, separate fax lines, and separate data communications hookups. Each of these requires a two-wire interface from the LEC. However, if a CATV provider has the technology installed, they can get a single coax (or fiber) to satisfy the voice, fax, data, and high-speed Internet access on a single interface, proving the convergence is rapidly occurring at the local loop.

It is far less expensive to install a coax running all services (TV, voice, and data) than multiple pairs of wire, so the topology is a dedicated local connection of one or more pairs from the telephone provider to the customer location or a shared coax from the CATV supplier. This is called a *star* and/or *shared star-bus configuration*. The telephone company connection to the customer originates from a centralized point called a *central office* (CO). The provider at this point might be using a different topology. Either a star configuration to a hierarchy of other locations in the network layout or a ring can be used. The ring is becoming a far more prevalent method of connection for the local Telcos. Although we might also show the ring as a triangle, it is still a functional and logical ring. These star/ring or star/bus combinations constitute the bulk of the networking topologies today.

Remember one fundamental fact: the telephone network was designed to carry analog electrical signals across a pair of wires to recreate a voice conversation at both ends. This network has been built to carry voice and does a reasonable job of doing so. Only recently have we been transmitting other forms of communication, such as fax, data, and video.

The telephone switch (such as DMS-100 or #a5ESS) makes routing decisions based on some parameter, such as the digits dialed by the customer. These decisions are made very quickly and a cross-connection is made in logic. This means that the switch sets up a logical connection to another set of wires. Throughout this network, more or fewer connections are installed, depending on the anticipated calling patterns of the user population. Sometimes there are many connections among many offices. At other times, it can be simple with single connections.

The telephone companies have begun to see a shift in their traffic over the past few years. More data traffic is being generated across the networks than ever before. As a matter of fact, 1996 marked the first year that as much data was carried on the network as voice. Since that time, data has continued its escalated growth pattern upwards of 30 percent, whereas voice has been stable at around a 4-percent growth.

The Local Loop

Our interface to the telephone company network is the single-line telephone line, which has been installed for decades and is written off after 30 or 40 years. Each subscriber or customer is delivered at least one pair of wires per telephone line. There are exceptions to this rule, such as when the telephone company might have multiple users sharing a single pair of wires. If the number of users demanding telephone service exceeds the number of pairs available, a Telco might offer the service on a party line or shared set of wires.

It is in this outside plant, from the CO to the customer location, that 90 percent of all problems occur. This is not to imply that the Telco is doing a lousy job of delivering service to the customer. In the analog dial-up telephone network, each pair of the local loop is designed to carry a single telephone call to service voice conversations. This is a proven technology that works for the most part and continues to get better as the technologies advance.

What has just been described is the connection at the local portion of the network. From there, the local connectivity must be extended out to other locations in and around a metropolitan area or across the country. The connections to other types of offices are then required.

The Telecommunications Network

Prior to 1984, AT&T owned most of the network through its local Bell operating telephone companies. A layered hierarchy of office connections was designed around a five-level architecture. Each of these layers was designed around the concept of call completion. The offices were connected together with wires of various types called *trunks*. These trunks can be twisted pairs of wire, coaxial cables (like the CATV wire), radio (such as microwave), or fiber optics.

As the convergence of voice and data networks continues, we see a revisitation to the older technologies as well as the new ones. Fiber is still the preferred medium from a carrier's perspective. However, microwave radio is making a comeback in our telecommunications systems, linking door-to-door private-line services. Carrying voice, data, video, and high-speed Internet access is a given for a microwave system. Light-based systems, however, are limited in their use by telephone companies. It has been user demand that has brought infrared light and now *Synchronous Optical Network*-based (SONET) infrared systems in place. Recently, the introduction of an unguided light introduced by Lucent Technologies operates at speeds up to 2.4 Gbps to 10 Gbps. This offers the connectivity to almost anyone who can afford the system, because the right of way is no longer an issue.

The Network Hierarchy (Post-1984)

After 1984, ownership of the network took a dramatic turn. AT&T separated itself from the *Bell Operating Companies* (BOCs), opening the door for more competition and new ventures. Equal access became a reality and users were no longer frustrated in their attempts to open their telecommunications networks to competition.

The Public-Switched Network

The U.S. public-switched network is the largest and the best in the world. Over the years, the network has penetrated to even the most remote locations around the country. The primary call-carrying capacity in the United States is done through the public-switched network. Because this is the environment AT&T and the BOCs built, we still refer to it as the Bell System. However, as we've already seen, significant changes have taken place to change that environment.

The public network enables access to the end office, connects through the long-distance network, and delivers to the end. This makes the cycle complete. Many companies use the switched network exclusively, while others have created variations depending on need, finances, and size. The network is dynamic enough, however, to pass the call along longer routes through the hierarchy to complete the call in the first attempt wherever possible.

The North American Numbering Plan

The network-numbering plan was designed to enable a quick and discreet connection to any telephone in the country. The North American Numbering Plan, as it is called, works on a series of 10 numbers. As progress occurs, the use of *Local Number Portability* (LNP) and *Intelligent Networks* (IN) enables the competitors to break in and offer new services to the consumer. Note that there have been some changes in this numbering plan. When it originally was formulated, the telephone numbers were divided into three sets of sequences. The area codes were set to designate high-volume usage and enabled some number recognition tied to a state boundary. With the convergence in full swing, the numbering plan became a bottleneck.

Now with the use of LNP, the numbering plan will completely become obsolete as we know it. No longer will we recognize the number by an area code and correlate it to a specific geographic area. LNP will make the number a fully portable entity. Moreover, 10-digit dialing in the age of convergence becomes the norm because of the multitude of area codes that will reside in a state.

Private Networks

Many companies created or built their own private networks in the past. These networks are usually cost-justified or based on the availability of lines, facilities, and special needs. Often these networks employ a mix of technologies, such as private microwaves, satellite communications, fiber optics, and infrared transmission. The convergence of the networks has further been deployed because of the mix of services that the telephone companies did not service well. Many companies with private networks have been subjected to criticisms because the networks were misunderstood. Often the networks were based on voice savings and could not be justified. Now that the telecommunications networks and systems are merging, the demand for higher speed and more availability is driving either a private network or a hybrid.

Hybrid Networks

Some companies have to decide whether to use a private- or public-switched network for their voice, data, video, and Internet needs. Therefore, these organizations use a mix of services based on both private and public networks. The high-end usage is connected via private facilities creating a *virtual private network* (VPN), while the lower-volume locations utilize the switched network. Installing private-line facilities comes from the integration of voice, data, video, graphics, and fax transmissions. Now VPNs are used on the Internet to guarantee speed, throughput, quality of service, and reliability. This new wave of VPNs takes up where the voice VPNs left off. Only by combining these services across a common circuitry will many organizations realize a savings.

Hooking Things Up

The Telco uses a variety of connections to service the customer locations. The typical two-wire interface to the network is terminated in a demarcation point. Normally, Telco terminates in a block; this can be the standard modular block. Another version of connector for digital service is an eight-conductor (four-pair) called the RJ-48X. When a Telco brings in a digital circuit, the four-wire circuit is terminated into a RJ-68 or a smart jack.

Equipment

Equipment in the telephony and telecommunications business is highly varied and complex. The mix of goods and services is as large as the human imagination, yet the standard types are the ones that constitute the ends on the network. The convergence and computerization of our equipment over the years has led to significant variations. The devices that hook up to the network are covered in various other chapters, but here is a summary of certain connections and their functions in the network:

- The *private branch exchange* (PBX)
- The modem (data communications device)

- The multiplexer (enables more users on a single line)
- *Automatic call distributor* (ACD)
- *Voice mail system* (VMS)
- *Automated attendant* (AA)
- Radio systems
- Cellular telephones
- Fax machines
- CATV connections
- Web-enabled call centers
- Integrated voice recognition and response systems

This is a sampling of the types of equipment and services you will encounter in dealing with telecommunications systems and convergence in this industry.

CHAPTER 3

Virtual Private Networks

The term *Virtual Private Network* (VPN) can have different meanings, but it usually refers to voice or Internet. In this chapter, we'll learn the meaning of the term in both environments.

History

As corporate communication volumes increased, organizations realized the cost of telephone service was escalating. Originally, all long distance service was charged on a per minute basis. AT&T introduced a volume discount outbound calling plan called *Wide Area Telephone Service* (WATS)[1]. For a monthly fixed payment, the organization got 240 hours of service to one of five bands across the country. Each band was priced, based on the distance from the originator's location. A typical company usually had a band 5 line and a band 1 or 2 to cover adjacent state calls. It took some analysis to determine the most cost-effective solution for each company's particular calling pattern.

Foreign exchange (FX) service provided a fixed rate calling plan if a company had a large call volume for in-state locations. This is essentially subscribing to telephone service at the foreign central office location and leasing an extension cord from the telephone company to the home location. Originally, there were no usage charges on this line so the more you used it, the less expensive it was. Of course, long distance calls made from the foreign exchange were billed at the long-distance rate. An FX line is needed to each high volume calling location.

Alternatively, a company could use a leased telephone line between locations. These lines went by several names: *Terminal Interface Equipment* (TIE) line, dedicated line, and a data line, when used for data. These are essentially point-to-point telephone lines that are available in two-wire or four-wire configurations. Because the difference in cost between two- and four-wire connections was small (relative to the cost of the line), the four-wire option was preferred unless the company needed many lines.

The next logical step was to use these TIE lines to connect *private branch exchanges* (PBXs) at the various locations. Here again, there were no usage charges on these dedicated lines. A company with locations in Seattle, Phoenix, Atlanta, and headquarters in Chicago might have a "hub and

[1]Some people refer to the term as Wide Area Telecommunications Services.

spoke" arrangement of TIE lines from their headquarters to each regional office. Each location then might have FX lines to adjacent cities; for example, a company based in Seattle might have an FX line to Tacoma, Kent, and Everett (see Figure 3-1).

There were corresponding inbound services where the called party paid. For example, the original Zenith operator provided toll-free calling in the days of manual switchboards. The inbound WATS service, now known as 800 service, was originally also structured in bands. Finally, for local toll service, *remote call forwarding* (RCF) allowed people to sign up for telephone service in a foreign exchange and have them make a long distance call from Tacoma, for example, back to Seattle at your expense. Although this was more expensive (depending on the number of calls) than FX, an advantage of RCF is that you can receive multiple calls at a time.

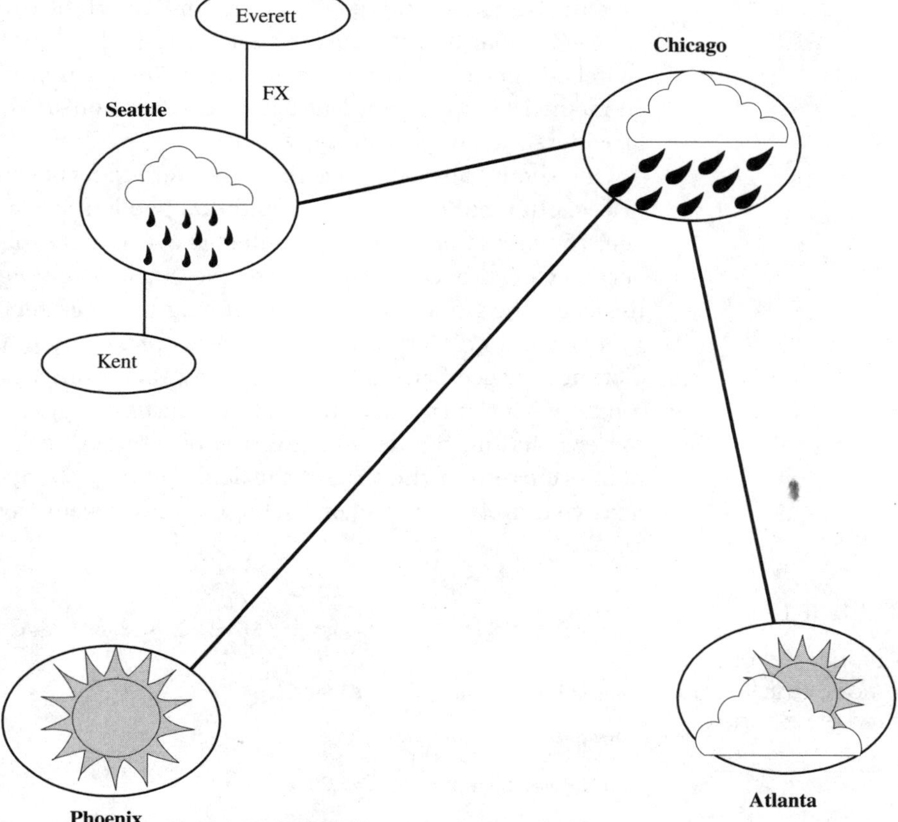

Figure 3-1
Hub and spoke arrangement for TIE lines

It soon became apparent to people working in the Phoenix location that they could call their uncle in Kent by first asking the company operator (later by dialing) for the TIE line to Chicago. They would then choose the TIE line to Seattle and finally dial across the FX line to Kent. The PBX, although not smart, did allow a person to dial up the TIE and FX lines.

The important fly in this otherwise ingenious solution (ointment) to high-cost long distance telephone service is that each TIE or FX line could only handle one call at a time. The challenge for the telecommunications manager was therefore to figure out the optimum number of TIE lines between locations to minimize cost and waiting time for the TIE line, while maximizing savings across the commercial long distance circuits.

About this time, AT&T noticed a small drop in its long distance revenue from such business and a sharp increase in the number of leased lines it was providing. Now, clearly it is much more profitable to rent a telephone channel out at $0.25 per minute than to lease that capacity to a corporation for $1,000 per month. Table 3-1 shows somewhat optimistically the amount of revenue that a normal telephone channel could return versus the lease line.

From Table 3-1, it is clear that the telephone companies much prefer switched service to dedicated service. (This thumbnail sketch focuses only on business hour revenue and ignores after hour revenue and the network providers' cost to provide the service.)

One should also be aware that the average corporation will not pay these prices, but smaller companies and independent contractors may! On average, 75 percent of the paying public is overpaying the cost of long distance because of the complexity and the various changes that take place. Recently, the three top providers of long distance service raised their rates by 7 percent (12/2001). The impact was primarily in the area of basic long distance service. This means that many small companies have subscribed to a plan with the carrier. The carrier selects the plan that best fits the customer's dialing habits and number of circuits used (lines). However, the plan is current at the time of the deal and may change several times in the next year. Better pricing or packaging may become available the very next

Table 3-1

Comparison of usage sensitive and fixed leased line costs

Usage	Cost per Month based on Usage
Leased line flat rate	$1,000
8 hr/day @ $.25/min.	$2,400
4 hr/day @ $.25/min.	$1,200

day. The consuming public may not realize that the new package is available and continue to pay the agreed to rates for the next *x* years, costing them hundreds to thousands of dollars extra per year.

To rectify the problem, many organizations periodically call the carrier and ask for the best plan to meet their dialing habits. Once again, the best plan is selected at the time of the call, not forever adjusted automatically.

Intelligent PBX Solution

Using these dedicated lines between locations, organizations created a private network. The next step in the evolution of private networks was to devise a corporate-wide numbering plan and have the now intelligent PBX determine the route to the dialed destination via its peers, just like the local telephone office does. After all, other than size, there is little difference between a PBX and a telephone company central office switch!

Virtual Private Networks (VPNs)

To get corporate America back on the switched network, AT&T devised a marketing strategy. The approach went something like this to the CEO/CFO: "Look, your primary business is banking [building airplanes, trading stocks, selling insurance or whatever], but it is not running a telephone company. Who knows better how to run a telephone system than we do? (You can substitute your favorite carrier here. AT&T is chosen here because they were the first to introduce this service.) You think you are saving money by using these dedicated lines. On the surface, it appears that you are. However, who is managing this network? What is it costing you to recover from outages? Do you have back-up facilities for each of your dedicated routes? Your dedicated team of telephony experts is costing you a bundle. Why are you doing this?"

The CFO and CEO look at each other and shrug their shoulders. "Our CIO or CTO[2] sold us on the idea for providing better service at a lower cost," they said in unison.

---

[2]CTO is the Chief Telecommunications Officer or Chief Technology Officer depending on the organization

"Look," said AT&T. "We have the ultimate (outsourcing) deal that will provide all your current capabilities for one low price. We will manage the whole network for you and give you all the service you currently enjoy with your private network with little or no hassle." Our product is called (somewhat obscurely) Software Defined Network™ because you can define the parameters of the network yourself," AT&T said proudly.

Sprint and MCI/WorldCom[3] offer essentially the same product and call it a virtual private network (VPN). We use VPN here because it is both the generally used term, and it is descriptive of the offering.

Here is how the deal works: The company defines the locations that will be part of the VPN as shown in Figure 3-2. The larger the average traffic commitment made between these locations, the lower the price per minute can be. (The catch is that if traffic falls below the average commitment, cost falls into the next higher rate category.)

The big advantage is that organizations no longer have to manage this far-flung network. The carrier will do it. Organizations can now lay off the telecommunications department. (Please note that the staff supporting the PBX in each location is still needed to handle moves, adds, and changes. In addition, the staff needed to maintain the dedicated data network is still needed. Even if the organization migrates to a Frame Relay network, some management of the vendor is always required).

All the calls to specifically defined locations (offices) in Chicago, Atlanta, Phoenix, and Seattle are known as on-net calls. These are priced at the reduced rate. Calls to business partners and customers are off-net calls and are charged at a higher rate. If the off-net call volume to these specific locations rises, the organization can still place FX lines into these areas. Again, there is no substitute for knowing the traffic distribution when evaluating any telecommunications plan.

As one can determine from the above description, it takes a sharp pencil to figure out if this is a good deal. It is definitely a good deal for the carrier who gets all those calls and minutes back on the switched network.

The VPN is more reliable than a dedicated, line-based network because calls are really riding over the *Public Switched Telephone Network* (PSTN), which is rich in multiple paths. One of the features of the private, line-based network was four- or five-digit dialing. This can be preserved intact if we want. Because the switches in the telephone network are computers that have access to a database, they can easily look up how to route a number

---

[3]MCI and WorldCom were different entities at the time of this offering, but for this book are updated to reflect current situations. ™Software Defined Network is a Trademark of AT&T.

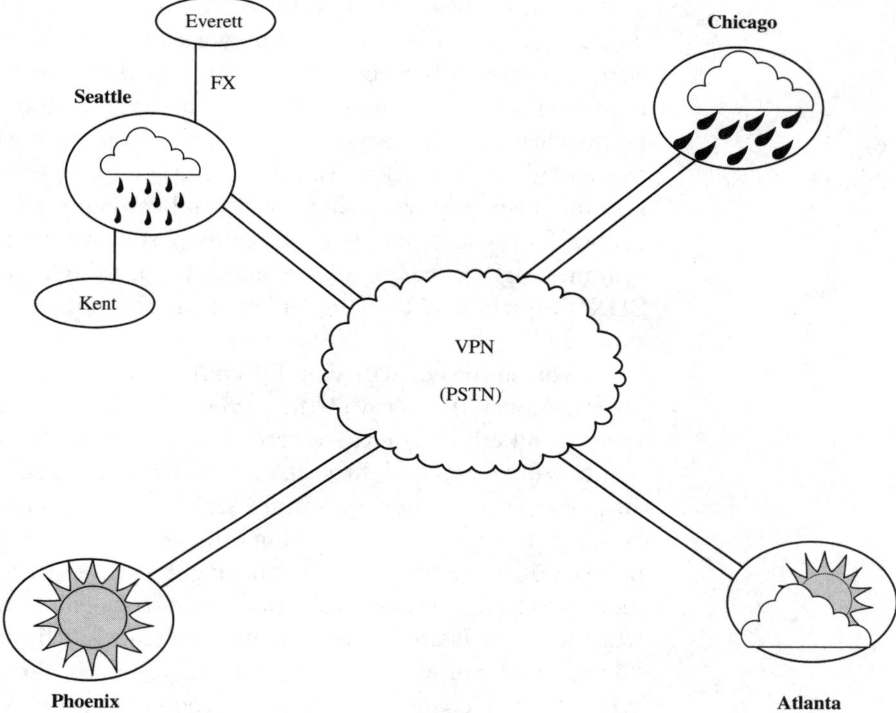

Figure 3-2 The VPN uses the PSTN as the backbone

based on the originating location and number dialed. The VPN then is a special discount-billing plan, with the carrier managing the network on which we can have a custom-dialing plan.

A caveat that should also be brought into the equation is that the large corporations will negotiate long-term SDN/VPN agreements with the carrier. Typically, the agreements will bear a 3 to 5 year term whereby the customer enjoys the benefits of the fixed pricing arrangement, with some caveats on usage such as minimums, numbers of locations, average revenue generated per month, and so on. If, however, the average volume falls below an agreed-to level, the carrier may charge a penalty. This penalty may be in the form of

- A minimum charge per site
- A minimum charge per month
- An averaged cost that is used on a quarterly basis (that is, they will bill the higher rate for an entire quarter if the customer does not achieve the minimum billing)

Any one of these charges may apply to the consumer's billing, depending on the agreement between the players. Incidentally, the customer and the carrier are usually sworn to secrecy regarding the rates and terms of the agreement, through some nondisclosure arrangement. The purpose of this nondisclosure is to keep the mass public coming back and asking for the same deal! Or is it? Sometimes the deal is not as good as it is supposed to be. One such case was a large financial company who had a deal with the carrier for 5 years, yet over that same period of time the costs were rapidly plummeting. The customer was actually spending more per minute for their SDN/VPN than if they just picked up the phone and made a long distance call.

Newer contracts will usually bear some terms that state if the costs decrease over the term of the agreement, then the carrier will annually review and adjust the rates accordingly. It may also state that the adjustments will be enacted if the costs drop by some fixed percentage point (like 10 percent). In either case, the carrier will also hook a contingency that because they are tied to reducing the costs in the contract period should the prices fall, they also reserve the right to raise the rates if their prices increase at greater than some tied percentage point (usually 10 percent). So what we have is an agreement that is somewhat fluid and can be modified during the term of the contract so long as both parties are in agreement. Where this is a benefit is when a company plans extraordinary growth over the term of the agreement, or when there is some speculation that some sites may be closed and contraction will drop the overall volumes.

Users May Not Like It

Without trying to throw a damper on the voice SDN/VPN, there are some conditions that may cause the end users to balk at its use. Many organizations' telecommunications management typically try to match the needs of the organization without causing undue stress on the user. However, the special dialing procedures necessary to use a SDN/VPN often got in the way. Let's use an example of a group with road warriors. The traveling person needs to use long distance to customers, contacts, and back to headquarters. Therefore, a special calling card is issued that has the caller go to a pay phone. From there the caller dials a special 800 number to call into the SDN/VPN (this requires 11 digits). This is nothing more than a switch that is keeping track of the traffic and usage verification. Once into the SDN/VPN, the caller then dials the 11-digit telephone number for a North

American location. The number of dialed digits may be higher for international calls. Finally, the caller must dial their user calling card number to validate it for authentication and billing purposes. This may be an additional 15 digits. So all told, the customer has just dialed 37 digits to make a call. This creates frustration for the caller, especially if they make several calls during the course of a day.

Let's complicate the above scenario a bit! After being frustrated by dialing all those digits, the caller gets a busy tone. This means that they have to start over. Now the frustration really starts to mount. Moreover, one may be reading this and saying "what is the author talking about? I can dial a number and if I get a busy tone, then I merely dial the pound key (#) and get my dial tone back."

That may be true for some calls and some phones, but this is not a guarantee. The individual phones at airports, hotels, and along the roadside may not allow this. Many may be phones that are used by a specific vendor/carrier (we have all seen the WorldCom and AT&T phone in the lobbies of hotels that only allow the features on their own specific network). So if the caller is using a WorldCom phone and calling an AT&T network, all bets are off. The service may require that the caller hangs up and starts over.

Moreover, when making a string of calls on a normal calling card, customers are able to use the # key to place the next call without entering the calling card number every time. This again is not necessarily true with the special SDN/VPN cards. Although the carriers have taken great strides in eliminating these problems, they still cannot guarantee that everything works at every phone. By the way, with the SDN/VPN, the carriers allowed stored numbers in the central switch so that a user could eliminate some of the dialing process by using a speed dialing arrangement. Corporate telecommunications personnel may have predefined calls to each office with a three- to five-digit speed number, thus the caller could eliminate some of the digits required. This is a noble gesture, but it does not always work the way it was planned, and therefore the end users begin to rebel against the amount of time they spend dialing digits to do their job.

Now back to the original purpose of the VPN—to save money and ease the process of communicating between and among users within an organization—the ease of use is not assured, as stated previously, so the goals are not met entirely. From there, however, the user can usurp the savings by doing many things:

- Reducing the amount of calls they make by refusing to dial the digits
- Calling around the VPN by using a separate calling card that is not billed under the special arrangement

- Placing operator assisted calls instead of dialing, thereby incurring a much higher cost per minute

Each of these situations complicates the overall purpose of using the VPN/SDN. One final comment here is that the users also begin to bemoan the use of the network to their superiors, who then begin a grass roots effort to override the VPN. What was planned as a cost containment tool, becomes a more expensive solution overall, and management really does not want to hear all the complaints about a system as mundane as the telephone. Bear this in mind as you look into the use of these systems.

This discussion so far has only considered the case where the corporation owns the PBX and connects it to the VPN. What if a Centrex system is provided by the *incumbent local exchange carrier* (ILEC) or leased from a reseller? The answer is that one can still implement all the above with a Centrex system at any or all locations.

Because Centrex is essentially a PBX that is physically resident at the local central office, it too can have TIE, FX, or RCF trunks. The long distance carrier supplying the VPN will be more than happy to terminate VPN trunks on a Centrex system.

In summary, the important points are as follows:

- Calls are carried over the PSTN.
- A custom dialing plan is used.
- Pricing is dependent on the locale.
- The number of locations.
- The projected or committed traffic volumes.

This is all achieved by computer databases in the network.

CHAPTER 4

Data Virtual Private Networks (VPNs)

Internet-Based VPN

One might say that these Internet-based data VPNs are the same as voice VPNs, but different at the same time. The philosophical point is that a dedicated network will be overbuilt in some areas and underbuilt in others. A shared network offers the hope that we can spread the overall cost out while getting the benefits of a private network. Historically, this accounts for the popularity of shared data networks beginning with X.25, Frame Relay, ATM, and now the Internet. The Internet has become a popular, low-cost backbone infrastructure.

Because of its ubiquity, many companies now want to use a secure *Virtual Private Network* (VPN) over the public Internet. The challenge in designing a VPN is to exploit the technologies for both intracompany and intercompany communication while still providing security. Of course the rule of thumb we now use in an *Internet Protocol* (IP) network is "IP on everything." A VPN is an extension of an organization's private intranet across a public network (that is, the Internet), creating a secure connection essentially through a tunnel. VPNs securely convey information across the Internet connecting remote users, branch offices, and business partners into the corporate network. Figure 4-1 is a graphic depiction of an Internet-based VPN.

VPNs are owned by the carriers, but used by corporate customers, as though the customers owned them. A VPN is a secure connection that offers the privacy and management controls of a dedicated point-to-point leased line, but actually operates over a shared routed network.

Figure 4-1 Tunnels provide secure access for VPNs.

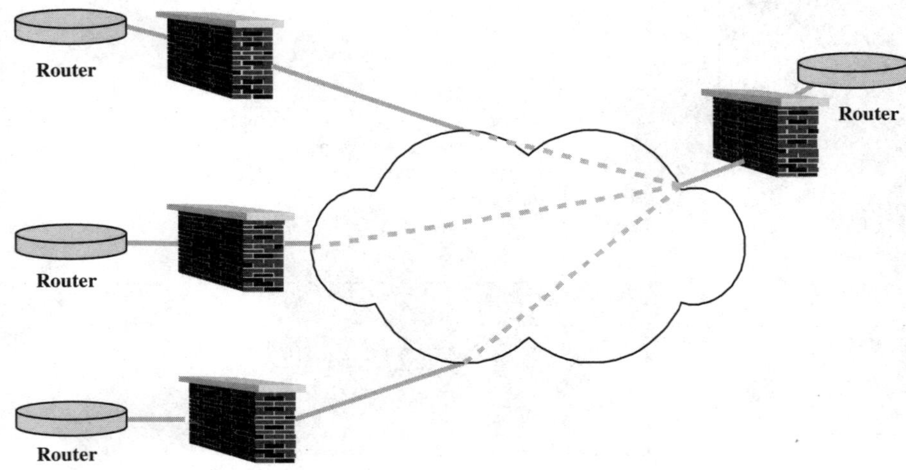

In the past we saw traditional networks being built as part of a leased line, point-to-point network. This was expensive and risky. A single link error brought the network down. Later a virtual networking scenario emerged using a packet-switching technology called *Frame Relay*. This demanded that presubscribed links were established by being premapped in logic.

VPNs are created using encryption, authentication, and tunneling, a method by which data packets in one protocol are encapsulated in another protocol. Tunneling enables traffic from multiple organizations to travel across the same network, unaware of each other, as if enclosed inside their own private steel pipe.

It is easy to jump to the conclusion that the Internet is free and, therefore, there are tremendous cost savings to be had from this free shared network. Later, we will explore some cost comparisons, but as one might guess, the relative cost benefit depends very much on each network's geography and traffic volume.

Goals

The goal of any network is to support users in a flexible, reliable, secure, and inexpensive manner:

- Network managers want the network to be flexible.
- Users want the network to be reliable and secure.
- Management wants the network to be inexpensive.

A balance of these often-competing goals can be achieved, provided a good dialog is maintained among the participants. Table 4-1 shows the network goals in terms of applications, users, potential network solutions, and access to the network. It is an exercise left to the reader to select from the list those applications and users who are to be served. The network list indicates that these users and applications could be interconnected by any of these network technologies. As indicated previously, dedicated networks are expensive and rarely fit the need perfectly. Frame Relay and *Asynchronous Transfer Mode* (ATM) are shared network technologies that can be very cost effective, depending on the geography and traffic volume. Dial-up telephony can be a networking technology for highly mobile, low-volume users. Normally, we would like to have a backbone network with direct access for various users and dial-up remote access for infrequent users. We will discuss these alternatives in the following sections.

Table 4-1
Mix of methods used to pick from

Access	Network	Users	Application
Dial-up	Dial-up	Road warriors	E-mail
ISP	Dedicated	Telecommuters	DB Access
xDSL	X.25	Branch office	Sales support
Cable modem	Frame Relay	Customer	Customer service
ISDN	Internet	Partners	E-Commerce
Dedicated	ATM		Order entry

Shared Networks

The advantage of shared networks is that organizations do not have to incur the entire cost of the infrastructure. For that reason, Frame Relay has been extremely popular. Because it (like X.25 before it) is virtual circuit based, there is little concern about misdirected or intercepted traffic. Still, Frame Relay service is not universally available and access charges to a *point-of-presence* (POP) can be expensive. However, compared to the cost of dedicated networks, shared networks offer equivalent performance and a much lower cost.

Internet

The next logical step is to use the Internet as the private network. It is almost universally accessible, minimizing access charges. From our discussion of the Internet in Chapter 29, "Synchronous Optical Network (SONET)," two things are clear:

- No one is watching the traffic or performance of the Net as a whole.
- The path our data takes across the network is quite unpredictable.

This leads to the conclusion that performance will be unpredictable and that our precious corporate data may pass through a router on the campus of "Den-of-Hackers University." (It is not the intent here to malign university students, but only to offer the observation that they are bright, curious, love a challenge, and may have time on their hands and access opportunity to do a little extra curricular research on the vulnerability of data on the Internet.) There are then two problems: performance and security.

Performance

The performance issue poses the problem of sizing the bandwidth on each link, which becomes a major task as the network grows. Unfortunately, few network managers have a good handle on the amount of traffic flowing between any given pair of locations. Typically, they are too busy handling moves and additions to the network, which frequently leads to performance problems. Because the network grew without the benefit of a design plan, invariably, it means that portions of the network, including servers, become overloaded.

A dedicated line network is expensive, requires maintenance, and necessitates a backup plan should a line or two fail. Using a shared network does not alleviate the problem of traffic analysis. On the contrary, we now have to worry about the capability of the Internet to provide the bandwidth we need when we need it. Selecting our ISP to provide the performance we need becomes an important issue.

Outsourcing

One solution is to outsource the network to a network provider (the analogy to a voice VPN here is strong). The most popular previous solution was to lease Frame Relay service. The benefit was that the network provider took care of the management of the network and even provided levels of redundancy (for which you paid) within its network. Unfortunately, to make most efficient use of this service, one still needed to have a handle on traffic volumes. For example, a *committed information rate* (CIR) that was too low resulted in lost data and retransmission, while a CIR set too high was a waste of money.

A national or international carrier with its own Internet backbone then becomes a good choice as a VPN provider. One negotiates *service level agreements* (SLA), which include *quality of service* (QoS) guarantees. Some ISPs even provide *Virtual IP Routing* (VIPR) in which they permit you to use internal, unregistered IP addresses.

If one builds a completely independent, internal (intranet) network, one could use any set of IP addresses one might choose. This alternative is attractive to large corporations that are constrained to using class C addresses. If these private addresses were to get out onto the Internet, chaos would quickly ensue. VIPR permits the flexibility to continue to use this unregistered set of addresses transparently across the Internet. This is strongly analogous to having one's own dialing plan on a voice VPN.

There are many possibilities and choices here. We can outsource the whole network, including the VPN equipment on each site, or outsource pieces.

Standard Outsourcing Issues A few points are worth making about outsourcing. One must take a realistic look at the task at hand:

- If the internal staff possesses the capability to implement the VPN, do they have the time?
- If you outsource the whole network, how permanent will the relationship be?
- To what extent will the internal staff become involved in the design and maintenance of the VPN?

Choose your vendor carefully. Evaluate responsiveness in the areas of presale support, project management, and post-sale support. As in any procurement process, writing a system specification and *Request for Proposal* (RFP) is essential. Also, make up the evaluation criteria ahead of time. You may (or may not) choose to publish the evaluation criteria in the RFP. Select the vendor who is most responsive to your requirements. Here is a good opportunity for the vendor to do the traffic analysis so that a traffic baseline for design can be established. Always include growth in the RFP.

Ongoing support will be critical. If the network spans multiple time zones, specify the minimum support requirements. For example, 9 A.M. to 5 P.M. CST is of little use to offices located in Taiwan. What training is offered as part of the package? The more knowledgeable the internal staff can be, the better they will be able to support the VPN—even when they are outsourcing support.

It is important to have a coordinated security plan so that we have an integrated and consistent view across our firewalls, proxy servers, and VPN equipment.

Security

The basic concept of a VPN is to provide a secure, point-to-point connection across the network between communicating entities. A couple of questions about security are important to keep our paranoia in check. The first question is how much security is enough? To answer that question, we must consider the impact on our business if the data we are sending is

- Simply lost. Is there a backup mechanism for sending or recovering the data?

- Found by another business (not a competitor).
- Found by a competitor.
- Actively pursued by a competitor as shown in Figure 4-2.

In the last case, we must ask how much effort is the competitor willing to invest to get our data? The answer to these questions will help us decided how much security is enough. Note that in the foregoing example, one can equally substitute the word hacker for competitor.

What About Security Issues? Turning to security, remote access to a system must have integral security to protect the network and users from unauthorized access and penetration. We have all heard about the teenaged hackers who have been creating havoc in the data processing and Internet business. These young hackers break into systems for the sheer pleasure of challenging the system and showing their prowess with the modem as shown in Figure 4-3. And it works, because they do it every day. We, therefore, have to consider these issues before opening a door.

We must start with different techniques such as VPNs, encryption, authenticating servers, and secure firewalls. The key technologies that compose the security component of a VPN are

- Access control to guarantee the security of network connections
- Encryption to protect the privacy of data
- Authentication to verify the user's identity as well as the integrity of the data

What Can We Do to Secure the Site? Remote access users sitting in a distant site need to know how to use the system, so training is important. Check the pieces of the puzzle as shown in Figure 4-4 to make sure that you

Figure 4-2 Competitors may actively pursue your data.

Figure 4-3
Hackers break in just to prove their prowess.

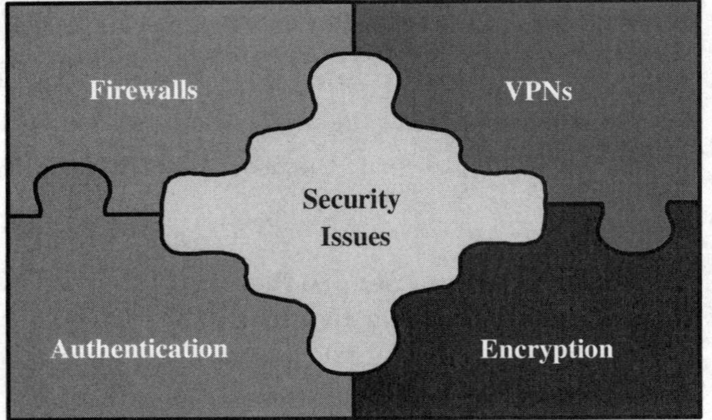

Figure 4-4
The pieces that must be considered for security

have a good solution provider to handle your needs. A company with salespersons who travel frequently would provide 800 number access. Hardware considerations vary, depending on what networking you're using, the number of users, and whether the users need desktops or laptops at the remote location. Standardization is essential—you don't want three or four different platforms, and you don't want to have to support 47 varieties of software. We want to leave the variety of flavors to the ice cream manufacturers!

Additionally, a firewall service will offer a bastion router capability to filter the packet, the protocol, or the user id and address. These systems will help to keep out unwanted guests. Firewalls can be in different places, as we will see. They can also be integrated or CPE solutions.

Security must also be ensured while the data is in transit. Therefore, we need to use a form of encryption so that an eavesdropper cannot listen in on our data and intercept it. By using *Internet Protocol Security* (IPSec) techniques, we introduce up to five different forms of encryption and digital signatures. These will be sufficient to delay any access to the data and by the time the code could be broken, the data will have little value.

Authentication is also a very effective tool that challenges the caller and requests a key-coded response. In a security dynamics environment, a challenge and response can be issued by default every 30 seconds or user variable to effectively manage the logged-on users.

What Are the Risks? The risks posed on data integrity and security take many forms. We usually think of data protection in terms of the corruption or total loss of data. However, other areas of concern may be from the undetected interception of the data by hackers or crackers. Moreover, the inaccessibility of our data from the denial of service attacks has become more prevalent in the security issues facing the IT manager. Lastly, there are also issues of invasions on our LANs or WANs when a promiscious device is attached to the network and picks off all data packets regardless of the addressee. These sniffers, as they are called, can capture all data packets from the network, usually undetected.

- Hackers
- Crackers
- Salami attackers
- Denial-of-service attacks
- Sniffers

Creating the VPN

There are five ways to create a VPN:

- Between desktops
- Between routers
- Between firewalls

- Between VPN-specific boxes
- With integrated boxes

Although not normally considered a VPN, one can certainly use desktop PCs to encrypt data and send it across the Internet securely. Additionally, software is available that runs on a desktop capable of creating a VPN to a firewall or stand-alone VPN device. Most VPN equipment vendors offer corresponding software that runs on a laptop or desktop in order to provide a secure path to the home office over the Internet. Most of the discussion then involves creating a VPN between business locations, branch offices, and road warriors.

Encryption

The first basic rule is the more secure it is, the less convenient it is to use and the greater impact (negative) it will have on overall system performance. The strength of an encryption mechanism is dependent upon the complexity of the calculation and the length of the key. The most popular mechanism for which hardware is readily available is *Data Encryption Standard* (DES), developed by IBM and now standardized. The basic key is 54-bits long. Triple DES involves simply running the algorithm with a 112-bit key. The question here is as always how secure do you need to be?

The more secure, the larger the key used (or the more times the algorithm is run with different keys). This all takes time to encode and to decode. Much has been made lately of the fact that by using thousands of computers, a DES-encoded message could be broken in 39 days.

Keep in mind that this is for one key. If we change keys, it would take the crackers and hackers another 39 days. Are they (hackers and competitors) motivated to do this? The method mentioned previously used the brute force attack of guessing keys. Changing keys often means that the attackers must start all over again. The other encryption standard (not widely supported) is *International Data Encryption Algorithm* (IDEA), which uses 128-bit keys.

The second basic rule is that encryption performed in hardware is much faster than in software.

Key Handling

A very important part (some say the most important) of an encryption is the mechanism used to disseminate keys. Here again, security is the inverse of convenience. True, keys can be sent in a multi-encrypted file. They can also be sent by snail mail or given over the telephone (not very secure). The problem with this private key system is that both communicating parties must have the same key. If all locations are talking to the home office, they all must have the same key, or the central office must keep separate key pairs for each location.

This key management nightmare can be handled in two ways. We could use the X.509 digital certificate system for key management. The other alternative is to use a public key system to encrypt the private key so that they can easily be exchanged.

Public Key Cryptography (RSA)

The layman's version (don't try this at home because it won't work as described here) is that each of us thinks up a couple of prime numbers (the bigger the better). One number we keep for ourselves and the other number we publish on our web site along with the product of the two prime numbers as our public key. Anyone wanting to send us something will use the public key to encrypt it with the public key, and only we can decrypt the message with our private key. We can authenticate the source if the sender used his private key to encrypt his signature because only his public key will decrypt his signature. The process is shown in Figure 4-5.

This system is secure because of the tremendous amount of processing power it takes to factor large prime numbers. (For example, if we could factor the product, we could determine the private key.) Unfortunately, performing the encryption and decryption are also processor intensive (slow).

Figure 4-5 Security key management is used for IPSec.

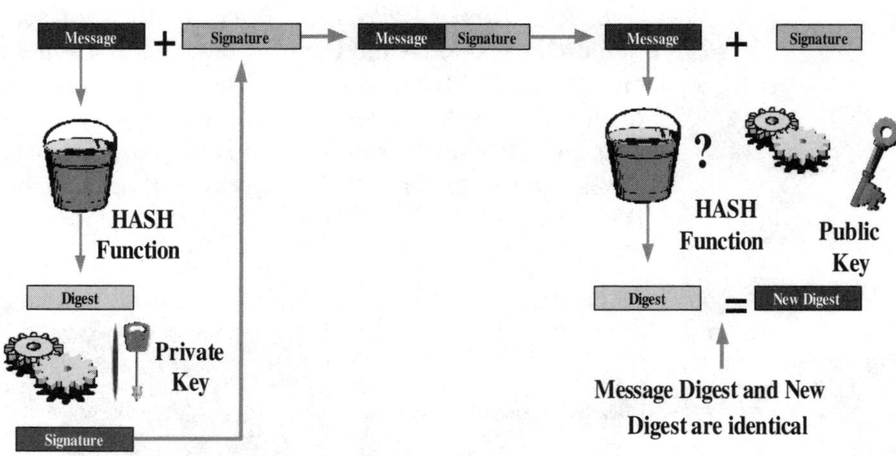

But it sure solves the key distribution problem. Therefore, we could use public key cryptography to encrypt and distribute the keys to all our VPN boxes.

Authentication

Authentication is the process of verifying that this is the party to whom I am speaking, and that they have authorized access. There are several ways of doing this; however, the most common way is to provide an authentication server that passes out authenticated certificates based on something the user has or knows.

User Level Authentication The user has or knows his/her account code (name) and password. User names are public, and passwords can be compromised. A more secure system is to use a type of secure ID card. These credit card sized devices are based on an internal clock that generates a different pseudo random code every minute. The authentication server is time synchronized with the card and therefore generates the same number at the same time. When the user calls in, he/she must enter his/her account code and the code from the card as the password. The IP has embedded in it a set of layer 2 protocols called the *Point-to-Point Protocol* (PPP). In PPP, the basic security methods used are *Password Authentication Procedure* (PAP) and the *Challenge Handshake Authentication Protocol* (CHAP). PAP and CHAP do little for security. In fact, PAP and CHAP are part of the basic PPP protocol suite and fall short in providing a true security procedure. These schemes do not address issues of ironclad authentication and integrity, or eavesdropping. The PAP and CHAP are rudimentary procedures used to log on to a network, but hackers and crackers easily defeat both.

Layer 2 Tunnel Protocol (L2TP) is another variation of an IP encapsulation protocol as shown in Figure 4-6. An L2TP tunnel is created by encapsulating an L2TP frame inside a UDP packet, which in turn is encapsulated inside an IP packet, whose source and destination addresses define the tunnel's ends. Because the outer encapsulating protocol is IP, clearly IPSec protocols can be applied to this composite IP packet, thus protecting the data that flows within the L2TP tunnel. *Authentication Header* (AH), *Encapsu-*

Figure 4-6
The L2TP packet

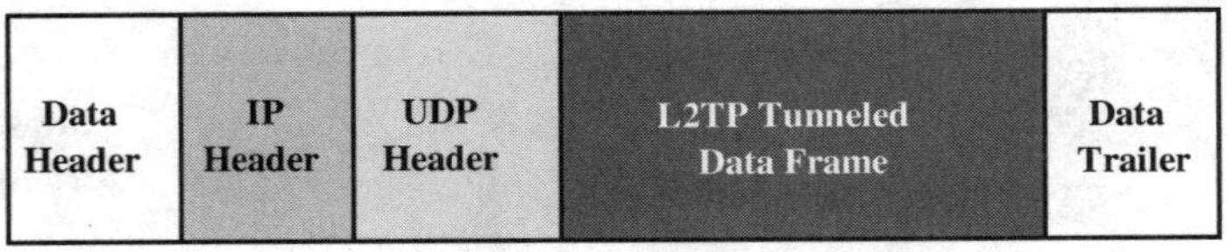

Data Header	IP Header	UDP Header	L2TP Tunneled Data Frame	Data Trailer

lated Security Payload (ESP), and *Internet Security Association and Key Management Protocol* (ISAKMP) can all be applied in a straightforward way.

L2TPs are an excellent way of providing cost-effective remote access, multiprotocol transport, and remote LAN access. It does not provide cryptographic robust security. L2TP should, therefore, be used in conjunction with IPSec for providing secure remote access. L2TP supports both host-created and ISP-created tunnels. A remote host that implements L2TP should use IPSec to protect any protocol that can be carried within a PPP packet.

Integrated at the VPN point of access, user authentication establishes the identity of the person using the VPN node, and this is because an encrypted session is established between the two locations. The user authentication mechanism enables the authorized user of the VPN system access to the system, while preventing the attacker from accessing the system.

Some of the common user authentication schemes are

- Operating system username/password
- S/Key (one time) password
- *Remote Authentication Dial-In User Service* (RADIUS)
- Strong two-factor token-based scheme

The strongest user authentication schemes available on the market are two-factor authentication schemes. These require two elements to verify a user's identity: a physical element in their possession (a hardware electronic token), and a code that is memorized (a PIN).

Some cutting-edge solutions are beginning to use biometrics mechanisms such as fingerprints, voiceprints, and retinal scans. However, these are still relatively unproven.

When evaluating VPN solutions, it is important to consider a solution that has both data authentication and user authentication mechanisms. Currently, there are VPN solutions that provide only one form of authentication.

Because of this, VPN solution providers that only support one of the two authentication mechanisms will typically refer to authentication generically, without qualification of whether they support data authentication, user authentication, or both. A complete VPN solution will support both data authentication (also known as the *digital signature process* or *data integrity*) as well as user authentication (the process of verifying VPN user identity).

Packet Level Authentication The IPSec standard provides for packet level authentication to prevent man-in-the-middle attacks. (This is where someone intercepts your packets and substitutes his/her own.) IPSec is a layer 3 protocol that enhances the use of the layer 2 underlying protocols. An authentication header is created for each packet. The layman's version of this is that a checksum is calculated and encrypted with the data. If the checksum calculated by the recipient doesn't match the one sent by the originator, someone has tampered with the data. The IPSec standard specifies two different algorithms for doing this MD-5 and SHA-1. If your vendor's equipment supports both algorithms, it improves the chances for intervendor compatibility. The other alternative is to simply not use packet level authentication.

In order to guarantee authenticity of the packets, a digital signature is required to authenticate the devices to one another. IPSec has included the X.509 digital certificate standard. Essentially, the X.509 certificate server keeps a list of certificates for each user. When you want to receive data from another device, you first ask for the certificate from the certificate server. The sender stamps all data with that certificate. Because this process is secure, you may be sure that these packets are authentic.

Your vendor then ideally supports both authentication algorithms and X.509. In any case, it is essential that someone in your organization understands in detail how each vendor supports the various levels of security that you intend to use. These authentication and encryption systems all have to work together flawlessly. If the vendors you choose stick to the standards, it improves the chances of, but does not guarantee, an integrated working environment.

IPSec offers a variety of advantages. Chief among those are

- IPSec is widely supported by the industry including Cisco, Microsoft, Nortel Networks, and so on.
- This universal presence ensures interoperability and availability of secure solutions for all types of end users. In addition, all IPSec-compliant products from different vendors are required to be compatible.
- IPSec provides for transparent security, irrespective of the applications used.
- IPSec is not limited to operating system-specific solutions. It will be ubiquitous with IP. It will also be a mandatory part of the forthcoming *Internet Protocol Version 6* (IPv6) standard.
- IPSec offers a variety of strong encryption standards. The key design decision to support an open architecture allows for easy adaptability of newer, stronger cryptographic algorithms.
- IPSec includes a secure key-management solution with digital certificate support. IPSec guarantees the ease of management and use. This reduces deployment costs in large-scale corporate networks.

IPSec used in conjunction with L2TP provides secure remote-access client-to-server communication. L2TP alone cannot provide for a totally secure communication channel due to its failure to provide per packet integrity, inability to encrypt the user datagram, and the limited security coverage only at the ends of the established tunnel. The major drawback to packet-filtering techniques is that they require access to clear text, both in packet headers and in the packet payloads.

There are two major drafts in IPSec: AH and ESP. They are defined as follows:

- AH is used to provide connectionless integrity and data origin authentication for an entire IP datagram (hereafter referred to as authentication).
- ESP provides authentication and encryption for IP datagrams with the encryption algorithm determined by the user. In ESP authentication, the actual message digest is now inserted at the end of the packet (whereas in AH the digest is inside the authentication).

AH provides data integrity only and ESP, formerly encryption only, now provides both encryption and data integrity. The difference between AH data integrity and ESP data integrity is the scope of the data being authenticated.

AH authenticates the entire packet, while ESP doesn't authenticate the outer IP header. In ESP authentication, the actual message digest is now inserted at the end of the packet, whereas in AH the digest is inside the authentication header.

The IPSec standard dictates that prior to any data transfer occurring, a *Security Association* (SA) must be negotiated between the two VPN nodes (gateways or clients). The SA contains all the information required for execution of various network security services such as the IP layer services (header authentication and payload encapsulation), transport or application layer services, and self-protection of negotiation traffic.

These formats provide a consistent framework for transferring key and authentication data that is independent of the key generation technique, encryption algorithm, and authentication mechanism.

One of the major benefits of the IPSec efforts is that the standardized packet structure and security association within the IPSec standard will facilitate third-party VPN solutions that interoperate at the data transmission level. However, it does not provide an automatic mechanism to exchange the encryption and data authentication keys needed to establish the encrypted session, which introduces the second major benefit of the IPSec standard: key management infrastructure or *Public Key Infrastructure* (PKI).

The IPSec working group is in the development and adoption stages of a standardized key management mechanism that enables safe and secure negotiation, distribution, and storage of encryption and authentication keys. A standardized packet structure and key management mechanism will facilitate *fully* interoperable third-party VPN solutions.

Other VPN technologies that are being proposed or implemented as alternatives to the IPSec standard are not true IP security standards at all. Instead, they are encapsulation protocols that tunnel higher level protocols into a link layer protocols. When encryption is applied, some or all of the information needed by the packet filters may no longer be available. There are many different forms of IPSec packets as shown in Figure 4-7. For example

- In transport mode, ESP will encrypt the payload of the IP datagram. In tunnel mode, ESP will encrypt the entire original datagram, both header and payload.
- In most IPSec-based VPNs, packet filtering will no longer be the principal method for enforcing access control. IPSec's AH protocol, which is cryptographically robust, will fill that role, thereby reducing the role of packet filtering for further refining after IPSec has encrypted the packet.

Basic Packet

Data Trailer | IP Header | TCP or UDP Header | Data | Data Trailer

Transport Mode Packet

Data Trailer | IP Header | Authentication Header | Encapsulating Security Payload | Data | Data Trailer

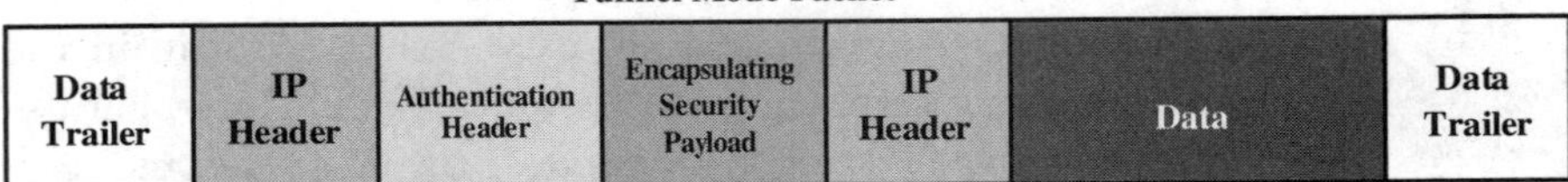

Figure 4-7
The various forms of IP packets

Moreover, because IPSec's authentication and encryption protocols can be applied simultaneously to a given packet, strong access control can be enforced even when the data itself is encrypted.

Router-Based VPN

Several router vendors offer VPN products based on the ability of the router to perform the requisite security functions. If your VPN is relatively small and the traffic volume not too heavy, then you might consider this option as a cost-effective approach. You need to have compatible routers at each location as shown in Figure 4-8. If there are individuals (for example, laptops or telecommuters) that don't have routers, they must have software that is compatible with that provided on the router. Make sure your vendor provides the compatible software that provides the level of security that you require for your VPN. The absence of a firewall in Figure 4-8 may be taken to mean that in this low-cost approach we are doing firewall functions on the router. In this case, the network would logically appear as in Figure 4-9.

The general admonition here is that you may be creating a bottleneck in the router. For large networks, let routers route.

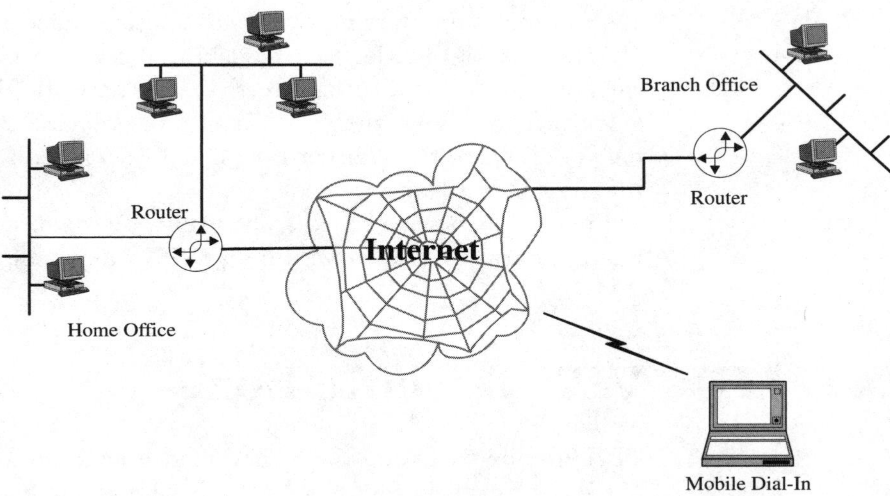

Figure 4-8 Compatible routers are used at each location for VPN services.

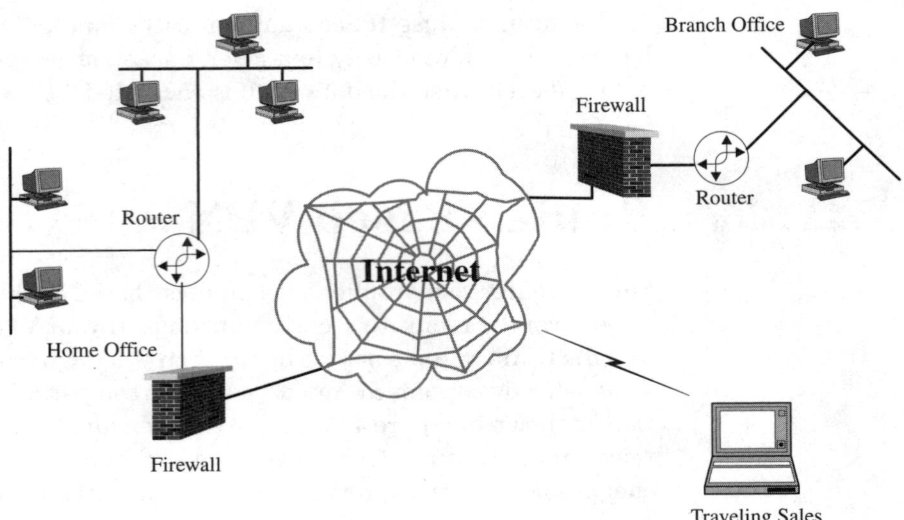

Figure 4-9
Stand-alone firewall

Firewall-Based VPN

The very same issues exist here as with routers. One needs to have compatible (preferably the same vendor's) firewalls at each location. Mobile users or telecommuters must have compatible VPN software. Firewalls are always potential bottlenecks, so asking them to perform VPN encryption can adversely affect all other access to your network. Here again, there is no substitute for traffic analysis. We only recommend this solution for small networks where the traffic through the firewall can easily be handled by the firewall hardware.

Figure 4-9 shows a stand-alone firewall hardware that filters all traffic into our network, while maintaining VPN functionality.

VPN-Specific Boxes

VPN specific boxes are the recommended solution for high volume, large networks. Several vendors offer these solutions in both hardware and software incarnations. The general rule is that hardware boxes will outperform software boxes and are theoretically more secure because they are based on proprietary technology that is harder to hack than publicly available operating systems. (A hardened Unix-based system is also extremely difficult to hack.) Traffic volume and feature support for remote terminals and industry compatibility will guide your decision here.

These boxes set up secure tunneling by using IPSec encryption and certificates as described previously. They are typically installed in parallel with your firewall. The firewall handles web (HTTP) requests, while the VPN box handles access to your internal database. Figure 4-10 shows the firewall and VPN box in parallel, reinforcing the division of labor between the two boxes. Because we now have two "holes" into our network, it is imperative that we have the permissions and access rights set up correctly. The firewall should not let users in who would be required to authenticate via the VPN box.

The integrated solution that some vendors are offering is an integrated custom box that does routing, firewall, and VPN all under one roof. This is an attractive option where traffic volume and performance is not going to be an issue. Again, Figure 4-8 or 4-9 might be used to depict this configuration.

Figure 4-10 The firewall and VPN box working in parallel

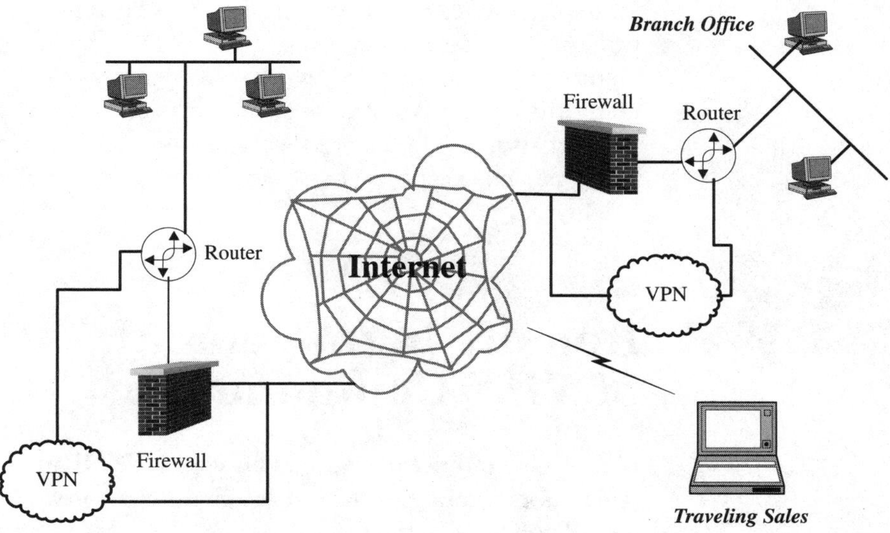

Throughput Comparison

Unfortunately, although there is compatibility testing, there are no consistent performance criteria across the industry. It, therefore, becomes difficult to compare the performance of different vendor offerings. Vendor claims tend to be exaggerated. They will measure their product in the best possible light (for example, maximum-sized packets and data compression turned on, using the simplest encryption algorithm). Our recommendation is to search the periodical literature for tests on the vendors you are considering as a starting point.

Then, in your *request for proposal* (RFP), specify a test sequence. With encryption and authentication, there is a lot of end-of-packet processing. This causes a significant performance hit when packet sizes are small.

The number of simultaneous sessions also affects performance. Vendors claim thousands of simultaneous sessions, but ask them how many they can set up or tear down at a time, and the number drops to fewer than 100. Notice also that during this peak-processing load of session setup, overall throughput will be affected.

Here again, having knowledge of how your users use the system, when the peak sign-on demand occurs, when the peak traffic occurs, and what kinds of response time you consider to be reasonable all influence your product selection. By the way, being able to set up 100 sessions/second is plenty in a 1,000-user network. (How many of these users are actually using the VPN?) Worst case (which statistically never occurs) means that the last user might have to wait 10 seconds to get a session setup. Most likely, no one except the network manager with the Sniffer will ever notice a delay.

Remote Management of VPN Components

If you have only two locations on your VPN, then remote management of policy is probably not an issue. For a large network, visiting each site to install policy rules becomes a burden. For larger networks then, look for the ability to provide remote policy management of not only your VPN devices, but also your firewalls and routers securely.

Cost Considerations

Figures don't lie, but liars know how to figure.

Although we're presenting some typical numbers here, you should run the numbers using your own particular configuration. The most beneficial comparisons of a VPN occur when compared to a dedicated, line-based network or one that makes extensive use of long distance dial-up lines. If you are already using a shared network (Frame Relay or ATM), the cost savings are not so striking.

Consider that a VPN box at each location might cost $5,000 including installation; multiplied by seven sites is equal to $35,000. Now, how long will it take to save this cost if you substitute your ISP charges for each location and subtract the cost of your existing T1 or Frame Relay network? If you had six T1s at $5,000/month, you might now have seven T1 access lines from your ISP at $3,000 or $4,000/month. The $7,000/month savings will pay off the $35,000 investment in 5 months. If your Frame Relay service is costing $1,000/month per location, the break-even point doesn't happen in any reasonable period.

Using remote access server and dial-up lines is cheaper to install, costing about $6,000 to $7,000 for about 20 users to install at the central location. Now comes the big bite, which is the long distance charge from all the remote locations. This could easily grow to $5,000/month if each of the users spent two hours online. Each working day at $0.10/minute is about $8,000/month. Plug in your own assumptions as to duration and cost of telephone calls here. (Even at 1 hr/day and $0.06/minute, that is $2,000/month for 20 users). A VPN system might cost $14,000 to install, including licenses for PC software at each location. The ISP charges that are $20/user/month, plus an ISDN line at the home shop for $100/month, means that we are saving $1,500 in monthly charges. We can pay off the system in 10 months. Again, do not assume that it will pay off in all cases. But, in all cases, it is worth the effort to perform the calculations.

Your VPN will definitely require more network management than a dial-up system, so the cost of perhaps an additional system administrator may have to be added.

Proprietary Protocols

Most VPN products are designed strictly around IP. They will often handle other protocols, such as AppleTalk and IPX, by tunneling them inside of IP packets. This introduces both overhead and delay. If the amount of "foreign" protocol traffic is small, then this is not significant. If the bulk of your network is IPX or Apple talk, we recommend you investigate VPN vendors who will support these protocols in native mode.

VoIP VPN

The justification for doing VoIP on a VPN is primarily security, along with the reduced cost of VoIP. Depending on usage, voice generates relatively large amounts of traffic. Be sure to include this additional traffic in your sizing estimates.

Our discussion of VoIP applies to whether we have a VPN or not. With a VPN, the delays due to encryption are larger, and therefore we would expect that the performance of voice over the VPN would be worse than VoIP. If we have chosen a network provider who will offer a SLA with QoS, there is a better chance for success, but the delays due to encryption and basic packet switching will still be there. With the exception of international calling, one must have a very large calling volume to make it worthwhile to put voice over the Internet and suffer the attendant quality reduction.

Summary

VPNs can provide a cost-effective solution to have secure communications across the Internet. Performance can be improved by utilizing a national/international ISP that will offer SLAs and QoS. Choosing hardware-based over software-based VPN equipment will generally provide better performance. Choosing VPN vendors who embrace standards and support multiple standards increases your flexibility to your vendor/equipment choices. Knowing your current and anticipated traffic volumes permits you to make improved cost performance studies.

CHAPTER 5

Advanced Intelligent Networks (AINs)

The *Intelligent Network* (IN) has been under development since Bell Communications Research first introduced it in 1984. The goal of intelligent networking is to integrate the features and benefits on the new-generation networks and to allow various types of information to pass through the telephone network without the need for special circuits. Data communications, Internet communications (using *Internet Protocols* [IPs]), and voice networking have converged to provide a new and exciting set of services. These services revolve around the backbone of the IN, which uses *Signaling System 7* (SS7). Network architects envision one network capable of moving any form of information, regardless of the bandwidth. Data and voice calls will traverse this network the same way—making communication as simple as placing a traditional telephone call.

Intelligent Networks (INs)

The INs consist of intelligent nodes (computer peripherals), each capable of processing and communicating with one another over low- to midspeed data communications links. All nodes in the intelligent SS7 network are called signaling points that work with packet transmissions. A signaling point has the capability to do the following:

- Read the packet address
- Determine if the packet is for that node
- Route the packet to another signaling point

Signaling points provide access to the SS7 network and the various databases on the network. They also act as transfer points.

More information will be explained in the SS7 chapter later. However, the switching network contains *Service Switching Points* (SSP) and provides the basic infrastructure needed to process calls and other related information:

- The SSP provides the local access because it emerged as the *Central Office* (CO). The SSP can also be other tandem points on the network or an ISDN interface for the *Signaling Transfer Point* (STP).
- The STP provides packet switching for message-based signaling protocols for use in the IN and for the *Service Control Point* (SCP).
- The SCP provides access to the IN database. The SCP is connected to a *Service Management System* (SMS).
- SMS provides a human interface to the database as well as the capability to update the database when needed.

- *Intelligent Peripherals* (IPs) provide resource management of devices such as voice announcers. IPs are accessed by SCPs when appropriate.

The IN enables customers to tailor their specific service requirements within hours instead of days. It is expected that full IN implementation will continue to evolve through the new millennium; however, the infrastructure has been laid and work continues. Some of the features available include the following:

- Find-me service
- Follow-me service
- Single (personal) number plans
- Call routing service
- Computer control service
- Call pickup service

Business is conducted by using a mix of public services, private networks, the Internet, wireless networks, and specialized carriers. Voice-processing requirements for a telephone switch were quite modest and could be satisfied by the limited announcement capabilities offered by the switch, usually as sequenced announcements of greetings, instructions, and terminating messages. No specialized service creation environment was required to develop and deploy these services. Complex voice-processing capabilities, such as CO-based voice mail were provided, but not as an integrated service of the switch. Complex voice-processing capabilities grew outside of the switching network in the form of interactive voice response systems, voice mail systems, automatic call distributors, and automated attendants. Over time, these systems embraced new technologies, including facsimile, speaker-dependent and speaker-independent voice recognition, text-to-speech, and voice identification.

Advanced Intelligent Networks (AINs)

AIN is a collection of components performing together to deliver complex call-switching and handling services. The SSP is the CO that provides robust, call-switching capabilities. When switching decisions require complex call processing, the SSP relies on the SCP, a subscriber database, and it executes service logic. The SSP uses SS7 signaling, specifically *Transaction Capabilities Application Part* (TCAP) messages, requesting the SCP to determine the best way to handle the call. The process supports telephony features, including 800 (888/877) and 900 calling, credit/debit card calling, call forwarding, and *virtual private networks* (VPNs). AIN has promised an architecture that is amenable to the rapid development and deployment of new services. How to maintain the stringent performance requirements of a CO service within this rapidly changing environment is a major challenge in the advancement of AIN.

IPs and *service nodes* (SN) are elements of the AIN and must be reliable to be deployed and used in the CO. IPs work in cooperation with SCPs to provide media services in support of call control. Service nodes combine the functions of the SCP and the IP. When viewed as point nodes in the network, these elements (IP and SN) are subject to failure and require redundant components and multiple communication paths. Software and procedures in support of CO reliability are also required.

When switching decisions require complex voice-processing services, the SCP cannot always provide all the required services to the SSP. The SCP also cannot provide termination of voice circuits and play recorded messages, collect touch-tone input, or perform other voice-processing services. The call must be redirected to an IP. The IP provides the voice-processing services not available from the SCP.

AINs provide more features and functions that are not provided by INs. AIN does not specify the features and services, but how end users use them. An essential component in AIN is the *Service Creation Element* (SCE). Today, Telco personnel at end offices handle service configuration and changes. The SCE specifies the software used to program end office switches. The single most important change is through a *graphical user interface* (GUI). Over time, the SCE will reside at the users' organization, allowing customers to tailor their services on an as-needed basis, without telephone company assistance.

Some of these enhanced features available include the following items:

- Calling name delivery
- Call rejection
- Call screening (visual or audio)
- Call trace
- Call trap
- Personal ID numbers (pins)
- Selective call acceptance

- Selective call forwarding
- Spoken caller identification

Some of these features are already available at certain COs, but they are not yet ubiquitous. Limitations to these services are based on the capabilities of end office switching equipment—service offerings and tariffs will not be consistent throughout telephone companies.

Information Network Architecture

Information Network Architecture (INA) is still in development and viewed by many as the successor to AIN. However, there is considerable controversy over this view, and some believe that two architectures will eventually develop. AIN is designed to facilitate the voice network, whereas INA will manage the broadband network. The common belief is that INA will provide better utilities for managing new broadband services offered by telephone companies.

Intelligent networking delivers computer and telephone integration capabilities inside the network. Two major market forces and architectural frameworks are merging to create the most explosive network services opportunities of the late '90s. Enterprise *Computer Telephony Integration* (CTI) applications and AIN services are being integrated to provide an array of advanced carrier-delivered services. Some of the features that will be available to the executive or professional, especially telecommuters, include the following items:

- Virtual call centers
- Consumer interactive applications
- Centrex productivity enhancements
- Formal and informal call centers
- Virtual *Automatic Call Distribution* (ACD)

Combining AIN and CTI Services

The evolution of AIN and CTI services underpins the marriage of these two architectural frameworks. AIN has its roots in the *Local Exchange Carriers'* (LECs) (both *Incumbent LEC* [ILEC] and *Competitive LEC* [CLEC]) and

Interexchange Carriers' (IEC) desire for vendor- and switch-independent network architectures.

Improving the speed of service provisioning and delivering advanced network services is crucial to maintain a competitive posture. As early as 1986, Ameritech began proposing a concept called *Feature Node Service Interface* (FNSI). Through successive industry efforts, AIN emerged as a network standard in the early '90s. Figure 5-1 is the basic framework of the AIN architecture.

CTI is a complement to AIN proposed by the *Private Branch Exchange* (PBX) manufacturers. The end user's desire is as follows:

- Advanced applications
- Faster feature delivery
- Control and customization of applications all driving the development of the CTI interfaces

Using an external processor, custom applications are configured to enhance the functionality of a PBX, especially in an ACD environment. Figure 5-2 shows the basic connections in a CTI application.

Both CTI and AIN use an external processor to deliver advanced complementary services to the switch. The switch controls call-processing

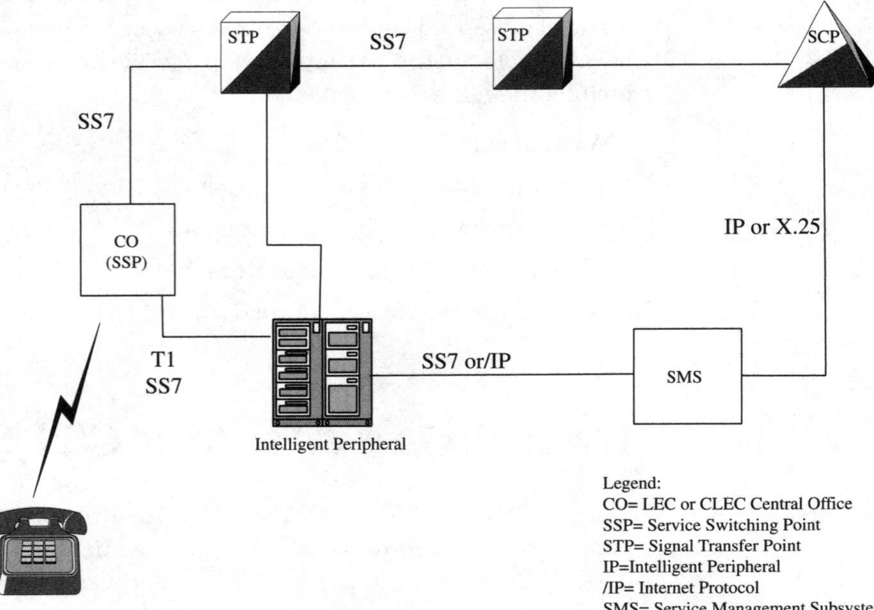

Figure 5-1 AIN architecture framework

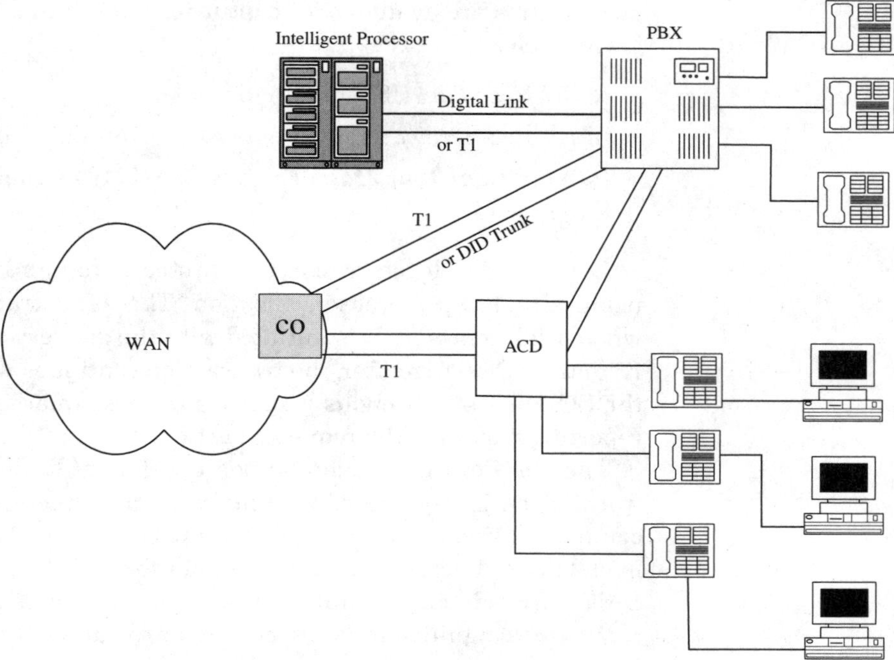

Figure 5-2
Basic CTI application

access to the external processor via an open interface. Both provide a GUI-based SCE for rapid service and application configuration. IPs provide additional context information for call treatment. Integrated voice response units are the most widely deployed IPs. Reporting, billing interfaces, and real-time monitoring tools are available in both.

With the introduction of the IN, these two technologies were brought together by necessity. Services rich in voice-processing content and requiring an abundance of digitized voice storage could no longer be created solely with the innate capabilities of the switch. The variety of service offerings, their complexity, and, in many cases, the requirement for multilanguage support, quickly outpaced the voice-processing capabilities and capacities of the switch. Voice-processing systems were integrated with switching systems to support these new services.

The Intelligent Peripheral (IP)

The *Intelligent Peripheral* (IP) must provide voice circuit support and some mechanism for obtaining information about an incoming call. Several mechanisms are available for capturing call information including the following items:

- *ISDN User Part* (ISUP) signaling
- *Integrated Service Digital Network* (ISDN) signaling
- In-band *Dual Tone Multi-Frequency* (DTMF) signaling
- DTMF

BellCore developed transaction protocols to handle the call processing using an 1129-protocol specification. The 1129 transaction is triggered when a call comes in. It is initiated with the delivery of a message from the IP to the SCP. Thereafter, the transaction continues with queries issued by the SCP and synchronous responses to these queries returned by the IP, reporting results of the requested action.

The BellCore recommendations call for multiple IPs within the network. Each is capable of working with multiple independent SCPs via communication across multiple data paths. Each IP operates as though it is distributed across multiple host platforms interconnected by multiple LANs. Introducing IP into an AIN environment is a major expense, requiring a significant investment in hardware, networking, and supporting software.

The BellCore philosophy is to provide redundant components and data paths, eliminating single points of failure wherever possible. However, many situations exist whereby an IP or SN provides a service, yet the service does not warrant a redundant infrastructure. Therefore, a solution is required for the IP or SN to provide suitable reliability inherently.

IP Services

The IP must be capable of establishing and maintaining communication with multiple SCPs. Furthermore, it handles the following functions:

- Encoding and decoding complex messages
- Interpreting these messages
- Performing the requested service

It must also be capable of switching functions including the following:

- Call setup
- Transfer

- Call teardown
- Detect call presentation and abandonment
- Process requests requiring service logic
- Access databases.

Finally, beyond the capabilities described in the 1129 interface specification, the IP needs support services such as logging, administration, alarm processing, statistics gathering and reporting, and database and network access. The IP falls into two categories of service: application processing and resource processing.

The application processor supports the call-processing logic and access to the databases and data networks. It initiates transactions, receives instructions, and reports success or failure of the action. The application processors also require connections to communication networks and hosts. An IP needs multiple data paths because it must be able to contact multiple hosts and must be able to survive the loss of a single communication pathway. It needs to communicate with SCPs to receive instructions and communicate with switching facilities in order to terminate circuits. It also needs to communicate with other network elements, such as the SMS (for provisioning), and to internal and external databases. This requires multiple data paths and alternate routing.

The resource processor manages the switching, voice channel facilities, and media-processing resources. The media processing normally requires multiple processors and disk storage to handle the media streams. It may also require switching, such as a *time division multiplexer* (TSM) or external switches.

These two elements are very different from each other. They perform different functions; consequently, they require very different hardware and software processing elements.

The architecture for AIN with the SSP, SCP, and IP in place on the stack is shown in Figure 5-3.

Software Architecture: Client, Router, Server

The objective is to build an IP in which application processors drive multiple resource processors. This requires software on the application processor capable of supporting the physical structure. A client-router-server scheme is used.

Figure 5-3
AIN protocol stack

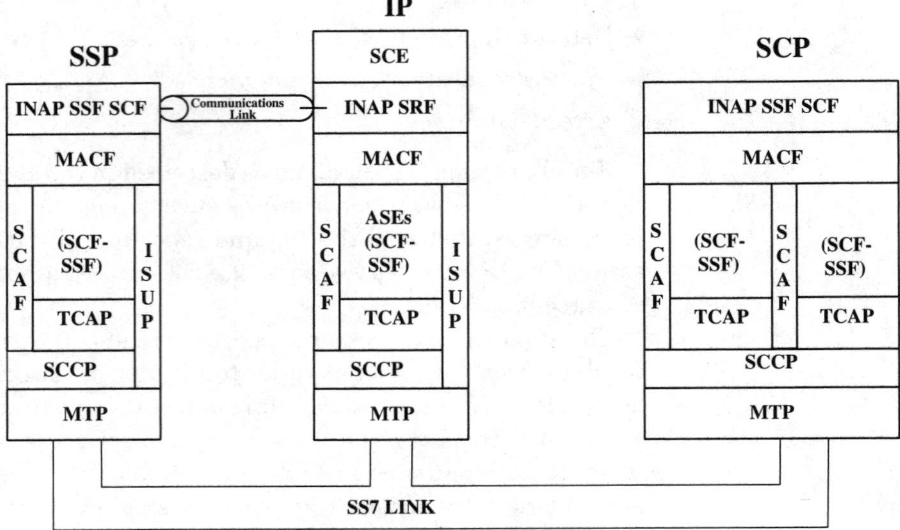

The client includes the application, media processing logic, and controls; manages the resources; and drives a state machine. The IP, driven by the 1129 message set, operates as a state machine. This state machine is driven by messages from the SCP and by trigger events from the network.

The router handles message routing and session management functions. It makes decisions about how to route traffic and load-balance.

The servers support both local and remote devices. The local devices include local file systems, databases, and locally attached devices such as encryption boxes. The remote devices are networks and hosts. The remote servers must be multiple processor capable.

The Application

The application must be capable of supporting real-time, online updates, which must be done without any disruption of service. The system administrator must be able to introduce new instructions and commands/ responses without affecting service. On-the-fly changes to parameters, such as timers and retry limits, are necessary. The state machine itself must remain operational without call loss while the application logic is being changed.

Results of AIN

It is possible to build IPs in support of the AIN. These peripheral devices must be economical and reliable enough to operate in a CO environment. Multiple hardware elements are required. Communication paths must be redundant.

Estimates are that by 2007, telephone industry spending will approximate $27 billion. Products and expenses supporting AINs will expand exponentially. Approximately $5 billion was spent on AIN products and services in 1998. Telephone industry spending on AIN will include STPs, SCPs, SSPs, IPs, SCE, and various hybrid products.

Recent studies indicate that the demand for AIN features and functions will be driven by the following factors:

- **Demand for more customer control and services** Businesses are placing more strategic reliance on telecommunications. They need features and functions tailored to their specific needs. The number of functions and services required will be dictated by the lifestyle and business-competitive environment of the end user. AIN is the only way to support such requirements. Services will have to be user-friendly and somewhat network-centric to meet the demands of the future user.
- **Greater geographical distribution and newer technologies** Future niche markets will probably emerge. They need to be widely available by the suppliers (LEC and IEC) to be effective and acceptable. AIN services must also become technology stable to support multiple vendor products and service on multiple switching systems. AIN-based services must find a way to work across carrier boundaries.
- **Mobility and mobile applications in a changing world** The business and personal use of communications support from cellular phones, voice mail, pagers, and e-mail is growing exponentially. The need to support a mobile user is now equally important. AIN-based services must cross vendor products and billing mechanisms to be effective. Wireless networks are among the fastest growing applications and services in the industry. The growth rate of 300 percent per year shows little sign of slowing. *Personal Communications Services* (PCS) networks are growing equally fast and demanding more services. Soon the AIN will have to support the single number for a user, regardless of where that user may reside.

- **Internet, Broadband, and Multimedia** A *Communications Industry Reports* (CIR)[1] report notes that most of the services that AIN is currently concerned with are narrowband and voice-oriented. However, the CIR report projects the belief that many of the services currently identified with AIN will migrate to the Internet. In addition, AIN concepts will increasingly be required for multimedia and broadband services. The market for the next few years around the world is shown in Figure 5-4.

Focus

SS7 is an essential technology supporting and developing ISDN, INs, mobile (cellular) telephony, PCS and information systems, personal communications, and many other applications. Currently, this technology is being pursued by the telecommunication industry, including carriers, large private networks, switch manufacturers, and an increasing number of software developers. The traditional methods of in-band signaling and other common channel signaling methods in the telecommunication networks have given way to an overlay network using a more capable, layered SS7 protocol.

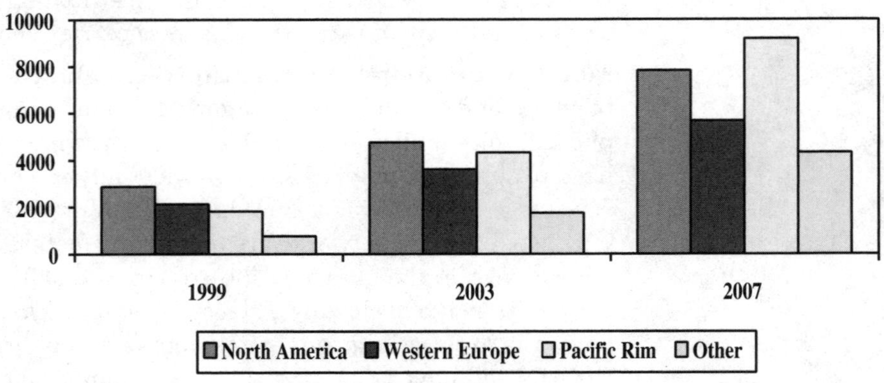

Figure 5-4 Growth in AIN services around the world

[1]CIR is a research firm specializing in market research and analysis of the telecommunications industry.

Although the primary function of SS7 is to handle call-control-signaling requirements for voice and data transmission services, it can provide a number of advanced services by use of network databases. These include toll-free and alternate billing, rerouting, virtual networks, and other highly sophisticated telecommunications services denoted by IN services. Additionally, the transaction capability of the SS7 protocol makes it applicable to a broad range of new services dealing with remote operations. In cellular telephony, SS7 is used for mobility management and handover functions.

An IN describes architecture designed to facilitate the implementation of highly sophisticated telecommunications services including the following:

- Features and functions
- Multivendor interworking
- Differing priority services offered by fixed networks, mobile networks, and personal communications systems

The new millennium will hold several benefits and service advantages unavailable in the past. These advantages will all be geared toward satisfying the changing needs and demands of the consuming public. As IN and AIN evolve, newer services will be introduced. The various new providers, such as the CLECs and the CATV companies, will compete to meet the one-stop shopping demands of the customer. AIN will be one of the deciding factors steering consumers to the various providers. The first providers to implement the AIN features will have the edge over taking the customer base.

The edge goes to the ILECs for now, because the infrastructure is theirs. However, as the new millennium rolls through, changes will occur. This is one of the critical components in being able to capture a niche in the market.

CHAPTER 6

Local Number Portability (LNP)

With the Telecommunications Act of 1996 in the United States (and the Telecom Act of 1997 in Canada), a series of competitive changes were required in the network. For years when competitors tried to compete with the Bell System, their efforts were stymied. Part of the reason was the competitor's ability to serve a new customer. Whenever the new competitor offered to serve a business or residential customer, customers were required to change their business or residential telephone number to use the new *Local Exchange Carrier* (LEC). *Local Number Portability* (LNP) is, therefore, an essential issue in the telephony business. Because of the need to change numbers, the *incumbent LECs* (ILECs) were able to prevent losing their customers to the competition.

Three Flavors of LNP

LNP gives end users the ability to change Local Service Providers without changing their telephone numbers. Three basic forms of LNP were introduced to the industry over the past few years (the only one implemented is service provider portability):

- Service provider portability enables subscribers to change Local Service Providers without changing their telephone number. This assumes that users can change suppliers and keep their existing telephone number. Still to be completed is the ability to provide LNP in a wireless world.
- Service portability enables subscribers to change from one type of service to another (for example, analog to digital—ISDN—without changing their telephone numbers) or to be served from a different central office (where the service is available) and not have to take a new telephone number.
- Geographic portability enables subscribers to move from one physical location to another (such as state to state) without changing telephone numbers.

The *Federal Communications Commission* (FCC) mandated service provider portability in July 1996. Service provider LNP involves a circuit-switched network capability, enabling users on one switching system to move or port numbers to a different switching system. Congress mandated LNP, set the regulations governing its implementation, and stated that any network modifications required to comply with these rules were the responsibility of the existing LECs.

In February 1996, President Clinton signed the Telecommunications Act into law. The single most significant characteristic of the new law is that it opens the local exchange market to competition. In an effort to eliminate technical as well as regulatory market entry barriers, the law requires that all LECs—both ILECs and new *competitive LECs* (CLECs)—provide LNP. LNP provides "users of telecommunications services with the ability to retain, at the same location, existing telecommunications numbers without impairment of quality, reliability, or convenience, when switching from one telecommunications carrier to another."

The Road to True LNP

LNP is germane to achieving true local competition. In November 1994, the industry began to seriously investigate methods of providing true LNP. Although the importance of retaining a telephone number was recognized in the early 1960s, it became a significant issue associated with 1-800 (and later 1-888, 1-877, 1-866, and 1-900) services during the 1980s. The need for portability among telephone companies and providers was never really an issue until the *North American Numbering Plan* (NANP) administration published a proposal for future numbering plans in North America. This NANP document was issued in January 1993. Of course, since that time, the level of interest has increased substantially. As equal access and competition catch up to deregulation, the use of LNP becomes critical.

In late 1994, MCI commissioned a study by the Gallup Organization to assess LNP and determine the following:

- The potential for businesses and consumers alike to switch local telephone service providers under various market scenarios.
- The perceived importance of various service factors regarding local telephone services.
- Not surprisingly, the results of the MCI study indicated the following for residential customers:
  - Nearly 2/3 of all customers were unlikely to switch providers if given the opportunity.
  - More than 3/4 stated that retaining their telephone number or numbers when switching carriers was very important.
  - Eighty percent were unlikely to switch service providers if they would have to change their telephone numbers.

Moreover, business customers showed the following results:

- Fifty-seven percent were unlikely to switch local telephone service providers if given the opportunity.
- Eighty-three percent said that retaining their telephone number or numbers when switching Local Service Providers was extremely important.
- Ninety percent were unlikely to switch providers if they would have to change their telephone numbers.

With business customers, the cost of changing numbers can be significant. When a business changes telephone numbers, the costs include reprinting stationary, business cards, advertisements, and literature. The statistics are shown in Figure 6-1.

These discussions led to an organization's development of the *Carrier Portability Code* (CPC) model, which was selected by the New York State Public Utility Commission LNP task force as the architecture to be used. MCI, with the support of several manufacturers, demonstrated the architecture to the FCC via a live test in May of 1995 to prove that true LNP was indeed technically feasible.

On June 27, 1996, the FCC adopted rules on LNP. The FCC required LNP availability on a permanent basis as of October 1, 1997. This availability must be complete for the top 100 *Metropolitan Statistical Areas* (MSA) by December 1998. After the December 1998 dates, all LECs (ILECs and CLECs) in other areas are required to provide LNP within six months upon request. The FCC actually adopted performance criteria, rather than a specific technology to meet the need. The criteria includes the following information:

- Support of existing services and capabilities
- Efficient use of numbering plans
- No change in numbers for end users
- No requirement to rely on databases of another carrier in order to process (route) calls
- No unreasonable degradation of service or reliability when implemented or when customers switch carriers
- No carriers' proprietary interest
- Ability to accommodate location and service portability in future years
- No adverse impact in outside areas where LNP is deployed

States will have flexibility if they meet the criteria listed previously. Wireless carriers have been granted a reprieve from the December dates,

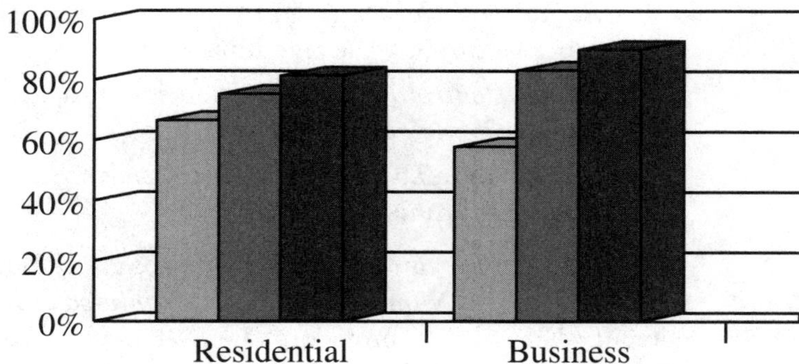

Figure 6-1 Comparison of business and residential user concerns over service provider change

originally enabling a June 1999 implementation date. This has since been postponed until 2002 and may be extended again. The *Telecommunications Industry Association* (TIA) continues to lobby against wireless carrier implementation of LNP. The obvious reasons are costs associated with the changes in the network. However, the wireless providers are the first to demand that the wireline carriers (that is, CLEC and ILEC) implement LNP so that the wireless provider can offer a single phone and number to their customers. The PCS service providers continue to offer customers a single number for their home and their cell phone. Moreover, the wireless carriers offer wider coverage areas that differ from the wireline carriers' coverage areas and the costs associated with placing and receiving a call.

Shortly after the studies were completed, several states began the process of officially selecting the architecture to be used for LNP in their respective states. The Illinois task force requested LNP solutions from a wide array of companies via a *Request for Proposal* (RFP) developed by the carriers that offered service in the state of Illinois at that time. An official voting body, which was comprised of those carriers, was established to select the architecture. After considerable discussion and deliberation, Lucent Technologies' *Location Routing Number* (LRN) architecture was selected.

Basic LNP Networks

The components of LNP are not that much different from the original *Signaling System 7* (SS7) networks used for years. The pieces serve different functions, which you can see by looking at the following components:

- The *Switching Service Point* (SSP) is the local CO or tandem switching office.

- The *Signal Transfer Point* (STP) is a packet mode handler that routes data queries through the signaling network.
- The *Signal Control Part* (SCP) is the database for features, routing, and *Global Title Translation* (GTT).
- The LSMS is the *Local Service Management System*; SOA is the *Service Order Administration*.

LSMS and SOA can be provisioned separately, but when combined it is referred to as the *Number Portability Administration Center* (NPAC) connectivity. For wireless providers, the LNP capabilities depend on the MSA served. Wireless providers' schedules keep changing because of the lobbying of the CTIA. See the network overview in Figure 6-2.

Figure 6-2
Basic LNP network

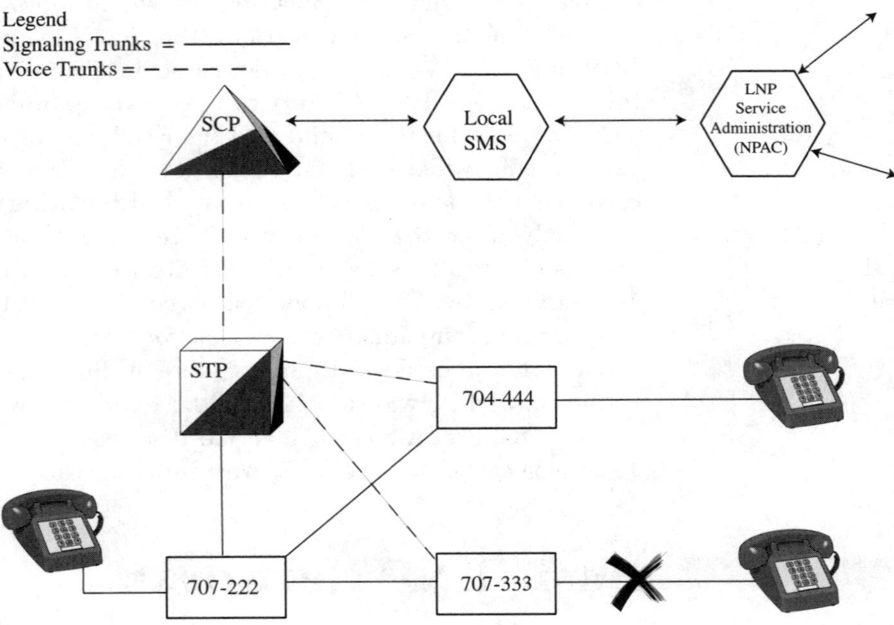

The Terminology

Several terms are introduced with LNP. Some of the more common ones are as follows:

- *Portable number* A number that is permitted to move between different service provider switches
- *Ported number* A *Directory Number* (DN) served by a switch other than the switch that has traditionally served that number
- *Nonported number* A DN that is portable but is currently served by the switch that is identified in the *Local Exchange Routing Guide* (LERG) as serving that number

These three concepts are shown in Figure 6-3 depicting the terminology.

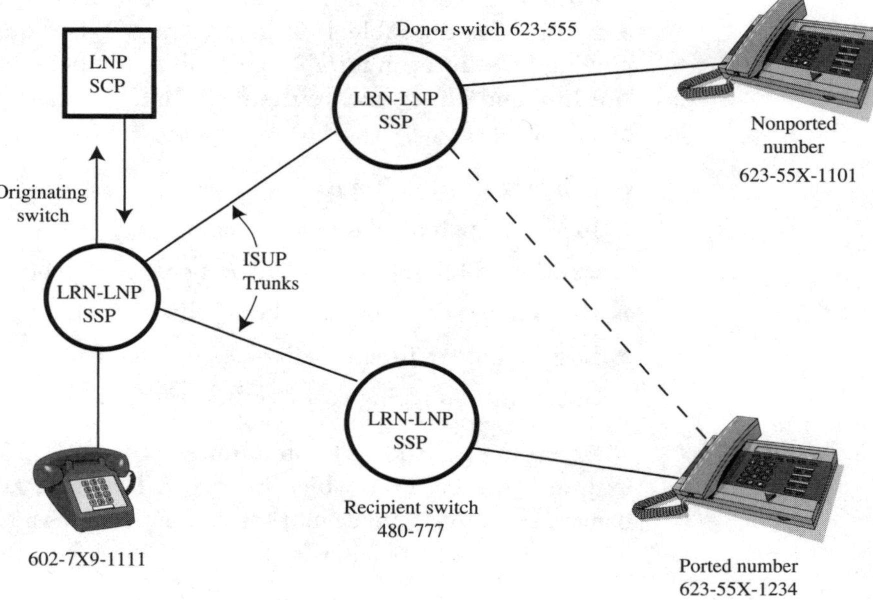

Figure 6-3 LNP terminology

Before LNP

In the earlier, non-LNP environment, a telephone number performed two functions:

- It identified the customer.
- It provided the network with information necessary to route a call to that customer.

LNP separates these two functions, providing the means for customers to keep their DN when changing Local Service Providers. By separating those two functions, LNP gives customers the flexibility to respond to pricing and service changes offered by rival carriers.

As we have seen, numerous studies conclude that most business and residential customers are reluctant to switch from one service provider to another if they must change their telephone numbers. Without LNP, new entrants (CLECs) would have to price their local exchange service 12 to 15 percent lower than the existing LECs in order to persuade customers to switch carriers. Although the degree to which the lack of LNP hurts competition is arguable, it is clear that LNP is required to provide a level playing field. Interim number portability methods, such as remote call forwarding and direct inward dialing, exist, but these methods have several disadvantages:

- Longer call setup times
- Increased potential for call blocking
- Continued reliance on the incumbent LEC's network
- Use of more directory numbers, which are fast becoming depleted
- Loss of feature functionality
- Substantial ongoing costs to the new Local Service Provider.

According to the Telecommunications Act, LNP will promote local exchange competition, which in turn will benefit all customers. As it has done in the long-distance market, competition in the local exchange market is expected to do the following:

- Drive down the cost of service
- Encourage technological innovation
- Stimulate demand for telecommunications services
- Boost the United States' economic growth

Number Administration and Call Routing in the Network

Telephone numbers in a pre-LNP environment have always been assigned to Local Service Providers' end offices on an area code and exchange code (*Numbering Plan Area* [NPA-NXX]) basis. Each NPA-NXX contains 10,000 telephone numbers. Because an NPA-NXX is only served by a single end office in the United States, the telephone number identifies the person as well as the actual end office that serves that person. In effect, the dialed NPA-NXX is the terminating switch's routing address to the rest of the network. With the implementation of LNP, which enables any number of Local Service Providers to serve the same NPA-NXX, this routing scheme can no longer be used.

LRN

With LRN, every switch in the network is assigned a unique 10-digit number that is used to identify it to the rest of the network for call routing purposes. An essential advantage of the LRN is that call routing is performed based on today's numbering format. LRN uses the strength of SS7 signaling with *Multi-Frequency* (MF) interworking and promises to be a long-term solution for LNP. LNP is SS7, *ISDN User Part* (ISUP)-oriented and therefore does not work well with MF interworking.

LRN-LNP on the *Switching Service Points* (SSPs) performs the integral part of the overall network LNP solution. The LRN is shown in Figure 6-2 with a new, 10-digit number assigned to the individual switching points. LRN depends on *Intelligent Network* (IN) or *Advanced Intelligent Network* (AIN) capabilities deployed by the wireline carriers' networks. LRN is a 10-digit number to uniquely identify a switch that has ported numbers. The LRN for a particular switch must be a native NPA-NXX assigned to the service provider for that switch.

LRN assigns a unique 10-digit telephone number to each switch in a defined geographic area. The LRN now serves as the network address. Carriers routing telephone calls to end users who have changed from one carrier to another (and kept their same number) perform a database dip to obtain the LRN corresponding to the dialed telephone number. The database dip is performed for all calls where the NPA-NXX of the called number has been flagged in the switch as a portable number. The carrier then routes the call to the new provider based on the LRN.

Shown in Figure 6-4 is the flow of LRN information, which is the same for both a wireline and a wireless provider (with one modification: the JIP is not used in the wireless network and LRN flow from wireless).

When the caller dials the number (333-3333 in this case), the originating switch (612-222) sends its signaling information (info dialed 333-3333) through the STP to the SCP, which analyzes the route and returns the LRN (612-444-0001). Next an IAM message is forwarded from 612-222 through the STP to the access tandem. The *access tandem* (AT) translates the LRN and sets up a speech path (trunk) from 222 to 444. Switch 612-444 detects the LRN (612-444-0001) as its address; therefore, the called number and the generic address parameter are swapped. From there, the call is connected (terminated) at 333-3333. The donor switch in this scenario has been uninvolved throughout the process.

Similar to the 800-number service, a database is used to store the routing information for subscribers who have moved or ported to another Local Service Provider. The LNP database contains the directory numbers of all ported subscribers and the location routing number of the switch that serves them. The LNP database can be accessed by switches using either the *Advanced Intelligent Network* (AIN 0.1) or *Transaction Control Application Part Intelligent Network* (TCAP IN) protocols. Each OLNP-capable

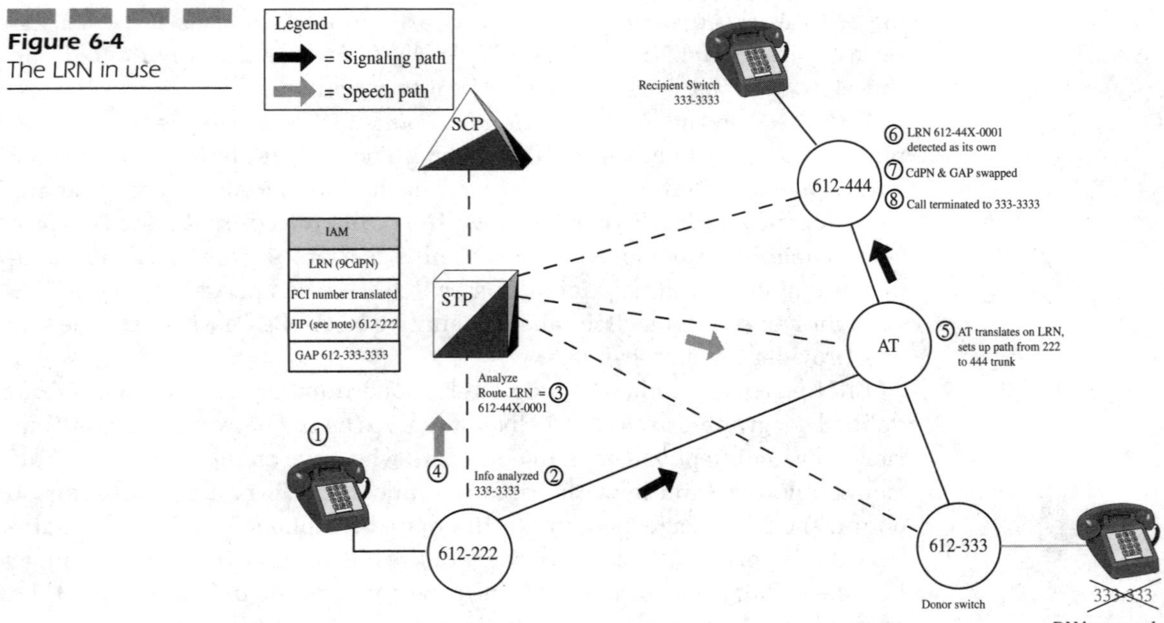

Figure 6-4 The LRN in use

switch in a portability area is assigned a unique 10-digit location routing number. A switch is defined as LNP-capable if it has the capability to launch LNP database queries and route calls based on the returned response. The 10-digit LRN is chosen from an NXX native to that switch (an NXX that was originally owned by that switch prior to LNP) and is used by other switches to route calls to that switch.

New routing and signaling methods are required to implement this feature because the LRN architecture requires the transport of the network routing address (or the LRN of the terminating switch) and the *called party number* (CdPN). Only the ISUP signaling is being modified to carry the additional information needed to support LNP. Workarounds to use MF trunks, such as sending only the dialed number when a MF route is used, have been included in the requirements.

The LRN feature uses the LRN returned from the LNP database to route the call and the CdPN to complete the call in the terminating switch. From a protocol perspective, the LRN of the end office serving the subscriber is placed into the CdPN and the actual CdPN is populated in the new *Generic Address Parameter* (GAP) field. As the call traverses the network, all switches will use the LRN to route the call. When the call is delivered to the terminating switch, the terminating switch will compare the LRN received in the CdPN field against its own LRN. If these numbers match, then the terminating switch will retrieve the CdPN from the GAP parameter and complete the call to the subscriber.

Once a switch performs an LNP database query, that switch must set a new bit called the *Translated Called Number Indicator* (TCNI) in the forward call indicator parameter of the *Initial Address Message* (IAM). This will indicate to other switches in the call path that a query has already been performed. This new bit and the corresponding call-processing logic ensure that multiple queries will not be performed for the same call.

Using a Database Solution

Many carriers felt that the best way to implement LNP is to establish databases that contain the customer routing information necessary to route telephone calls to the appropriate terminating switches. The LNP database, similar to that already used by the telecommunications industry to provide 800-number portability, will use intelligent network or AIN capabilities. These capabilities separate the customer-identification information from the call-routing information.

Rather than establishing one national database, similar to the 800-number database, carriers will provide LNP via a system of multiple, regional databases. A national LNP database is not feasible simply because one database could not store the telephone numbers for the entire United States population or even the subset of the largest metropolitan areas.

Further, a regional database system offers specific advantages for carriers deploying LNP. Regional databases effectively reduce the distance over which carriers must transmit routing information. By minimizing that distance, a regional system reduces the associated routing costs incurred by the carriers. A regional database system also ensures that carriers will have all the number portability routing information they need to route calls between carriers' networks for that regional area.

Because many of the major carriers install their own SCP where the LNP database resides, a single access point must be provided to effectively manage and distribute updates to the common regional LNP database. These updates will provide all carriers with the changes made when end users port from one Local Service Provider to another. Most states have decided that an NPAC, similar to the arrangement used for 800-number portability, should be used. The LNP task force in Illinois issued an RFP to the industry to solicit technical (and financial) proposals for the NPAC. Lockheed-Martin was chosen to develop and administer the NPAC in Illinois. Other companies were selected by different regions, but complications arose, and Lockheed-Martin was selected to replace those who could not deliver. Management of the LNP database by a third party will ensure the security of all carriers' customer bases. The full scenario of a call process is shown in Figure 6-5, using the components of the network and the databases.

Triggering Mechanisms

Using LNP requires end office switching systems to determine if a dialed NPA-NXX has been declared open for portability. The switching systems must set triggers on a portable NPA-NXX to cause, or "trigger," a query to the LNP database and retrieve the LRN of the dialed number. Most switching vendors are using AIN 0.1 triggers. AIN 0.1 triggers were defined by industry requirements well before the development of LNP. These triggers are administered on the NPA-NXX digit string by using the administration capabilities of the switching systems. Unconditional trigger mechanisms

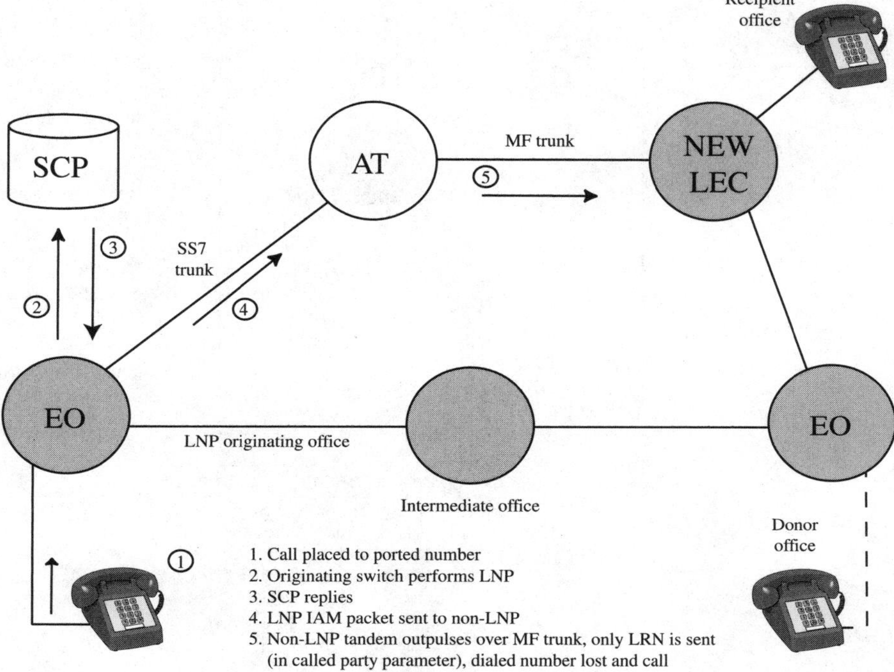
Recipient
office
SCP
AT
MF trunk
NEW
LEC
SS7
trunk
EO
LNP originating office
EO
Intermediate office
Donor
office
1. Call placed to ported number
2. Originating switch performs LNP
3. SCP replies
4. LNP IAM packet sent to non-LNP
5. Non-LNP tandem outpulses over MF trunk, only LRN is sent
(in called party parameter), dialed number lost and call

Figure 6-5
LNP scenario

are used during the transition from one service provider to another, such as the following:

- From a wireline to a wireless provider
- From an ILEC to a CLEC
- From a wireless to a wireline provider

These triggers can be assigned at the donor switch (the old provider) and the recipient switch (the new provider). By using an unconditional trigger, the trigger is used at the switch even if the directory number is present (see Figure 6-6).

The industry requirements for the location routing number model state that either a TCAP IN "800-like" or AIN 0.1 query can be used. The protocol defines the information and structure of the query between switching systems and the LNP database. Both of these protocols, defined within industry standards, are deployed in the network today. The AIN model is shown in Figure 6-7.

Figure 6-6
Unconditional trigger

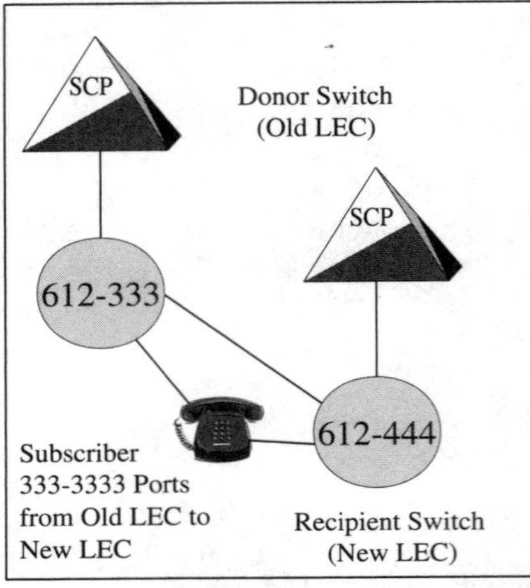

10 Digit Unconditional Trigger

Assigned to a line during the transition from the old to new service provider (LEC).

Can be assigned at both the Donor Switch (Old LEC) and the Recipient Switch (New LEC).

Unconditional trigger: triggers at switch even if DN is present, ESCDN ignored.

Triggers at Donor or Recipient ensuring no subscriber downtime or service outage during service provider transition.

Old LEC		New LEC	
Pre-SCP Update	Post-SCP Update	Pre-SCP Update	Post-SCP Update
Dialed number returned, terminate on this switch (Donor)	LRN returned, route to New LEC (Recipient)	Dialed number returned, route to (Donor) switch	Home LRN returned, terminate on this switch (Recipient)

How Is a Telephone Number Ported?

Many steps are required to move or port an end user from one Local Service Provider to another. Database changes will be required in the following locations:

- Old Local Service Provider's central office
- New CLEC central office
- LNP database

Beyond that, the end user's copper pair must be disconnected from the old service provider's central office equipment and connected to the new Local Service Provider's central office equipment or some patch field in a co-located office. All of these activities are controlled by the NPAC.

Several important steps must proceed the actual porting of a subscriber:

1. The NPA-NXX must be opened for porting by the NPAC. This drives the update of the *Local Exchange Routing Guide* (LERG) database, global title translations in the STPs, and end-office translations (to

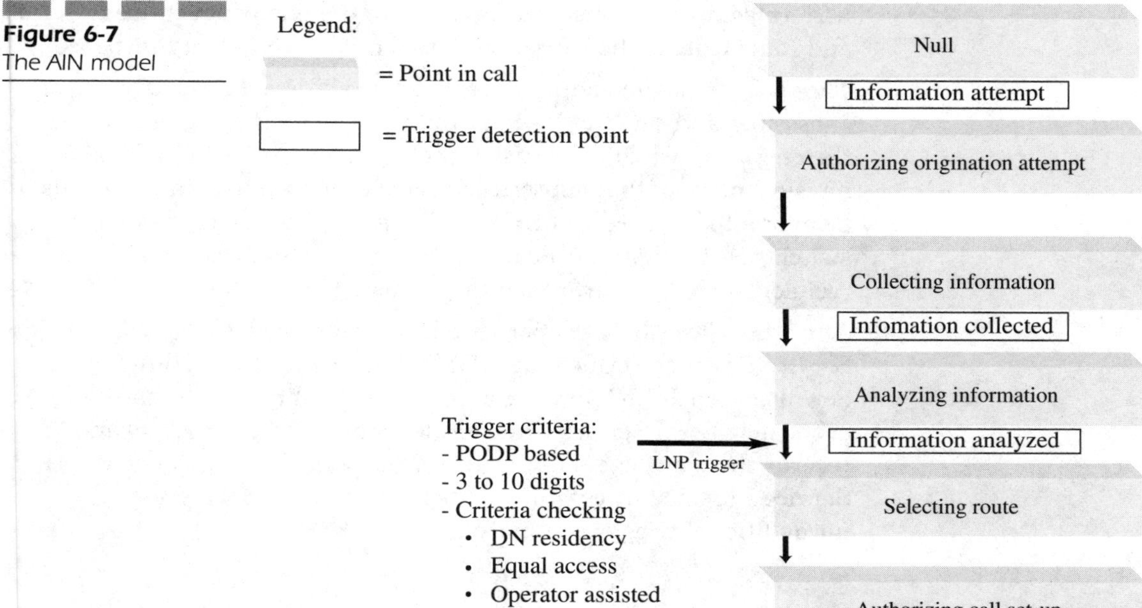

Figure 6-7 The AIN model

set the LNP trigger for the NPA-NXX) for all Local Service Providers in the portability area.

2. The two service providers must arrange the physical transfer of the subscriber's copper pair. Both providers must have implemented the 10-digit unconditional trigger function in their respective end offices.
3. Once all the pre-move actions have been performed, a subscriber can be ported. The availability of the unconditional trigger in both the new and old Local Service Provider's end offices will preclude the need to synchronize every step of the porting process.
4. The donor switch administers the 10-digit unconditional trigger on the porting subscriber's directory number. This will cause an LNP query under all conditions (even when the subscriber is still served by the old service provider's end office), thereby eliminating the critical timing coordination between the donor and the recipient (new) Local Service Provider. This enables the activities at the donor switch to be

performed autonomously with respect to the recipient service provider and route calls to the subscriber based on the LNP query response.

5. The recipient switch administers the translations for the new subscriber and sets the 10-digit unconditional trigger. This enables the recipient switch to provision the subscriber prior to the actual physical move of that subscriber's copper pair without causing calls to be misrouted during that time period. As in the case of the donor switch, the 10-digit unconditional trigger also enables activities at the recipient switch to perform autonomously.
6. After the subscriber's copper wire has been moved, the new Local Service Provider notifies the NPAC of the change. The NPAC then downloads each LNP database (in the portability area) with the LRN of the new Local Service Provider (for the ported subscriber) and records the transaction (for example, date/time). After the new Local Service Provider successfully tests the new subscriber, the unconditional trigger is removed.

Other Issues

The following topics are but a few of the effects that LNP brings to the telecommunications network.

Switching Systems

Substantial changes to call-processing logic and administration software are required in all switching systems in use in today's telecommunications network in order to implement LNP. The cost of the system upgrades and changes has been accumulated in a pool by the RBOCs and is now being passed on to consumers. The costs typically range from $.27 to $.54 per line per month. This pass-through cost will continue for five years (supposedly ending in 2004) as the plan calls for today. Additional effects will be evident when a *Line Information Database* (LIDB) requires a dip to determine the 0 plus calling card model, as shown in Figure 6-8. It is interesting to note that the fees for LNP have been in place since the year 2000, yet the LNP availability across North America is still primarily limited to the major downtown metropolitan areas. The consumer is paying for a service that is significantly limited and not being pushed.

Billing, Administration, and Maintenance Systems

Because LNP removes the direct association of a subscriber's directory number to a central office, substantial changes will be required in most of the systems in use today in local telephone company networks.

Signaling

LNP will require an LNP database query for every call to a ported subscriber that is not served by the originating switch. This will require capacity increases in the number of SS7 links to the signaling transfer points and the deployment of new service control points to run the LNP database application. These costs have also been accumulated, and they are now being passed through as part of the fee discussed previously.

Operator Services

Operator calls from subscribers (for example, 0-, 0+, and so on) are routed directly to the operator services system where an LNP query must be performed to determine the Local Service Provider to which the call must be routed. Thus, significant modifications are required in operator services systems (see Figure 6-9).

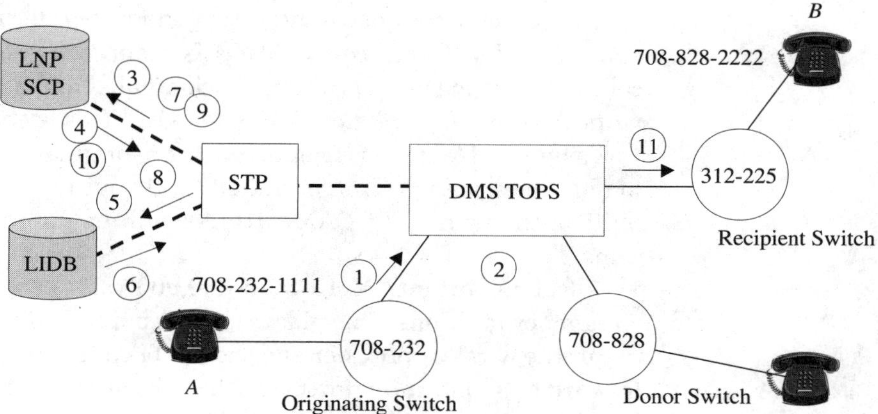

Figure 6-8 LIDB model for 0 plus calling card

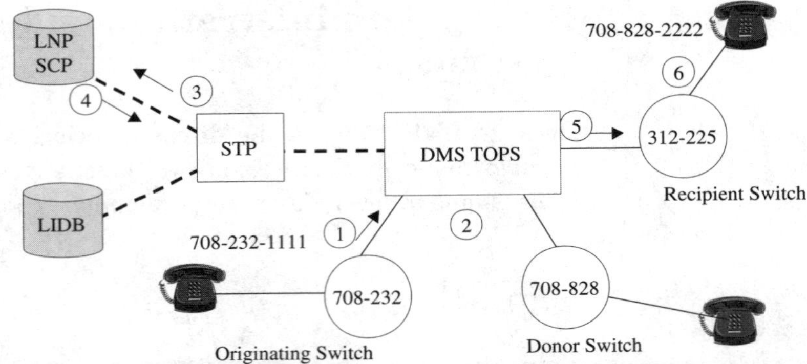

Figure 6-9 Operator service effects

Call flow for 0-Call Model

1. Subscriber 708-232-1111 dials 0- and is connected to TOPS operator.
2. Operator receives request to connect to 708-828-2222 and bill to originator's station.
3. TOPS determines that requested number has been ported and sends an LNP query to LNP SCP.
4. LNP SCP responds with the LRN (312-225-0000) of Recipient Switch.
5. TOPS routes call to Recipients Switch.
6. Recipients Switch terminates call to ported number 708-828-2222.

911 Services

Maintaining the 911 and enhanced 911 databases create another impact on the processing of calls to emergency response agencies. When we think about LNP in a wireline environment, the process is straightforward. The caller places an emergency call from a fixed location, which is easily identifiable. Routing of the E-911 call is to the proper *Public Safety Answering Point* (PSAP), based on a known, prearranged location for each telephone. *Automatic Number Identification* (ANI) is mapped one-to-one for all wireline calls by using the appropriate callback number and a location from an *Automatic Location Information* database (ALI). This is shown in Figure 6-10.

Using CAMA trunks (trunks that were originally used for cost accounting and messaging call information), 911 routing is built on limited capability of the trunk. The CAMA trunks will remain in use for some time to come.

Is 911 important? More than 300,000 calls per day are made to 911 emergency locations. This places a great burden on the network.

For the wireless provider and the E-911 caller, things are not as straightforward as shown in Figure 6-11. The calls are placed through the network, but it is processed through the MSC where a substituted number is used.

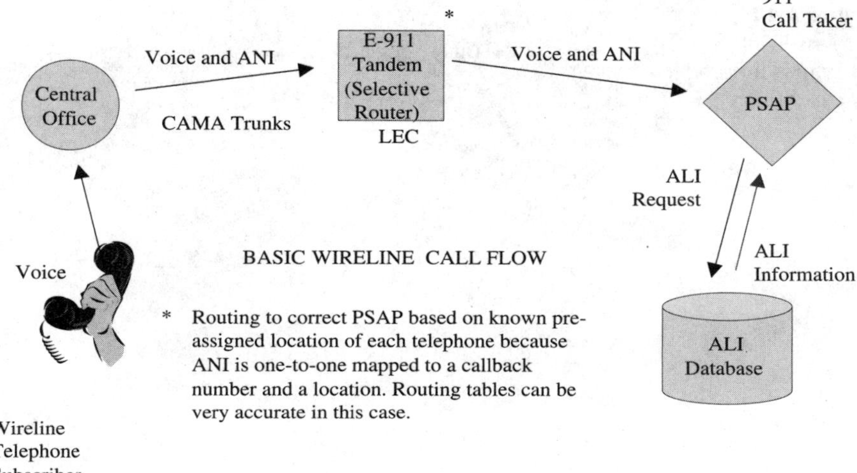

Figure 6-10 E911 calling with LNP

Routing to the PSAP based on a telephone number substituted by the MSC is handled by the end-office as though the call came from a fixed wireline location. Problems crop up because the substituted number is assigned in different ways:

- One per system
- One per several cell sites
- One per cell

The arrangement of how the calls are processed is agreed to in advance by the various carriers involved and placed in the ALI database. Routing is to the PSAP on the calling number (not the wireless number) and may cause inaccuracies due to the configuration or due to crossing PSAP jurisdictions.

Simplifying the Wireless E-911 Call

Today things are changing. By using sectorized cells, the providers can be more specific. When a wireless subscriber makes a 911 call, each sector in a cell site is assigned a pseudo-ANI, a fictitious nondialable subscriber telephone number assigned to the cell. This enables the call routing to the

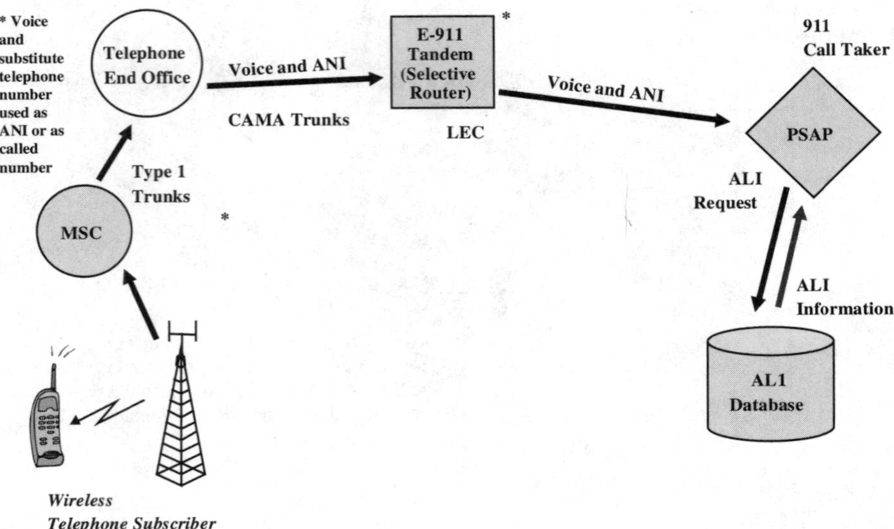

Figure 6-11 Wireless E911 substitutes a number for call routing.

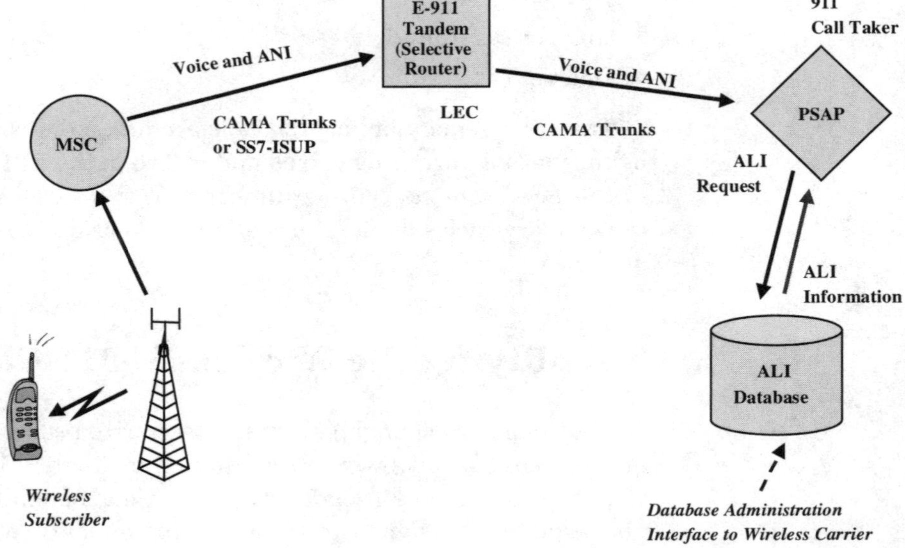

Figure 6-12 Simplifying the wireless E911 call

PSAP based on a fixed number and a location to route. Using an ANI within the wireline networks, the end office can more specifically process the wireless 911 call. This is shown in Figure 6-12 as the simplification process continues.

The overall benefit will be to use the LNP for both wireline and wireless organizations. Each of the services will bear transparency in the services, number assignments, and the operation of a single network strategy. Concurrently, the use of the LNP services will prevent the "disconnect" of a wireless user who is a moving target, so that the network will be used in such a way as to locate the LNP user within a matter of three to five meters using a GPS location and triangulation. The benefits are immense, the services will increase, and the features will be transparent. The only limiting factors right now include the timing for implementation, the demand from end users, and the willingness of the carriers to cooperate among themselves.

CHAPTER 7

Computer Telephony Integration (CTI)

Throughout the past two decades, business users have sought to find a means of integrating their telephony and computing systems. One can only imagine the great demand for this, assuming the computer and communications budgets are in the same domain. If one were to come up with a totally integrated package, there would be unimaginable savings (or so the story goes). It has been the goal of many organizations to reap the benefit of a single infrastructure.

In the past, the *Management Information Systems* (MIS) department and the voice communications department were separate entities, yet their operating budgets and staff overlapped in certain areas. The degree of overlap varied by organization, but the overlap itself was the concern. Since the convergence began nearly a decade ago, the fundamental shifts in these operations were somewhat dramatic. These shifts were not explosive in themselves, but more in line with a slowly rolling wave. In many camps within the industry, however, these changes were virtually unnoticeable.

The driving forces behind this convergence combined with voice communications for a commodity-priced service, data communications for an open standard in the *Transmission Control Protocol/Internet Protocol* (TCP/IP) world, and the preponderance of *Local Area Network* (LAN) technologies overtaking the desktop. From the very beginning of these three disparate technologies, end users looked to the manufacturers to come up with an integrated scheme. It was through the initial offerings of AT&T, IBM, and DEC (albeit all proprietary) that the computer and communications convergence began. Past examples of the manufacturers' attempts to provide an integration scheme included *computer-to-phone integration* (CPI) from both IBM and AT&T and the *computer-integrated telephony* (CIT) from DEC. Unfortunately, the proprietary nature and the blocked architectures of the manufacturers met with disappointing results. Users were not willing to pay for another proprietary solution as they had in the past. Moreover, the nature of the evolving LANs was leaning towards a new set of proprietary solutions from companies like Novell and Banyan, far from the openness the industry was busy praising.

Today two technologies are revolutionizing the way the world communicates: telephones and computers. Telephones are everywhere. For decades, they have been prevalent in offices, public establishments, and homes. And computers are catching up. These two technologies historically have remained separate, however. Often the only thing they share in common has been the desk on which they both sit, but the computer telephony industry has been combining the best of telephone and computer technology to let people exchange information more quickly and easily.

The power driving the computer telephony industry is telephone network access to computer information through almost any convenient, easy-to-use, and available terminal device, including

- Telephones (pulse dial, touch-tone, and wireless)
- Facsimile (fax) machines
- *Personal computers* (PCs)

These terminal devices access a multiuser computer telephony platform that supports applications that process the information within the call and/or route and monitor the progress of the call. These functions of the computer telephony platform are an integral part of enterprise information systems.

Until recently, our two most widely used business tools, the telephone and the computer, were separated by many factors. For decades, the telephone has been our primary means of business communication. However, the majority of our necessary business data has been accumulated on mainframe, mini-, and personal computers. To make matters even more confusing, we have been bombarded with a multitude of alternative communication methods (such as fax, e-mail, and voice mail), each accessible in different ways from different places.

Computer telephony integration (CTI) is the merging of the computer and telephone, which will transform the personal computer from being an information-processing device to being a powerful platform for communications. Computer telephony is the art of intelligently linking and combining these tools to create systems that enable us to use technology to our advantage. The goal, of course, is to dramatically increase the access we have to the information we need, when we need it. A variety of cost-effective solutions is available to businesses of all sizes.

Linking telecommunications to the processing capabilities and *graphical user interface* (GUI) of the computer enables new forms of communications and more robust access to existing types of communication, including

- Voice
- Asynchronous data
- Fax
- Remote access to LANs
- Internet access
- Online services

The communicating PC will redefine how we share ideas and information and will provide a portal to other people, computers, and network services anywhere in the world. CTI will also transform the telephone into an information-processing device. By giving the telephone access and control of nonvoice media, users will enjoy the convenience and economics of multimedia communications anywhere and at any time.

The Computer World

What had been the bastion of the big iron providers (mainframe and midrange computers) quickly eroded in the early 1980s to a desktop PC-based architecture. Surely, legacy systems, such as the departmental computers, became servers, and the mainframe for the entire organization remained as the mainframe (or became a major server) because of the investments and the nature of the data stored. One cannot just walk in and trash what has been in place for years. Instead, the integration of the computer and the telephony world requires a slower process that includes a methodology to preserve the legacy data.

Over the years, much time and effort has been spent in trying to develop interfaces that would tie the computer and the telephone network together. As technology has changed, many new interfaces and applications can be created through the integration of these two techniques. Consequently, a whole new industry has emerged. The ability to link computer systems and voice systems together offers some new possibilities on how we approach the office. When one thinks of the *automatic call distribution* (ACD) systems with the capability to link the *automatic number identification* (ANI) and a database to provide screen-popping capabilities, it becomes an exciting opportunity.

Figure 7-1 is a representation of the CTI capability using a screen-popping service. In this case, as a call comes into the building, it is initially delivered to an ACD. At this point, the ACD captures the ANI of the calling party. When the called party's number is packaged into a small data packet, it is then sent to the computer as a *structured query language* (SQL) inquiry into a computer database. Assuming the database already exists for the established client, the SQL opens the client file and responds to the query. In this particular case, the telephone number of the calling party is being used to create the query from the database. Once the database entry is recognized based on telephone number, a computer screen will "pop" to the agent's desk. This occurs at the same time the telephone call is being deliv-

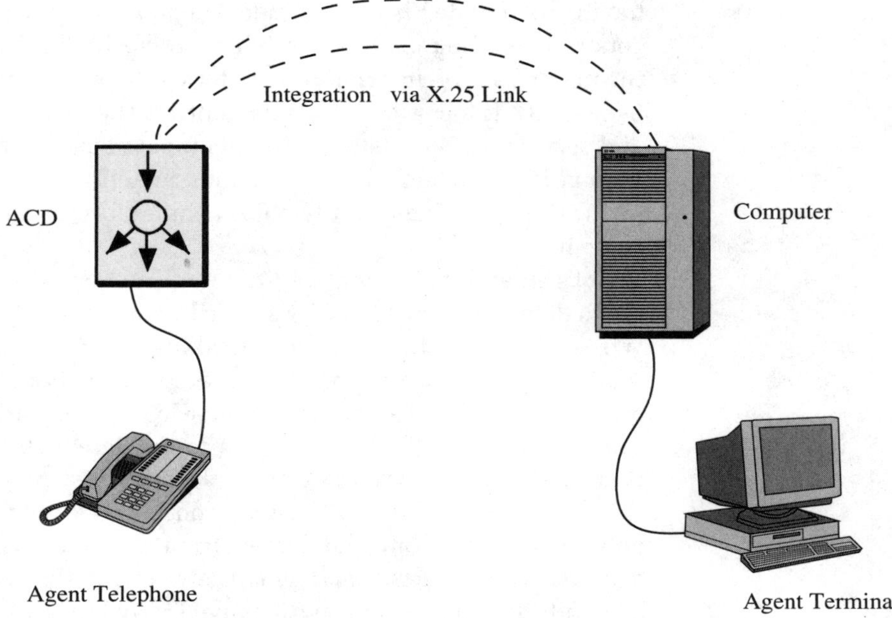

Figure 7-1 The CTI capability using a screen-popping service

ered to the agent's desk. Now the agent taking a telephone call does not have to ask the caller for all of the associated and applicable information, such as

- Name
- Address
- Telephone number
- City and state
- Account number
- Any other pertinent information

By reducing the time spent gathering this information, because it is already available in the database, savings of 20 to 30 seconds of call-processing time can be achieved. However, the 20 to 30 seconds is not the critical part. Instead, the ability to satisfy the customer and provide better service levels is what is important.

When one thinks of other alternatives that might exist here, a couple of thought-provoking ideas come to mind. For example, assume that an agent has received a call and a screen of information from an established customer. In this regard, the established customer has saved time by not having to provide his or her name, telephone number, and so on. The agent gets right to the task at identifying and verifying the customer. Now that the agent has the information, historical files on such things as buying or usage patterns can be developed so that the agent can suggest products and services to the customer based on past buying experiences. Moreover, if there is a problem with a particular customer, such as a delinquent payment, the agent can readily obtain this information while talking to the customer.

Let's extend that thought a little further. Assume that the agent encounters a delinquent customer who hasn't paid his or her bill for three months. While talking to this customer or taking an order, which is what the customer called for, the agent sees on the screen that the customer is delinquent and that collection action must be taken. Rather than becoming a collections manager, the agent can immediately suggest to the customer that the call must be transferred to the accounts receivable manager. When the agent transfers the call, not only does the call go to the accounts receivable department, but the screen from the database follows. When the accounts receivable department receives the call, the manager there will have a full screen of information about why that call has been sent to him or her. Answering the call, the accounts receivable manager can then immediately and proactively get to the job at hand, collecting the money. Instead of picking up the telephone and asking for information that has already been determined, totally frustrating the customer, the accounts receivable manager can immediately get to the problem at hand, collecting the money. When the call goes back to the agent, the screen with all of the notes that the accounts receivable manager has entered returns as well.

Again, the agent sees who is on the telephone and what has transpired. This prevents the replication of information gathering and keeps the entire conversation in a proactive mode. It is this positive handling of calls, particularly those of an unpleasant nature, that helps to boost customer service and influence the continued buying relationship.

The previous scenario assumes that certain things have taken place. It must be assumed that the customer is calling from a known location and that the database already exists. If, in fact, the database does not exist, the initial data gathering can be accomplished at the receipt of the first call. Thereafter, any time the customer calls from the same telephone number, the database will be activated and brought up with the screen-pop. The first time is the most laborious. Each successive contact between the customer and the organization is much simpler. Using this database and screen-pop arrangement, companies have successfully achieved several things, including

- Improved customer service
- Improved customer satisfaction
- A reduced number of incoming lines
- Reduction in the number of agents necessary to handle the same volume of calls

Through the combined savings of 20 to 30 seconds for each successive attempt, the average hold time (talk time) can be reduced significantly.

Other Possibilities

Screen popping in an ACD is one of CTI's primary uses today, but other applications can take advantage of CTI. The newer integrated *private branch exchange* (PBX) will become a voice server on the LAN. In Figure 7-2, the integrated PBX is shown as a voice server residing between two separate LANs.

Now that the various communications vendors have produced an open applications interface, the ability to use a PBX for several components exists. One of its primary applications is to install a third-party bridge or router

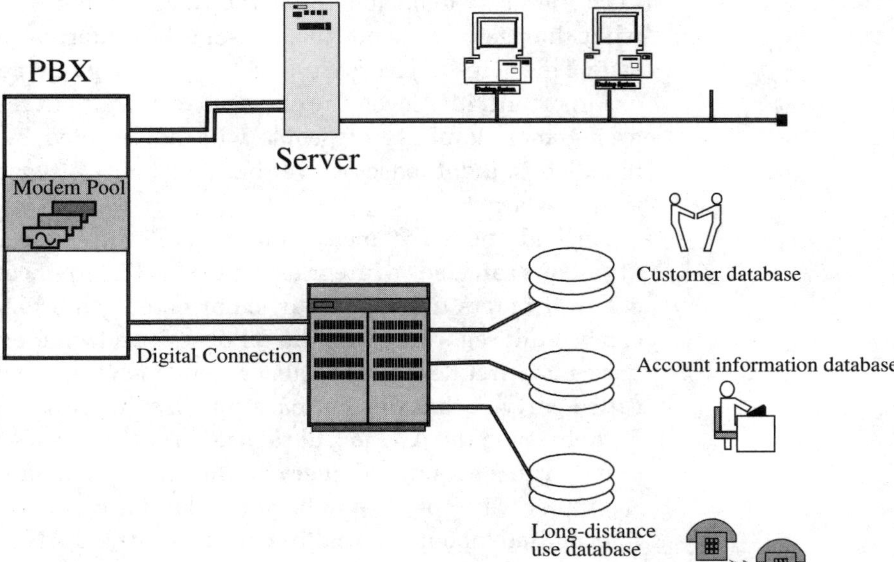

Figure 7-2 The integrated PBX as a voice server between two separate LANs

card inside the PBX. The PBX now serves as the bridging function between the two LANs. Using the PBX also enables an extension of the PBX access to other LAN services. Inside the architecture of the PBX is a high-speed communications infrastructure such as *Asychronous Transfer Mode* (ATM), *Synchronous Optical Network* (SONET), or *Fiber Distributed Data Interface* (FDDI), depending on the direction chosen by the PBX manufacturer. The PBX manufacturers have devised means of running 10 Mbps or 100 Mbps to the desktop. Now they are looking to implement an IP-based card, creating a true data application in the telephony architectures. Newer applications will also include the gigabit Ethernet over twisted-pair wires to the desktop (1000 base T). When dealing with the high-speed communications to the desktop, the user will have unlimited access to voice features, LAN features, and multimedia applications running inside the organization. The combined investment in computer and telephony integration offers substantial opportunities for high-speed broadband communications to the desktop. As a true communications server, the PBX equipped with an IP card (router) inside can bring other services such as Internet or intranet access to the PBX.

Still other features can be included with this integration of CTI. Voice messaging, as integral or peripheral to the PBX, can now become available as a feature or a function hooked to the LAN. As an outside caller dials into an organization's PBX, the capability to capture the ANI and hook to the database server on the LAN allows for the delivery of the caller ID to the desktop. This makes it possible for a user who has a screen (PC or other) on his or her desk to be able to see the voice messages in the queue. In reality, what the user sees is not the message, but information regarding the message. This helps to prioritize, or select from a queue, calls that are necessary or important. Moreover, the end user can print out a list of voice messages, a customer database of people who have called, ANI information to be included in a database not yet built, or many other useful and productive types of data.

Beyond the voice messaging capability, connecting to an automated attendant can also be integrated into this CTI application. When an outside caller attempts to reach a particular extension or department, a database can be built regarding who the callers are and what extensions are typically used most frequently. This also extends ANI access to the desktop of the called party so that the database can also link to some form of a screen-pop. By delivering the ANI to a database server on the LAN, the user can immediately run a structured query on the database to pull up the customer file. A simpler way would be to have the database automatically deliver all the information about the calling party to the LAN terminal device. This doesn't work in all applications, but it does offer some unique capabilities and opportunities that never before existed.

Each of the discussed services can all be coupled in the combination of computer and telephony integration. CTI offers the capability to link multiple databases, multiple communication channel capabilities, and the LAN-to-PBX server functions. The PBX becomes the voice server, whereas the LAN server becomes the PBX's database. Corporate directories, departmental lists, product lines, and other types of usable information can all be made available to the desktop device when a called or calling party needs access. Although the cost is incremental, the overall increases in productivity and satisfaction from both employees and customers may well be worth the effort. This is one of those decisions that a user must make.

In the same manner, you take higher risks by putting all of your eggs in one basket and having a single-server environment. However, the distributed computing architectures that are emerging minimize the risks. Distributed computing architecture can also add to the depth of the PBX by using servers on a LAN that can be voice servers; the functions and the real estate required for a PBX architecture can now be combined and reduced. A large room with special air conditioning is no longer required. The servers can just be deployed wherever they are needed to serve their users.

We can see the excitement and the benefits of CTI when we look beyond the basic functionality of each individual box. Hence, the decision-making process becomes a little more complicated than the purchase of a PBX or a server when thinking about acquiring a combined infrastructure to support the communication needs of an organization.

Why All the Hype?

A lot has already been said about what is going on within the industry and organizations today. Business has reached an intensity and a pace that is extraordinary. The competition and changes in technology are driving the pace exponentially. All of these factors have driven organizations to seek an improved means of dealing with their customers and to answer the needs for increasing productivity, decreasing costs, and enhancing the competitive edge over the existing market segment. Keeping pace with the industry and maintaining the competitive edge has become a crucial element in the survivability of an organization. Therefore, the demands on organizations have escalated dramatically to provide for

- The ability to deal with corporate decision-making processes at a rapid pace
- Readiness to deal with issues, regardless of the time of day, day of week, or availability of executives

- The continued restructuring and reengineering of the organization, which requires fewer people to perform more functions
- The changing workforce, in which talent and skill sets are no longer as available as they were, making technology a necessary solution
- Preparing the administrative assistants who have supplanted the administrative and support staff of old (secretarial support) to deal with issues when the senior staff is not available
- Ensuring the availability of the information that is needed to achieve these goals

Therefore, in order to effectively use the resources that remain within the organization, employees must have information readily available at their fingertips. Using technology of both computers and communications provides the edge to maintain the link between the employee and the customer. It is not necessarily a given that these technologies will always be the solutions, but the right implementation can have significant benefits if done correctly. Several examples in this book show how organizations have tied their computers and communications together through various network technologies to facilitate the timeliness and usability of information. Using a CTI system, employees have the ability to share information on the spot by using updated information on a call-by-call basis. Each of these tools and capabilities is an absolute must if a business is to remain competitive.

The reengineering of an organization must focus on serving the needs of their customers. Examples of this are clearly demonstrated in both the financial community and the airline industry. Customers can no longer be looked on as a nuisance; rather, they must be seen as the most valuable resource and asset that an organization has (excluding its own internal personnel). Providing the necessary treatment and care of the customer becomes one of the primary goals of the organization. It should have been the goal all along, but the focus on driving costs down and increasing productivity began to erode the customer and organizational relationships.

Now that CTI has begun its movement within the industry, it is much more feasible to restructure and reaffirm the relationship between the two parties. Customers expect to be treated as individuals who are special for the size and volume of the business that they do. Competition in any segment of industry is fierce. Therefore, if the necessary care and treatment is not provided, customer loyalty shifts dramatically. Changing market demands due to the rightsizing, downsizing, and capsizing in external organizations place more demand on your organization to meet the customer's needs. In planning our technological innovations, we must therefore look to how we can better satisfy the customer's needs by the following tasks:

- Giving our customers unlimited access to our employees to place orders, check on status, or make inquiries.
- Assuming that the employee/agent will have information readily available and not put the customer on interminable hold.
- Anticipating the demand for products based on sales projections and customer demands so that out-of-stock or backordering can be curtailed.
- Having technical support available to the agent so that either the technical or the administrative/sales person can easily answer a question.
- Considering that the customer's success in business and/or product lines can be matched to your ability to meet the customer's demands for products and services.
- Anticipating customers' needs and satisfying their demands so they will not be exposed by today's diminishing inventories and just-in-time manufacturing processes.

If these conditions can be satisfactorily met, the customer will feel better about placing orders or checking on the status and availability of products with your organization. This equates to better sales. Telecommunications and computer integration can therefore increase productivity and assist in increasing sales. Using the right technology in the right mix at the right time can enhance a relationship that is ever so fragile in today's competitive marketplace.

Linking Computers and Communications

The architecture of using a three-computer technology is the result of many years of evolution and revolution within the computing platforms. The LAN emerged in the early 1980s as a means of moving away from control by a single entity known as the MIS department. As this evolution began, users were empowered to manipulate their own data, handle their own information, and share that information with others on a selective or exclusive basis. Because of this movement, a lot of pressure was placed on the communications side of the business to provide interconnection using a cabling infrastructure to the various desktop computers and departments that would be interrelated. Management saw an opportunity to produce ad hoc reporting, allow flexible data manipulation, and provide more timely information to managers and departments alike.

As LANs emerged, they were primarily tied to a desktop and an individual server within a department. Later in its evolution, the client/server architecture started to emerge more strongly by allowing multiple departments within a single organization to fulfill their data communications and data access needs by accessing each other's databases or sharing the files they created with their customers. It is this sharing that created additional appliance affiliations within the different groups. The PC, as it attached to the network, enables users to store their applications and files on a single device. By linking these computing systems together with a communications medium, the user can best seek the information wherever it resides within the network. If a new application is needed, either a server can be adapted to fit that need or a new server can be added.

From a communications perspective, devices have emerged that include e-mail within departments, fax servers for transmitting information directly from a desktop to a customer, and voice recognition and response systems that enable customers to call in and literally talk to the server, as opposed to having to talk to a person. Furthermore, the linkage of all of these systems through a communications medium is the glue that binds everything together.

The PC of old, although structured as strictly a desktop device, has changed dramatically. Now, new features include the following:

- Extensive processing power
- Sophisticated operating systems
- Integrated voice boards or sound cards
- Integrated video-conferencing capabilities
- The client/server software that enables all of the pieces to be pulled together

With the use of CTI, an organization's needs are easily met using the buildout of the desktop devices connected to the LAN. As already mentioned, the client/server architecture has the capability to bring together the computers that reside on the LAN and the telephony services that can be built into either the PCs or the servers that reside on the LAN. This all becomes possible by bonding two technologies that began in the mid-1980s and matured in the mid-1990s. The architecture within an organization has dramatically changed. In place of a single provider of information, such as an administrative assistant or a particular sales group, there now exists an integrated homogenous workgroup that can share information, transfer calls among members, or add other departments on calls as needed. This has already been shown in our examples.

While this information process was taking place, organizations began to connect to the outside world and to their customers with their telephony services. In the early 1990s, a new explosion took place within the industry. The emergence of the World Wide Web and the use of the Internet as a commercial resource added to computer telephony development. Attaching an organization to the outside world through the Internet allowed customers to use their telecommunications services to enter the organization and check inventories, catalogs, and other services that would otherwise not normally be available to them. The ability to introduce fax-back servers also enabled customers to obtain catalog information or technical brochures immediately, instead of having to wait for an agent to mail products or brochures through the mail. This increased the availability of information by linking the communications' infrastructure to the computing architecture. Moreover, as organizations started to use *electronic document interchanges* (EDI), they created the ability to place orders, perform functions as dynamic workgroups, and facilitate the ordering and payment process. These services enhanced the true organizational capability.

The mid-1990s has also marked the innovation and growth of this computer and telephony integration with intranets. For now, think of it as the internal Internet to an organization. Using the intranet and communications infrastructures within an organization's client/server architectures, internal organizations can access the same information that might be available to a customer. Furthermore, internal organizations can share files, technical notes, and client notes regarding their customers, as well as other valuable information that would not be readily available to the masses. Specific files or bulletin boards can be set up within the organization to achieve that result. Therefore, it is through the integration of our communications systems and the capabilities of the far more powerful and processor-intense desktop devices that the pieces are coming together quickly. No distinction can be made anymore between computing and communications, because the two really draw on each other's resources to facilitate the organization's day-to-day mission: to serve the customer and maximize shareholder wealth. Serving the customer increases sales, which hopefully decreases costs and raises profits. This maximizes the shareholder's wealth, which is the charter for all business organizations.

The Technology Advancement

Through the integration of computer and telephony-type services, the old philosophy of telephony being a necessary evil has passed. Advances in computing technology have enhanced the telephony world. The age of mass pools of telephone operators is gone. Newer technologies enable customers to call and proceed through an organization without ever communicating with an operator or a secretary. The customer can get directly to the person or information that is desired.

In the key system marketplace, the old labor-intensive electromechanical system is now a computer-driven telephone system. However, these computer-driven systems are now rich in features and laden with capabilities undreamed of in the past. The integration of voice messaging, ACDs, and automated attendants in key systems is now commonplace. Key systems are also emerging as voice servers within the organization.

The PBX, serving hundreds if not thousands of employees within an organization, enables an internetwork of services and capabilities to be spread throughout an organization. What was once just a stand-alone telephone system is now the high-end computer that just happens to handle voice. Today's technology provides full-digital transmission systems or services and permits linkage to all devices that were exclusively used by the elite. PBXs can now be integrated tightly into your computing systems.

The PBX also acts as the high-end server on a digital trunk capability to the outside world. By linking to a high-speed digital communications trunk from the outside world, users can now access ANI. Using inbound 800/888/877/866 services and a common channel-signaling arrangement enables the delivery of caller ID directly to the called party. Moreover, enhancements in software (such as call forwarding), overflow arrangements when agents or other individuals are busy, and other services can all be chartered into a single PBX architecture. Another service called *directory number information service* (DNIS) can be used on the 800/888/877/866 service to direct the call to a specific group, such as technical support or marketing.

The Final Bond

As the client/server architecture and the communications systems were developing throughout the 1980s and into the 1990s, it was the software vendors who became the aggressors. With operating systems that could work on a LAN-based server, the software vendors were aggressive in finding the integration tools necessary for the organization. Microsoft developed a de facto protocol called the *telephony application program interface* (TAPI) in a Windows environment to bring the telephony services right to the desktop LAN-attached device. A TAPI interface on a Windows platform enables users to access information through what is called a GUI. Throughout the entire architecture, a server application can handle the distribution of calls to members within workgroups or departments. This includes such services as screening calls, rerouting calls to new agent groups if the primary agent group is busy, or routing calls to the voice messaging and voice response systems as necessary.

Similarly, Novell and AT&T developed what is known as the *telephony services application programmers interface* (TSAPI) as a means of providing the computer with telephony integration capabilities from a LAN. TSAPI works with Novell's NetWare telephony services. The basis for using this particular product is to gain what is known as *third-party call control*. Third-party call control uses the CTI application on behalf of any clients in a workgroup or department. The application is running a shared environment, typically on a server, so there is no direct contact or connection between the user's PC and the telephone interface. What happens, then, is that a logical connection is produced when the PC applications talks to the server, which in turn controls the telephone switch. The server then is the controller and sends the order to the PBX to make calls or connections on behalf of the end user. The shared-server environment can handle individual as well as dynamic workgroup applications, such as directories, individual *personal information managers* (PIMs), and other workgroup functions that occur within a larger organization. Therefore, the server provides the linkage for all calls being handled within a dynamic workgroup. This is a more powerful arrangement regarding call control. A central server can handle the distribution of calls to any member in a workgroup and provide such services as call screening, call answering groups, backing up agents, or routing calls to a supervisory position in the event of overflow. Many of the CTI vendors have seen these capabilities as the benefits and strengths of the CTI applications.

These are the primary applications that have been used within the CTI environment. Thus, developers and manufacturers alike have seen the application working as a telephone set. This is instrumental in designing the capabilities of a call answering/call processing environment, but more can be done through the use of the computer-based technology that is available on the market today. The applications that run in server or computing platforms can use call-monitoring features within the PBX and collect information of any type. Using the call-monitoring features, a CTI can watch every keystroke that an agent group enters. This can include such things as

- Dialed digits
- When agents "busy out"
- Answering a call
- How long the call is off hook

Additionally, by monitoring the trunk groups within the PBX, the CTI application can see all the incoming calls and collect the data associated with each call. This application uses ANI and DNIS to see where the call was directed and when it was answered.

The selectivity available in a CTI application therefore enables the supervisor to monitor the activity of each agent within a workgroup and get a clear picture of what transpires during the course of a day. This helps in determining workload effort, staffing requirements to meet a specific demand, or any seasonal adjustments that must be made based on time of year. About 65 to 70 management reports can be generated on an ad hoc basis. Supervisory personnel can monitor the work flow, the productivity of each agent in a group, abandoned calls that were not answered soon enough, or any other anomalies that might occur within a given day. Through these useful tools and reporting structures, the supervisors know whether they have enough, too few, or too many agents on board at any one time. This can aid in workflow scheduling as well as in determining the productivity of the individual agents or of the workgroup as a whole. In the event an agent is not performing satisfactorily, management can then take whatever corrective actions are deemed necessary.

If abandoned calls are escalating because the agents are on the phone too long, several other activities might result. These could include such things as new training, analysis of the call type and the information requested, or simply determining the morale and productivity within the workgroup.

Beyond the call-processing and the call-monitoring capabilities, CTI can be implemented to integrate the facilities and capabilities of the PBX as well as the computer systems into one homogenous unit. The typical PBX today has features that are rarely used. There can be as many as 300 to 400 of these features that are designed to either improve productivity or make the job simpler. Although the average user typically activates only three or four of the normal features, many of the functions are available but never used. The CTI applications can customize features and functions for individual users within a workgroup and provide more powerful interfaces using point-and-click GUIs. Through a CTI application, a PC can be used for dialing, activating features, conference calling, call transfers, or any other necessary feature. Simply clicking a telephone set icon saves the end user from the risk and the unfamiliarity of the PBX features, preventing cutoffs or lost calls and facilitating better utilization of the PBX features.

Taking this one step further, each individual user within a group supported by a CTI application can customize his or her own features and functions in personal folders. These can either be stored on a PC or a server. Therefore, using the CTI application, individuals can select different forms of call screening, call forwarding, or call answering, according to the applications and the individualized services they prefer.

CHAPTER 8

Signaling System 7 (SS7)

The ability of a caller to go off-hook in a telephone world, dial digits, and then miraculously talk to someone anywhere in the world is still a mystique to many. The network's capability to set the call up almost instantly and then tear it down just as fast is what really carries the mystique. How can the network figure out where to send the call, get the connection, and ring the phone on the other end so quickly? All of this happens in under a second, and the user is oblivious as to the intricacies of what occurs. What happens behind the scenes constitutes the backbone of the signaling systems. The networks are now dependent on the capability of handling subsecond call set-ups and teardowns.

Several signaling systems have been introduced to the telecommunications networks. The current one in use is called SS7 in North America. In the rest of the world, this is referred to as *CCITT Common Channel Interoffice Signaling System 7* (CCS7 for short). Although the names are different, the functions and the purposes of the two systems are the same. As always, the North Americans do things one way, and the rest of the world does things a different way. This is an age-old problem, but one that we have learned to deal with and adjust to.

The essence of the signaling system boils down to many different factors, but one of the most significant reasons the carriers employ these systems is to save time and money on the network. Following that fact, the carriers are also interested in introducing new features and functions of an intelligent network, as discussed in an earlier chapter. The best signaling systems are designed to facilitate this intelligence in the network nodes that are designated as signaling devices, separate and distinct from the switching systems that carry the conversations.

Presignaling System 7

Prior to the implementation of SS7, *per-trunk signaling* (PTS) was used exclusively in the networks. This method was used for setting up calls between the telephone companies' exchanges. PTS continues to be used in some parts of the world where SS7 has not yet been implemented. Admittedly, the exchanges using the PTS method are declining, as SS7 is gaining in its deployment worldwide. However, the network is always in a state of change, and this is no exception. PTS sends tones or *multiple frequencies* (MF), as they are called, to identify the digits of the called party. The trunk also provides all of the intelligence for monitoring and supervision (call seizure, hang up, and answer back) of the call. Telephone systems at the customer's location (PBXs) that are not *Integrated Services Digital Networks* (ISDN) *Primary Rate Interface* (PRI)-compatible use the per-trunk signaling method.

On a long distance call when a call set-up is necessary, each leg of the call repeats the MF call set-up procedure until the last exchange in the loop is reached. In essence, the call is being built by the signaling as the progress is occurring on a link-by-link basis. As each link is added to the connection, the network is building the entire circuit across town or across the country. Each leg of the call set-up takes approximately 2 to 4 seconds, with the configuration shown in Figure 8-1, or a total call set-up takes approximately 6 to 12 seconds (at a minimum) from end to end.

This method works but is an inefficient use of the circuitry in both major and minor networks. Although the call gets to its end destination, several complications could arise, causing extensive delay or incomplete calls. Regardless of the complications, the outcome is the same; the carrier ties up the network and never completes the call. Hence, no revenue is generated for the use of the circuits or the network. This inefficient use of the network costs the carriers a significant amount of money. Therefore, something has to be done to improve this method of call establishment. The call establishment part of the connection could take as much as 24 seconds, then time out, and never get to its end point. However, the carrier ties up parts of the

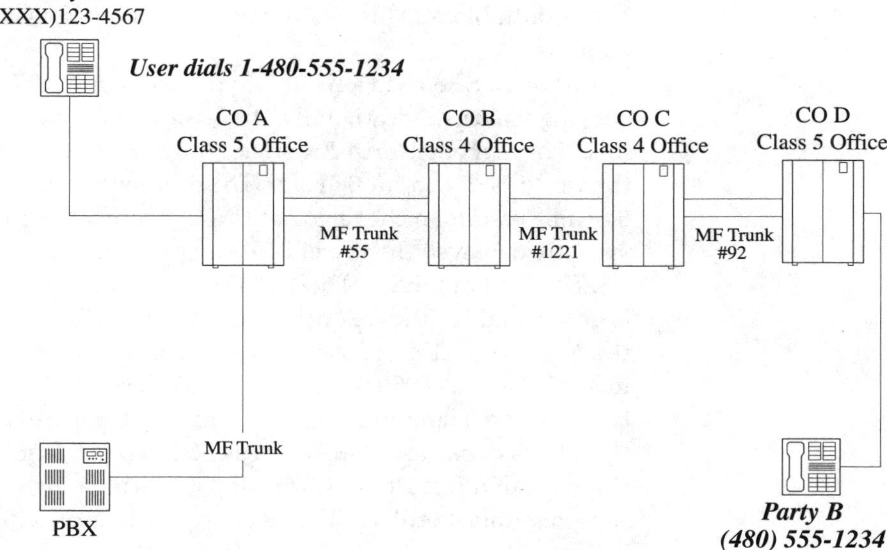

Figure 8-1 Per-trunk signaling preceded SS7 but was slow

network without getting a completion. This is no big deal when discussing one call, but when a network carries hundreds of millions of calls per day, this accumulated lost time is extensive and expensive.

Introduction to SS7

The ITU-TS (once called the CCITT) developed a digital signaling standard in the mid-1960s called *Signaling System 6* (SS6) that would revolutionize the industry. Based on a proprietary, high-speed data communications network, SS6 later evolved to SS7. SS7 has now become the signaling standard for the world.

The success of the signaling standards lies in the message structure of the protocol and the network topologies. The protocol uses messages much like the X.25 and other message-based protocols to request services from other entities on the network. The messages travel from one network element to another, independent of the actual voice and data that they pertain to, in an envelope called a *packet*.

The first development of the SS6 in North America was used in the United States on a 2400-bit-per-second data link. Later these links were upgraded to 4800 bits per second. Messages were sent in the form of data packets and could be used to request connections on voice trunks between Central Offices, placing 12 signal units' (of 28 bits each) assembled packets into a data block. This is similar to the methods used today in SS7 architectures.

Although SS6 used a fixed-length signal unit, SS7 uses signal units with varying lengths. Additionally, the later version of SS7 uses a 56 Kbps data link. Throughout North America 56 Kbps are used, whereas in the rest of the world, SS7 runs at 64 Kbps. The difference in the speeds between 56 and 64 Kbps results in the fact that the *local exchange carriers* (LECs) have not yet fully deployed the use of B8ZS on the digital circuits (For a discussion of B8ZS, see Chapter 28, "The T Carrier Systems"). Consequently, the 56 Kbps is an anomaly in the SS7 networks. Further, SS6 was still being installed by the North American carriers up through the mid-1980s (even though it was invented in the 1960s), while the SS7 deployment began in 1983, leaving two separate signaling systems in use throughout North America.

SS6 networks are slow, whereas SS7 is much faster; 64 Kbps and the possible use of a full DS-1 (1.544 Mbps) is still being considered in the North American marketplace. This is an evolutionary service that is continually being modified.

Purpose of the SS7 Network

The primary purpose of SS7 was to access remote databases to look up and translate information from 800 and 900 number calls (now the addition of 888 and 877 area codes are included). There were several benefits to using this lookup process, such as that carriers do not have to maintain a full database at each switching node but know how to get to the remote database and find the information quickly. The second purpose of the SS7 network and protocols was to marry the various stored program controlled systems throughout the network. This enables the quick and efficient call set-up and teardown across the network in 1 second. Moreover, this integration provides for better supervision, monitoring, and billing systems integration. Additional benefits of the SS7 network were geared to replacing the SS6 network, which, as of today, is well over 30 years old. Like anything else, the networks have served us well but need upgrading on a regular basis due to technology changes and demands for faster, more reliable services.

SS7 networks enable the introduction of additional features and capabilities into the network. This makes it attractive to the carriers so they can generate new revenues from the added features. SS7 also enables the full use of the channel for the talk path, because the signaling is done out-of-band on its own separate channel. This is more efficient in the call set-up and teardown process.

What Is Out-of-Band Signaling?

Out-of-band signaling is signaling that does not take place in the same path as the conversation. We are used to thinking of signaling as being in-band. We hear dial tone, dial digits, and hear ringing over the same channel on the same pair of wires. When the call connects, we talk over the same path that was used for the signaling. Traditional telephony used to work this way as well. The signals that set up a call between one switch and another always took place over the same trunk that would eventually carry the call.

In early days, the out-of-band signaling was used in the 4 kHz voice grade channel (see Figure 8-2). The telephone companies used band pass filters on their wiring to contain the voice conversation within the 4 kHz channel. The band pass filters were placed at 300 Hz (the low pass) and at 3,300 Hz (the high pass). The range of frequencies above the actual filter was 700 Hz (4,000 − 3,300 = 700). In this additional spectrum, in-band signaling was sent down the wires outside the frequencies used for conversation. Actually, the signals were sent across the 3,500- and 3,700-Hz frequencies. Although these worked and were not in the talk path (out of the band), they were limited to the number of tones that could be sent. The result was also a limit to the states that could be represented by the tones.

Out-of-band signaling has evolved to a separate digital channel for the exchange of signaling information. This channel is called a *signaling link*. Signaling links are used to carry all the necessary signaling messages between nodes. Thus, when a call is placed, the dialed digits, trunk selected, and other pertinent information are sent between switches using signaling links, rather than the trunks that will ultimately carry the conversation.

It is interesting to note that although SS7 is only used for signaling between network elements, the ISDN D channel extends the concept of out-of-band signaling to the interface between the subscriber and the switch. With ISDN service, signaling that must be conveyed between the user station and the local switch is carried on a separate digital channel called the *D channel*. The voice or data that comprise the call is carried on the B channel. In reality, the out-of-band signaling is virtual because the signaling information is actually running on the same path as the B channels. Time slots on the same physical paths separate the signaling and the conversational data flows. Therefore, it is virtually out-of-band, while it is physically in the same bandwidth.

Figure 8-2
Out-of-band signaling used the high frequencies

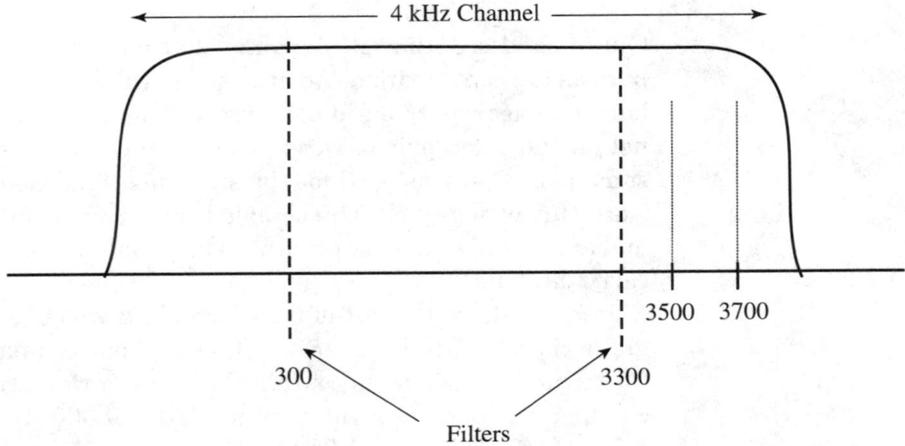

Why Out-of-Band Signaling?

Out-of-band signaling has several advantages that make it more desirable than traditional in-band signaling:

- It allows for the transport of more data at higher speeds (56 Kbps can carry data much faster than MF out-pulsing).
- It allows for signaling at any time in the entire duration of the call, not only at the beginning.
- It enables signaling to network elements without a direct trunk connection.

The SS7 Network Architecture

If signaling is to be carried on a different path than the voice and data traffic it supports, then what should that path look like?

The simplest design would be to allocate one of the paths between each interconnected pair of switches as the signaling link. Subject to capacity constraints, all signaling traffic between the two switches could traverse this link. This type of signaling is known as *associated signaling*. Instead of using the talk path for signaling information, the new architecture includes the connection from the *Signal Switching Point* (SSP) to a device called the *Signal Transfer Point* (STP). It is then the responsibility of the STP to provide the necessary signaling information through the network to affect the call set-up.

When necessary, the STP sends information to the *Signal Control Point* (SCP) for translation or database information on the routing of the call. The pieces that form the architecture of the SS7 network are described in Table 8-1 and are shown in Figure 8-3 with the connection of the overall components.

The drawing shows a typical interconnection of an SS7 network. Several points should be noted:

- Paired STPs perform identical functions and are redundant. Together they are referred to as a mated pair of STPs.
- Each SSP has two links (or sets of links), one to each STP of a mated pair. All SS7 signaling to the rest of the world is sent out over these links. Because the STPs are redundant, messages sent over either link (to either STP) will be treated equivalently.

Table 8-1
Components of the SS7 networks

Component	Function
SSPs	SSPs are the telephone switches (end offices and tandems) equipped with SS7-capable software and terminating signaling links. They generally originate, terminate, or switch calls.
STPs	STPs are the packet switches of the SS7 network. They receive and route incoming signaling messages toward the proper destination. They also perform specialized routing functions.
SCPs	SCPs are the databases that provide information necessary for advanced call processing capabilities.

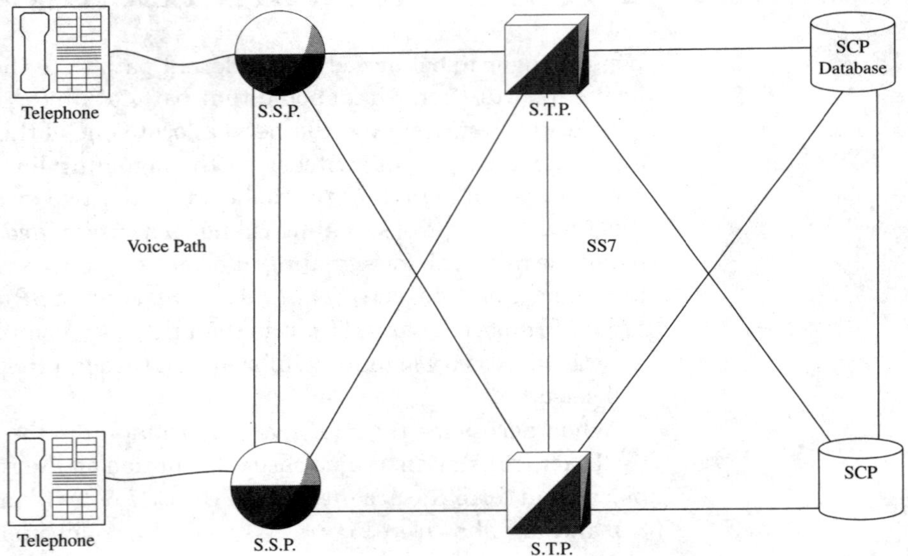

Figure 8-3 SS7 architectural beginnings

- A link (or set of links) joins the STPs of a mated pair.
- Four links (or sets of links) interconnect two mated pairs of STPs. These links are referred to as a quad.
- SCPs are usually (though not always) deployed in pairs. As with STPs, the SCPs of a pair are intended to function identically. Pairs of SCPs

are also referred to as mated pairs of SCPs. Note that a pair of links does not directly join them.

- Signaling architectures such as these that provide indirect signaling paths between network elements are referred to as providing *quasi-associated signaling*.

SS7 Interconnection

The actual linkage enables the local exchange offices to send the necessary information out of band across the signaling links. SS7, therefore, uses messages in the form of packets to signal across the network through the STPs. This enables the full use of the talk path for information exchange and the messaging paths for informational dialogue between the switching systems and the transfer points.

The links are used to pass control and billing information, network management information, and other control functions as necessary without interfering with the conversational path.

Basic Functions of the SS7 Network

The basic functions of the SS7 network include some of the following tasks:

- The exchange of circuit-related information between the switching points along the network
- The exchange of non-circuit-related information between the databases and the control points within the network
- The facilitation of features and functions by marrying the stored program control systems together throughout the network into a homogenous network environment

Further, the SS7 network enables these features to be put into place without unduly burdening the actual network call path arrangements. SS7 also accomplishes the following tasks:

- It handles the rerouting of network traffic in the event of circuit failures by using automatic protection-switching services, such as those found in SONET or Alternate Routing information.

- Because it is a packet-switching concept, the SS7 network prevents misrouted calls, the duplication of call requests, and lost packets (requests for service).
- It enables the full use of out-of-band signaling using the ITU Q.931 signaling arrangements for call set-up and teardown.
- It allows for growth so that new features and functions can be introduced to the network without major disruptions.

Signaling Links

SS7 signaling links are characterized according to their use in the signaling network. Virtually all links are identical in that they are 56 Kbps or 64 Kbps bidirectional data links that support the same lower layers of the protocol; what is different is their use within a signaling network. The bidirectional nature of these links enables traffic to pass in both directions between signaling points. Three basic forms of signaling links exist, although they are physically the same. They all use the 56 Kbps DS0A in North America and 64 Kbps DS0C data facilities in nearly every other portion of the world (except Japan where they still use a 4.8 Kbps link). The three forms of signaling links are as follows:

- **Associated signaling links** This is the simplest form of signaling link, shown in Figure 8-4. In associated signaling, the link is directly parallel from the end office with the voice path for which it is providing the signaling information. This is not an ideal situation, because it requires a signaling link from the end office to every other end office in the network. Some associated modes of signaling are in use, but they are few and far between.

  The most associated signaling is deployed at the end user location, using a single T1 and common channel signaling. Channel number 24 on a T1 is the associated out-of-band signaling channel for the preceding 23 talk channels.

Figure 8-4
Associated signaling

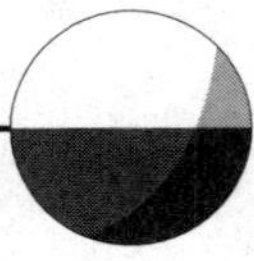

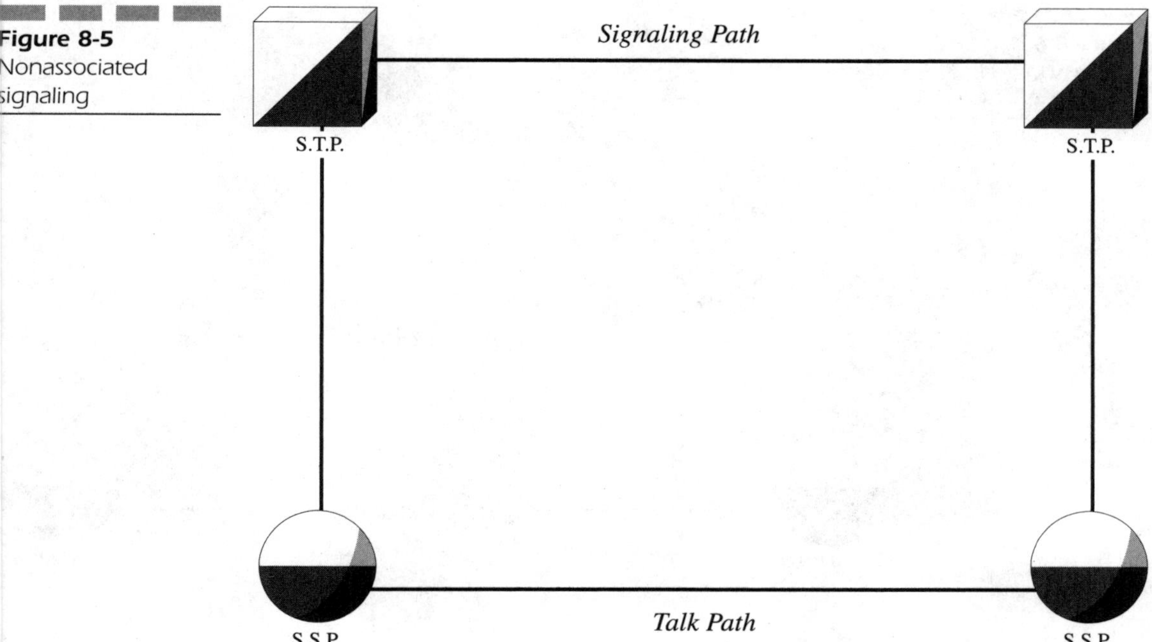

Figure 8-5 Nonassociated signaling

In some cases, it may be better to directly connect two SSPs together via a single link. All related SS7 messages to circuits connecting the two exchanges are sent through this link. A connection is still provided to the home STP using other links to support all other SS7 traffic.

- **Nonassociated signaling links** In this signaling link arrangement, there is a separate logical path from the actual voice path, as shown in Figure 8-5. Usually, multiple nodes reach the final end destination, while the voice may have a direct path to the final destination. Nonassociated signaling is a common occurrence in many SS7 networks.

  The primary problem with this form of signaling is the number of signaling nodes that the call must use to progress through the network. The more nodes used, the more processing and delay that can occur. Nonassociated signaling involves the use of STPs to reach the remote exchange. To establish a trunk connection between the two exchanges, a signaling message will be sent via SS7 and STPs to the adjacent exchange.

Figure 8-6 Quasi-associated signaling

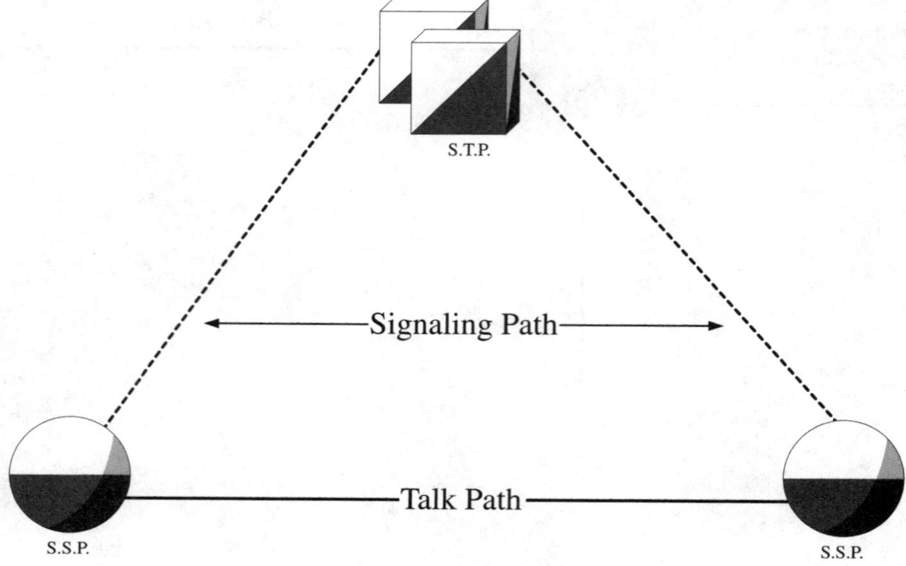

- **Quasi-associated signaling links** In quasi-associated signaling, a minimum number of nodes is used to process the call to the final destination, as shown in Figure 8-6. This is the preferred method of setting up and using an SS7 backbone because each node introduces additional delay in signaling delivery. By eliminating some of the processors on the set-up, the delay can be minimized.

SS7 networks favor the use of quasi-associated signaling. In quasi-associated signaling, both nodes are connected to the same STP. The signaling path is still through the STP to the adjacent SSP.

The Link Architecture

Signaling links are logically organized by link type (A through F), according to their use in the SS7 signaling network. These are shown in Figure 8-7 with the full linkage in place.

- **A link** An *access* (A) link connects a signaling end point (SCP or SSP) to a STP. Only messages originating from or destined to reach the signaling end point are transmitted on an A link.

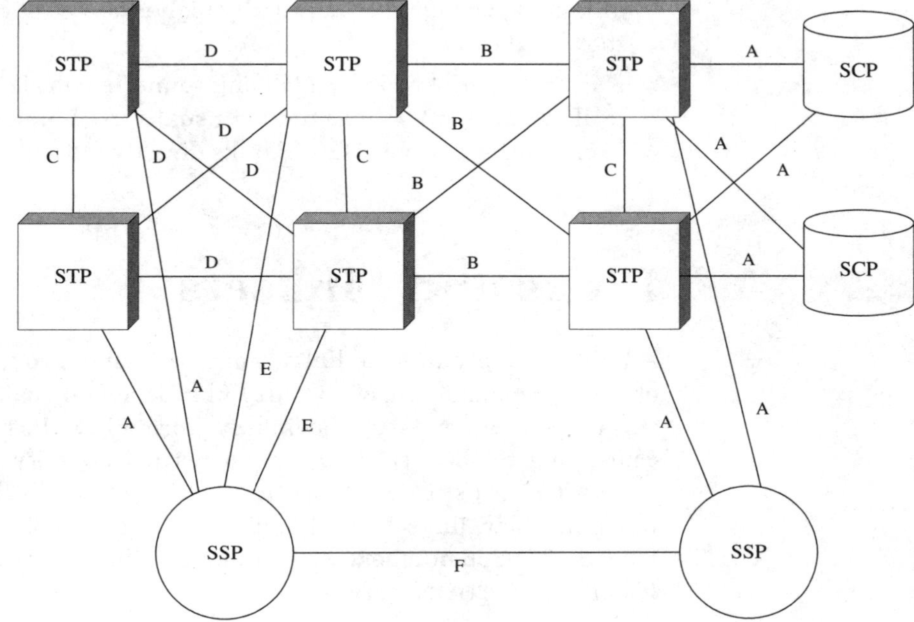

Figure 8-7 The signaling link architecture

- **B link** A *bridge* (B) link connects one STP to another STP. Typically, a quad of B links interconnects peer (or primary) STPs (the STPs from one network to the STPs of another network). The distinction between a B link and a D link is rather arbitrary. For this reason, such links may be referred to as B/D links.
- **C link** A *cross* (C) link connects STPs performing identical functions into a mated pair. A C link is used only when an STP has no other route available to a destination signaling point due to link failure(s). Note that SCPs can also be deployed in pairs to improve reliability. Unlike STPs, however, signaling links do not interconnect mated SCPs.
- **D link** A *diagonal* (D) link connects a secondary (local or regional) STP pair to a primary (internetwork gateway) STP pair in a quad-link configuration. Secondary STPs within the same network are connected via a quad of D links. The distinction between a B link and a D link is rather arbitrary. For this reason, such links may be referred to as B/D links.
- **E link** An *extended* (E) link connects an SSP to an alternate STP. E links provide an alternate signaling path if a SSP's "home" STP cannot be reached via an A link. E links are not usually provisioned

unless the benefit of a marginally higher degree of reliability justifies the added expense.

- **F link** A *fully associated* (F) link connects two signaling end points (SSPs and SCPs). F links are not usually used in networks with STPs. In networks without STPs, F links directly connect signaling points.

Links and Linksets

A linkset is a grouping of links joining the same two nodes. A minimum of one link to a maximum of 16 to 32 links (depending on the part of the world this is used) can make up the linkset. Normally, SSPs have one or two links connecting to their STPs based on normal capacity and traffic requirements. This constitutes a one- or two-link linkset. SCPs have many more links in their linksets to handle the large amount of messaging for 800/888/877/866 numbers, 900 numbers, calling cards, and *Advanced Intelligent Network* (AIN) services.

Combined Linksets

A *combined linkset* is a term that defines routing from a SSP or SCP toward the related STP where two linksets share the traffic outward to the STP and beyond. The requirement is not that all linksets are the same size, but the normal practice is to have equally sized groupings of linksets connecting the same end node. Using a linkset arrangement, the normal number of links associated with a linkset is shown in Table 8-2.

Linksets are a grouping of links between two points on the SS7 network. All links in a linkset must have the same adjacent node in order to be clas-

Table 8-2

The configuration of linksets

Type Link	Number of links
A links	Maximum of 16–32 links
B/D links	Installed in quads up to a maximum of eight links
C links	Installed individually up to a maximum of eight links

sified as part of a linkset. The switches in the network alternate traffic across the various links to be sure that the links are always available. This load spreading (or balancing) serves many functions and can help you do the following:

- Be aware when a link fails.
- Recognize when congestion is occurring in the network.
- Use the links when traffic is not critical to know when a link is down before it becomes critical (see Figure 8-8).

Routes and Routesets

The term *routeset* refers to the routing capability of addressing a node within the SS7 network. Every node within the network has a unique address that is referred to as a *point code*. The addressing scheme or point code is the major routing characteristic of the CCS7 (SS7) network. The terms routeset and point code are somewhat synonymous.

Figure 8-8
Linksets combined

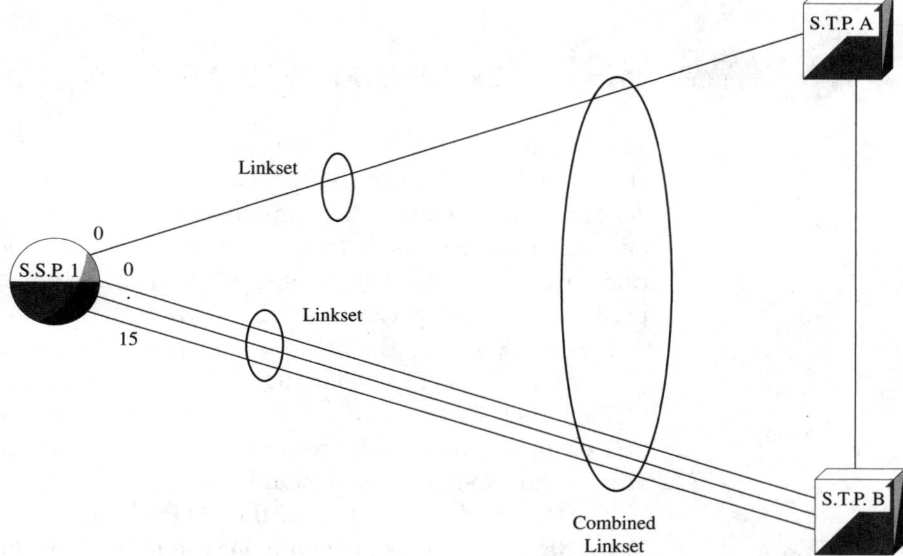

The point code is made up of nine digits broken down into three, three-digit sequences. An example of this is 245-100-000. Reading the point code from left to right, we find that

- The first three digits refer to the network identifier (245).
- The next three digits refer to the cluster number (100).
- The final three digits refer to the member number (000).

In any given network, there can be 256 clusters with 256 members. The network number in this case is for Stentor Communications in Canada.

The routing of SS7 messages to a destination point code can take different paths or routes. From the SSP perspective, there are only two ways out from the node, one toward each of its mated STPs. From that point on, the STPs decide which routes are appropriate, based on the time, resources, and status of the network. From the SSP, various originating and terminating (destination) addressing scenarios are defined as follows:

- If the route chosen is a direct path using a directly connected link (SSP1-STPA), then the route is classified as an *associated route*.
- If the route is not directly connected via links (SSP1-SSP2), the route is classified as a *quasi route*.

All routing is controlled by nodal translations, providing flexible and network specific routing arrangements. This is shown in Figure 8-9.

SS7 Protocol Stack

The SS7 uses a four-layer protocol stack that equates to the seven-layered OSI model (see Figure 8-10). These protocols provide different services, depending on the use of the signaling network. The layers constitute a two-part functionality; the bottom three layers are considered the communications transmission of the messages, whereas the upper portion of the stack performs the data process function.

The stack shows that the bottom three layers make up the *Message Transfer Part* (MTP) similar to the X.25 network function. At one time, the SS7 messages were all carried on X.25. Now newer implementations use SS7 protocols, yet in older networks or third-world countries the X.25 may still be the transmission system in use.

The SCCP is used as part of the MTP when necessary to support access into a database and occasionally for the ISDN User Part. This extra link is

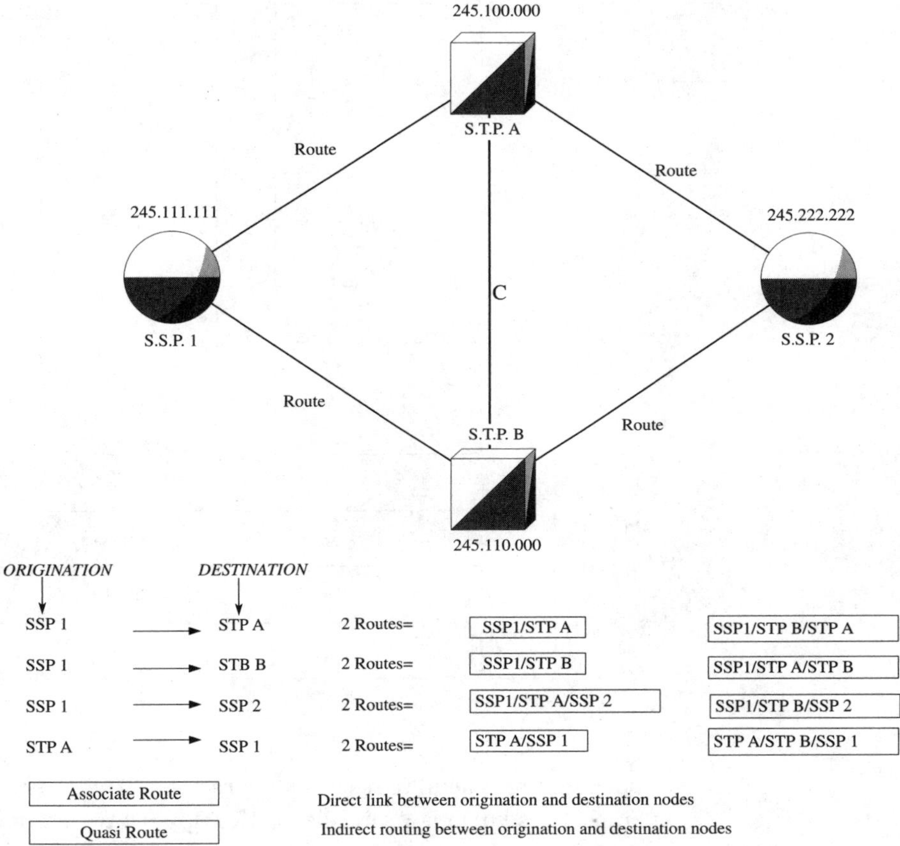

Figure 8-9
Routes and routesets

the equivalent of the transport layer of the *Open Systems Interconnect* (OSI) model supporting the TCAP.

The SS7 network is an interconnected set of network elements that is used to exchange messages in support of telecommunications functions. The SS7 protocol is designed to both facilitate these functions and to maintain the network over which they are provided. Like most modern protocols, the SS7 protocol is layered. Functionally, the SS7 protocol stack can be compared to the OSI reference model. Although OSI is a seven-layered stack designed to perform several communications and transparent functions, the SS7 protocol stack is similar but different.

Like any other stack, the SS7 protocol stack is specifically designed for the reliable data transfer between different signaling elements on the network. The guaranteed delivery and the prevention of duplication or lost

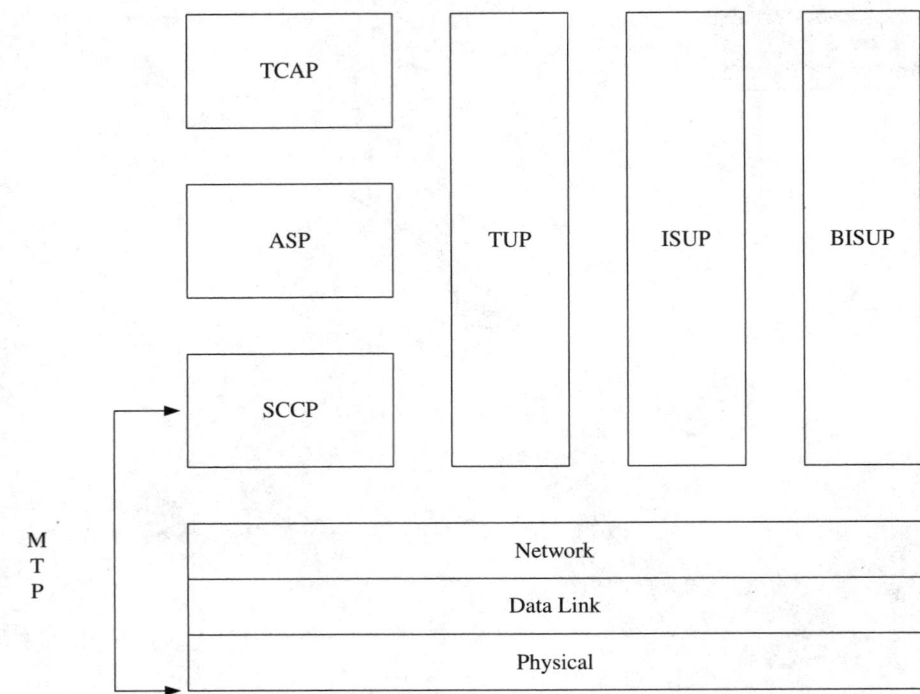

Figure 8-10
SS7 protocols

packets are crucial to network operations. To satisfy differing functions, the stack uses various protocols on the upper layers but consistently uses the same lower layers.

Basic Call Setup with ISUP

The important part of the protocols is the call set-up and teardown. This next example is shown in Figure 8-11 and is described in the following section.

When a call is placed to an out-of-switch number, the originating SSP transmits an ISUP *initial address message* (IAM) to reserve an idle trunk circuit from the originating switch to the destination switch (1a). The IAM includes the originating point code, destination point code, circuit identification code dialed digits, and, optionally, the calling party numbers and name. In the following example, the IAM is routed via the home STP of the originating switch to the destination switch (1b). Note that the same signaling link(s) are used for the duration of the call unless a link failure condition forces a switch to use an alternate signaling link.

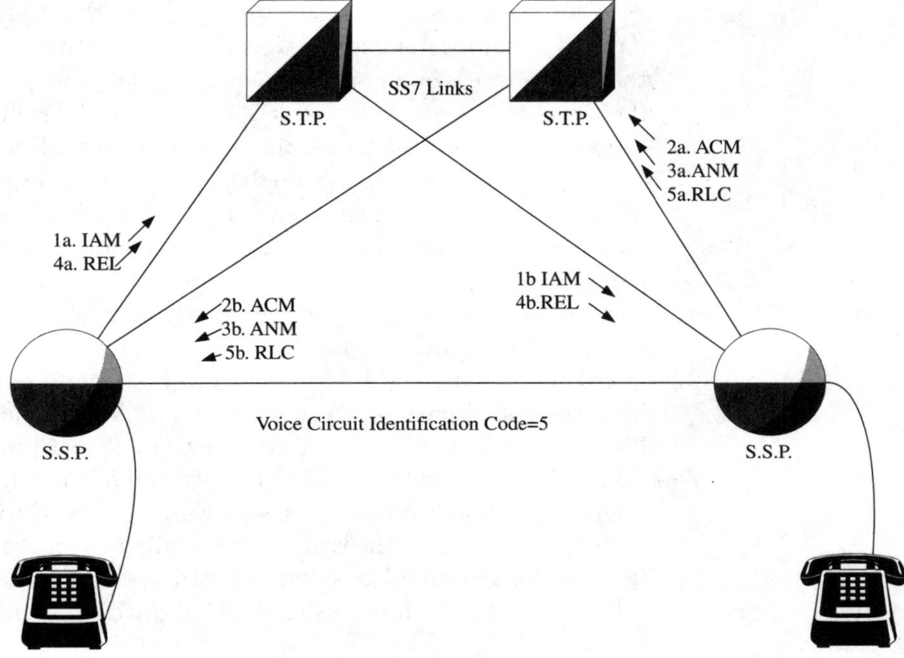

Figure 8-11
Call set-up with ISUP

The destination switch examines the dialed number, determines that it serves the called party, and that the line is available for ringing. The destination switch transmits an ISUP *address complete message* (ACM) to the originating switch (2a) (via its home STP) to indicate that the remote end of the trunk circuit has been reserved. The destination switch rings the called party line and sends a ringing tone over the trunk to the originating switch. The STP routes the ACM to the originating switch (2b), which connects the calling party's line to the trunk to complete the voice circuit from the calling party to the called party. The calling party hears the ringing tone on the voice trunk.

In the previous example, the originating and destination switches are directly connected with trunks. If the originating and destination switches are not directly connected with trunks, the originating switch transmits an IAM to reserve a trunk circuit to an intermediate switch. The intermediate switch sends an ACM to acknowledge the circuit reservation request and then transmits an IAM to reserve a trunk circuit to another switch. This process continues until all the trunks that are required to complete the connection from the originating switch to the destination switch are reserved.

When the called party picks up the phone, the destination switch terminates the ringing tone and transmits an ISUP *answer message* (ANM) to the originating switch via its home STP (3a). The STP routes the ANM to the originating switch (3b), which verifies that the calling party's line is connected to the reserved trunk and, if so, initiates billing.

If the calling party hangs up first, the originating switch sends an ISUP *release message* (REL) to release the trunk circuit between the switches (4a). The STP routes the REL to the destination switch (4b). If the called party hangs up first or if the line is busy, the destination switch sends an REL to the originating switch indicating the release cause (such as a normal release or a busy signal).

Upon receiving the REL, the destination switch disconnects the trunk from the called party's line, sets the trunk state to idle, and transmits an ISUP *release complete* (RLC) message to the originating switch (5a) to acknowledge the release of the remote end of the trunk circuit. When the originating switch receives (or generates) the RLC (5b), it terminates the billing cycle and sets the trunk state to idle in preparation for the next call. ISUP messages can also be transmitted during the connection phase of the call, such as between the ISUP ANM and REL messages.

SS7 Applications

At this point, we switch gears and look at some of the applications that are possible because of SS7 implementations. The use of AIN features, ISDN features, and wireless capabilities all are a reality as a result of the functions of SS7 integration. Some of the features are listed here, but remember they are formulated as a result of SS7, even though they may be part of other systems or concepts. These include the following:

- 800/888/877/866/900 services
- Enhanced 800/888/877/866 services within call centers
- 911 enhancements
- Class features
- Calling card toll fraud prevention
- Credit card approvals and authentication
- Software/virtual defined private networks
- Call trace
- Call-blocking features

SS7 and IP

Much of the growth in SS7 networks requires that the carriers add dedicated 56 KB or 64 KB circuits between and among the nodes. As already discussed for reliability, most carriers add these circuits in redundant pairs. The voluminous growth of database dips and SS7 queries due to network expansion, AIN and LNP, means that more dedicated circuits must be installed. Network operators can no longer accurately predict the volume or growth rates of their SS7 circuits and networks. When we view the amount of traffic required in the wireless networks (that is, GSM networks with GPRS and SMS service growths approaching explosive proportions), the carriers cannot even begin to predict the volumes they will see on their networks. Usage has been increasing at over 400 to 500 percent annually, causing the carriers and operators to fail in their predictions. To circumvent the problem, carriers have actually been over-provisioning to ensure that network blockage will not occur.

As more SS7 links are being installed in the SS7 network, devices like STPs, SCPs, and HLRs are increasing in size and complexity. Moreover, they require faster processing speeds to handle the loads being generated. These high-speed database engines and processing units are expensive. Because the systems are installed in redundant architectures, they become less efficient because operators must connect myriad devices in the network. Each time a new STP or HLR/ VLR is added to the network, a reconfiguration of the entire network occurs, which also results in additional network management costs.

Network planners are anxious to find alternatives that can reduce their STP or HLR port consumption and delay major unit replacements. Today's SS7 network planners face the following obstacles:

- Their SS7 networks are growing exponentially.
- Dedicated 56–64 Kbps SS7 link solutions are expensive when we contrast that to the shared resources of other networks.
- Ports on the nodes (that is, STP, HLR/VLR) are being rapidly consumed.
- New technology may be required; replacing existing STPs is expensive.

The carriers began to look for other solutions. One of the initial questions that they considered are as follows:

- With all the emphasis on moving to shared packet networks, would it be possible to reliably transport SS7 messages on an IP networks?

- Can the cost advantages be maximized through a shared IP network?
- Is there a way to ensure the reliability of an IP datagram if it is carrying SS7 traffic?
- Can other solutions be used to minimize the load on the existing nodes in the SS7 network?

The availability of cost-effective hardware and the growing global knowledge of IP networking has led many of the carriers to reconsider the way they deploy the SS7 networks. Advancements in reliable IP communication (using various tools such as MPLS, *quality of service* [QoS], RSVP, and others) and the market successes of *Voice over IP* (VoIP) enable the carriers to consider the next logical step—the convergence of SS7 and IP networks. For the network planners, any means or device for off-loading SS7 traffic must be

- Capable of handling carrier-grade traffic loads
- PSTN/SS7 network transparent—there must be no additional point codes and no network reconfiguration requirements
- As reliable for message transfers as the current *Public Switch Telephone Network* (PSTN)
- Remotely manageable with support for existing operations standards, like *Simple Network Management Protocol* (SNMP)

The industry response to their dilemma is the development of a new standard protocol for routing SS7 messages over IP—the *Stream Control Transmission Protocol* (SCTP). This *Internet Engineering Task Force* (IETF) standard ensures the reliable transmission of SS7 messages routed over IP networks.

SCTP

SCTP is an IP transport protocol developed by the *Signal Transport* (SIGTRAN) working group of the IETF. The basic structure of the SCTP stack is shown in Figure 8-12 with its sublayers defined. It is used to replace the *User Datagram Protocol* (UDP) and *Transmission Control Protocol* (TCP) transports for performance and security critical applications, such as voice signaling where protocols like SS7 or ISDN are carried over IP. To place it on an equivalent level of the SS7 protocol, it is more related to the MTP-2 layer as shown in Figure 8-13.

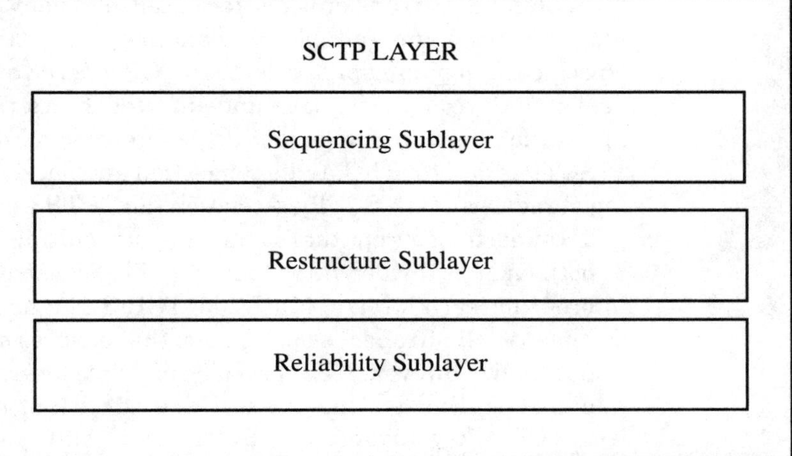

Figure 8-12
The SCTP sublayers

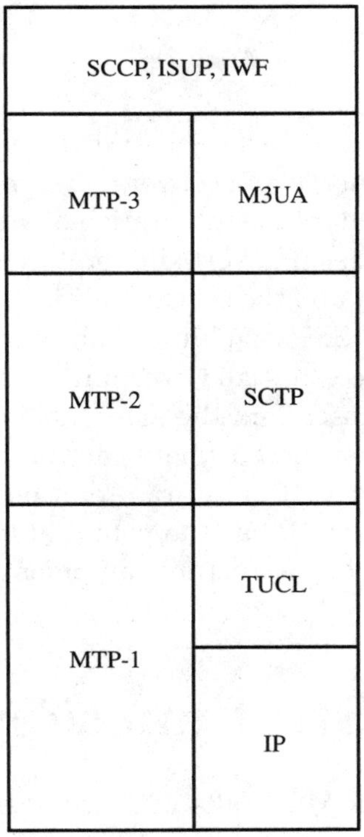

SCCP, ISUP, IWF
MTP-3
M3UA
MTP-2
SCTP
TUCL
MTP-1
IP

Figure 8-13
The SCTP relates to MTP-2

In SCTP, data transfer is packet-based, and delivery is guaranteed. This makes SCTP more suitable for handling transaction-based applications (specifically, signaling protocols) than TCP, where an application is forced to deal with the complexity of an undelineated data stream, or UDP, where an application needs to implement its own retransmission algorithms.

SCTP is designed to handle congestion and packet loss better than existing standards. Each SCTP association (an SCTP association is similar to a TCP connection, except that it can support multiple IP addresses at either or both ends) is divided into a number of logical streams. Data is delivered in order for each stream. Much like TCP, SCTP uses a message acknowledgement and retransmission scheme that ensures message delivery to the remote end. However, SCTP provides multiple message streams in order to minimize the head-of-the-line blocking effect that can be a disadvantage with TCP. A key advantage of SCTP is its ability to support multiple network interface controllers that enable applications to dynamically determine the fastest and most reliable IP network for message transmission.

VoIP Impacts

What this all means is that the convergence is rapidly being adopted so that SS7 and IP networks can merge and converge. The need to converge these services is also a direct result of the integration of VoIP. With VoIP, the use of session initiation and session advertising protocols (SIP and SAP) take advantage of the call set-up and the service notifications. This involves the transparent transport of SS7 signaling information between circuit-switched networks that are connected over an IP network. The goal is to provide voice telephony subscribers the same ubiquitous access and features regardless of whether the backhaul for the call is over a circuit-switched network or over a VoIP network. Additionally, infrastructure to provide transport of SS7 over IP has the potential to be significantly less costly than traditional SS7 infrastructure equipment.

Overview of SIP Functionality

The IETF describes SIP for VoIP calls in a document that discusses the overall concept and the supporting protocols necessary. The *Session Initiation Protocol* (SIP) is an application-layer control protocol that can establish, modify, and terminate multimedia sessions or calls. These multimedia sessions include

- Multimedia conferences
- Distance learning
- Internet telephony
- Similar applications

SIP can be used to initiate sessions as well as invite members to sessions that have been advertised and established by other means. Sessions can be advertised using multicast protocols such as SAP, e-mail, news groups, web pages, or directories (LDAP). SIP transparently supports name mapping and redirection services, enabling the implementation of ISDN and Intelligent Network telephony subscriber services. These facilities also enable *personal mobility.* Personal mobility is the ability of end users to originate and receive calls and access subscribed telecommunication services on any terminal in any location and the ability of the network to identify end users as they move. Personal mobility is based on the use of a unique personal identity (personal number). Personal mobility complements terminal mobility, that is, the ability to maintain communications when moving a single end system from one subnet to another.

SIP supports five facets of establishing and terminating multimedia communications:

- **User location** Determination of the end system to be used for communication
- **User capabilities** Determination of the media and media parameters to be used
- **User availability** Determination of the willingness of the called party to engage in communications
- **Call set-up** "Ringing," establishment of call parameters at both called and calling party
- **Call handling** Including transfer and termination of calls

SIP can also initiate multi-party calls using a *multipoint control unit* (MCU) or fully meshed interconnection instead of multicast. Internet telephony gateways that connect PSTN parties can also use SIP to set up calls between them.

SIP is designed as part of the overall IETF multimedia data and control architecture currently incorporating protocols such as RSVP for reserving network resources, the *real-time transport protocol* (RTP) for transporting real-time data and providing QoS feedback, the *real-time streaming protocol* (RTSP) for controlling delivery of streaming media, the *session announcement protocol* (SAP) for advertising multimedia sessions via multicast, and the *session description protocol* (SDP) for describing multimedia sessions. However, the functionality and operation of SIP does not depend on any of these protocols.

SIP can also be used in conjunction with other call set-up and signaling protocols. In that mode, an end system uses SIP exchanges to determine the appropriate end system address and protocol from a given address that is protocol-independent. For example, SIP could be used to determine that the party can be reached via H.323, obtain the H.245 gateway and user address and then use H.225.0 to establish the call. In another example, SIP might be used to determine that the called party is reachable via the PSTN and indicate the phone number to be called, possibly suggesting an Internet-to-PSTN gateway to be used.

SIP does not offer conference control services, such as floor control or voting, and does not prescribe how a conference is to be managed, but SIP can be used to introduce conference control protocols. SIP does not allocate multicast addresses. SIP can invite users to sessions with and without resource reservation. SIP does not reserve resources but can convey to the invited system the information necessary to do this.

In VoIP networks, packetizing the voice occurs in real-time. VoIP also decreases the bandwidth utilized significantly because multiple packets can be transmitted simultaneously. The SS7 and TCP/IP networks are used together to set up and tear down the calls. *Address Resolution Protocol* (ARP) is also used in this process.

The process of creating IP packets works as follows:

1. An analog voice signal is converted to a linear *pulse code modulation* (PCM) digital stream (16 bits every 125 µ-sec).
2. The line echo is removed from the PCM stream. It is further analyzed for silence suppression and tone detection.
3. The resulting PCM samples are converted to voice frames, and a vocoder compresses the frames. G.729a creates a 10 ms long frame with 10 bytes of speech. It compresses the 128 Kbps linear PCM stream to 8 Kbps.
4. The voice frames are integrated into voice packets. First, a RTP packet with a 12-byte header is created. Then, an 8-byte UDP packet with the source and destination address is added. Finally, a 20-byte IP header containing source and destination gateway IP addresses is added.

5. The packet is sent through the Internet where routers and switches examine the destination address and route and deliver the packet appropriately to the destination. IP routing may require jumping from network to network and may pass through several nodes.
6. When the destination receives the packet, the packet goes through the reverse process for playback.

The IP packets are numbered as they are created and sent to the destination address. The receiving end must reassemble the packets in their correct order (when they arrive out of order) to create voice. The IP addresses and telephone numbers must be mapped properly.

VoIP Telephony Signaling

Telephony signaling functions include the following:

- **Call processing** Performs the state machine processing for call establishment, call maintenance, and call teardown. This also includes Address Translation and Parsing, which determines when a complete number has been dialed and makes the dialed number available for address translation.
- **Network signaling** Performs signaling functions for establishment, maintenance, and termination of calls over the IP network. There are two widely used standards: H.323 and SGCP/MGCP.
- **H.323 Protocols** H.323 is an ITU standard that describes how multimedia communications occur between user terminals, network equipment, and assorted services on local and Wide Area IP networks. The following H.323 standards are used in VoIP gateways:
  - **H.225** Call Signaling Protocols. Performs signaling for establishment and termination of call connections based on Q.931.
  - **H.245** Control Protocol. Provides capability negotiation between the two end-points such as voice compression algorithm to use, conferencing requests, and so on.
  - **RAS** *Registration, Admission, and Status Protocol*. Used to convey the registration, admissions, bandwidth change, and status messages between IP Telephone devices and servers called *gatekeepers*, which provide address translation and access control to devices.

- **Real-time Transport Control Protocol (RTCP)** Provides statistics information for monitoring the QoS of the voice call.
- **Simple Gateway Control Protocol (SGCP)/Multimedia Gateway Control Protocol (MGCP) Protocols** is a standard that describes a master/slave protocol for establishing VoIP calls. The slave side or client resides in the gateway (IP telephone), and the master side resides in an entity referred to as a *call agent*. SGCP has been adopted by the cable modem industry as part of the DOCSIS standard. SGCP is evolving to the MGCP.

SS7 and Wireless Intelligent Networks

The *wireless intelligent network* (WIN) mirrors the wireline intelligent network model. The distinction between the wireless and wireline network is that many of the wireless call activities are associated with the end user's movement, not just the actual phone call. In the WIN, more call-associated pieces of information are communicated between the *Mobile Switching Center* (MSC) and the *Service Control Point* (SCP) or *Home Location Register* (HLR). The WIN moves service control away from the MSC and up to a higher element in the network, usually the SCP.

- **MSC as SSP** In the intelligent network, the SSP is the switching function portion of the network. The MSC provides this function in the WIN.
- **SCP** This device provides a centralized element in the network that controls service delivery to subscribers. High-level services can be moved away from the MSC and controlled at this higher level in the network. It is cost effective because the MSC becomes more efficient and does not waste time processing new services and simplifies new service development.
- ***Intelligent peripheral* (IP)** The IP gets information directly from the subscriber. This can be in the form of calling card or credit card information, a PIN number or voice-activated information. The peripheral gets information, translates it to data, and hands it off to

another element in the network—like the SCP—for analysis and control.

- **STP** This is a packet switch in the signaling network that handles distribution of control signals between different elements in the network such as MSCs and HLRs or MSCs and SCPs. The advantage of an STP is that it concentrates link traffic for the network. It can also provide advanced address capabilities like global title translation and gateway screening.
- **Location registers** These are used to supplement MSCs with information about the subscriber. The number of subscribers that the switch supports changes as roamers move in and subscribers move to other switches. The database of active subscribers changes very dynamically. Each MSC cannot have the database for all potential users of that switch. Location registers help to get around that problem.
- ***Visitor location register* (VLR)** Within an MSC, there is a VLR that maintains the subscriber information for visitors or roamers to that MSC. Every MSC or group of MSCs will have a VLR.

GSM Network Connection to SS7 Networks

The MSC is the Central Switching function of the GSM network. The MSC is connected to a SS7 network for the purpose of signaling and performing database queries. The SS7 network uses a network node called the STP, which is a packet switching node (can be SS7, IP, or X.25). Using a 64 Kbps channel connection between STPs, the network can process its signaling information.

Next in a SS7 network is the use of the SCP, which houses the databases congruent to the network. In many cases these databases interact with the HLR, VLR, EIR, AuC, and PSTN nodes. The SCP is used whenever a Global Title Translation is required, which converts numbers (800-322-2202 equates to 480-706-0912) and whenever the *Mobile Application Part* (MAP) is used . These services link across an SS7 interface. The GSM architecture using the SS7 protocol is shown in Figure 8-14.

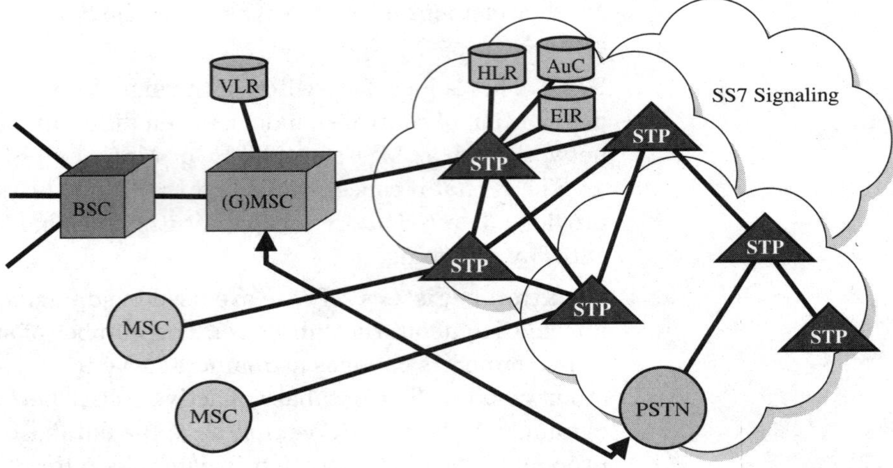

Figure 8-14
SS7 protocol stack and GSM

The Signaling Protocol Stack for GSM

The normal SS7 network uses the bottom three layers in what is called the *Message Transfer Part* (MTP) 1-3. These parts use a different layer of the OSI model to provide the routing and data link layers across the physical link. Between the layers three and applications, is the *Signaling Connection Control Part* (SCCP), which is used when database queries are required and when providing both connection and connectionless access to the SS7 networks. The combination of the MTP1-3 and SCCP creates what is called the *actual MTP*.

When looking at the upper layers, the SS7 protocols support the use of the following protocols shown in Figures 8-15 and 8-16.

- ***Telephony User Part* (TUP)** For a voice circuit-switched call across the PSTN (refer to Figure 8-15).
- **ISDN User Part** A newer implementation and replaces the TUP (refer to Figure 8-15).
- ***Transaction Capabilities Application Part* (TCAP)** An application layer that supports the features and functions of a network (refer to Figure 8-15).
- **MAP** Sits on top of the TCAP as a means of supporting the difference application service entities for mobile users (refer to Figure 8-15).

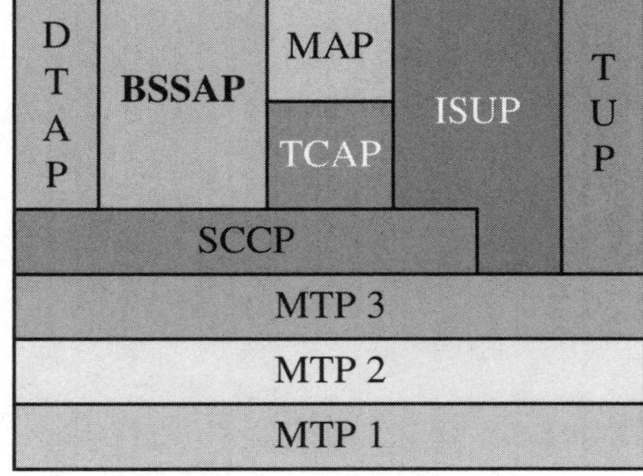

Figure 8-15 The protocols for GSM and SS7 networks

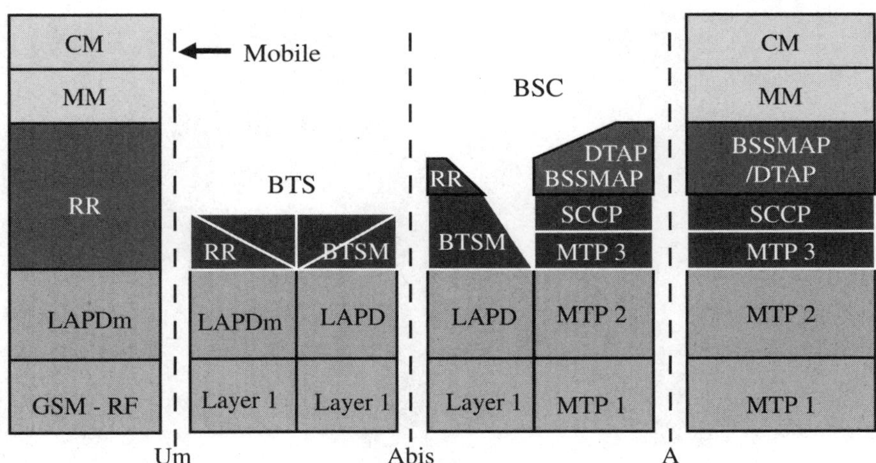

Figure 8-16 The protocols for the wireless GSM architecture

- ***Base Station Systems Application Part* (BSSAP)** A combination of the BSSMAP and DTAP (refer to Figure 8-16).
- **BSSMAP** Transmits messages that the BSC must process. This applies generally to all messages to and from the MSC where the MSC participates in RR management (refer to Figure 8-16).
- ***Direct Transfer Application Part* (DTAP)** Transports messages between the mobile and the MSC, where the BSC is just a relay function, transparent for the messages. These messages deal with MM and CM (refer to Figure 8-16).

CHAPTER 9

CTI Technologies and Applications

Computer telephony systems can range from simple voice mail to multimedia gateways. The equipment used in these systems includes *Voice Response Units* (VRUs), fax servers, speech recognition and voice recognition hardware, and intelligent peripherals deployed by telephone companies and service bureaus.

Today's businesses need to leverage the power of these diverse, multiuser computer telephony systems to improve productivity, give users more access to information, and provide communication options and services to both customers and employees. Business customers can use the telephone to automatically receive information about a product through a fax machine, while employees can access computer-managed voice, fax, and even data through telephones and computers to connect offsite workers to the office and expand relationships with outside enterprises.

The computer telephony industry offers the power of sophisticated telephone systems to any size business the same way the PC industry exploded in the 1980s. In just over 10 years, the computer telephony industry has grown to encompass many diverse applications and technologies. In 1999 alone, analysts estimate the revenue from multiuser computer telephony applications, development toolkits, and services and technologies to be $10 billion worldwide.

Understanding Computer Telephony Technologies

Many manufacturers and *value-added resellers* (VARs) are committed to providing technologies and products to customers for achieving success with automated call-processing applications. These products take advantage of technologies and enable users to store, retrieve, and manipulate computer-based information over a telephone network.

Voice Processing

Voice is the fundamental technology at the core of most computer telephony systems. It encompasses both the processing and the manipulation of audio signals in a computer telephony system. Common tasks include filtering, analyzing, recording, digitizing, compressing, storing, expanding, and replaying signals.

Telephone Network Interfaces

Network interfaces enable computer telephony systems to communicate electrically within specific telephone networks. Calls arriving from the *Public Switched Telephone Network* (PSTN) can be carried on a variety of lines, including the following:

- Analog loop start
- Analog ground start
- Direct Inward Dial (wink start) lines
- T-1/T3
- E-1/E3
- *Integrated Services Digital Network* (ISDN) *Primary Rate Interface* (PRI) and *Basic Rate Interface* (BRI) lines

Network interfaces interpret signaling, provide data buffering, and include surge protection circuitry.

Tone Processing

Tone processing includes the capability to receive, recognize, and generate specific telephone and network tones. This facilitates an application placing a call and monitoring its progress. Tones that are processed include

- Busy tones
- *Special information tones* (SIT)
- No answer (RNA)
- Connection
- Ringing
- No ringing
- Dial tones
- Fax tones
- Modem tones

Facsimile (Fax)

Facsimile lets you transmit copies of documents and images over telephone lines to another location. To transmit and receive electronic faxes, PC-based systems use fax boards. This computer-based fax technology can improve productivity by enabling documents to be sent through a broadcast fax to a large number of people in a short period of time. People can also retrieve any number of documents on demand (FaxBack) that reside on a fax server.

Automatic Speech Recognition (ASR)

Automatic speech recognition (ASR) technology (also known as voice recognition) reliably recognizes certain human speech, such as discrete numbers and short commands, or continuous strings of numbers, like a credit card number. ASR can be divided into two groups:

- Speaker-independent ASR, which can recognize a limited group of words (usually numbers and short commands) from any caller.
- Speaker-dependent ASR, which can identify a large vocabulary of commands from a specific speaker. This is popular in password-controlled systems and hands-free work environments.

Text-to-Speech (TTS)

Text-to-speech (TTS) generates synthetic speech from text stored in computer files. TTS provides a spoken interface to frequently updated information and information stored in extensive computer databases. TTS is an economical way of giving customers telephone access to information that would be too expensive or impractical to record using voice technology.

Switching

Switching technology handles the routing, transfer, and connection of more than two parties in a call. Once the domain of *private branch exchanges* (PBXs) and proprietary switches, switches are now available on boards that can be easily installed in PC-based computer telephony systems.

Understanding Computer Telephony Solutions

Customers use building block components to develop open systems products that are sold in all of the major computer telephony markets. These markets can be grouped into four major sections: Information Access and Processing, Messaging, Connectivity, and *central office* (CO)/*Advanced Intelligent Network* (AIN).

Information Access and Processing Applications

Businesses worldwide are expanding their corporate communication systems to automate employee access to information and the capability to process that information. For example, a customer may want to check on the balance of a loan and then get a fax report of the adjusted interest charges for early payment.

Information access and processing applications can improve communication and increase customer service levels. These systems include AudioText, *fax-on-demand* (FOD), *interactive voice response* (IVR), *interactive fax response* (IFR), and simultaneous voice and data.

AudioText

AudioText provides prerecorded information to callers. Businesses can offer callers a single message or a choice of messages through touch-tone or ASR.

Voice Recording for Transaction Logging

With the pace of communications escalating and the demand for real-time information increasing, most of us are faced with the challenge of better managing our time and resources by establishing priorities.

One of the toughest elements to manage is inbound telephony, mostly because we do not know who is calling and why. Therefore, we have choices on how to handle the incoming calls:

- Screen the call by a personal assistant
- Route the calls to voice mail
- Answer the calls

Another problem is dealing with our calls while away from the office. Callers are forced to either leave a message or to page us. Unfortunately, we spend a large portion of our valuable time playing "telephone tag." This costs time, opportunities to complete transactions, and money.

Newer systems can handle the phone process for us when they become available. Some of these features include the following:

- **Voice announce** Knowing who is calling is the most important criteria in call management. By knowing who is calling, one can decide the priority and nature of the call before taking action. Positive Caller Identification enables us to know who is calling, allowing us to manage and prioritize calls. Callers can be identified by various methods:
  - Matching the *automatic number identification* (ANI) with a database or *personal information manager* (PIM).
  - Having the caller speak their name.
  - Having the caller input their telephone or account number prior to transferring the call.
- **Follow-me** Regardless of where you may be, you can be available to take that important call. Call Management enables you to take calls you want when you want them.

  If an important customer calls your office while you are traveling, they will be identified by the Call Management system and your "Virtual Assistant" will try to locate you by calling previously programmed telephone numbers in the Call Management system. You may have designated your cellular telephone as your primary call-back choice. Your cellular telephone rings, you answer, and your Virtual Assistant says, "You have a call from (caller's voice plays). Would you like to take the call?" At this point, you can take the call, elect to send it to voice mail, or reroute to another associate. You can even choose to play one of many prerecorded greetings to the caller.

  While using your network PC, you can see the name of the caller on your screen and select the appropriate call-handling action. You can connect your PC to the Call Management system remotely via *Remote*

Access Server (RAS) or the Internet and manage your calls remotely as if you were sitting at your desk. The Call Management system can enable you to uniquely handle calls from important business associates and customers. Let's suppose you are expecting a call from your boss and he is expecting you to provide him with an update on a particular project. The Call Management system would enable you to record a message, giving him the desired update, so that when he calls, the update would be played for him.

- **Single number availability** Rather than giving out numbers for your office telephone, fax, cellular, and pager, you (and your customers) can enjoy the simplicity and convenience of a single number. The Call Management system can recognize whether a person or fax is calling and can handle the call accordingly. Your Virtual Assistant can provide your callers with options to contact you, page you, or simply leave you a voice message with a call-back number.

  Your Virtual Assistant can also be programmed to locate you and deliver faxes, voice mail, and e-mail messages. You can even have your Virtual Assistant make calls for you when you travel, so you do not have to deal with calling cards. The Virtual Assistant keeps your telephone directory and you can make calls from your directory.

Technology Enhancements

When CTI was first delivered, it was done through a large computing platform, either a mainframe or a high-end midrange computing system. These systems included such things as the IBM mainframe, AS/400, DEC VAX, or HP systems. Although they worked, they were expensive and sophisticated, requiring an extensive investment as well as application programming interfaces that made the service available only to large organizations.

The implementation of server-based or LAN-based platforms, however, has trickled down to the very small organizations. No longer can one determine or assume that a company using CTI applications is very large. As a matter of fact, many of the CTI applications are now rolling out on PC-based platforms for small organizations. Companies with 3, 5, or as many as 10 call-answering or telemarketing positions have implemented CTI very effectively. These lower-cost solutions have made CTI a reality within organizations around the world.

It was through LAN technology and not PBX technology that CTI actually got a foothold within organizations. Imagine if we had waited for the

PBX manufacturers and the telephone companies to roll out CTI integration for us. We would probably still be waiting. In the past few years, however, the computer manufacturers have provided the push and the software developers have contributed the innovations to make CTI a reality.

This relatively new yet rapidly accepted approach to using the server as the instrument to provide CTI has thrust CTI into the forefront of telecommunications technology. As mentioned earlier, the voice server on a LAN is designed to connect directly to the public switch telephone network, handle calls coming into the group, and then process those calls directly to the desktop. Priority customers and special handling arrangements enable specific users to work around the high-end PBX and Centrex platforms and go directly to the individual department or customer service group without proceeding through the corporate platform. This in turn changes the architecture, because of the CTI applications that can work on a server platform.

PBXs, once known for their large investments and proprietary nature, can now remain single-line telephony service providers. When higher-end features and functions are necessary, the end user merely has to buy computer-based software, as opposed to high-end PBX architectural software. Of course, the PBX manufacturers have recognized this shift in the technology implementation and all of the PBX manufacturers are now developing the CTI interfaces or the software with third-party developers to reclaim customers. Because the PBX need only be an uncomplicated telephone system for the masses within an organization, the technology can last significantly longer. In the old days of the PBX, the plan was to keep the system for a period of about 10 years. However, reality dictated that the PBX was changed on a basis of about five to seven years. This involved major investments and changes within architecture, causing a significant amount of corporate stress.

As a quick side note, whenever a new PBX was installed, it usually meant that the Telecom manager within a corporation would be leaving soon. Regardless of how many technological advancements or enhancements were installed, users' expectations were never met. The frustrations of the users and the complaints made to management usually led to the Telecom manager's demise. Now with the features and functions moved to a PC-based platform, the Telecom manager can breathe easier. Without the need to upgrade the PBX or change an entire infrastructure, the Telecom manager can implement which features or functions are necessary and available on a department-by-department basis. On a large scale, all PBX features would be available to all users. By using the CTI implementation on a server, since features are purchased on a department-by-department basis, they are subsequently less expensive. This has been the boon of the 1990s.

Other Technologies

Because of the innovations in the telecommunications environment as well as the server marriage, many other applications and features can be made available by other producers. For example, through the use of the Telecom server on a LAN, the automated attendant, voice messaging, ACD, and IVR functions can all be united in a single server-based platform.

The developers of voice messaging systems' automated attendants recognized this opportunity several years ago. They leapfrogged the market, bypassing the PBX manufacturers, and developed single-card processing systems that could use high-end digital trunking capabilities directly into the servers. Using microprocessor control devices, these companies were able to write the necessary software that would provide the capabilities of all of these features. No longer would an organization have to buy a room-sized voice messaging system; this function can now be performed on a PC-based platform.

When voice messaging was first introduced, the size, heat, and cost of systems were exorbitant. Now, using PC-based systems, just about every vendor offers the capability of allowing several hundred to several thousand active users on a voice messaging system or a call-processing system at a much lower cost. Actually, it's becoming much more difficult to tell a PBX performing CTI applications from a CTI server performing PBX capabilities. This convergence is blurring the lines between the various departments. Many of the organizations that now produce systems with these capabilities may have once been niche market providers, but are now moving across the border that once separated these two technologies. Just about every feature, function, and capability can now be had using a very low-end server platform at a very reasonable price.

The integration includes features and functions such as

- Voice messaging
- Automated attendant
- IVR
- Text-to-speech
- Speech-to-text
- Directory services
- Fax services
- Fax-back services
- Intranet access for catalogs

Taking these one at a time, we'll see how they can all play together and provide unified messaging and integration capabilities. The capabilities of the integrated messaging and unified messaging services enable the desktop user to functionally perform all day-to-day operations at a single interface device, now the desktop PC.

Automated Attendant

With technology moving as quickly as it is, the use of single processing cards in a PC can deliver a combination of voice messaging and automated attendant functions directly to the CTI application. A digital signal-processing capability can literally compress voice calls so that they can be conveniently stored on a hard disk drive. The voice is already in a digital form when it arrives from a digital trunk or digital line card; therefore, storage is a relatively simple technique. The application software used in a voice mail system is basically a file service in which storage and retrieval can be easily accommodated.

Integrated Voice Recognition and Response (IVR)

IVR enables customers to manipulate information in a computer database, such as retrieving an account balance and transferring funds from one account to another. These applications range from AudioText and pay-per-call information systems that deliver a single audio message or a selection of messages to transaction-based systems that enable callers to access accounts and update information on a LAN-based or host-based database. AudioText entertainment lines are popular applications in computer telephony markets.

IVR systems are primarily based on the same type of technology as the auto attendant and voice mail. Using a single digital-processing card, the capability now enables users to arrange for prescripted capabilities that will actually walk a caller through a menu. The IVR will play digitally stored messages and solicit a response from the caller at each step, usually in the form of a touch-tone from a telephone set. The response from that tone will then cause the next step of the message to be played in accordance with the script. This is useful when a user is trying to access information from a host-based system, for example, and a played-back message will enable the user to retrieve any form of information.

IVR has been used by several medical providers and insurance providers, enabling a caller to dial in and access information regarding a payment or the processing of claims by merely using a touch-tone telephone. When dialing into the IVR, the user is prompted each step of the way by the system. As the user enters an ID number, a query is sent to a database in a host-computing platform. The appropriate information is then retrieved and played back. Using this CTI application saves an immense amount of time for an organization, because this normally labor-intensive activity can now be achieved through technology.

Fax-Back and Fax Processing

The *digital signal processor* (DSP) card can also be programmed to function as a fax modem that can provide for the sharing of fax services within a single-server environment. A fax image can be downloaded across a LAN, converted by a fax card, and transmitted across the network over a digital trunking facility. In the reverse direction, if an incoming fax is received from the network, it is then converted back to a file format that is easily usable within the PC environment. This can then either be stored in a fax server file for later retrieval by the individual recipient of the message or redirected by a fax operator. Some of these systems and services take more effort to implement and facilitate, but they may well be worth the effort in terms of an organization's needs.

A fax-back capability means that when a user dials into an organization equipped with CTI applications, that user can be directed to a fax server that has a numerical listing of specific documents that the user can retrieve by keying in a telephone number. When the user enters the telephone number, the server retrieves the fax from the file and then automatically transmits it to the designated telephone number the user has just entered. Through this application, catalog information or specific customer information can be retrieved without the use of human intervention. One can see how much time using these types of services could save.

Fax-on-Demand (FOD)

Fax capabilities are indispensable in the business world today. Businesses with dedicated fax-based systems or with fax as an enhancement to their existing communications systems can automatically deliver information on demand to their customers. For example, customers can dial in and listen to a menu telling them which documents are available by fax. They can make a selection by speaking or by pressing a touch-tone digit and then enter the number of their fax machine to receive the document.

Interactive Fax Response (IFR)

Interactive Fax Response (IFR) enables customers to automatically receive a fax in response to a transaction performed through either the telephone or a computer. For example, a customer may receive a printout of an account balance after having transferred funds.

E-mail Reader

An e-mail reader resides on a media server that uses TTS technology. E-mail readers translate the ASCII text of an e-mail message (stripping out unnecessary header information) into voice that can be retrieved by callers through any analog device, such as a telephone.

Text-to-Speech and Speech-to-Text

In speech-to-text applications, a prestored pattern of words can be used through the CTI application to enable a highly mobile workforce to dial into a server-based platform and literally speak to the machine, as opposed to using touch tones. Speech from the callers, whose voice patterns are already stored in the computer, can then be converted into usable text using a server-based CTI application. This is instrumental when the user cannot access a touch-tone telephone. Without the touch-tone telephone, the user would have to carry a portable touch-tone pad generator, which is very inconvenient. Inevitably, the batteries on these devices die at the very moment the user needs access to information. Consequently, the use of voice patterns or speech patterns that have been prerecorded with a series of words, such as get, save, retrieve, file, and so on, can be used to facilitate and walk through a computing system.

The TTS applications are comparable in that when a user accesses a particular file, again without a terminal device, for example, the system can convert the text into a speech pattern. What this effectively means is that e-mail and other documents can literally be read back to us no matter where we are. This is exciting because an end user might well dial into the

CTI application while traveling on the road and learn that he or she has six voice messages and four e-mail messages waiting. Rather than that user being forced to log on with a different form of terminal device, these e-mail messages can be read right down the telephone line to the end user, facilitating the easy retrieval, storage, or redirection of messages. It is through these types of services that the CTI applications are drawing so much excitement.

Optical Character Recognition (OCR)

Another form of DSP technology can convert scanned images into text. When used with fax machines or fax images, *optical character recognition* (OCR) can change an incoming fax into a document that can easily be edited or incorporated into other types of applications. This would include editing a fax and placing it into a word processing document for easy editing capabilities. The additional storage capabilities could then convert the OCR, which usually would be a file of significant size, into a text-based document, which would be much smaller. Furthermore, by using the OCR to scan a pre-typed document, for example, the application could transform the scanned document into a TTS application, which could be read aloud. One can just imagine the uses and applications of some of these technologies.

Summary

Hopefully, this discussion of CTI will provide you with some appreciation of the capabilities and features that contribute to the merger of computing and PBX architectures. The use of an onscreen interface at a desktop PC enables users to manage and maintain their mailboxes for voice messaging as well as e-mail. Beyond that, a visual display can be received directly to the desktop, outlining the number of faxes, e-mails, or voice messages waiting to be retrieved. With the integration of the voice and text applications, the user can also see who the messages are from and prioritize the receipt of each of these messages based on some preconditioned arrangement. All of this facilitates the integration of the computer and telephony capabilities onto a single, simple platform that empowers the end user to access information more readily.

Moreover, with the implementation of CTI as a frontend processor for the organization's telemarketing or order-processing departments, customers have the ability to retrieve information at will. This use of touch-tone or voice response systems enables a customer to literally walk through catalogs, check the status of orders, check inventories, or even check the process of billing information, all without human involvement. It is not the intent of this discussion to rule out the use of all humans, but rather show how humans can be more productive in performing the functions for which they were initially hired. By taking the repetitive "look-up"-type applications along with the data applications and allowing them to be controlled by the end user (or customer), the organization can save a significant amount of time and money and better utilize the human resources they have. The industry, however, is now facing a severe shortfall of skill sets and talents that could facilitate some of these functions. With the use of the CTI application, this human resources shortfall can easily be supplemented through technology. As things progress even more, additional applications such as video servers may well be added to this architecture and enable callers to view displays on a downloadable file, so that catalog information could be easily retrieved with a video clip that would show exactly what the customer is ordering or buying. Moreover, as the video clips and the fax services and voice messaging capabilities all become integrated into one tightly coupled architecture, customers could see the article, place the order, and literally "construct" the order customized to their needs. One can only imagine some of the possibilities of these features and functions that will be available in the future, but as with anything else, the first steps must be implemented.

In short, an organization must recognize the potential benefits that can be derived from a CTI application. Its capabilities are exciting because of all of the different ways CTI can be used. With a GUI-based system, a point-and-click, mouse-driven application on a desktop enables end users as well as customers to literally walk their way through all catalogs and information.

CHAPTER 10

Integrated Services Digital Network (ISDN)

This chapter will describe the concept of the *Integrated Services Digital Network* (ISDN) and will focus specifically on the following topics:

- Its original goals
- How it can be used
- Some of the alternatives

The world's telephone companies conceived ISDN in the early 1980s as the next generation network. The existing voice networks didn't deal well with data for the following reasons:

- One had to use modems to transmit data.
- The data rates were around 9600 bps.
- Connections (worldwide) were unreliable.

Not only would the connection drop without notice, but also the error rate was high enough to require a complex protocol to recover from errors. The back end of the network, that is, the interoffice trunks, were practically all digital, with more being installed daily. The switching systems were becoming digital just as quickly. It was expected that the only part of the network that would still be analog was the local loop (our infamous last mile). It was a logical step to provide digital capability already in the network directly to the customer.

This created what is always called the local loop problem. The problem is that much of the local loop plant was installed between 40 and 60 years ago and had been designed for normal voice communications only. The local loop problem therefore is how to run high-speed digital data on the local loop. Several solutions to the local loop problem are discussed in this book, and the ISDN solution is a little different from the solutions that will be presented in Chapter 16, "xDSL."

Although the digital network exists and the digital switching systems exist, ISDN has not made large inroads into the customer premises. Once heralded as the solution for Internet access, it has been overtaken by xDSL (although there is an ISDN-like DSL service called IDSL) and cable modems. It has found some success in the business community, specifically for telecommuting and teleconferencing.

Although greeted with enthusiasm by many equipment makers, ISDN has been treated coolly by several of the major North American carriers. The European monopoly carriers implemented ISDN in the major cities, while their rural telephone systems still use electromechanical switching. The committees that defined ISDN concentrated on the interface.

Origins of ISDN

ISDN was a concept developed by the *Consultative Committee on International Telegraph and Telephone* (CCITT). The CCITT has since changed its name to the ITU-T, but it is still the same folks, made up primarily of representatives of the world's government-owned monopoly carriers. Recall that in 1980 all telephone companies were monopolies and all were government owned and run by the post office (thus, the name *Post Telegraph and Telephone* [PTT]). Canada and the United States were exceptions because most of the Telcos were private companies that had been granted monopolies.

Although there were about 1,500 Telcos in the United States, AT&T, GTE, and ITT owned the bulk of the important ones that covered the large population centers. In Canada, the provincial governments owned the telephone company. Exceptions were in British Columbia, Ontario, and Quebec. Canada also has many small independent Telcos in the outlying areas.

Origins of the Standards

The CCITT is a consultative committee to the *International Telecommunication Union* (ITU) and have recently changed their name to ITU-T. The ITU-T (CCITT) is a UN treaty organization and, as such, each country is entitled to send representatives to any committee meeting. The representative typically comes from the government-run PTT monopoly. The world's Telcos are becoming privatized and competition is being permitted. This creates an interesting struggle within each country to determine who will represent that country's interest at the ITU-T. Note that when discussing the organization's historical actions and composition, we call it the CCITT. When we discuss its current actions, we call it ITU-T.

The name ITU-T came about due to the privatization trend separating telephone business from the post office and the general elimination of telegraph service. Since its members were no longer PTTs, the organization couldn't be called CCITT. The CCITT was a consulting committee to the ITU, so the *ITU-TSS* (ITU-T) is the *Telecommunications Standard subsection* (TSS) of the ITU.

The CCITT is comprised of *study groups* (SG). Each SG has its own area of expertise. Here are some of the better known ones related to ISDN:

- SG VII on public data networks (X.25) X-series standards
- SG VIII terminal equipment for telematic services

- SG XI ISDN and telephone network switching and signaling
- SG XII transmission performance of telephone networks and terminals
- SG XV transmission systems
- SG XVII data transmission over public telephone networks
- SG XVIII digital networks, including ISDN

Although we are calling them standards, technically the CCITT and ITU-T publish recommendations. The philosophy of the ISDN committee (essentially the Telcos) is to specify the customer interface first and then figure out how to support it in the network. The theory being that if we can get all the peripheral or end equipment makers to make equipment based on our specifications, then when we get ready to roll out the service, the store shelves will be stocked with inexpensive ISDN interface equipment.

Interfaces

The customer interface I.45x specifies the *Basic Rate Interface* (BRI). It was intended to become the standard subscriber interface.

The BRI specifies two bearer channels and a data channel. The two bearer channels would bear the customer's information. The initial concept had this as being everything from analog telephone calls (digitized) to teleconferencing data and these would be switched channels. The only difference between a conventional telephone circuit and a bearer channel is that the bearer channel would be 64 KBps all the way to the customer. (Note in the current network, your analog telephone circuit is digitized to 64 KBps at the local Telco office before being switched across the network. It is then turned back into analog at the far end before being delivered to the called party.)

Now with BRI, we have not one, but two such telephone circuits. Since it is digital, we have switched digital 64 KBps to (theoretically) anywhere in the world.

The problem with the existing Telco network is that the signaling information shares the telephone channel with the user information. With plain voice circuits, a customer doesn't notice or care. With the advent of modems, this represents a loss of channel bandwidth, and with digital transmission, it means a loss of usable bits per second. The customer is therefore stuck with 56 KBps, instead of the actual channel rate of 64 KBps.

The BRI interface therefore specifies a multifunctional data channel at 16 KBps that could handle signaling (its primary function) and network data (X.25) when not needed for signaling. BRI is therefore referred to as 2B + D, two bearer channels and a data channel.

Figure 10-1 shows the BRI graphically and indicates the bandwidth allocation on the ISDN interface. Remember that this is a *time-division multiplexed* (TDM) interface where the B, D, and overhead bits are interleaved.

In Figure 10-2, the BRI is created by the *Network Terminal type 1* (NT1). The NT1 creates a four-wire bus called the T interface onto which each ISDN device is connected. We will discuss the interface designations *user* (U), *terminal* (T), *system* (S), and *rate* (R) later. Two points should be kept in mind. First, the boxes shown in Figure 10-2 can be combined in any reasonable way. Second, it is not necessary to have an NT2 element. This means that the S and T interfaces are logically and physically identical. They have separate identities to allow us to describe the functionality of the NT2 element. The NT2 could create multiple S interfaces and perform the switching to adjudicate access to the B channels on the T interface. Since the two interfaces are the same, they are frequently referred to as the *S/T interface*. As described below, but not shown in Figure 10-2, up to eight devices can be connected to the bus.

Figure 10-1 BRI bandwidth allocation

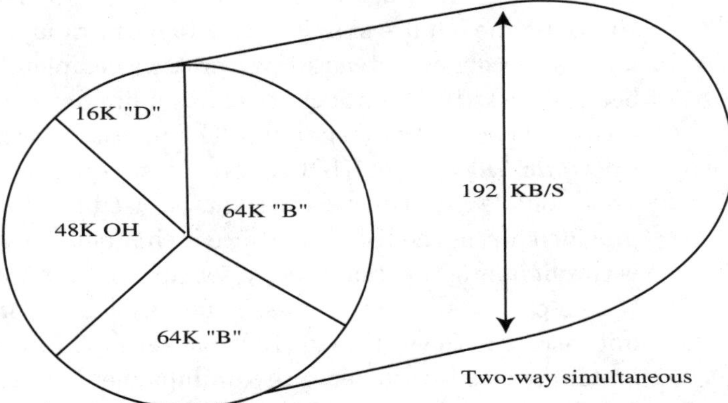

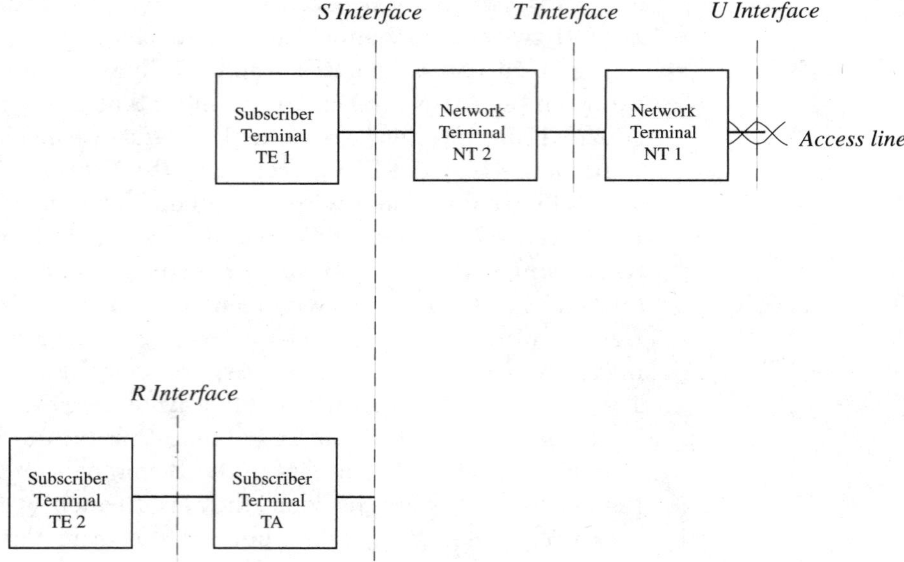

Figure 10-2
NT1 creates the BRI.

It is helpful for the following discussion to have a picture of the ISDN interface architecture handy, and Figure 10-3 serves this purpose. We have outlined the basic access. It provides 2B + D. The two bearer channels shown on the left are simply access to network layer services. Both the link layer protocol and network layer protocol completely depend on the service being accessed. The data channel has a link layer protocol defined as Link access protocol—Data (channel). The link layer (which is essentially *high-level data link control* [HDLC]) provides sequenced acknowledged delivery. This reliable link layer service can be used by the network layer services provided across the ISDN interface D channel. The most important of which is the signaling function used to set-up and teardown the B channels.

The packet-switching access is not universally implemented, but it permits access to the worldwide X.25 packet-switching network. The telemetry system access is also generally unimplemented. Several telemetry experiments have been done in Europe and in the United States, but few resulted in a cost-effective solution. We may have to wait for new lower-cost technology. The concepts, discussed in the following paragraphs, are sound ones.

The higher-layer services could be almost anything. A typical example might be video conferencing. Here the higher-layer services would be the functions of the Codec.

If the ISDN B channels are used for Internet access, then IP would be the network services and TCP, HTTP, and so on would be the higher-layer ser-

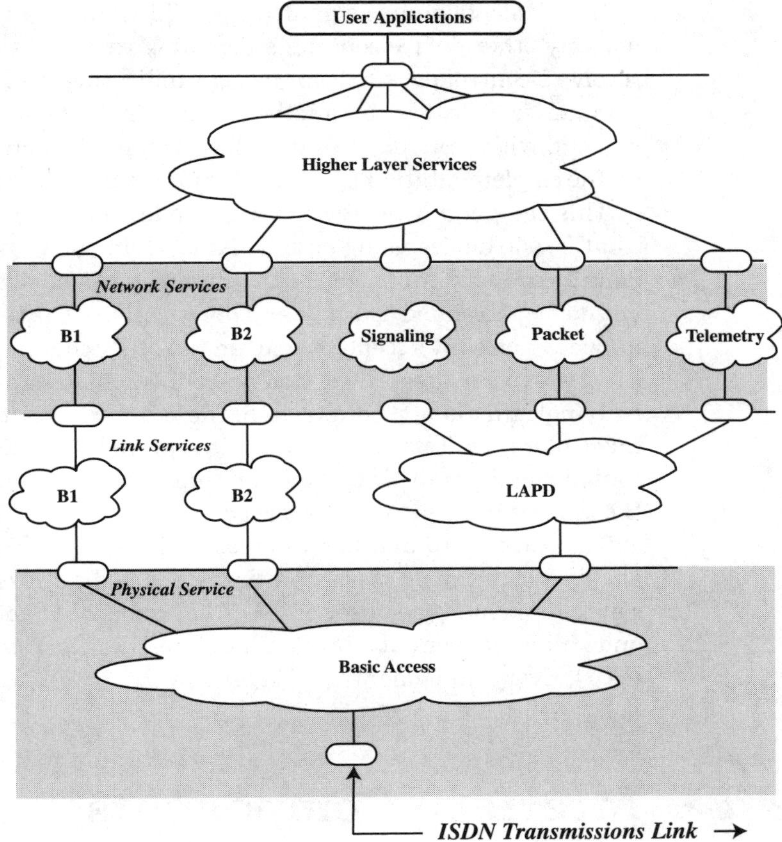

Figure 10-3 Architecture of the ISDN interface

vices. The B and D channels are then just timeslots on the bus that can be grabbed by any of the connected devices. The trick is how the telephone, for example, keeps the computer from grabbing its timeslot while it is off-hook. The answer lies in the 48 KBps of overhead. In this 48 KBps, the NT1 provides timing to all the devices on the bus. Remember this is a bidirectional channel, so there is 48 KBps of overhead going back as well. The station that wants to grab the D channel effectively puts its address in the contention slot in the 48 KBps inbound bus. These bits are echoed by the NT1 as a confirmation of success. Effectively, the highest address will win since the one bits would overlay any zero bits of a device with a lower address. When the terminal sees its address echoed in the outbound overhead channel, it has won the right to send a packet (signaling or otherwise) on the data channel.

If the telephone has signaled the NT1 channel and grabbed a B1 channel, any other device sending a signaling request for that B1 channel will receive a control packet from the NT1 indicating that it is busy. Since one B channel is identical to another B channel, the NT1 will theoretically respond with "you may use the B2 channel." The actual operation depends on the implementation of the NT1 and terminal equipment.

This can get a little sticky when bonding channels. Essentially, bonding (that is, one device using both 64 KBps channels to get 128 KBps) must be done at call initiation. One of the channels can be dropped during the call, but it can't be added again later. This is because the called end (such as ISP) may have already accepted a call on the other channel. The logical question is, why can't it accept a mated or bonded channel on any other channel? Although this is a good question, the answer is that when the rebonding takes place, the Telco network treats it like a brand new call. Thus, it can be routed via different offices and the timing relationship between the bonded B channels would not be preserved.

This brings up another interesting issue: Telco implementations vary widely. The number of possibilities is nearly limitless. Each Telco has chosen to implement a subset of ISDN based on what they think they can sell and still implement at a profit. Therefore, there is no such thing as standard ISDN. Every implementation is unique.

Interface Components

The previous interface doesn't allow for switching on premises, which would be necessary if there were multiple telephones.

The European and North American Telephone systems are quite different in operation. In North America, extension telephones simply bridge onto the line, or connect in parallel. Adding a party to a call or transferring from one extension to another is as simple as picking up one telephone and putting down another. In the European system, specific actions (pushing the designated button) are required on the part of the parties to transfer a call. This also implies on-premise switching.

NT1

We have mentioned the function of the NT1. It creates the T interface for premise devices (from the U interface). In the original CCITT concept, the

NT1 was provided by the Telco as part of the ISDN service. The U interface was therefore only the concern of the Telcos with open networks (which concerned North America at the time). It is now important to understand on general principles.

NT2

This device would do the switching, permitting more than the standard eight devices to share the T bus by creating perhaps multiple S buses. Therefore, an ISDN *terminal equipment* (TE) device can't really tell if it is connected to an NT1 or NT2.

TE1

The *terminal equipment type 1* (TE1) is a standard (there is that word again) ISDN terminal that is capable of dealing with the B and D channels. In other words, it can interface with the S/T bus.

TE2

The *terminal equipment type 2* (TE2) is a standard device having an RS-232 or V.35 interface. (In ITU parlance, this is called a V-series interface.) It may be intelligent, but it doesn't have an ISDN interface capable of handling the D and B channels.

TA

The *terminal adapter* (TA) is the semi-intelligent device that lets a TE2 connect to the S/T ISDN interface. The primary function of the TA is to run the ISDN interface for our TE2. The functionality varies widely due to the manufacturers. Some are simple and support only one TE2; others support two TE2s and an analog telephone. The TA need not be a stand-alone box; it can come on a PC card and plug into the computer's internal bus. With the proper software to run the card, this instantly creates a TE1 out of your computer.

Thus, the BRI then was designed for *small offices or home offices* (SOHOs).

Physical Delivery

One of the more interesting parts of the ISDN service is the solution to the local loop problem. Remember our local loop is ancient and designed for voice. From Figures 10-1 and 10-2, we expect it to support 192 KBps bidirectionally! This would not be a problem if we only had to go a few hundred feet. Unfortunately, there aren't that many customers within a few hundred feet of the central office.

The problem is approached from the other end actually. If we want to reach 95 percent of the customers, what is the average length of the cable to them and what are the quality and characteristics of that cable?

Some of the answers to these questions you might find surprising. In cities, 95 percent of the customers are within 15,000 cable feet of the central office. When we go to suburbs and the rural areas in particular, then all bets are off.

Figure 10-4 shows a typical local loop layout with emphasis on the fact that local loops, particularly those in older parts of a city, are comprised of different gauges of cable and may have several bridge taps. Although

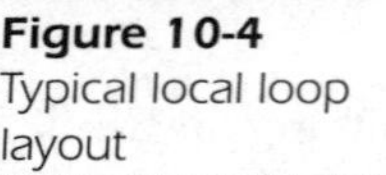
Figure 10-4 Typical local loop layout

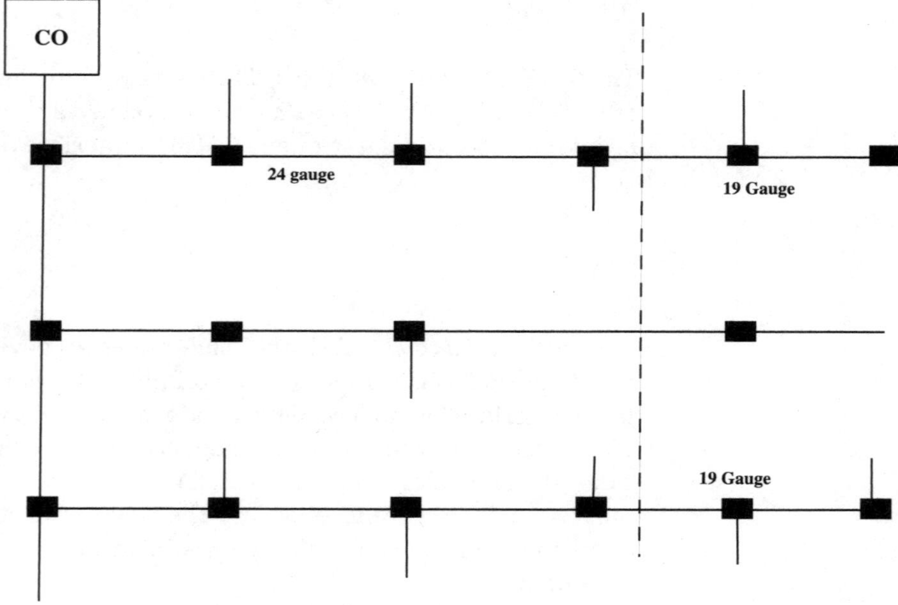

slightly exaggerated, one could imagine that all of the cable taps are on one pair. That is, at one time or another during the 60-year life of this particular cable plant, that pair was used to provide telephone service to each of those different locations. Note also the gauge change from 24 to 19 gauge. This was obviously not a problem for the analog telephone network because it worked fine for 60 years on that cable.

High-speed digital transmission presents a whole new set of problems. We will outline just the main issues here. (See Chapter 16 for a more in-depth discussion of the transmission problems.) The fundamental problem centers on the fact that a wire is an antenna. This is OK if we are in the radio or TV transmission business, but it is an unfortunate side effect of the telephone transmission business. The problem is worse at higher frequencies that characterize digital signals. If we use our local loop to send digital signals containing high frequencies, the low frequency portion of the signal goes a relatively long way. The high frequency portion of the digital wave is radiated off the wire and is delayed by the characteristics of the wire. The longer the wire, the worse the problem. The result is that the signal dribbling out the end of the wire is weak and distorted because all the components didn't arrive in time or with the right strength. Adjacent wires in the cable pick up the components of our digital signal that are radiated. We call this *crosstalk* in the telephone business and the unwanted signal picked up is noise. The brute force method can be employed to remedy this. Sending a stronger signal makes more signals available at the destination, but also increases the amount of energy radiated.

As the telephone company, we must make sure that any new service added to our outside plant (local loops in aggregate) does not interfere with any existing service. Our goal of wanting to provide more distance must be tempered by the realities of crosstalk.

Interestingly enough, the European Telcos ignored the question of ISDN delivery, preferring to concentrate on defining the S/T interface instead. Their theory was that "we're selling a service and how we deliver it to the customer isn't important—we can always figure that out later." (Also since they were government monopolies, if the cost of delivery was high, they could always raise the tariff.)

The North American Telcos had a different situation. They were not allowed to sell or lease the NT1 interface as part of the ISDN service and had to specify the electrical and logical interface at the Telco demarcation point. Therefore, the signaling on the local loop had to be worked out and standardized before ISDN could be deployed.

The U Interface

The U interface is unique to North America and the open telephone network interconnection. Figure 10-5 shows the U interface connecting to the NT1 and the NT1 in turn creating the internal S/T bus to which up to eight ISDN devices can be connected. This U interface can be either a two-wire or a four-wire connection. In the following discussion, we concentrate on the two-wire connection because it is the more technologically challenging. The four-wire interface requires much less technology and can be delivered over a greater distance. Many early BRI interfaces were installed using the four-wire interface for just these reasons. For wide-scale deployment, however, the two-wire interface has to be perfected, since the Telcos are not about to double the size of their already extensive outside plant.

AT&T (Bell Labs) came to the rescue with a technique known as 2B1Q (At this time, this was all part of AT&T. Since then, Bell Labs was spun off

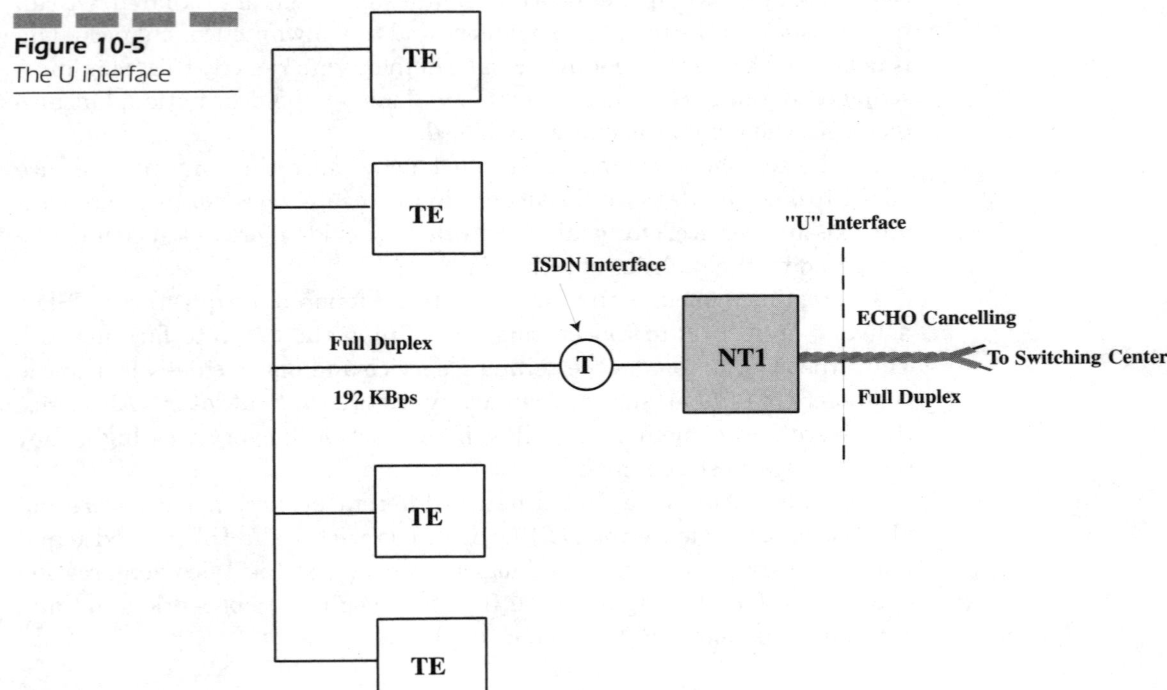

Figure 10-5 The U interface

from AT&T as part of Lucent Technologies). Figure 10-6 shows the 2 B1Q signal. Although it appears (and is) simple, it solves several problems:

- It is easy to generate (that is, it is a simple wave form).
- It minimizes crosstalk.
- It will work on most local loops.

Notice that each of the four levels contains two bits. Therefore, the signaling rate (baud rate) is one-half the bit rate. From our previous discussion of the S/T interface, we might conclude that the bit rate must be 192 KBps. Fortunately, those 48 KBps of overhead are only needed on the S/T interface to provide timing and priority (refer to Figure 10-1).

On the Telco side of the interface, we only need 2B + D and some overhead for timing and control. The total is 160 KBps. This means in engineering terms that the primary spectral peak is at about 80 kHz. In layman's terms, it means we are only trying to send 80 kHz down the old local loop twisted pair wire. What we have effectively done is halved the bandwidth requirements of the line. This trade-off isn't free though; it means that the signal-to-noise ratio has to be better than if we were to use a simpler encoding system.

The next problem is that the old local loop has different gauge wire and has bridge taps (refer to Figure 10-5). Unfortunately, when sending pulses down the line, we are going to get reflections from these gauge changes and bridge taps. These reflections show our transmitted signal much lower in amplitude and delayed in time (Figure 10-6 shows this reflection as a dashed line). The reflections will always be

- Of the same magnitude and
- At the same relative time from each of the cable plant anomalies

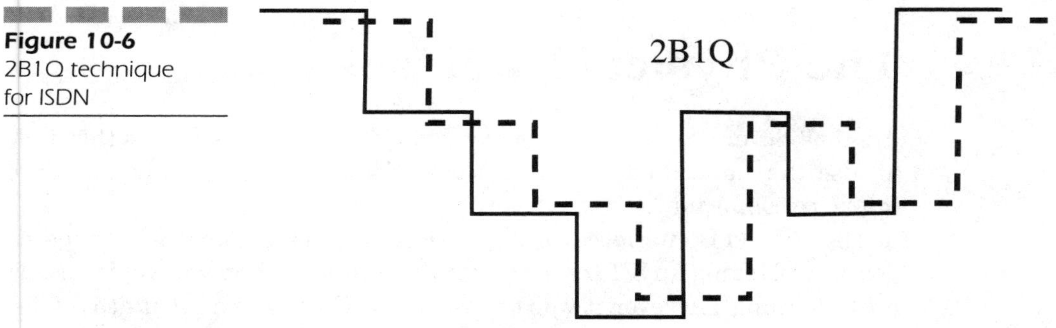

Figure 10-6
2B1Q technique for ISDN

Therefore, when the ISDN NT1 goes off-hook, it transmits a known pattern. That pattern contains all possible bit combinations (there are only 16 combinations). The receiver at the transmitting end then monitors the resultant complex signal. Since it knows what it sent and can subtract that signal, it memorizes the resulting reflected energy. The NT1 stops and lets the central office end do the same thing. Each end has learned the reflection characteristics of the local loop.

Now the clever part: Both ends can now simultaneously send data to the party at the other end. However, it is hard to listen to the relatively weak signal from the other end when you are also talking. This is where the learning comes in. Each transmitter can subtract its own transmitted signal *and* the reflections, which it knows to be there. After subtracting our transmitted stuff (and reflections), whatever is left over must be the data from the far end. Yes, it really works!

Unfortunately, we aren't quite home free. This old local loop doesn't handle all frequencies equally, so some of the signal components arrive out of the precise time (or phase) with the other parts of the signal, as mentioned above. This effectively distorts the signal. The U interface hardware therefore uses the old modem technique of equalizing the line. While each end trains itself on reflections, the opposite end receives the known signal, recognizes the distortion, and tunes its equalizer to make the signal appear correct. If we were to independently measure the line characteristics (amplitude and phase) and the equalizer, we would (as you might expect) find them equal and opposite. The result is that the equalization process takes the distortion out of the line.

The *2 binary 1 quaternary* (2B1Q) works so well that this basic technique is now the primary technology used by the local Telcos to deliver T-carrier service, instead of the old *alternate mark version* (AMI) technique. It will go twice as far before repeaters are required. Of course, the name has been changed to confuse us and is now called *High bit-rate Digital Subscriber Line* (HDSL).

The Physical Interface

Another clever design feature of both the S/T and U interfaces is that they all use the same RJ-45 type connector. Figure 10-7 shows the standard interface connector.

The S/T and U interfaces carefully select the pin assignments so that accidentally plugging an S/T connector into a U interface and vice versa doesn't hurt anything. However, it won't work either. There was a great deal of dis-

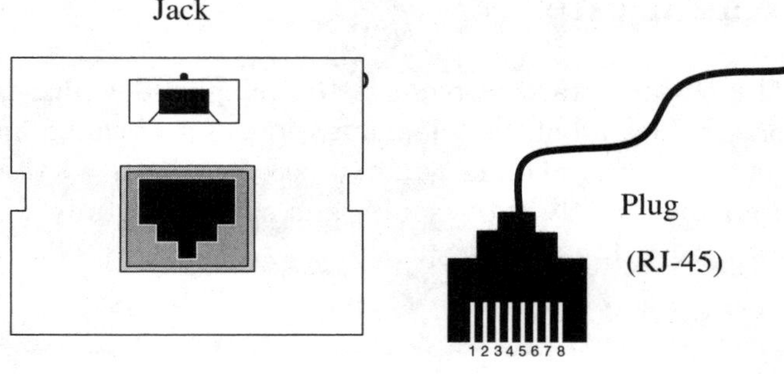

Figure 10-7 The standard interface connector

cussion about the customer interface concerning how and whether the carrier should provide power. In the current analog system, the carrier provides power to the telephones so that they work, although commercial power is off. Shall this capability be preserved for ISDN? Although the augment lasted several years, three powering mechanisms are provided across the interface:

- The customer provides power to the NT1.
- The carrier provides power to the customer from the NT1.
- The carrier provides a small amount of keep-alive power on the actual bus leads.

Here is another opportunity for the implementations of the carriers to diverge. Most carriers don't provide power. The safe bet is for the customer to be able to power his own equipment in case of a power failure.

Applications of the ISDN Interface

The following section describes the areas in which ISDN functions, including multiple channels, telephone services, digital fax, analog fax, computer/video conferencing, signaling, telemetry, and packet switching.

Multiple Channels

Figures 10-1 and 10-2 display the logical BRI interface. The plan is to provide access to every possible home device. The original concept was for up to eight devices. After all, you only have two B channels and one D channel to share among eight devices.

Telephone

The obvious starting point is the telephone, which is now a digital telephone. Instead of the telephone conversation being analog from the handset to the central office where it becomes digitized, the conversation can be digitized directly at the source and passed digitally all the way through the network to the other end.

Digital Fax

Fax machines now have to be digital. Therefore, the Group IV fax standard specifies 64 KBps fax operation.

Analog Fax

Analog fax machines use a modem, so it has to plug into the telephone (or similar device) that would take the analog modem tones and digitize them at 64 KBps. This would provide compatibility with all existing Group III fax machines.

Computer/Video Conferencing

Our computer or video conferencing equipment can use one of the 64 KBps or bond both bearer channels together for a 128 KBps digital channel across the network.

Signaling

The primary function of the data channel is to provide for signaling, that is, the setting up and tearing down of the switched bearer channels. At 16 KBps, the data channel has more bandwidth than is needed for signaling alone. Therefore, when it is not being used for its primary and high-priority signaling function, it could be used for other things.

Telemetry

This feature has never been well defined. The concept is that many household devices can be connected to the data channel. This can include an energy management system that would let the power company selectively turn off the refrigerator or air conditioner for an hour or so at peak usage time. The concept also includes connecting the utility meters to permit remote monitoring and billing. Although several proof-of-concept trials of this technology have been conducted, apparently the cost of implementation outweighed the potential savings.

Packet Switching

The 16 KBps data channel has bandwidth to spare. Therefore, the local carrier can provide a data service on this excess bandwidth. X.25 is just maturing and is the logical packet-switching technology to offer. As it turns out, all the data on the data channel, whether it is signaling data, telemetry data, or X.25 data, are always sent in packets anyway. (This packet-handling channel was the logical genesis of Frame Relay. If we can distinguish different kinds of packets, why not frames too?)

These devices are all connected to the same 2B + D interface; therefore, three of them could be in operation at one time. For example, the telephone could be using a B channel, the fax could be using a B channel, and the computer could be doing X.25 packet switching. (Today we run all these services over our Internet connection.)

Primary-Rate ISDN

The BRI interface offers 2B + D. The *Primary Rate Interface* (PRI) provides 23B + D and, in this case, all are 64 KBps channels. This should sound familiar. What technology provides 24 64-KBps channels? Of course, the answer is T1. The physical interface for PRI then is simply a T1. Technically, it is a T1 with *extended superframe framing* (ESF). The obvious question is, how is PRI different from an ordinary T1?

The answer is that at the physical interface it isn't any different. Channel 24 now becomes the signaling channel. One might say that we have added common channel signaling to the T1. PRI then is frequently used for a PBX interface where the full signaling capability of the D channel is needed.

Figure 10-2 shows the S/T interface components for PRI. This same figure is then used to describe the BRI and PRI interfaces, but because it is a T1, we can add familiar labels to the diagram components. The NT1 is now a *Channel Service Unit* (CSU), and the NT2 is now a *private branch exchange* (PBX). It could also be a router, but typically routers are members of dedicated networks and don't use switched channels. The Telcos charge more for a PRI because it offers many more switched-channel features. If a router needs to be connected to another router, an ordinary T1 will do fine and cost less. Circumstances will occur when we need both our router and another device on the same interface and will periodically adjust the amount of bandwidth to each. For this unique application, PRI is ideal.

If we stick with our PBX example, the NT2 creates multiple S/T interfaces for TE1 or TA devices. TE1s are ISDN BRI telephones. What is interesting about the definition of the PRI interface is that it isn't limited to a single T1. Multiple T1s can be added to the PRI, so that up to 20 T1s can be part of a single interface and the single 64 KBps D channel on the first T1 would control all. That is, a single D channel could control 479-switched B channels.

H0 Channels

Equally interesting is that aggregated channels have been defined. These are known as H channels. An H0 channel is 384 KBps or six B channels treated and switched as a single channel. This means our PRI NT2 could call for an H0 to Atlanta, another to Phoenix, and one to Chicago and still have five B channels to individually switch.

H11 Channels

The H11 channel is a 1.536 Mbps (T1) switched channel too. If our PRI NT2 had, let's say, 5 T1s, it could configure any one of them as an H11 channel and another as 2 H0 channels with 12 B channels. This essentially gives us dial-up T1, 384 KBps, and 64 KBps clear channel service.

H12 Channels

The Europeans use an E1 system that has 32 channels, each 64 KBps. One channel is used for timing and alarms and one is used for common channel signaling. In the PRI, channel 16 becomes the D channel, so the E1 version gives 30B + D. The H12 channels are only in the E1 system and are 1,920 KBps. Again; theoretically this is switched E1, 384 KBps, and 64 KBps service.

Signaling on the D Channel

The signaling packets on the D channel are the same for BRI and PRI. In the early days of defining the interfaces, an argument arose about whether to grant direct access to the signaling system by the customers. As we indicated, *Signaling System 7* (SS7) is the mechanism for managing the network and it's logical to simply let the D channel use SS7 packets. Unfortunately, the more paranoid faction won the argument, so the D channel signaling packets require a small amount of conversion to change them from D channel signaling packets to SS7 packets. This is OK, because the D channel link layer frame is standard HDLC (*Link access protocol for the D channel* [LAP-D]), which can carry the packets on the D channel whether they are signaling packets or X.25 packets.

Figure 10-8 shows a D channel packet. If it has the look of an X.25 packet, it is not by accident. Since both X.25 and signaling packets are handled (time interleaved) on the D channel, it was thought that a protocol discriminatory byte would be a good idea just in case the packet was received by the wrong entity. (The link layer protocol makes this virtually impossible anyway, but just in case . . .) The protocol discriminator byte pattern never occurs in X.25 and would be immediately recognized as an illegal packet and discarded.

The call reference value is essentially a random number chosen to identify a particular call. Because there could be many (479) calls in progress at once, you might need more than a fixed-length value. (In order to prevent confusion about which call is doing what, you don't want to reuse those numbers too quickly.) All signaling packets associated with a given call will have the same call reference value.

The message type indicates the format of the packet. The setup message, for example, would have the called and calling telephone numbers in the

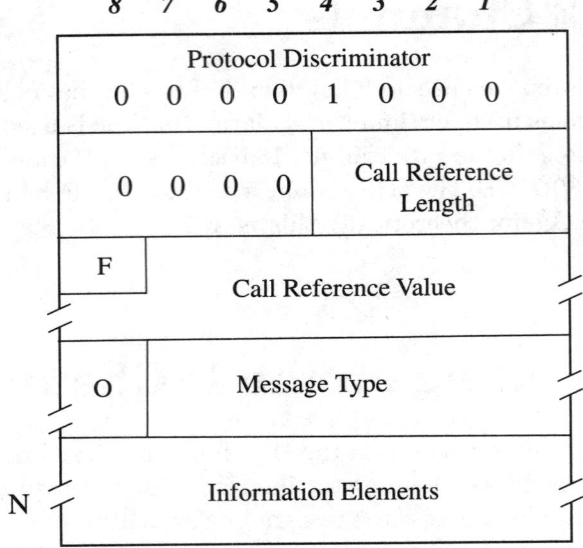

Figure 10-8
D-channel packet

information elements field. The call-clearing message would have a single-byte cause code in the information element field.

Of course, all of this wonderful capability isn't of much value if the service provider hasn't implemented the backroom functions to enable these features for the customer. The carriers have simply implemented that subset that they individually think they can sell. It should come as no surprise that the offerings vary widely and all features are not generally compatible across carriers.

Installation Problems

It is not our intent here to denigrate the carriers, but rather to point out that installing ISDN for other than plain vanilla applications can be frustrating and time consuming.

The carriers come from a long tradition of providing voice services. Their entire management system and technical system is designed around providing highly reliable (and profitable) voice services. ISDN is digitally based, the local distribution is digital, and the switching is digital. Historically, the Telcos have strictly partitioned the knowledge of each technical specialty. The installer knows about local drop wires, pedestals, and telephones. The switchman knows about switching. The outside plant folks know about cabling in the streets and on poles. The transmission expert knows about carrier systems. Moreover, no one knows it all!

Therefore, specially trained ISDN technicians are required to deal with the local loop. They must know the parameters for operation and be able to test it out. In the old days, the installer would hook his butt to the line and listen. He could tell a lot about the local loop by the background noise. Sorry, you can't hear digital. Therefore, the technician now has to be trained in handling a special piece of test equipment to make sure the line will support the data rate.

Because PRI is simply T1, the tuning of the local loop is understood; BRI, however, is a different issue. We are now dealing with local loops. T1 is not normally delivered to residential loops, but it could be. T1 is delivered on business loops. There is no real difference between business and residential loops, but the Telcos do tend to run data circuits (T1s) in separate binder groups and they tend to avoid the problems of many bridge taps, shown earlier in Figure 10-4.

Another complicating factor is that each carrier implements its own version of ISDN, making for a frustrating experience getting it installed and working. These different versions often lead to compatibility problems end to end. As a rule, you can count on the 64 KBps B channels working end to end, and in most cases they can be bonded to 128 KBps. On the PRI, using switched H0 and switched H11 channels is very iffy. The carriers just haven't implemented the switching algorithms to enable this feature.

BRI Application

One of the major uses of ISDN is in video conferencing. This is normally done by installing three BRI lines. The video conferencing equipment has a built-in inverse *multiplexer* (mux). Figure 10-9 shows a typical configuration. The three BRI interfaces are connected to the inverse mux and the control panel enables the operator to specify how much bandwidth is to be used for the videoconference. Although one could use 64 KBps for a videoconference, it is quite impractical. A minimal usable videoconference is 128 KBps, and as long as someone else is paying for the bandwidth, 384 KBps is acceptable. This is the crux of the problem: How much bandwidth are we willing to afford for the videoconference? The three BRI interfaces give you a choice in increments of 64 KBps to 384 KBps.

The inverse mux takes the commands from the control panel and provides the appropriate packets on the D channels to set up the amount of

Figure 10-9
Typical configuration of the inverse mux

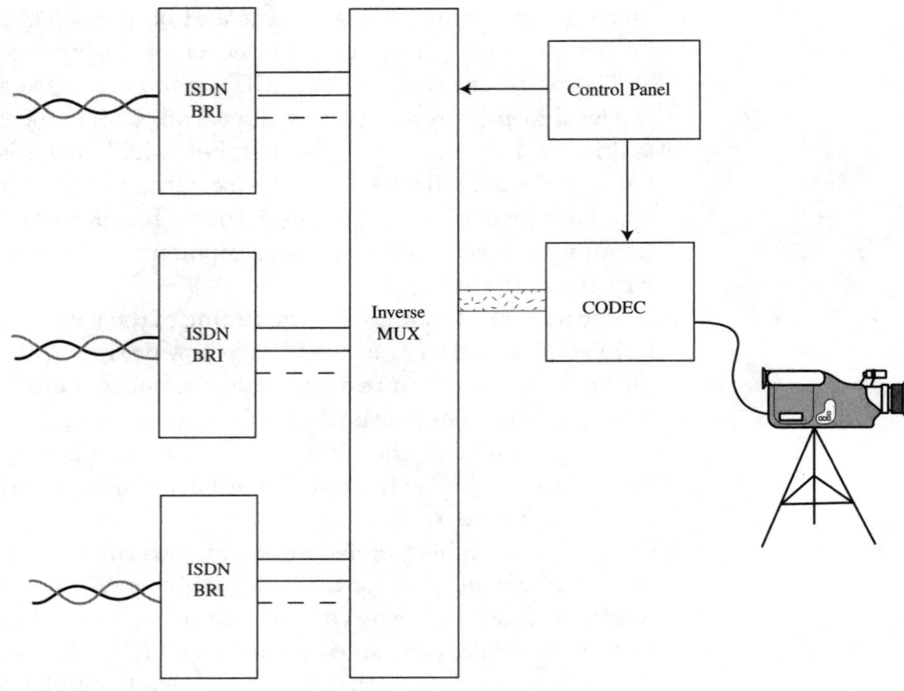

bandwidth required. The inverse mux is needed because it is essentially making up to six independent 64 KBps calls. The network routs these calls independently. Therefore, if we are in Phoenix making conference calls to Atlanta, one of the calls might go through Denver, Minneapolis, and Atlanta. Another might go to Albuquerque, Kansas City, St. Louis, and Atlanta. A third might go through El Paso, Dallas, Birmingham, and Atlanta. The point is that each of these paths has a different length and delay. The inverse mux plays "scatter gather" by sending the digitized video in packets (actually frames) alternately over each of the circuits. The peer unit at the other end puts the frames back in order with the proper timing to provide the 384 KBps channel to the *coder / decoder* (Codec). If we could afford a PRI, we could simply set up an H0 channel to Atlanta.

Broadband ISDN

There has been much confusion over exactly what is broadband ISDN. This confusion stems from at least two sources. First, the definition of the word broadband, and second from the fact that even after we are over that hurdle, no one knows what it means.

Definitions

The original meaning of broadband started with high-bandwidth as compared to low-bandwidth telephone lines. The Telcos used the word wideband to describe some of their analog carrier systems that used frequency division multiplexing to put multiple telephone channels on a single transmission facility. So when the Telcos were looking for a new name for their high-data-rate digital service, they had an existing meaning for wideband. They chose the word broadband, despite the fact that this word (outside the Telcos' world) meant a high-bandwidth analog multiplexed system. A cable TV system is a broadband system. In the early days of *Local Area Networks* (LANs), Wang and several others used the analog cable TV technology (using wideband modems) to provide computer connectivity. These were therefore broadband networks. Along comes broadband ISDN and everyone is naturally confused.

One could think of it as a high-bandwidth digital service. Actually, this is exactly how the Telcos would like to sell it. Don't ask what it is or how it works. It is simply a high-data-rate-capable transport service.

Originally then, the Telcos intended to offer this high-rate telephone service. How high a rate was in doubt. What did the customers need? Clearly, we can't sell something that the customers don't need.

There were a couple of stumbles along the way. First, there was a lot of hype about broadband ISDN, yet there was precious little supporting evidence in the Telco infrastructure. Second, it was the intent from the beginning to base the service on a statistical multiplexing type of service, a very high-rate X.25 packet switching technology. It was clear to everyone at the time that such a shared network would offer the customer more bandwidth while requiring less infrastructure bandwidth than the existing circuit-switched, time-division multiplexed network based on T1s, T3s, and T4s.

One of the stumbles grew out of the popularity of LANs. Why not become citywide or even countrywide LAN providers? AT&T invented a thing called *Switched Multimegabit Digital Service* (SMDS). The LAN committee in the *Institute of Electrical and Electronics Engineers* (IEEE) grabbed the idea and after about 10 years of study accepted the standard 802.6 as the *Metropolitan Area Network* (MAN). This was an ingenious concept that used an access method known as *distributed queue dual bus* (DQDB).

Figure 10-10 is a diagram of the DBDQ system. Although it looks like a ring, it is really two buses (one side of a T1) running in opposite directions.

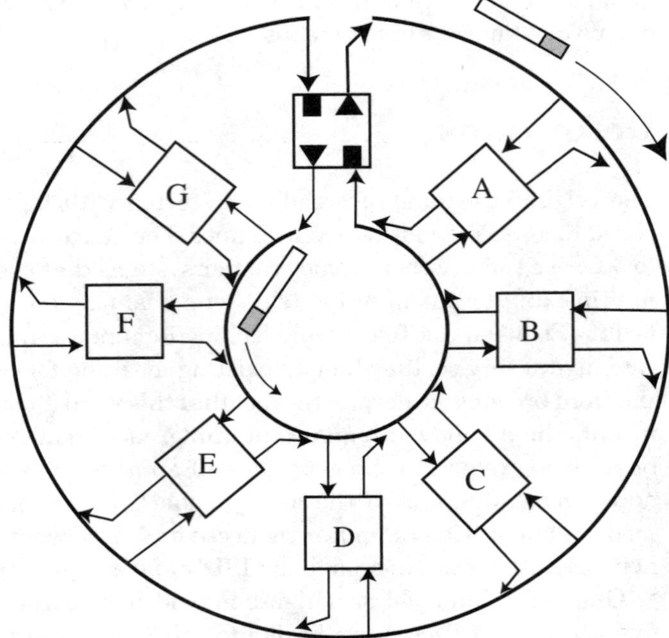

Figure 10-10 The distributed queue dual bus (DQDB) system

The Telco is the bus controller. It is well known that token passing busses are not efficient, since they effectively poll each station. The time taken in polling when a station has no data is time that a station with data cannot lose. DQDB solves this problem by having a continuous set of packets flowing in both directions, which could be assured by the Telco. These packets are small and are called *cells*. Sending empty cells doesn't hurt the efficiency. Each cell has a header that contains a busy or free bit and a request field. If station D wishes to send something to station B (or to another station) serviced by the Telco on the *clockwise* (CW) bus, D places a reservation request on the *counterclockwise* (CCW) bus in the reservation field. (It has been watching the CCW bus for reservations, so it has a queue of prior reservations.) Each free cell arriving on the CW bus causes one to be decremented from the queue. When the queue is empty, the next free cell on the CW bus can be used for transmission. Yes, this system is a little complex. (If you really want to challenge your powers of visualization, try to envision this happening in both directions at the same time.) The point of the discussion is that it is sort of a demand-based system and there are no collisions. If no one else has requested cells, then you can have them all.

The problem is that the rest of the protocol layer built upon the basic DQDB access scheme makes it very inefficient. The service is known as SMDS. Unfortunately, out of the 1.5 Mbps of a T1, you can only use a little over 1 Mbps. Very few customers found that the service was cost effective. Another unfortunate aspect of the system is that the cells are not identical to ATM cells. Thus, to use ATM for transport, these cells have to be repackaged into ATM cells. The best part is that the Telco can implement it using existing T1 technology.

SMDS, then, was part of the broadband ISDN service offering. What about long-haul transport? That was left to ATM, which intended not to be a service per se, but as the implementation technology for broadband ISDN. You can now see that ATM was intended as the core technology (invisible to the customer) to support the Telco network of the future. Just as narrowband ISDN (PRI and BRI) were intended to provide the network connection of the future, broadband ISDN service was intended to be the high-rate interface. ATM would be the core technology supporting all of the services from circuit switching to packet switching.

ATM as the supporting technology was not ready when broadband ISDN was announced, so the carriers couldn't provide the service. All eyes turned to the new kid on the block, ATM. Although not well defined, like a kid, it had lots of promise. Who was to do the heavy lifting until ATM was mature?

Frame Relay was developed to fill the void. The idea for Frame Relay came from both X.25 concepts and from the D channel packet handling of ISDN. Frame Relay filled the need for broadband service. Few customers actually needed the nosebleed speeds of ATM, and Frame Relay was a cost-effective replacement for dedicated lines. Being packet switched, it provided bandwidth on demand. ATM also provided bandwidth on demand at much higher rates.

Conclusion

ISDN, therefore, was a great technology-driven service that didn't really solve a business (or home) need. It is little wonder that ISDN is not widely implemented or used, but there are, as we have noted, some clear exceptions. The most notable is video conferencing. Internet access is also a possibility, but ISDN can't compete with xDSL technology in performance for the cost. The ISDN primary rate is used extensively in call centers, utilizing computer telephony integration to maximize their efficiency. PRI is also used in PBX applications, where the digital PBX can make use of the network control and status information provided by the PRI.

CHAPTER 11

Frame Relay

Frame Relay is a fast packet-switching technology introduced in 1992. Its installed base has skyrocketed since its development and introduction in the industry. No one could have ever predicted just how popular this technique would become, but the final outcome is that Frame Relay has become the "bread and butter" service for many carriers. The reason is not as simple as one might believe; therefore, this chapter will explore the details of what has led to the popularity in Frame Relay installations.

Moreover, some carriers have encountered problems with their installations because they did not understand the benefits or the operations of a Frame Relay service. Unfortunately, this is the epitaph of the telecommunications industry: the carriers (and specifically the long-distance carriers) do not understand data communications! Yes, they have some talented people on their staff, but they have always stifled these individuals who knew what was happening. Instead, the carriers let the voice and engineering folks become the spokespersons for the carriers in a data communications arena. How sad, they just never really got the point.

Even to this day, the voice people have not taken the time to learn data communications. More specifically, the carriers (*Incumbent Local Exchange Carriers* [ILECs] and *Interexchange Carriers* [IECs]) feel that their sales people do not need to know much about the technology, just how to sell the product. Interestingly, this leads to the carrier sales representatives being less aware of their own product than the customers they are attempting to sell to. Sales people do not need to be specific engineers, but if you want to "walk the walk" you have to "talk the talk." How can the carriers sell a product when they do not know the basic concepts of data and, more specifically, Frame Relay?

Now in an era when all the voice carriers are trying to become data-literate and the data carriers are trying to become voice-literate, the voice people stand to lose in the overall transition because of the innate ability to mess up the data installation. Whenever we hear about a carrier's capability to handle data traffic, we seem to be mesmerized by the carrier's overall capabilities, instead of understanding specific characteristics for processing and delivering data communications. Many of the more trusted names in the industry continue to tout their products and services, but fail to deliver on the promise of data communications because they churn their people over too often. The training they provide their sales representatives is enough for these people to barely get by in discussing the product, but fails to cover the application of the product and the benefits that the customer can gain.

Frame Relay Defined

First, a definition of packet switching is in order because Frame Relay falls into the category of a packet-switching family.

Packet switching is a store and forward switching technology for queuing networks where user messages are broken down into smaller pieces called *packets*. Each packet has its own associated overhead containing the destination address and control information. Packets are sent from source to destination over shared facilities and use a statistical *time-division multiplexing* (TDM) concept to share the resources. Typical applications for packet switching include short bursts of data such as electronic funds transfers, credit card approvals, point of sale equipment, short files, and e-mail.

Fast packet switching is a combination of packet switching and faster networking using high-speed communications and low-delay networking. Fast packet is a "hold and forward" technology designed to reduce delay, reduce overhead and processing, improve speed, and reduce costs. It is designed to run on high-speed circuits with low (or no) error rates. Errors are corrected at the two ends, instead of every step along the route.

Frame Relay, as stated, is a fast packet-switching technology used for the packaging and transmission of data communications. Moreover, Frame Relay packages the data into a data link layer frame (LAPF-Core Frame) used to carry the data across the network on a *permanent virtual circuit* (PVC) without all the handling of the X.25 networks. Although X.25 acknowledges every packet traversing the network, Frame Relay does not use *acknowledgments* (ACKs) or *negative acknowledgments* (NAKs). Also, when an X.25 packet is corrupted, the network node requests a retransmission, which is not so on Frame Relay. Both of the services do, however, use a statistical TDM concept. Table 11-1 is a summary of the comparison of X.25 and Frame Relay services.

By design, Frame Relay is focused on eliminating several of the older networking problems, yet in reality it does more to move the responsibility to others than to solve any of the older network problems. What it does, however, is come up with a streamlined type of communications transmission system, eliminating the older overhead. In the original days of data communications, networks were highly unreliable, yet today newer networks are very reliable due to the increased use of fiber in the backbone networks. Consequently, the entire older overhead dealing with error recovery is

Table 11-1

A summary of the X.25 and Frame Relay services

Service	X.25	Frame Relay
Statistical TDM	Yes	Yes
OSI layer used	Layer 3 (Network)	Layer 2 (Data link)
ACK and NAK	Yes, extensive	None
Retransmissions	Yes, extensive done at each node on the network	None done by the Frame Relay nodes; retransmissions are requested by higher-level protocols at the end
Packet/frame size	Up to 128 bytes average network; up to 512 bytes in some implementations	Up to 1,610 bytes in networks; up to some 4,096 bytes in some vendor products
Speed of transmission	Up to 64 Kbps	Starts at 56 Kbps; up to 50 Mbps, depending on the vendor products

somewhat superficial. Frame Relay can eliminate this overhead and use the saved capacity to carry more data. Frame Relay also assumes that if an error occurs, the higher-level protocols at the end-user level will be intelligent enough to correct the problem or request a retransmission. This can be a variable, depending on the implementation and what the end user is willing to pay for these services.

What Can Frame Relay Bring to the Table?

Frame Relay in itself is merely a communications protocol designed to eliminate the overhead discussed previously. What one can expect from Frame Relay services is the use of the higher-speed communications, the basis of the newer fiber-based *Wide Area Networks* (WANs). Taking advantage of the capacity improvements, Frame Relay can use the *bandwidth on demand concept* to get faster data across the network. The use of excess capacities in a Frame Relay network also brings newer ideas to the forefront. By allowing the end user to use extra capacity, the bottleneck of the older networks goes away. This excess usage is called the bandwidth on demand concept.

Where People Use Frame Relay

Frame Relay is designed as a WAN technology primarily for data. When the deployment began, end users and carriers alike all felt that digital voice (data) could ride on Frame services. However, that aside, the network and protocols were designed to carry data traffic across the WAN. More specifically, Frame Relay was developed to carry data traffic across the WAN and link *Local Area Networks* (LANs) to other LANs, as shown in Figure 11-1. The transmission of data across the local loop to the local telephone company's central office that is connected to the interexchange carriers' network switching system is handled by a leased T-1 or T-3 link. In Figure 11-1, a T-1 provides the connection. Note also in this figure, the access device is through a dedicated Frame Relay router on both ends of the connection.

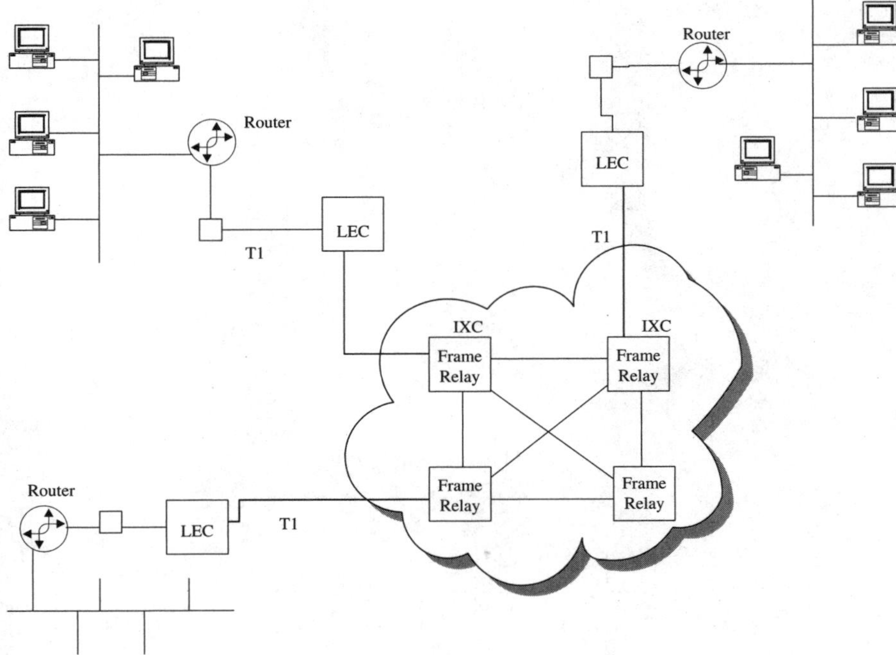

Figure 11-1 A typical Frame Relay connection

When Frame Relay was first introduced in 1992, the speeds were limited from 56 Kbps up to 1.544 Mbps in North America and 64 Kbps up to 2.048 Mbps in the rest of the world. However, as time wore on, a small upstart company named Cascade Communications (acquired later by Ascend, who was then acquired by Lucent) decided to set the world on fire by increasing the access speeds and information throughput to approximately 50 Mbps. The industry quickly jumped on this speed and made Cascade the number one supplier in the industry at the time. Figure 11-2 shows the connection at T-3 speeds on one end and *Synchronous Optical Network* (SONET) OC-1 on the other end of the connection. Note this was still used just for data transmission across the network. A higher-end router is installed on each end of the connection to facilitate the data throughput of up to 50 Mbps.

The next step was to use a different device to access the high-speed connection on a Frame Relay network. Devices known as *Frame Relay Access Devices* (FRADs) were introduced to provide the access. This was done through a high-speed CSU/DSU, through a multiplexer, or some form of a switching system. The FRAD enabled the access to be simplified for the end user and the network provider alike. In Figure 11-3, the FRAD is shown. Note the FRAD used here is through a CSU/DSU on one end and a high-

Figure 11-2 A higher speed Frame Relay connection

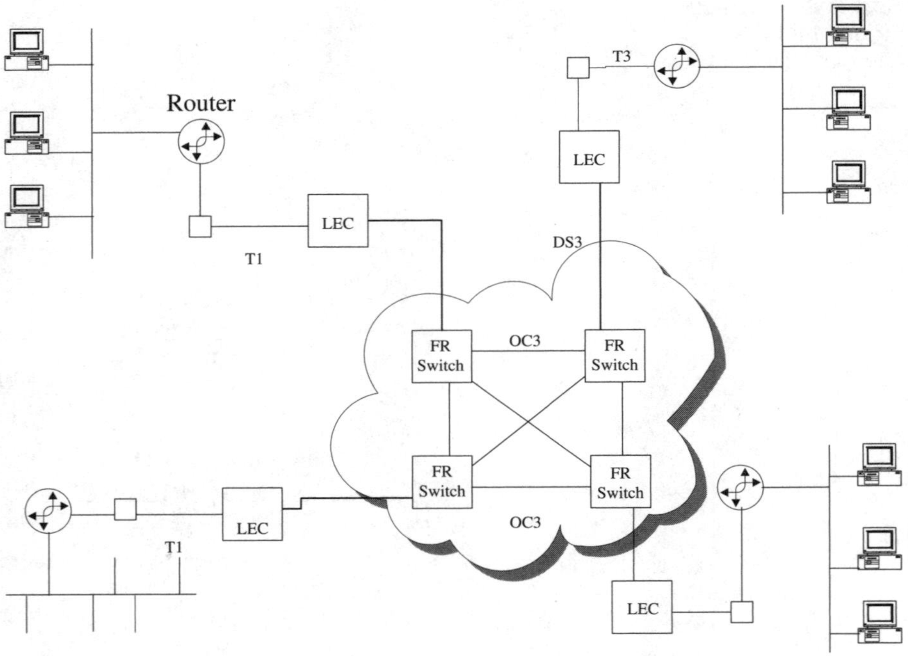

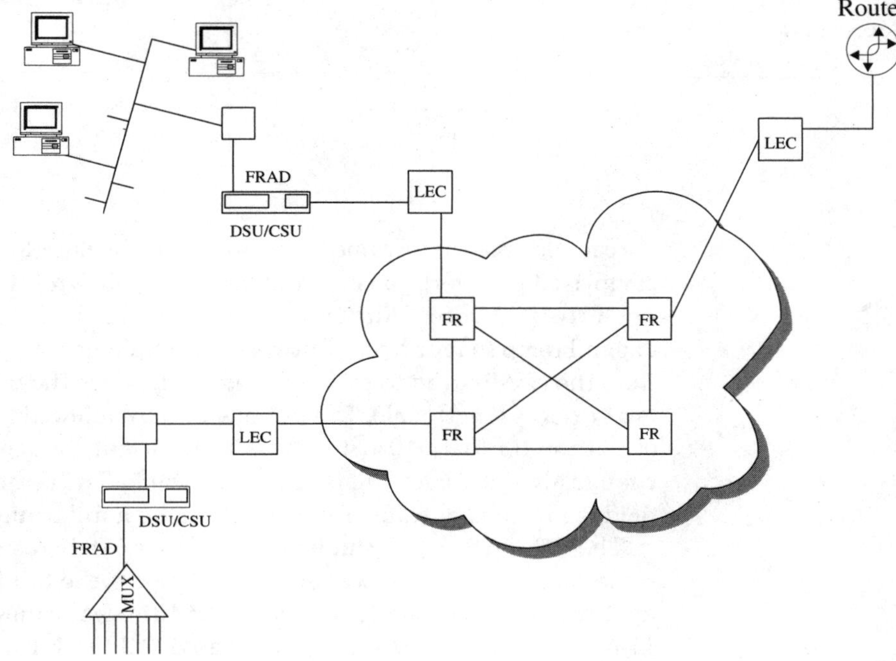

Figure 11-3 A FRAD used on each end of the circuit at high speeds

speed T-3 multiplexer on the far end. The choices vary for the end user, so the flexibility of the access is one of the strong points for the Frame Relay network.

The Frame

When Frame Relay was developed, the important part of the data-carrying capacity was the use of the frame to carry the traffic and not have the same overhead as an older technology (such as X.25). The frame was filled with data as necessary, but it handled the speed and throughput via the high-speed communications and lower overhead.

In Figure 11-4, the frame is shown. The frame is a *High-level Data Link Control* (HDLC)-framed format, as shown in this figure. The beginning of the frame (as with most HDLC formats) starts with an opening flag. Next, a two-byte sequence defines the addressing of the frame. This is called the *Data Link Connection Identifier* (DLCI). By very nature of the title (DLCI),

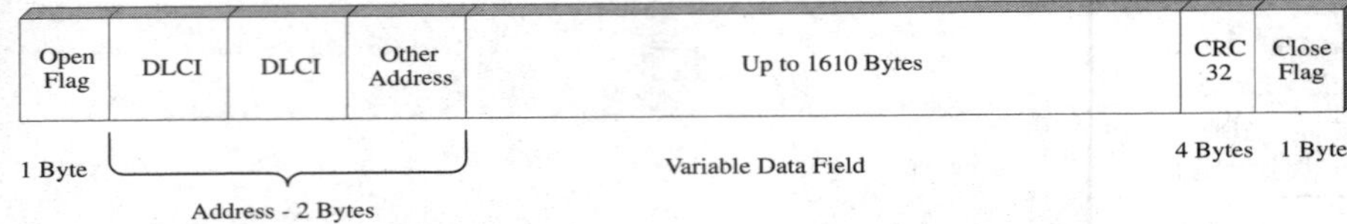

Figure 11-4
A Frame Relay frame

we can assume that Frame Relay works at the data link layer. The DLCI is comprised of several pieces of information, shown later, but is normally a two-byte sequence. Provisions have been made to enable the DLCI to expand to up to four bytes, but very few implementations occur using more than the two-byte address. Following the DLCI is the information field. This is a variable length field. The initial standard allowed for a variable amount of data is up to 1,610 bytes. This is sufficient for most installations, but change always occurs when things are stable. The reason for the 1,610-byte field is to handle a frame from a LAN using a full frame of Ethernet traffic.

The Ethernet frame can be as large as 1,518 bytes (with overhead) and some *subnet access protocol overhead* (SNAP); the full frame should, therefore, accommodate the 1,610 bytes. The Ethernet frame is shown in Figure 11-5. This frame is the same size for an 802.3 IEEE frame or for a DIX Ethernet frame. Therefore, the variable data frame is sufficient to carry the traffic loads necessary.

Following the data field in the Frame Relay frame is a *cyclic redundancy check* (CRC) used only to check for corruption. The CRC determines if the frame or the address information is corrupt. If so, the frame is discarded; if not, the frame is forwarded. There is no ACK or NAK in the Frame Relay transmission along the route. Lastly, there is a closing flag on the frame, indicating that the transmission of the frame is ended and the switching system can then process the entire frame. In many cases when a variable data field is used, the switches must allocate enough buffer space to hold a full frame, regardless of how full each frame is. This is somewhat wasteful across the WAN, but does provide the necessary flexibility to handle the traffic.

Shortly after Frame Relay was introduced with the 1610-byte information field, a new issue cropped up. What about the clients who use an IBM Token Ring? The Token Ring LANs can carry a variable amount of information up to 4,068 bytes. This means that a frame in the Frame Relay world is not large enough to carry a full token and the data must be truncated into three tokens to accommodate the Token Ring. To solve this problem, the frame was expanded to accommodate up to 4,096 bytes in the information field. Not all suppliers supported this change, but the two major suppliers of Frame Relay products (Cascade and Nortel) both

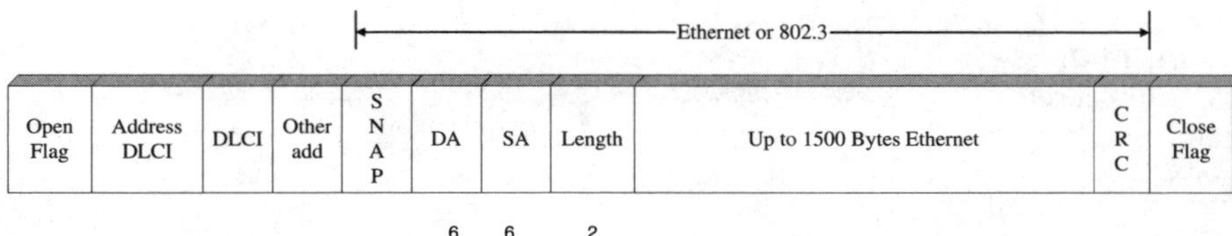

Figure 11-5 Ethernet and IEEE 802.3 frames fit into the Frame Relay frame.

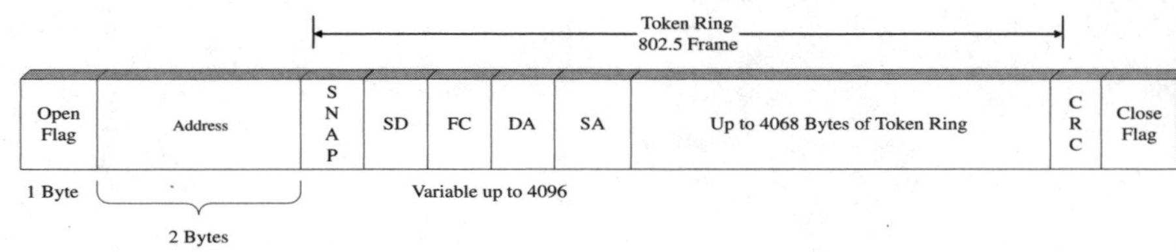

Figure 11-6 Modified frame size of the Frame Relay information field

adjusted their systems to accommodate the larger frame. This frame is shown in Figure 11-6 with a variable frame size of up to 4,096 bytes.

The OSI Protocol Stack and Frame Relay

When we discuss the use of the data link protocols, one always compares the *Open Systems Interconnection* (OSI) to whatever other protocol is being discussed. This book is no different because one needs to understand where Frame Relay falls on the OSI stack and what Frame Relay's purpose is. The development of any new set of standards is usually done to improve network performance. Frame Relay works at the data link layer to reduce the overhead associated with the movement of data across the wide area. Because we refer to Frame Relay as a WAN technology, it is natural that the protocols will work with the improvements made in the network over the past decades.

In the older days, data was shipped across the layer three protocols (such as X.25) to assure the reliability and integrity of the data. This is because the networks back in the 1970s were unreliable, so the protocols were put in place to accommodate this network flaw. The X.25 protocol worked at layer three, as shown in Figure 11-7. The overhead associated with the transmission and reception of the data on the X.25 networks was inordinate. To facilitate better data throughput and eliminate some of the

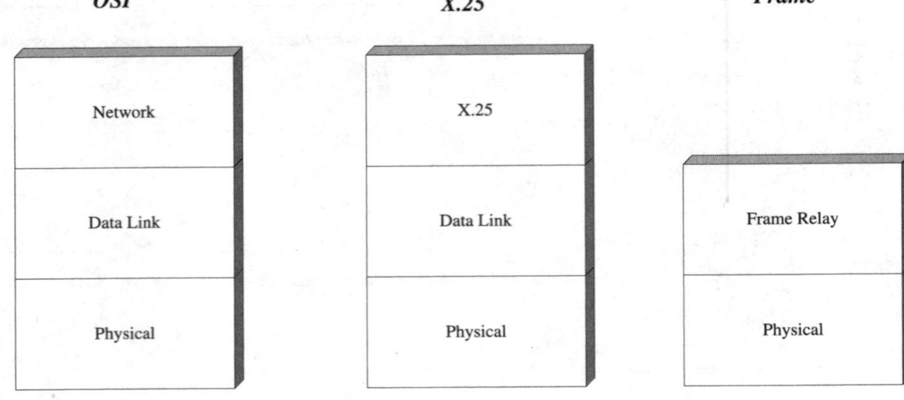

Figure 11-7 OSI compared to Frame and X.25 stacks

overhead, Frame Relay was developed. The comparison of Frame and X.25 to the OSI model is shown in the figure. Two things were in place, however, to enable the use of Frame Relay instead of X.25:

- The networks were now improved through the mass deployment of fiber-based networking technologies and the use of SONET protocols.
- The networking strategies of many end users were based on router technologies and LAN-to-WAN communications instead of the older terminal-to-host intercommunications.

These two changes actually revamped the way we communicate. No longer did we have to use a timing relationship, as in the older data networks. Any form of data transmission could be accommodated across the newer improved techniques and protocols.

With this comparison in mind, one will note that the Layer 2 protocol (in this case, Frame Relay) eliminates some of the overhead associated with the transmission of data. The need for network addressing using Layer 3 is reduced because many of the link architectures are based on point-to-point circuits or private networking techniques. Moreover, where the network address was required to send the data to its end destination, Frame Relay uses the DLCI as the PVC connection. Therefore, by using PVCs, the routing of the data traffic is predetermined to occur across a highly reliable direct connection to the far end. Switching and routing decisions are not required once the connection is established because all the traffic for this connection between two end nodes follows the same path. This mapping is done through the use of the DLCI address to presubscribe the connection in a virtual circuit connection, as shown in Figure 11-8. The only time the traffic might traverse some other route is when a link failure occurs, but this is already mapped in the logic for the connection. This is shown in the routing example in Figure 11-9 where the end nodes are premapped in the Frame Relay switches across the network.

By using this arrangement of DLCI mapping across the network, the network can also accommodate various other types of traffic, such as IBM *Systems Network Architecture/Synchronous Data Link Control* (SNA/SDLC) traffic, which is very time-sensitive and times out if the traffic does not arrive in time. Moreover, if a customer is using an older form of interactive terminal traffic using some older 2770/3770 bisynchronous protocols, these can be placed into the frame.

Figure 11-8 Mapping using the address and DLCI

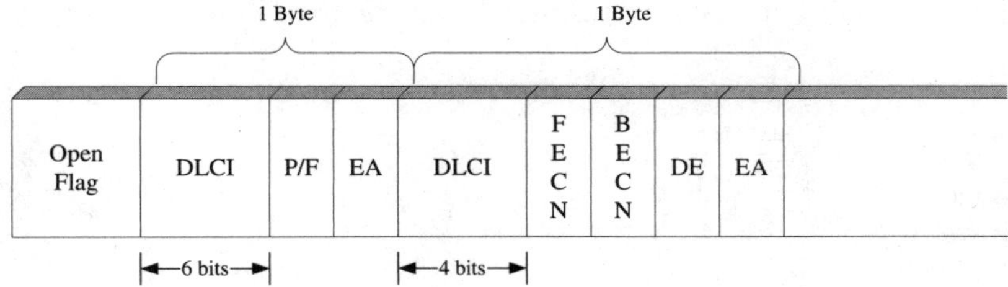

Figure 11-9 A typical mapping of the DLCI in a Frame Relay network

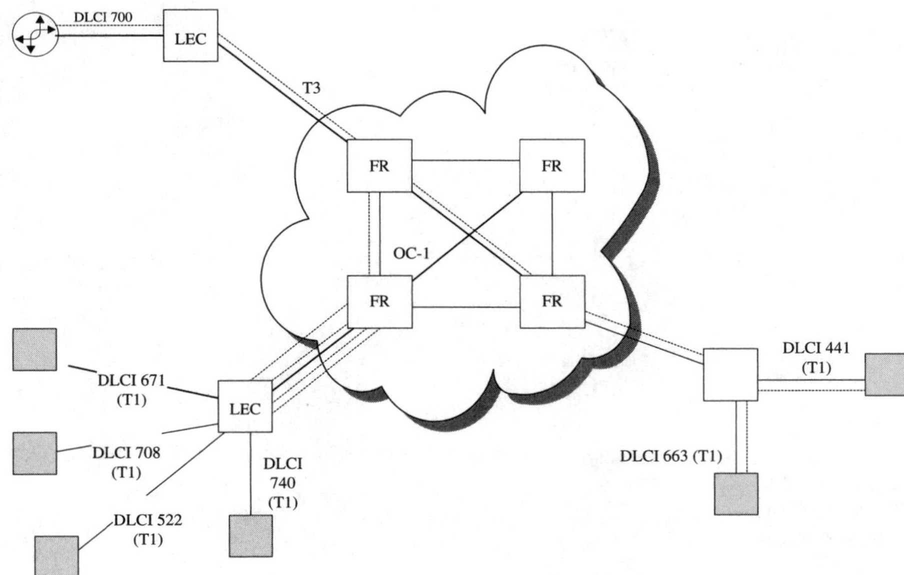

The concept of placing the traffic inside these frames is called *tunneling*. Clearly, what happens is the traffic is sent inside the frame transparent to the network. No checks or validations are made on the data through the network, reducing some of the delays of data handling. Also, while tunneling through the network, the data is actually encapsulated inside the Frame Relay frame, so the only place where the data is actually enacted on is at the two ends. There is some minor overhead associated with the use of these other protocols, called SNAP, but the overhead is minimized. This tunneling concept is shown in Figure 11-10 where the data is encapsulated inside the frame (tunneled).

The term tunneling is getting a lot of press these days because of the Internet and the *Internet Protocols* (IPs) using *Virtual Private Networking* (VPN) by tunneling through the Internet with a private link. This will be discussed in a later chapter in greater detail, but is shown in Figure 11-11 with TCP/IP tunneled into a Frame Relay connection.

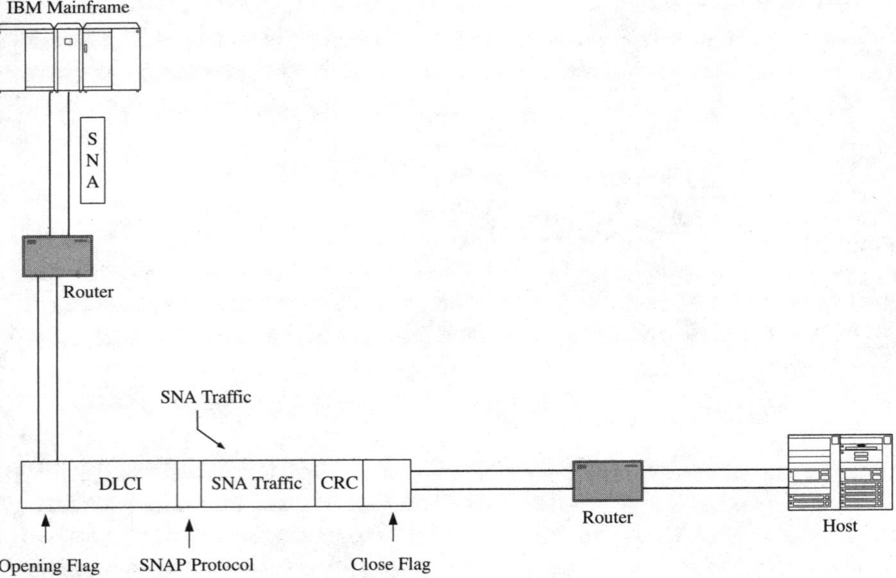

Figure 11-10 Traffic is tunneled in the Frame Relay frame using a small amount of overhead.

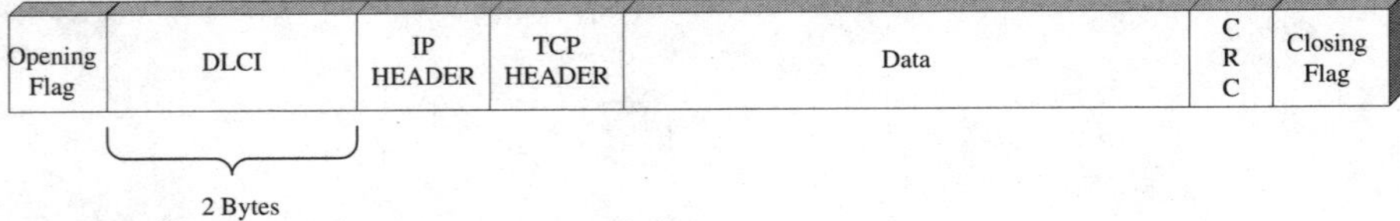

Figure 11-11 TCP/IP traffic is tunneled into a frame.

Frame Relay Speeds

As a means of keeping everything in order, it is appropriate to discuss the speed that can be achieved with the use of Frame Relay. Although in the beginning of this chapter, it was stated that Frame Relay was designed for speeds up to T-1/E-1 (1.544–2.048 Mbps); it later evolved to speeds of up to 50 Mbps. Actually, few end users have ever implemented Frame Relay at the higher speeds; this is more of a speed for the carrier community, but the need for stepped increments has always been a requirement for data transmission. Therefore, Table 11-2 is used to show some of the speed increments typically used. Other rates are possible in 4 Kbps increments, but implementations are normally done at the speeds shown in Table 11-2.

Table 11-2

Typical speeds used in Frame Relay

Frame Relay Access	Typical Committed Speed	Average Additional Burst Speeds
56 Kbps (DS0 or ISDN)	32 Kbps	24 Kbps
64 Kbps (Clear channel DS0 or ISDN)	32 Kbps	32 Kbps
128 Kbps (ISDN)	64 Kbps	64 Kbps
128 Kbps (ISDN)	128 Kbps	0
256 Kbps	128 Kbps	128 Kbps
256 Kbps	192 Kbps	64 Kbps
384 Kbps	256 Kbps	128 Kbps
512 Kbps	384 Kbps	128 Kbps
1.544 Mbps	512 Kbps	256 Kbps
1.544 Mbps	1.024 Mbps	512 Kbps
2.048 Mbps[1]	1.024 Mbps	1.024 Mbps

[1]In most implementations, when a customer exceeds 256 Kbps access, the normal installed link for access is a T-1 in North America at 1.544 Mbps. This is a pricing and an availability situation.

Frame Relay Access

A link is installed between the end-user location and the network carrier's node. The normal link speed is T-1, although many locations can and do use *Integrated Services Digital Network* (ISDN) or leased lines at lower rates. Some customers may choose to install a local loop at speeds up to T-3 (45 Mbps approximately) to support higher-speed access and faster data throughput. (In most implementations, when a customer exceeds 256 Kbps access, the normal installed link for access is a T-1 in North America at 1.544 Mbps. This is a pricing and an availability situation).

In many cases, the use of the T-3 will also allow for consolidation on the same link. Many of the carriers (and in particular the LECs) will offer the T-3 access and enable Frame Relay throughput at rates up to 37 or 42 Mbps. Now the ILECs are offering flat-rate services and bundled capacities to be more attractive to their end users. The ILECs are the local telephone companies with their installed base of services and facilities. Often these ILECs are offering high-speed Frame Relay services *Local Access and Transport Area* (LATA)-wide (or statewide, depending on the geographic topology of the state and LATA boundaries) all for competitively priced services. Better yet, some of the ILECs offer high-speed Frame Relay at speeds up to 10 Mbps across the LATA at competitive rates on either T-3 or an OC-1. The 0 to 10 Mbps bursts of data are designed for the very large customer, but they may fit smaller organizations needing broadcast (or near-broadcast) quality for voice and video applications in the future.

Figure 11-12 is an example of the connection installed at a large organization yet fed across the network by lower-speed feeds from branch offices at T-1 and lower rates. This scenario is likely the most common implementation for the near term, but will shift as the pricing model becomes more conducive to the smaller organization.

Overall Frame Relay Core Protocols

When the Frame Relay specification was developed, the primary goal was to carry data over the WAN. To handle this form of wide area communications, the core protocols for Frame Relay were established using the revised version of the data link protocols. Instead of using the network layer protocols, Layer 3 was gleaned down to efficiently carry the traffic while per-

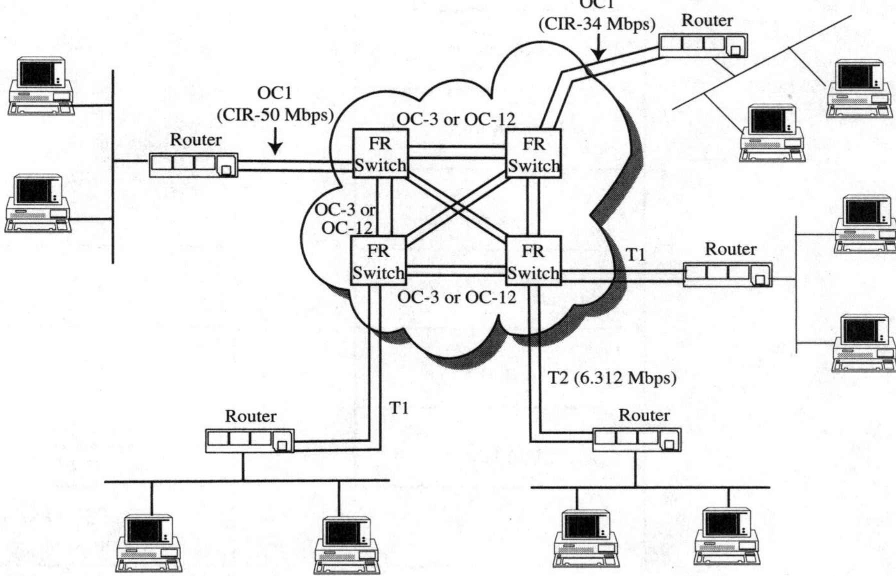

Figure 11-12 Examples of connections at higher speeds

forming the same function as the network layer. Moreover, the data link layer was also streamlined to offer less overhead and processing on a link-by-link basis. Because the circuits across the wide area are much more reliable and error-free (thanks to fiber optics), the ACK and NAK functions can be eliminated. Furthermore, the use of PVCs in the connection eliminates the need for the sequence numbering on the link. One can assume that if we send multiple frames onto a circuit between two end points (even if it is a virtual circuit), the data will come out in the same sequence that it went in on the other end. Unless a frame is discarded, there should be no way that the data will arrive out of sequence. Because the data should not arrive out of sequence, there should not be a need to do the counting.

If, however, something goes wrong on the circuit, how then do we recover? The answer is that we rely on the upper-layer protocols on both ends of the circuit (the transport layer) to recognize if the data is missing. If a frame is lost, then a transport will request a retransmission from the sender. This eliminates much of the processing and checking at each node across the link. In Figure 11-13, the core protocols are shown for Frame Relay using a subset of the Q.922 data link layer for the actual link protocols.

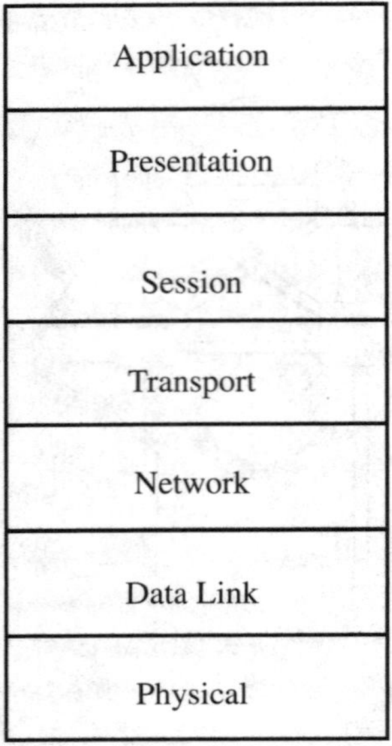

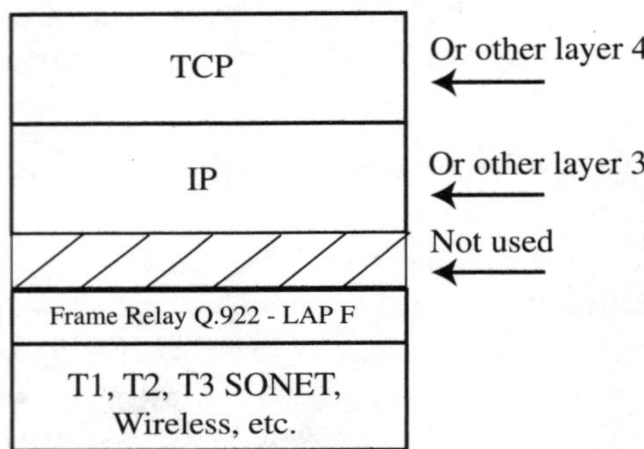

Figure 11-13 The core protocols for Frame Relay mimic the ISDN standards.

Q.922 is an ITU-specified protocol for ISDN. Frame Relay was developed as an offshoot of the ISDN protocols because of the amount of work invested in the technologies and standards. As a result, many of the core protocols for Frame Relay mimic the ISDN standards. This is helpful in understanding why and how the networks have evolved throughout the past 10 years to improve native throughput and increase the acceptance of the standards.

Carriers' Implementation of IP-Enabled Frame Relay

Carriers are now offering IP-enabled services, enabling a customer to use an existing Frame Relay access link to tap into a connectionless *Multiprotocol Label Switching* (MPLS)-based IP backbone or a private IP network. The primary benefit is that achieving mesh connectivity within a customer's VPN requires just a single "access" PVC from each remote site.

Many companies have already configured their networks in a hub-and-spoke configuration. Thus, it would appear that there is no benefit to consider the IP-enabled offering. With traditional PVCs in a hub-and-spoke configuration, a PVC runs from each remote site to the central site. With IP-enabled services, the customer installs one PVC at each site to the carrier network and an IP-enabled PVC at the central site.

Other organizations are moving to at least two PVCs per site, each terminating at a central point. This covers a few key areas within the IT group. First, the dual homing allows for load balancing. Next, by feeding the second PVC through a different route, the risk of downtime is minimized by alternate routing. Still an option is to have a regular site handling the routine traffic but a hot site for recovery purposes located in a different location. In another instance, the company might run two major processing centers, such as one for back office functions and another for front-office functions such as customer service. Yet all sites typically need access to both locations. To facilitate this, two PVCs per remote site are needed, creating the opportunity to justify IP-enabled Frame Relay. The break-even point occurs at low *committed information rate* (CIR) speeds where one IP-enabled PVC reaches both sites. The price of the IP-enabled PVC is usually twice the cost of a normal PVC. When using higher-speed access and PVCs, the value is far more dramatic. As the speed of the CIR increases, the IP-enabled PVC costs are more closely aligned to the traditional PVC costs. As the speeds increase to the point where the prices are the same, you have twice the connectivity for the same price.

Frame Relay Versus IP

We now compare the pros and cons of Frame Relay and IP. IP and applications such as *Virtual Private Data Networks* (VPDNs), intranets, and extranets have garnered a lot of mind share in the industry today. Although these services show a lot of promise, today's services lack comprehensiveness and robustness. They lack much of the potential functionality and service guarantees they can eventually deliver. Security and performance top the list of areas in which network managers will need to see improvements prior to seriously considering these services en masse. Frame Relay's support for multiple protocols, its predictable and reliable performance, and the wide availability of network management tools and service-level guarantees make it the most logical choice for a majority of the business applications used today. In Table 11-3, we draw a comparison of the Frame Relay and IP from a logical progression.

Frame Relay has evolved to address the mass market's increasing bandwidth demand by supporting connectivity up to DS-3 speeds. However,

Table 11-3
Comparing Frame Relay and IP

Frame Relay	IP
Multiprotocol support	Any-to-any connectivity
Predictable and reliable performance	Limited *quality of service* (QoS)
Robust network management capabilities	Security concerns abound
Primarily intracompany connectivity	Linking intercompany business partners
WAN only	LAN or WAN
Focus is on logical connections versus intelligence	Dialup and international access available

niche markets exist where end-user IP applications require much more bandwidth. IP over SONET is one such solution. IP over SONET's appeal for many users has been its management simplicity and transport efficiency when compared to alternate solutions such as IP over ATM.

Voice over Frame Relay (VoFR)

Just as the industry was getting used to the idea of reduced overhead for data transmission, some radical thoughts began to surface in the industry. In the past, all data ran over voice networks, adjusted, and accommodated according to voice standards. But what if voice could run over a data network instead? By using the capability of reduced overhead, more reliable circuits, and faster throughput, the network could be tuned to accommodate voice in the form of packets of data.

Much of the pressure for voice over any data technology has been based on cost in the past. Newer ideas are based on efficiency and the convergence of the network protocols and services. Voice is fairly inefficient! Actual voice traffic is carried only about 25 percent of the time we are on a connection. The rest of the time, we are sending silence (no information). If we can integrate the two networks and carry interleaved traffic (voice or data), then we can efficiently fill the network with traffic all the time, instead of just sending idle conditions.

Thus, a new concept was born that could be accomplished through compression and interleaving the data. Therefore, on a digital circuit, data is data and voice is too! The voice is just a data stream of ones and zeros. The Frame Relay Forum has been busy defining the standards for three service offerings, one of which is the VoFR specification. In conjunction with VoFR offerings, the forum was busy developing other protocols and specifications to support some of the unique challenges with real-time data on a network. This includes the use of PVC fragmentation protocols and multilink Frame Relay services.

The fragmentation protocols are necessary to support the different types of delay experienced on the network for time-sensitive traffic (that is, voice). Interleaving the voice communications (using small frames) onto a high-speed data link with larger frames is one way to handle this need. This sharing of the same physical link enables both real-time and nondelay sensitive traffic to coexist yet receive separate treatment as it moves across the network. The fragmentation of the traffic enables variability, depending on the speed of the link, the congestion, local timing needs, and the type of service being used. This makes the implementation of fragmentation available at various interface points. In Figure 11-14, the use of a fragmentation procedure is shown in three different places: at the *User-to-Network Interface* (UNI), at the *Network-to-Network Interface* (NNI), and on an end-to-end basis. These three ways enable enough flexibility to accommodate the different types of service being delivered across the Frame Relay network.

With the implementation of VoFR, particularly with carrying international telephony traffic, the Frame Relay Forum introduced the specification FRF.11 that deals specifically with the voice side of the business. This specification goes beyond the possibilities of fragmenting the data and

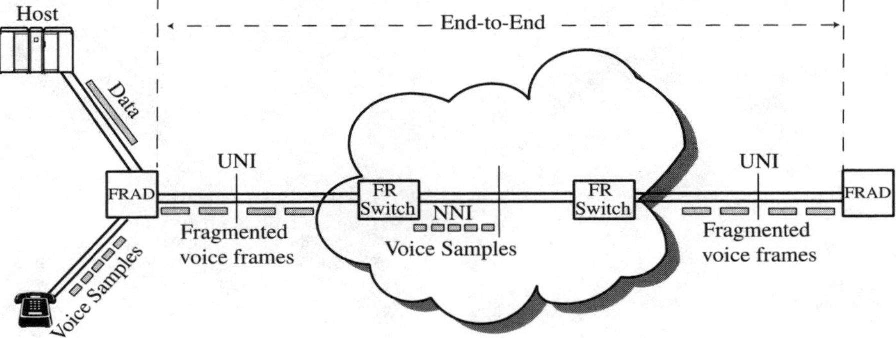

Figure 11-14 Three places where fragmentation can take place in support of VoFR

incorporates the necessary steps of call setup and call teardown. The other major issues that the FRF.11 deals with are as follows:

- Analog-to-digital conversion
- Digital-to-analog conversion (back to analog)
- Compression techniques
- The sizing and transmission of a frame of traffic

Many of the specifications involve more sophisticated steps, such as handling various forms of analog traffic, including voice, fax, and compressed video communications. To accommodate that need, the FRF.11 handles a specification dealing with multiservice multiplexing. This covers the ability to multiplex multiple voice and data channels on a single Frame Relay connection (or PVC). A gateway function is used in many cases to handle the various informational streams multiplexed onto a single PVC, as shown in Figure 11-15.

Compressing the Information on VoFR

By using some industry-accepted standards for compression techniques, such as *Adaptive Differential Pulse-Coded Modulation* (ADPCM) or *Code-*

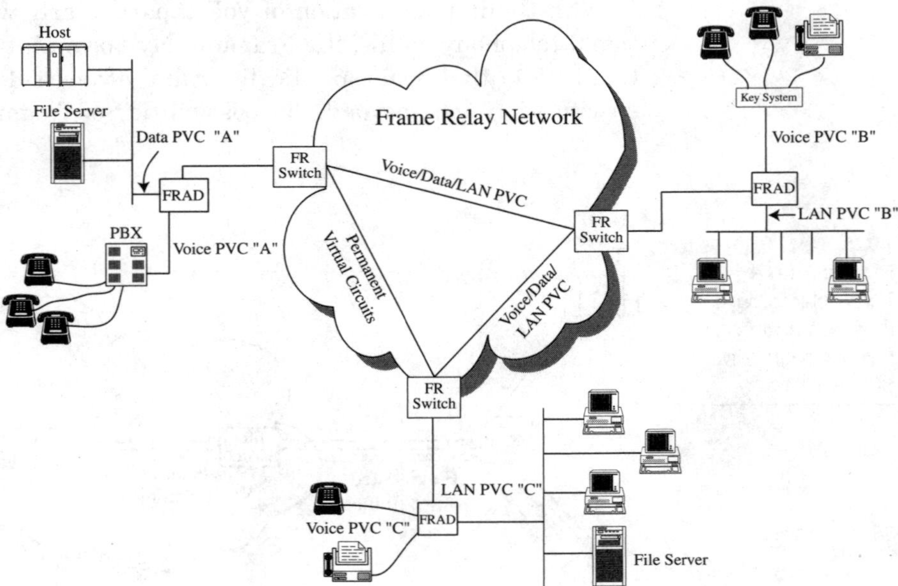

Figure 11-15 A Frame Relay gateway acts as the multiplexer of different services onto a single PVC.

Excited Linear Predictive Coding (CELP), the conversation can be compressed from 64 Kbps to a data stream of 40, 32, 24, or 16 Kbps using ADPCM compression techniques and down to 5 to 8 Kbps using the newer CELP standards.

Following this idea, the next step is to view how a service bureau approach might work using Frame Relay to act as an international callback service, or in the case of a corporation with multiple international locations, this service can be used for the intranet calls. One cannot underestimate the robustness and power of the Frame Relay networks. In many cases, the carriers are offering throughput across their backbone in the proximity of 99.99 percent of the *Committed Information Rate* (CIR) and with an availability of 99.5 percent or better. These two statistics absolutely beat anything we have ever seen from voice or data networks in the past. Notwithstanding the capability to achieve the throughput and availability statistics, the carriers will sign a *service-level agreement* (SLA) with these guaranteed throughput and availability characteristics without hesitation. Never before have we seen such confidence and acceptance of a single-standard interface and network topology to carry the WAN traffic.

Still, when one considers the possibility of running voice data across a network, it is uncomfortable to think that the voice networks have truly not kept pace with the developments of the data networking strategies. Do we really need another packet-switching technology, rather than improve the ones we already have in place? The answer can be found in the overall characteristics of the Frame Relay networks already discussed here.

Provisioning PVCs and SVCs

The primary difference between PVCs and *switched virtual circuits* (SVCs) is whether the connections are provisioned or established. Both types of connections need to be defined. The difference is when the connections are defined and resources allocated.

The network operator typically provisions PVCs. The network operator can be the carrier (public services) or the MIS manager (private networks). Once the PVC is provisioned, the connection is available for use at all times unless there is a service outage. On the other hand, the end user, not the network operator, establishes SVCs. Prior to each use, an SVC is established to the destination end user. The connection is cleared after use.

SVCs are ideal for networks with highly meshed connectivity, highly intermittent applications, remote site access, and interenterprise communications. Each of these applications will be discussed in more detail in the succeeding slides. SVCs are also ideal for networks that are not primarily dependent on resources housed at a single location such as the headquarters site or at regional offices. In a nonhierarchical networking environment where there is a need to communicate with many locations, SVCs can offer a viable solution. The advantages of SVCs are magnified as the number of locations and the degree of connectivity requirements increase. Highly meshed networks are becoming more common as more and more companies deploy intranets. It is conceivable that all end users will have their own Web page within the corporation. This will increase the amount of peer-to-peer intracompany traffic. Additionally, it can offer a cost-effective solution for occasional inter-company connections to suppliers, partners, and even customers, provided that they all subscribe to the same public Frame Relay service.

Because SVCs only consume network bandwidth when there is information to send, it is a good solution for short duration applications. (PVCs also only consume bandwidth when there is information to send. The difference is in the amount of bandwidth that needs to be reserved for the connection. In a PVC-based environment, the network operator must ensure that CIR can be guaranteed to all connections transmitting at any given time. This means that the network may have to design for the worst case regardless of the applications' CIR utilization. In an SVC environment, the appropriate amount of CIR is allocated during the call.) Intracompany voice calls are an excellent example. The average telephone call is only three minutes. SVCs are also ideal for scheduled, time-bound applications like videoconferencing. You may only need videoconferencing capabilities every Monday morning from 9 to 10.

SVCs can be used in conjunction with PVCs for traffic overflow during peak traffic periods. Traffic overflow is highly intermittent because, hopefully, overflow doesn't occur on a regular basis. When it occurs, it only happens for a short time. If end users notice that overflow is occurring more frequently, then it might be more cost-effective to increase the CIR of the PVC to accommodate the overflow. SVCs can also provide a backup connection to a secondary host location if the primary host location fails or is unavailable.

Benefits of SVCs

Some public Frame Relay services' pricing structures are forcing end users to build star networks even when the underlying traffic patterns warrant more meshing. Some pricing structures incant high subscription rates on ports, which result in star and hub-and-spoke configurations. For example, all remote offices may have direct connectivity to the headquarters location only. Remote-office-to-remote-office communication happens by tandeming through the headquarters router. The headquarters router and port connection can become bottlenecks. Network latency increases with tandeming.

With SVCs, a direct connection is established between the caller and the called party. There is then no need to tandem through one or more nonterminating or nonoriginating end user locations for that particular transmission. SVC usage can be tracked on a call-by-call basis. The information can be used to further optimize the network for billing purposes.

Frame Relay Selected for Wireless Data on GPRS

Newer wireless services known as the *General Packet Radio Services* (GPRSs) are designed around the movement of IP datagrams (always-on Internet access) from a cell phone or *personal digital assistant* (PDA) to the public Internet or a VPN connection to an intranet. Regardless of the direction that the data is going to flow, the use of the IP services from the handset enables us to use the network ad hoc. When an IP is created, it is packaged in a radio message. Once this message gets to the base station, it is then encapsulated into a Frame Relay frame to be carried across the wireless carrier's network to a router. The *Packet Control Unit* (PCU) is, therefore, a form of a FRAD. Our packets are sent to the PCU where it is slotted into the Frame Relay service and carried through the cellular network.

One might think about this and wonder why Frame Relay is used. First, it is widely deployed as a networking architecture. Second, it is based on the PVC from the PCU to the network device called a *Serving GRPS Support Node* (SGSN), as shown in Figure 11-16, which operates at speeds up to 2.048 Mbps (the standard for the E-1). By connecting across this architecture, it is a virtual private line adding some degree of security. Third, the standards allow for the sharing of the circuitry from many devices by interleaving the data frames on the same physical channel. Fourth, it does minimize the overhead on the channel.

In general, the use of Frame Relay has been continually climbing due to the robustness, industry acceptance, and wide availability. Many

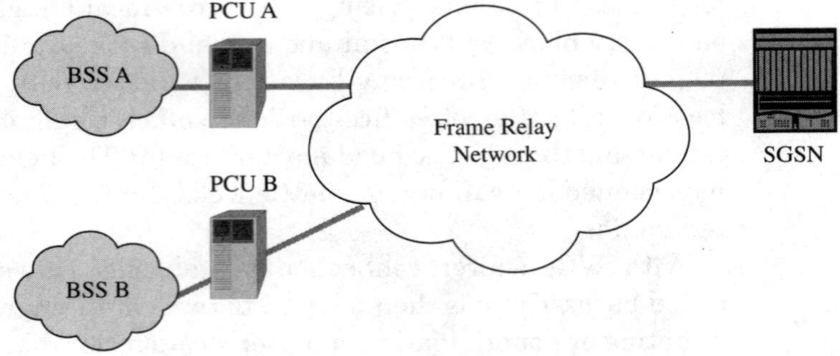

Figure 11-16 The PCU uses Frame Relay to connect to the SGSN.

organizations are not ready to displace their networks by moving to newer or different services. However, where the customer has used an IP-based network, the use of managed services, burstable data rates, and inexpensive access of PVCs and SVCs combined now with IP-enabled Frame Relay continues to lend credibility and acceptance in this networking standard. We can expect to see this around for a long time to come.

CHAPTER 12

Asynchronous Transfer Mode (ATM)

In 1992, a group of interested parties developed a set of standards-based specifications called the *Asynchronous Transfer Mode* (ATM). This was a step at developing a single set of standards for the integration of voice, data, video, and multimedia traffic on a single backbone network. Prior to this development, the industries offered separate standards and networks for voice, others for data, and still others for video communications. Above and beyond that, data networks were also treated differently with separate data networks for point-to-point needs, dialup data, and packet-switched data transmissions.

Beyond being expensive, this concept also proved to be confusing to users of the networks. Should separate networks be used or should an integrated approach be sought? Even beyond that, those few brave souls who tried the integrated approach were destined to more confusion because of the carrier offerings being different. What the industry needed was a single way of handling all forms of traffic and one network service that could carry the different forms of traffic. The end user was looking for something that would clear up the confusion and make life simple. Hence, ATM was born.

Unfortunately, ATM can cause as much confusion as the older techniques, strictly by virtue of its name. Why is ATM called asynchronous? Wasn't it designed to run on a synchronous networking platform and support synchronous traffic? The answer is yes! One can see that confusion reigns in this industry just because of the naming conventions used to describe the various protocols and specifications.

What Is ATM?

ATM is a member of the fast packet-switching family called *cell relay*. As part of its heritage, it is an evolution from many other sets of protocols. In fact, ATM is a statistical *time-division multiplexed* (TDMed) form of traffic that is designed to carry any form of traffic and enables the traffic to be delivered asynchronously to the network. When traffic in the form of cells arrives, these cells are mapped onto the network and are transported to their next destination. When traffic is not available, the network will carry empty (idle) cells because the network is synchronous. In Figure 12-1, a representation of the traffic delivery is shown, and Figure 12-2 shows the carrying of idle cells across the network.

Therefore, what we can derive is that the ATM technique is a combination of TDM, with cells using preassigned slots, and Statistical TDM, with

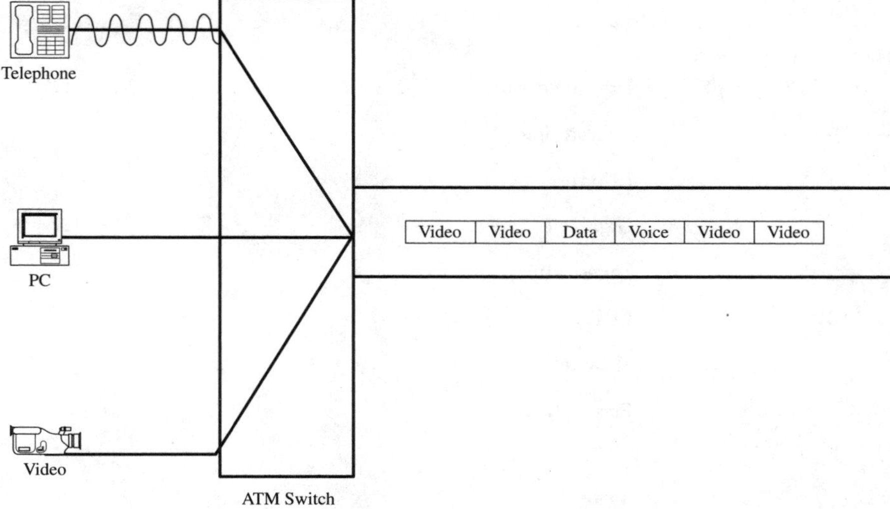

Figure 12-1 Traffic is mapped onto the network as it arrives.

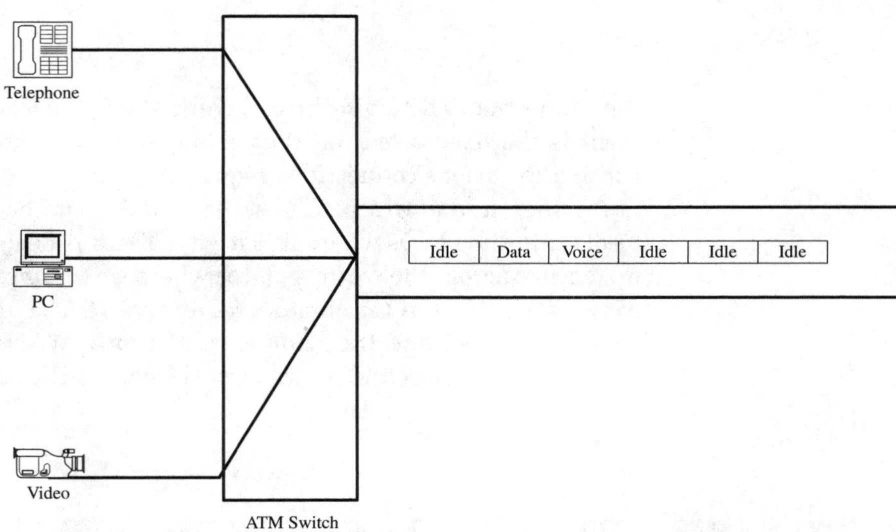

Figure 12-2 When no cells arrive, idle cells are transported across the link.

cells using whatever slots are available or needed to handle a particular traffic flow. This offers the carrier and the network user the best of both worlds. It is also a connection-oriented protocol much the same as dialup voice communications services, but it uses virtual circuits, such as *permanent virtual circuits* (PVCs) and *switched virtual circuits* (SVC), to handle

Table 12-1
Summary of where various protocols are used

Technology	LAN	CAN	MAN	WAN
Dialup network			X	X
Leased line			X	X
Ethernet		X	X	
Gigabit Ethernet	X	X	X	
Token Ring	X			
FDDI		X		
X.25 networks				X
Frame Relay				X
SMDS			X	
ATM	X	X	X	X

the connection. The main thrust behind the fast packet-switching arrangement is the fixed sized cell that is employed to carry the end user traffic across the various connections required.

Another area in which ATM stands out from other technologies and protocols is in the places where it is used. ATM was designed from the ground up to work across the various places where we communicate: the *Local Area Network* (LAN), the *Campus Area Network* (CAN), the *Metropolitan Area Network* (MAN), and the *Wide Area Network* (WAN). Table 12-1 summarizes the differences and locations of the capabilities of ATM in comparison to other sets of protocols.

Why the Interest in ATM?

When one considers the disappointing capacities of past technologies, we can see why there is the hype for ATM. ATM will be the basis of many of our future broadband communications systems; as such, it starts where other technologies stop. Many organizations have escalated their demands and needs for raw bandwidth, yet no single entity has emerged as a clear-cut

Table 12-2

Summary of speeds for various technologies

Technology	Application	Speeds
Dialup analog switched	Modem communications	Up to 56 Kbps circuit using current technologies. Stops at 33.6 Kbps using older modems.
Leased line	Point to point	T1 @ 1.544 Mbps normal, T3 @ 44.736 Mbps occasionally.
Ethernet	LAN shared/switched bus	10 to 100 Mbps normal, 1,000 and 10,000 Mbps emerging.
Token Ring	LAN shared ring switched	4 to 16 Mbps normal, or 1,000 Mbps under consideration.
FDDI X.25	Shared ring dialup or leased line	100 Mbps normal. 56 Kbps top end in North America, 64 Kbps top end in rest of world.
SMDS Frame Relay	Shared/switched bus point-to-point PVC	Up to 34 Mbps common. Starts at 56 Kbps normally, top end is 50 Mbps.
ATM	Switched PVC	Starts at 1.544 Mbps, top end is 622 Mbps today, 2.488 Gbps for future application.

winner to deliver the services necessary to support the demands of today's multimedia applications. Table 12-2 compares the capacities of ATM to the other techniques we used in the past. This will give the reader a chance to see what the excitement is all about.

A lot of discussion is always bantered in the industry regarding the best way to handle data and voice communications. There is a continual discussion on the difference between using ATM for the higher-bandwidth-intensive applications and the use of gigabit Ethernet for the same applications. One cannot predict what the outcome will ultimately be, but it keeps the end user totally confused and uncertain. What could happen is that a complement of both techniques will come rolling out to serve the application and support each other.

The use of ATM over other technologies is, however, an attractive alternative when considering the various aspects of the options. For example, ATM offers the following types of service, shown in Table 12-3, which may or may not be defined in other protocols.

Table 12-3
Summary of ATM features and functions

One technology for voice, data, video, and multimedia
Bandwidth on demand as needed
Scalable as needs dictate
Quality of service (QoS) is well defined
Management systems and services prebuilt into ATM
Hardware-based switching instead of complicated routing and software schemes

ATM Protocols

It takes many protocols to support an ATM network, which is one of the issues that continually comes up as a negative from the supporters of the gigabit Ethernet crowd. To develop the necessary interfaces in support of the various points within a network (networks are pretty complex in themselves), different protocols are necessary. The actual protocols needed depend on where the traffic originates, what transport mechanisms must be traversed, and where the traffic will terminate. To see this in a clear picture, a summary of protocols for the ATM user is shown in Table 12-4 and is also shown graphically in Figure 12-3.

Table 12-4
Summary of where protocols are used for ATM

Location	Protocol/Specification
End user to LAN	Private *User to Network Interface* (Private UNI)
End user to WAN	Public *User to Network Interface* (Public UNI)
Between network nodes in a WAN	Public Network Node to *Network Node Interface* (Public NNI)
Between nodes in a interprivate WAN	Private Network Node to Network Node face (Private NNI)
Between carriers in a public network	The *Intercarrier Interface* (ICI)
Between a Frame Relay and an ATM interface device	*Frame User Network Interface* (FUNI)
From a legacy router to the network	*Data Exchange Interface* (DXI)

Table 12-4 cont. Summary of where protocols are used for ATM

Location	Protocol/Specification
On a LAN-to-LAN interconnection	*LAN Emulation* (LANE)
Between LAN nodes and other network UNIs	Next Hop Routing (Resolution) Protocol
On a LAN, PBX, or CAN interface	*Multi-protocol over ATM* (MPOA)

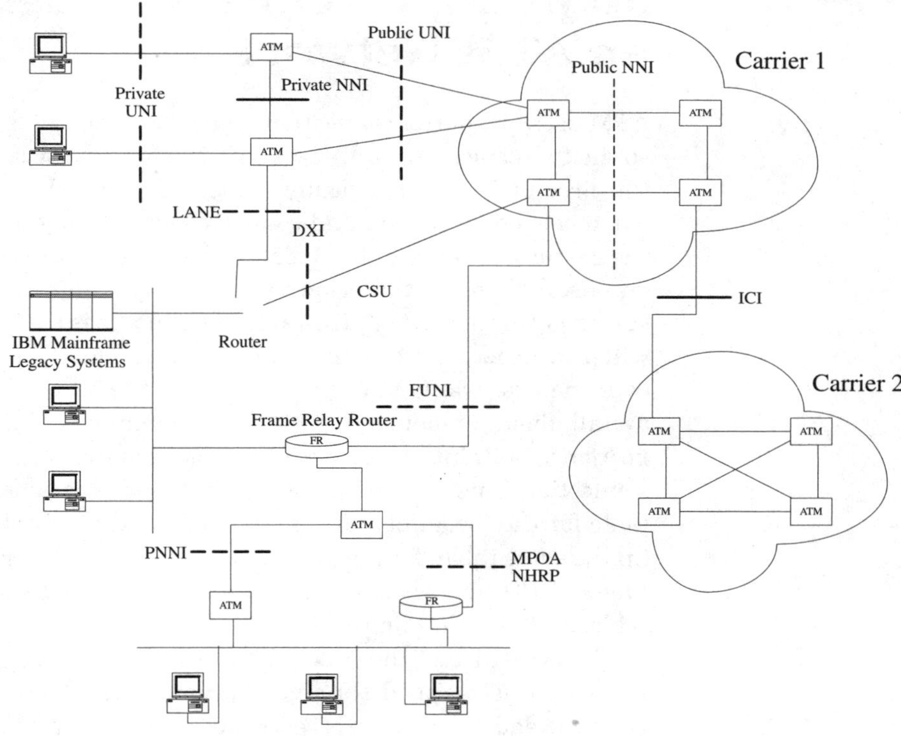

Figure 12-3 Graphic representation of the ATM protocol interfaces

One can see the reason why the gigabit Ethernet proponents support a simpler and less expansive set of protocols when dealing with LAN technologies, now moving toward a WAN protocol with Layer 3 switching being developed. The problem with ATM is that in order to support the older legacy systems, many protocol points and interfaces are necessary. To get around the problem of "forklift" changes, the necessary protocols have been developed. Due to its protection mechanism for existing systems, ATM is its own undoing from the opponents of the technique. Too often the opponents cite the different protocols needed instead of what they are actually doing. However, the use of each of these protocols satisfies a specific need and should not be taken strictly as an overhead problem, but as an interworking function. If all one can do is complain about the various protocols, then the true benefit of ATM is lost in the mire.

Mapping Circuits Through an ATM Network

ATM uses one of two connection types. The protocol is connection-oriented, so the two choices are a PVC or a SVC. There is actually no permanency to the circuits. They are logically mapped through the network and are used when needed for PVC or dial-connected when using the SVC. In either case, the carriers promise only to make a best attempt to serve the needs of the end user when the time is appropriate. With no true guarantees, the consumer is at risk (sort of). However, the concept is that the network provider will provide a committed bandwidth available to the user on demand whenever the user wants to use it. This forms the basis of what ATM networks are all about: on-demand, high-speed communications networks. The connection is built into a routing table in each of the switches involved with the connection from end to end. As such, the switches only need to look up a table for the incoming port and channel and then determine the mapping (in the same table) for the output port and channel. Using *virtual path identifiers* (VPI) and *virtual channel identifiers* (VCI), the carrier maps the table, as shown in Figure 12-4.

The structure of the link is shown in Figure 12-5, where a full virtual connection is mapped through the various switches across the network. Here the end-user device is connected across an ATM access link through a switch. The switches provide the cross connection and link to the next downstream node. Note that the connection from the end user to the network may be on a T1, T3, or OC-n. From the first switch out, the network will use *Synchronous Optical Network* (SONET) or *Synchronous Digital Hierarchy* (SDH) capabilities possibly mapped onto a *Dense Wave Division Multiplexer* (DWDM). The network carrier will use whatever services and bandwidth is available at the connection points.

In Figure 12-6, the network switches handle the mapping on the basis of VPI switching. VPI switching means that the switches use the virtual path

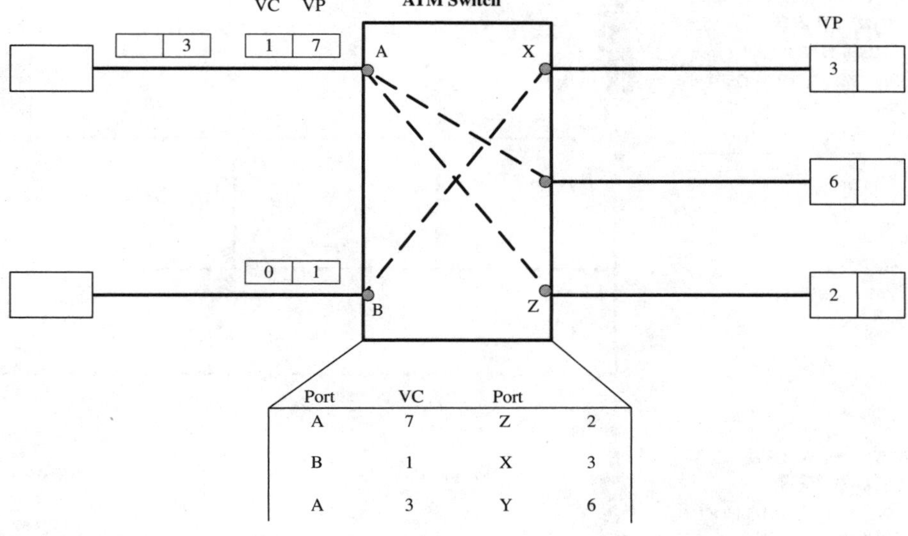

Figure 12-4 ATM table lookup maps the input and output channels.

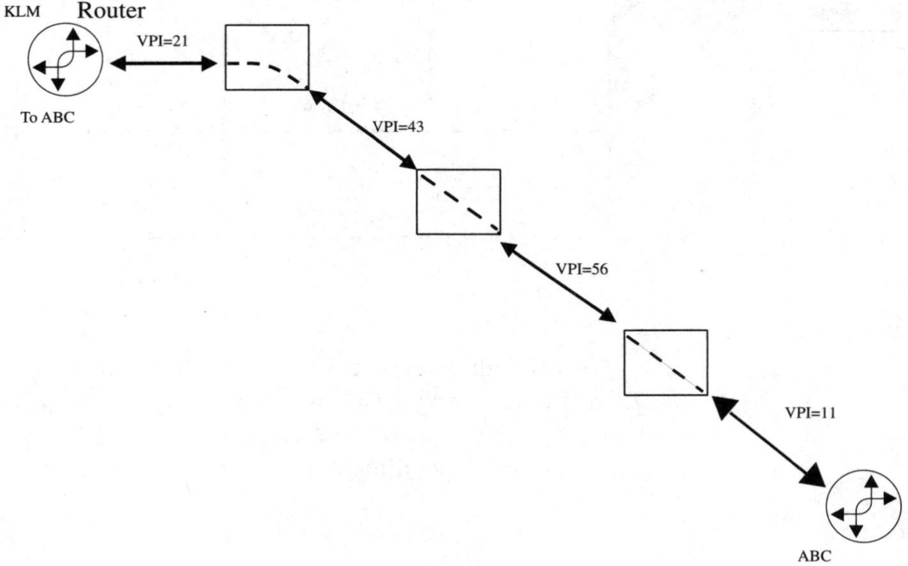

Figure 12-5 The end-to-end connection through the network

for mapping through the network and will remap from one virtual path to another, while the virtual channel number is held consistent through the entire network.

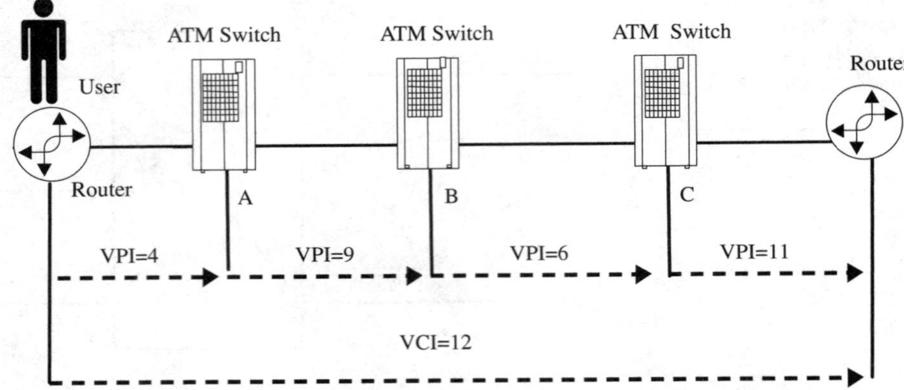

Figure 12-6 Virtual path switching remaps the path, but keeps the channel the same.

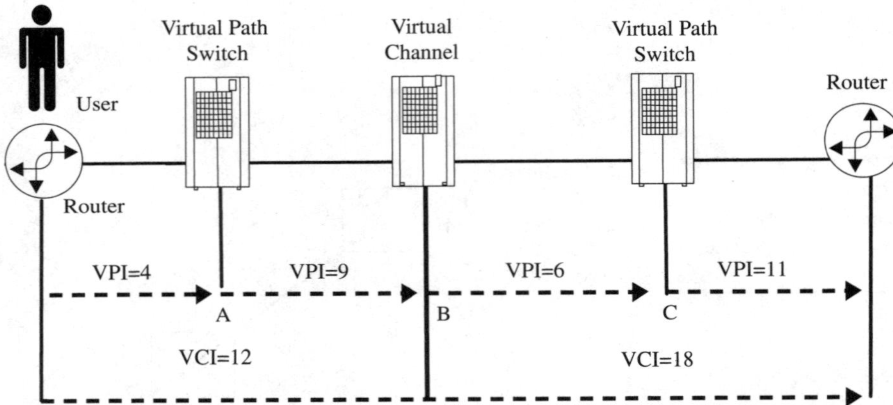

Figure 12-7 The virtual path and virtual channel switches process and remap both elements.

A second alternative is to use VPI/VCI switching, as shown in Figure 12-7. In this case, the ATM switches along the route will switch and remap both on a virtual path and a virtual channel. This is an installation-specific arrangement depending on how the carrier chooses to handle the switching.

The ATM Layered Architecture

Nearly all documents describe the layered architecture of every protocol against the OSI model as a reference only. The ATM architecture is no different when trying to compare what the protocol is doing. Using the OSI model as a base reference, the ATM layers fall typically in the bottom two layers (data link layer and physical layer) of the architecture. ATM has been designed to run on a physical medium such as SONET.

In Figure 12-8, the ATM layer is shown as the bottom half of Layer 2 in its equivalency. There is no real way to draw true one-to-one mapping of the ATM and OSI models, but for purposes of this document, it is done that way. Now the bottom half of the data link layer is ATM, but below the ATM layer is the physical layer such as SONET or some other physical media dependent layer (SDH, DS3, and so on).

Moving up the architectural map, we have the upper portion of Layer 2, the LLC equivalent, called the *ATM adaptation layer* (AAL). Within this portion of the Layer 2 protocol stack, several sublayers are seen, depending on the services required. For example, using Figure 12-9 as a reference, the uppermost portion of the layer is called the *service specific convergence sublayer* (SSCS). This portion of the protocol stack is used when mapping Frame Relay, *Switched Multimegabit Data Service* (SMDS), or another protocol to the ATM adaptation process. Under the SSCS is the *common part convergence sublayer* (CPCS). The combination of the SSCS and the CPCS make up the *convergence sublayer* (CS). Convergence, as the name implies, is the changing and melding of the data into a common interface for the ATM networks. Following the CS portion of the upper layer is the next

Figure 12-8 Comparing the OSI and ATM layered models

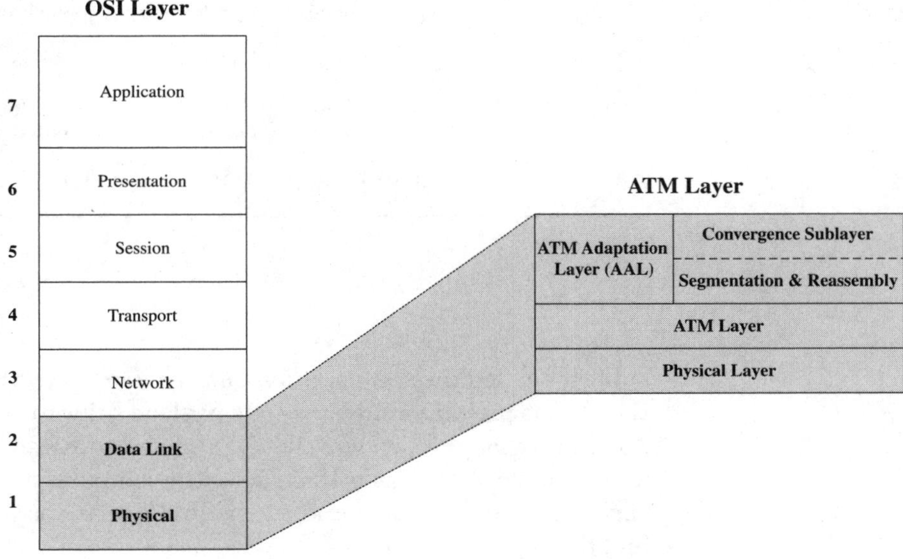

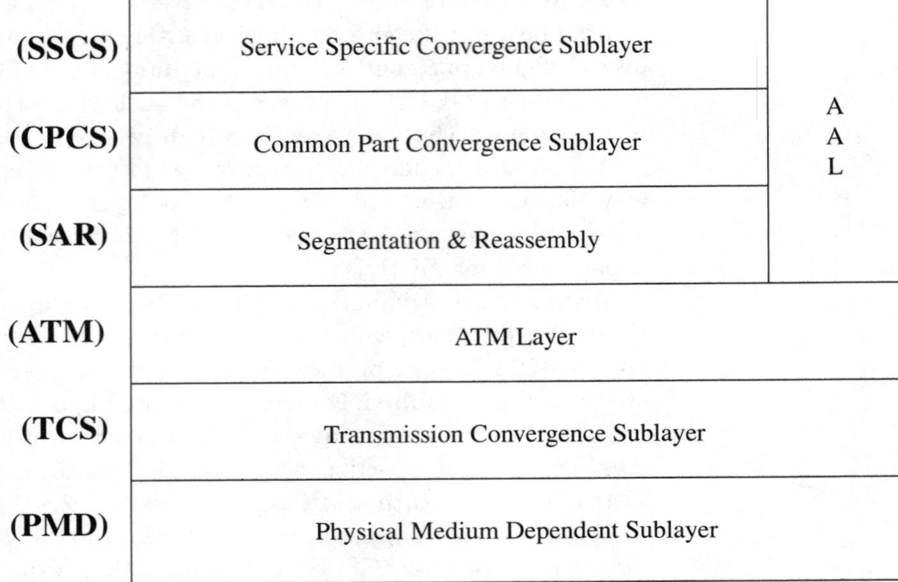

Figure 12-9 Upper-layer services of ATM

Table 12-5 Types of AAL and services offered

Class of Service	Type of Service A	B	C	D
Timing	Synchronous	Synchronous	Asynchronous	Asynchronous
Bit rates	Constant	Variable	Variable	Variable
Connection type	Connection oriented	Connection oriented	Connection oriented	Connectionless oriented
AAL	Type 1	Type 2	Type 3-4	Type 3-4-5

sublayer called the *segmentation and reassembly* (SAR). The SAR is where the data is prepared into a 48-byte payload prior to being submitted to the ATM layer for the header. The CS and the SAR combine to form what is known as the AAL. Table 12-5 shows the combination of the AAL types, the services being provided, and the working relationships between the AAL and the type of service.

ATM Traffic Management

When dealing with traffic management, some of the goals of the ATM Forum and other developers include the following:

- *ATM must be flexible.* It must meet the constantly changing demands of the user population. These goals mean that the demands for traffic will rise or fall as necessary, and therefore managing this traffic is of paramount importance.
- *ATM must meet the diverse needs of the end-user population.* Many users will have varying demands for both high- and low-speed traffic across the network. Using a QoS capability throughout the ATM network, a user can determine the performance and the capabilities of how the ATM network will meet their demands. These demands must be met in terms of the delay or the actual delivery of the cells across the network. Diverse needs are always going to be different, depending on the type of service (voice, data, video, or other traffic). Meeting these diverse needs for multiple users across the single threaded network is a major goal in traffic management and traffic delivery.
- *Cost efficiency is a must.* If ATM is truly to succeed, traffic management must also include the effective usage of all of the circuitry available. ATM is designed to reduce the inefficient circuit usage by efficiently mapping cells into dead spaces, particularly when data is involved. In the past, variable amounts of data would be sent across the network. Although it is good for data because variations exist, such as the use of a fixed cell size and managing that fixed cell throughout the network in terms of its performance, buffering and delay become the crucial issues addressed.
- *Robustness in the event of failure or in the event of excess demand is a requirement of the traffic management goals.* If the network is to be readily available for all users to be able to transmit information on demand, then the network must be very robust to accommodate failures, link downtime, and so on. Through this process, the managing of traffic must accommodate such diverse needs on a WAN.

Fair allocation of traffic capacity is essential on a fair and equitable basis. The goal of the traffic management scenario is to ensure that no one user would dominate the network; rather, all users would have equal access and an equal shot at using the capacity on demand. Specific goals of delivery can be achieved through committed information capabilities, but the intent is really to fairly arbitrate the traffic capacity and divide it up among multiple users. Traffic management is a set of actions taken by network devices to monitor and control the flow of traffic on the network. A highway, for example, is built to carry a certain amount of traffic. Any more traffic at a given time (a one-hour period, for example) causes congestion. Congestion causes frustration and forces some traffic to overflow to other roads. Each stream of traffic onto a network can carry a finite amount of flow. If the flow exceeds the capacity (bandwidth), then actions must be taken to minimize delays (red lights and green lights are used for flow control), control losses, redirect or discard traffic, and prevent collisions.

When using ATM networks, traffic management becomes critical. Too often the networks are built around fixed resources, which are finite and must be managed to provide equitable access and bandwidth to the end user. Network suppliers and carriers are, therefore, under constant pressure to get the best utilization from the networks.

Contention Management

Traffic in the form of asynchronous bursts of information (cells) enters the network at random times. This randomness is what causes the confusion and the unpredictability of the data. To manage the traffic flow, buffers are used to enable the flow and ebb of traffic volumes. Because data tends to be very bursty, it is extremely difficult to predict the demands of the network and the capacity needed at a given time. Therefore, when sufficient resources are not available, the use of buffers helps to offset the immediate demands. It is this bursty traffic that produces a contention for the network resources.

The use of "leaky buckets" in the buffering of the traffic helps to manage and control the flow of traffic onto and through the network. The leaky bucket, as the name implies, is a buffer that is constantly flowing. In Figure 12-10, a leaky bucket concept is shown with the two stages of buffering. Traffic enters into the buffers and is tagged, based on the amount of cells enabled by the carrier. If the user exceeds the amount of cell flow per increment (per second, and so on), then the buffer is filled and begins to empty out the bottom side. If more cells enter the buffer than are allowed, the cells are flagged for discard. A *first in, first out* (FIFO) concept is normally used to handle the traffic as it flows, but the end user may flag specific traffic according to application, priority, and the like.

ATM involves finding values that the network needs and making decisions about how the network will perform. These decisions involve taking

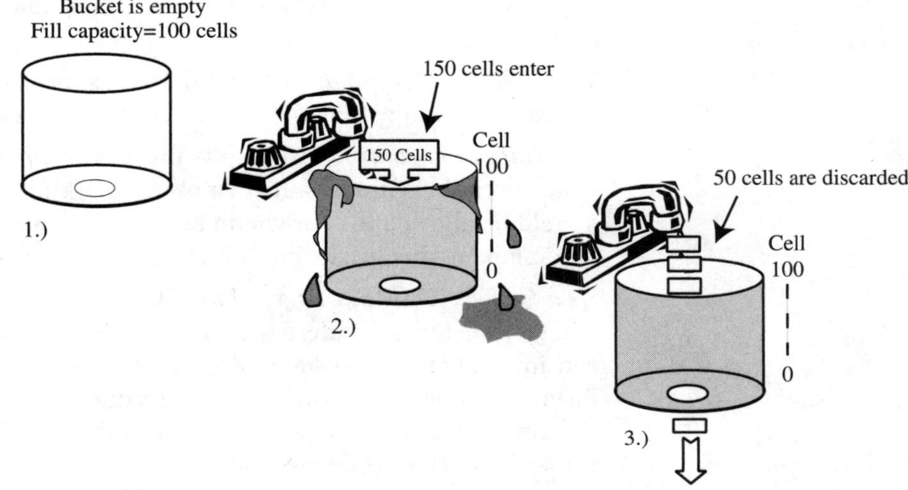

Figure 12-10 Leaky buckets allow for buffering of the traffic.

action to ensure its availability to users and to ensure that all current connections are receiving adequate service based on their class of service. To facilitate these decisions, the following performances are examined:

- *Cell loss ratio* (CLR) The ratio of lost cells to the sum of the total number of lost and successfully delivered cells.
- *Cell insertion rate* The number of cells inserted into an ATM network within a specific period of time.
- *Severely errored cell ratio* The ratio of severely errored cells (more than one bit of an error in the header) to the number of successfully delivered cells.
- *Cell transfer capacity* The maximum number of successfully delivered cells over a specified ATM connection during a period of time (one second).
- *Cell transfer delay* (CTD) The average delay and the arithmetic delay of a specified number of cell delays. The *cell delay variation* (CDV) is the difference between a specific cell delay and the average. This variation causes the most problems, especially with real-time voice and video.
- *Priority control* Networks must adequately service buffers in the network nodes under all kinds of conditions. When congestion occurs

(too many cells in the network), a priority mechanism can be used in the following ways:

- *To remedy the situation* Some cells can be discarded under congestion circumstances.
- *For congestion control* Networks must prevent congested conditions from spreading throughout the network. In this case, no sender should be allowed to overwhelm any receiver, as the network will accommodate by discarding cells.
- *As a Generic Cell Rate Algorithm* (GCRA) The GCRA is an example of a generic term. It refers to a virtual scheduling algorithm or a continuously leaky bucket and expresses a complex series of formulas. In most network implementations, it is widely known that the double leaky bucket algorithm is what we call the GCRA. Using the leaky bucket as a means of describing the GCRA is a good way to figure out exactly what is happening. The GCRA functions like a bucket with a hole in the bottom. The bucket leaks at a steady rate, no matter when water is poured into the bucket. The bucket may be initially empty, partially full, or full to the brim. As water pours in, it may splash or completely overflow the bucket if poured in from a much larger container. Water emerges from the hole in the bucket at precisely the same rate at all times. (Of course, everyone knows there are no real buckets in computers and networks, although the term "bit-bucket" is constantly used.)
- *As a counter or buffer* The simplest leaky bucket implementation is a counter. The counter has a minimum value (usually 0) and a maximum value. With the empty bucket, it holds up to 100 cells. All of a sudden, 150 cells are delivered to the network. This means that the bucket will overflow, or the 50 cells will be discarded because the buffer can only hold 100. At the bottom, the cells enter the network at a ratio of 1 over the *sustained cell rate* (SCR). Cells will exit at a steady rate.

The Double Leaky Bucket

In this double leaky bucket case, which was used in most of the early implementations of ATM, cells arrive from the CPE across the UNI totally uncontrollable in time. All cells leaking into the network are sent with a *cell loss priority* (CLP) equal to zero. As long as the arriving cells do not exceed a given rate, they are admitted to the network unchanged. If the cell arrival rate exceeds the limit imposed by the network, namely the SCR of the connection, the cells in excess of this limit have their loss priority bits changed to one. The cells with a CLP equal to one are subjected to another leaky bucket with another limit. Any cells sent with a CLP equal to one already set are also added to the cell stream. However, this time the limit corresponds to the maximum burst size of the connection. Cells under the MBS are admitted onto the network, while any others are discarded. This double leaky bucket is shown in Figure 12-11.

Here are two functions in ATM network implementation that perform traffic control:

- *The connection administration control* (CAC) Networks must set aside the proper amount of resources to service a connection at the

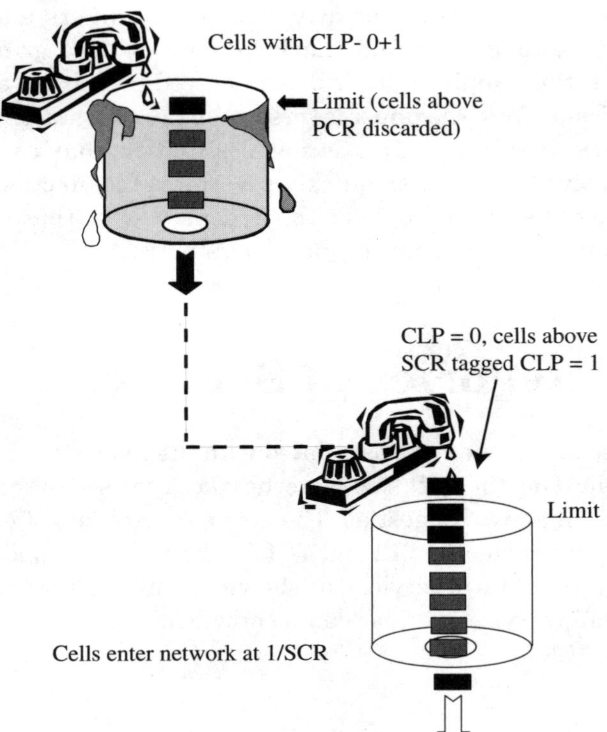

Figure 12-11 A double leaky bucket

time of the connection, whether it was set up as a service provision time on a semipermanent basis, or by means of a signaling protocol dynamically.

- *The usage parameter control* Networks must police the UNI to make sure cell traffic volumes do not affect overall network performance.

Traffic control and management can also be based on a credit- and debit-type system. As traffic enters the buffers, a certain amount of cells are authorized to enter the buffer. As the cells fill the buffers, they use up the credits allowable. When any extra cells are introduced into the buffers, then a second stage buffer can be used for the traffic as being eligible for discard. If, in fact, the eligibility for discard is exceeded, any new cells are automatically flushed away. These mechanisms are used to control the flow of data into the network. If some type of control is not used, the switches across the wide area can get into trouble. Buffer overruns and system overloads can also occur. If this continues to occur, the network suppliers cannot provide or live up to their guarantees for service levels. The continuous overruns across buffers and switches will cause severe congestion and snarl the network to a grinding halt. Data from real-time applications will be delayed, requiring a request for retransmission from the upper layers in the protocol stacks, thereby causing even more congestion. This is not a situation that can be taken lightly. To resolve the situation quickly, several traffic functions are used to manage the network and deliver the promised QoS, that elusive term everyone banters around, but no one understands.

Categories of Service

Because of the risks associated with the congestion and the impairments to delivering the QoS, several techniques are used to shape the traffic and prevent massive congestion. The network providers offer specific service offerings based on the VPI and VCI for the end-user applications. These come in several types of services, as shown in Table 12-6 and constitute the way the traffic service offerings can be provided.

Table 12-6

Comparing the categories of service

Service Category	Equates to
Constant bit rate (CBR)	The equivalent of a dedicated point-to-point leased line. This type of service is used when a specific amount of throughput is required, but it changes infrequently. The end user can commit to a fixed bandwidth and throughput and pay accordingly for the sustained throughput demands. The user will get and use a *peak cell rate* (PCR) from the carrier. This type of service can be used for real-time applications such as videoconferencing, sustained file transfers, telephony applications, audio applications, and the like.
Variable bit rate in real-time (VBR-rt)	This service is a variable rate of throughput typified by some of the same real-time applications listed previously but not always the same sustained rate of throughput. This service may also apply to applications that cannot tolerate lengthy delays such as real-time SNA traffic, financial transactions, compressed and packetized video or voice applications, and some forms of multimedia. In this service, the customer will buy into a PCR, a *Sustainable cell rate* (SCR), and a *Maximum Burst Size* (MBS) for the traffic.
Variable bit rate (VBR-nrt)	This service category is characteristic of nonreal-time bursts of data, but requires delivery guarantees from the carrier. This type of connection uses similar agreed-to definitions as listed in the previous VBR-rt mode. This includes the PCR, SCR, and MBS modes of data transmission. The real application here may be data processing, transaction processing, banking and credit card processing, and airline reservation services. Some process control applications also fall into this category.
Unspecified bit rate (UBR)	UBR does not specify any form of QoS guarantees from the network. Applications here can handle delay and latency without too much trouble. The network provider will make a best effort to deliver the traffic within a reasonable amount of time. This can apply to applications such as e-mail, file transfers, remote printing, LAN-to-LAN interconnections, and telecommuting services for a *small office / home office* (SOHO).
Available bit rate (ABR)	The ABR is suited for applications that can definitely tolerate longer delays on the network. The ABR adapts to whatever resources the network has available, rather than a specified amount of throughput. Typically, the ABR also uses a *minimum cell rate* (MCR) as a means of controlling the traffic. The network continually feeds back information on the regular traffic and therefore allocates the MCR to hold the connection alive, while other traffic runs across the network in a priority over the ABR rate.

Getting to the Elusive QoS

Above and beyond the ways that the ATM networks allocate resources as shown in the categories of service described previously, other methods are used to get to the elusive QoS guarantees. The PVCs in an ATM network can be provisioned with certain throughput parameters. The primary parameters are listed here:

- Cell loss ratio (CLR)
- Maximum cell transfer delays (maxCTD)
- Cell delay variations (CDVs)

Each of these methods is described briefly here for the sake of knowing what the accomplished goal is:

- The CLR is a means of determining the ratio of lost cells to the total number of cells that have been transmitted. There are many reasons why cells get lost, yet the goal is to hold the amount of lost cells to a minimum. An ATM switch can discard calls that have been corrupted, especially the header. Moreover, cells can be explicitly marked (tagged) as eligible for discard if the network gets congested. The formula for the ratio is as follows:

  CLR = Lost cells/total cells transmitted

- The CTD is calculated as the total amount of elapsed time from when a cell enters a network (a switch) until it exits the network. This takes into account the total amount of internodal processing time, buffering time, and propagation delays across whatever the medium is used.

  CTD = Node processing 1…n + buffer 1…n + propagation n

- The third formula is used to calculate the average amount of variations used in the cell delays. Often this is called *jitter on the network* where timing is lost or skewed a little to the left of right of the average. The mean of the variations is the overall delay variation. This is typically used for applications requiring some real-time consequence, such as packetized voice or video.

Shaping the Traffic

The ITU-TSS defines four possible situations when a cell enters an ATM network:

- *Successfully delivered* The cell arrives at the destination with less than time T-cell delay.
- *An errored cell occurs* A cell arrives with at least one detected bit error in the information field in the cell. Another possibility is the severely errored cell with information bits errors equal to n or n>1.
- *Lost cells* A cell either never arrives or arrives after the time T-cell delay, in which case it is discarded at the destination.
- *An inserted cell* A cell contains an undetected error or is misdirected by an ATM node and therefore shows up at the wrong destination.

In the bandwidth, the user is allocated a certain amount of capacity across the network. Clearly, the network will perform in different peaks and valleys, as shown in this picture. As the increase of cells per second grows and exceeds the network available rate for normal traffic over a period of time, the network will begin to discard cells. Although buffering might take place for certain periods of time, cells will likely be discarded because the network won't be able to process all of them.

If the user exceeds the network rate, the cells will be discarded, and the user will, therefore, have to retransmit at a later time, as shown in Figure 12-12.

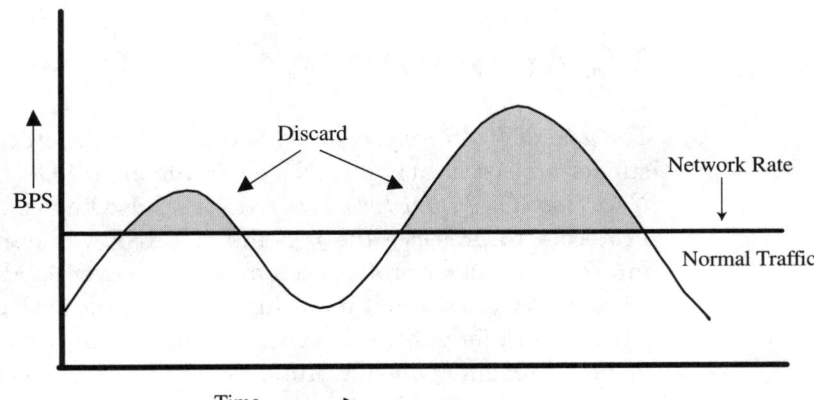

Figure 12-12 When a user exceeds the network rate, then cells are discarded.

Normal Bandwidth Allocation

If, in fact, the user does *not* exceed the network rate or the agreed-to rate the network will provide, and the user traffic is shaped, under the capacity, then the available capacity left over will be provided to other users. Allocating this capacity on a fair and equitable treatment can far more efficiently use the ATM network.

Once again, as the users exceed their network thresholds, the network will discard cells, but if they do not exceed the network rate, the network should deliver all available cells delivered to the network, as shown in Figure 12-13.

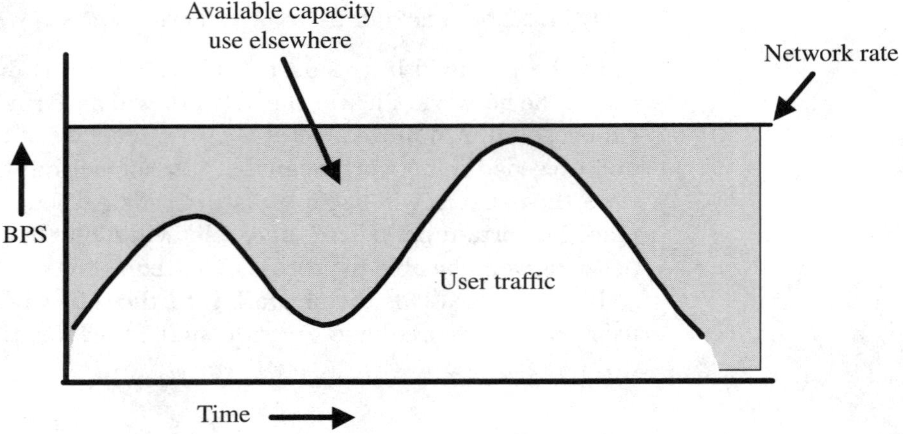

Figure 12-13 Cells enabled through the network when the user does not exceed the agreed-upon throughput

What Is MPOA?

The goal of *Multiprotocol on ATM* (MPOA) is the efficient transfer of intersubnet unicast data in a LANE environment. MPOA integrates LANE and *Next Hop Resolution Protocols* (NHRPs), also known as Next Hop Routing Protocols, to preserve the benefits of LANE while enabling intersubnet, internetwork layer protocol communication over ATM *Virtual Circuit Connections* (VCCs) without requiring routers in the data path. MPOA provides a framework for effectively synthesizing bridging and routing with ATM in an environment of diverse protocols, network technologies, and IEEE 802.1 *virtual LANs* (VLANs). This is designed to provide a unified paradigm for overlaying internetwork layer protocols on ATM. MPOA is capable of using both routing and bridging information to locate the optimal exit from the ATM cloud.

MPOA enables the physical separation of internetwork layer route calculation and forwarding, a technique known as *virtual routing*. This separation provides a number of key benefits:

- It enables efficient intersubnet communication.
- It increases manageability by decreasing the number of devices that must be configured to perform internetwork layer route calculation.
- It increases scalability by reducing the number of devices participating in internetwork layer route calculation.
- It reduces the complexity of edge devices by eliminating the need to perform internetwork layer route calculation.

MPOA provides *MPOA Clients* (MPCs) and *MPOA Servers* (MPSs), and defines the protocols that are required for MPCs and MPSs to communicate. MPCs issue queries for shortcut ATM addresses and receive replies from the MPS using these protocols. MPOA also ensures interoperability with the existing infrastructure of routers. MPOs also make use of routers that run standard internetwork layer routing protocols, such as *Open Shortest Path First* (OSPF), providing a smooth integration with existing networks.

LANE

Whenever we deal with the choice of service offering, we wind up with confusion. Such was the case when ATM was introduced as a means of replacing LANs such as Ethernet and Token Ring. Users were confused as to when this would be done and why it should be done. Moreover, the cost of such a change was not inconsequential. An Ethernet card sells for $10 to 20 and will support 10/100 Mbps. However, an ATM LAN card will cost $200+ with some being offered at $400. Thus, the industry decided to resolve the confusion. A standard has been developed for encapsulating and transmitting Internetwork protocols such as TCP/IP over ATM networks. One such protocol is *LAN Emulation* (LANE). To make it possible to continue using existing LAN application software while taking advantage of the increased bandwidth of ATM transmission, standards have been developed to enable the operation of LAN layer protocols over ATM. LANE is one such method, enabling the replacement of 10 Mbps Ethernet or 4/16 Mbps Token Ring

LANs with dedicated ATM links. It also enables the integration of ATM networks with legacy LAN networks. This software protocol running over ATM equipment offers two major features:

- The capability to run all existing LAN applications over ATM without change. The immediate benefit is that it is not necessary to reinvest in software applications.
- The capability to interconnect ATM equipment and networks to existing LANs and also the capability to link logically separate LANs via one ATM backbone. The benefit is that ATM equipment may be introduced only where it is needed.

The function of LANE is to emulate a LAN (either IEEE 802.3 Ethernet or 802.5 Token Ring) on top of an ATM network. Basically, the LANE protocol defines a service interface for higher-layer protocols, which is identical to that of existing LANs. Data is sent across the ATM network encapsulated in the appropriate LAN *Media Access Central* (MAC) packet format. Thus, the LANE protocols make an ATM network look and act like a LAN, only much faster.

Therefore, for LANE to operate as a traditional LAN, there must be the appearance of a connectionless service. This is the most important function for LANE. The main objective of the LANE service is to enable existing applications to access the ATM network by way of MAC drivers as if they were running over traditional LANs. Standard interfaces for MAC device drivers include *Network Driver Interface Specifications* (NDIS), *Open Data-link Interface* (ODI), and *Data-link Provider Interface* (DLPI). These interfaces specify how access to a MAC driver is performed. Although the drivers may have different primitives and parameter sets, the services they provide are synonymous. LANE provides these interfaces and services to the upper layers.

LANE was designed to enable existing networked applications and network protocols to run over ATM networks. It supports using ATM as a backbone for connecting "legacy" networks. It was also designed to support both directly attached ATM end systems and end systems attached through Layer 2 bridging devices. In addition, because one of the goals of ATM is to provide complete worldwide connectivity, it is important to enable multiple emulated LANs to exist on the same physically interconnected ATM network shown in Figure 12-14. The choices for connecting the two ends together include the following:

- Ethernet-attached end systems may communicate with other Ethernet-attached end systems through bridges across the ATM network in a backbone-type configuration.

Figure 12-14
The configuration of Ethernet and ATM combined

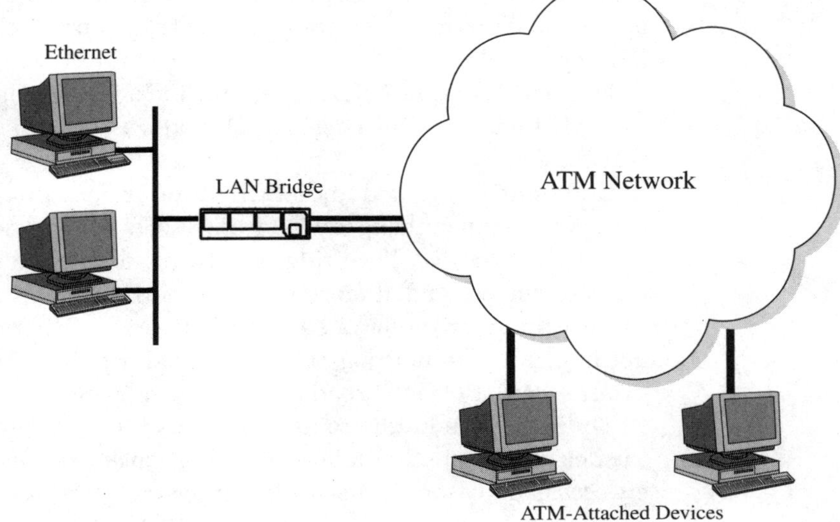

- ATM-attached end systems may communicate with ATM-attached servers and both may communicate with Ethernet-attached end systems via the bridges.

Significant portions of the LANE protocol were designed specifically to address the bridging-related issues. The specification refers to bridges generically as proxies. A proxy is any edge device that needs to forward LAN traffic, but may not have definitive information about the stations that are located on the legacy side.

The LANE configuration server takes requests from LANE clients and provides information on the LAN type being emulated and which LANE server to use. In its request, the LANE client must provide its ATM address as well as its 6-byte LAN address. The configuration server then provides the ATM address of the LANE server, to which the client connects in order to join the emulated LAN.

End stations and proxies are treated the same until they start communicating with the LANE server. The problem is that if the proxy is a bridge, it cannot know the MAC address of all the stations it may serve. The bridge learns about stations over time.

An end station only needs to join the emulated LAN and set up the virtual circuits to communicate and the virtual circuits used to control the connection. It must inform the LANE server about such things as its desired maximum frame size and LAN type. A proxy client must register as a proxy.

One of the jobs a LANE server may tackle is translating MAC addresses into ATM addresses or passing MAC addresses onto some device that can translate them.

One solution allowed under LANE for bridges involves a special virtual circuit for transmitting all address resolution requests that the LANE server can't resolve. The bridge can respond if the address is already in its address tables, and then employ its normal mechanisms to attempt to resolve the address for the emulation server. Otherwise, the bridge does nothing; if it does nothing, the client must use the *broadcast and unknown frame server* (BUS) to broadcast unknown frames.

The BUS, in addition to assisting in address resolution, helps facilitate the delivery of small numbers of unicast packets—for example, those packets occasionally sent by network management stations, where a virtual connection isn't really warranted. The BUS also delivers broadcasts and multicasts.

What this means is that although the service is available for high-speed LANs and high-speed WANs, LANE melds the two services together into a single homogenous networking strategy for an organization. A lot of hype was built into the delivery of LANE, yet the acceptance of LANE has met with only mild enthusiasm.

Voice over DSL and over ATM (VoDSL and VoATM)

In general, a VoDSL system functions as an overlay solution to a DSL broadband access network, enabling a LEC or CLEC to extend multiline local telephone service off of a centralized voice switch. For example, some VoDSL solutions enable up to 16 telephone lines and high-speed continuous data service to be provided over a single *digital subscriber line* (DSL) connection. A VoDSL solution typically consists of three components:

- First, a carrier-class voice gateway that resides in the *regional switching center* (RSC) and serves as a bridge between the circuit-based voice switch and the packet-based DSL access network.
- Second, an *Integrated Access Device* (IAD) resides at each subscriber premises and connects to a DSL circuit. It also serves as a

circuit/packet gateway and provides the subscriber with standard telephone service via up to 16 analog *plain old telephone service* (POTS) ports and Internet service via an Ethernet connection.

- Third is the management system.

With VoDSL solutions, DSL broadband access networks now have the coverage, capacity, and cost attributes to enable LECs and CLECs to deliver local telephone services as well as data services to the small and midsize business markets as shown in Figure 12-15.

It has already been established that DSL access networks have the right bandwidth to serve the data needs of small and midsize businesses. With VoDSL access solutions, this is true for serving the local telephone service needs of those subscribers as well. Some VoDSL solutions are capable of delivering 16 telephone lines over a DSL circuit along with standard data traffic. Because 95 percent of small businesses use 12 or fewer telephone lines, a single DSL circuit provides sufficient bandwidth to serve the voice needs of the vast majority of the market. In addition, if more than 16 lines are required, most VoDSL solutions enable a provider to scale service by provisioning additional DSL connections. In addition to providing the right

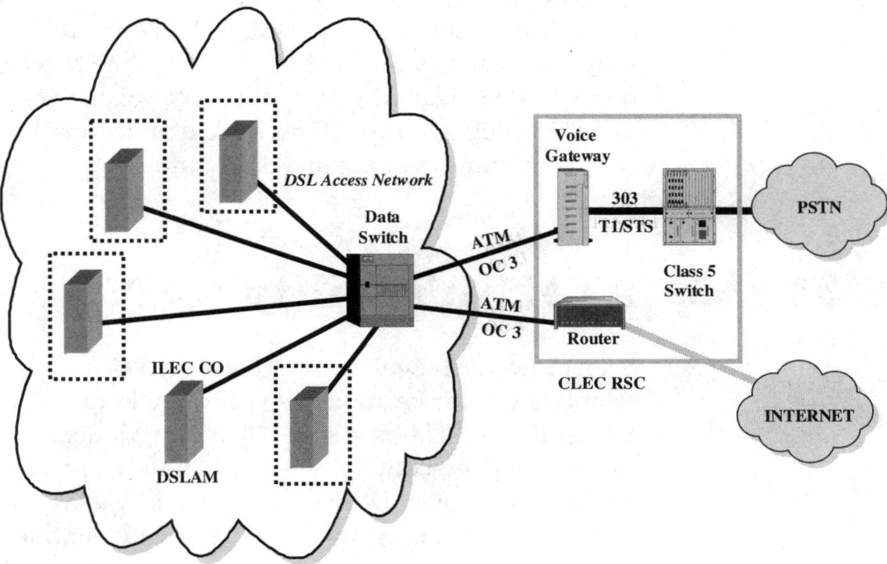

Figure 12-15 VoDSL can be provided easily.

capacity for providing local telephone service, DSL broadband access networks are very efficient in the way they deliver service. TDM-based transport services, such as a T1 line, require the bandwidth of the line to be channelized and portions dedicated to certain services, such as a telephone line. Even if a call is not active on that line, the bandwidth allocated to that line cannot be used for other purposes. DSL access networks are packet-based, allowing VoDSL solutions to use the bandwidth of a DSL connection dynamically. VoDSL solutions only consume bandwidth on a DSL connection when a call is active on a line. If a call is not active, then that bandwidth is available for other services, such as Internet access. This dynamic bandwidth usage enables providers to maximize the potential of each DSL connection, delivering to subscribers the greatest number of telephone lines and highest possible data speeds.

Because telephony traffic is more sensitive to latency than data traffic, VoDSL solutions guarantee the quality of telephone service by giving telephony packets priority over data packets onto a DSL connection. In other words, telephony traffic always receives the bandwidth it requires, and data traffic uses the remaining bandwidth. Fortunately, telephony traffic tends to be very bursty over the course of a typical business day, so the average amount of bandwidth consumed is minimal. For example, over a single 768 Kbps symmetric DSL connection, a LEC or CLEC could provide 8 telephone lines (serving a PBX/KTS with 32 extensions) and still deliver data service with an average speed of 550 Kbps.

VoATM unites ATM and DSL technologies to deliver on the promise of fully integrated voice and data services. VoATM meets all requirements in terms of QoS, flexibility, and reliability because the underlying technology is ATM, a highly effective network architecture developed specifically to carry simultaneous voice and data traffic.

ATM Suitability for Voice Traffic

Sometimes mistakenly associated with *Voice over IP* (VoIP), VoATM is a completely separate technology that predates VoIP. In contrast to IP and Frame Relay, ATM uses small, fixed-length data packets of 53 bytes each that fill more quickly, are sent immediately, and are much less susceptible to network delays. (Delays experienced by voice in a Frame Relay or IP packet network can typically be 10 times higher than for ATM and increase on slower links.) ATM's packet characteristics make it by far the best-suited packet technology for guaranteeing the same QoS found in "toll-quality" voice connections.

The part of ATM responsible for converting voice and data into ATM cells, the AAL, enables various traffic types to have data converted to and from the ATM cell and translates higher-layer services (such as TCP/IP) into the size and format of the ATM protocol layer. A number of AAL definitions exist to accommodate the various types of network traffic. Those AAL types most commonly used for voice traffic are AAL1, AAL2, and AAL5.

VoATM with AAL1 is the traditional approach for CBR, time-dependent traffic, such as voice and video, and provides circuit emulation for trunking applications. ATM with AAL1 is still suitable for voice traffic, but is not the ideal solution for voice services in the local loop because its design for fixed bandwidth allocation means network resources are consumed even when no voice traffic is present. AAL5 is used by some equipment manufacturers to provide VoATM, it provides support for VBR applications, and it is a better choice over AAL1 in terms of bandwidth used. However, the means for carrying voice traffic over AAL5 is not yet fully standardized or widely deployed, and implementations are usually proprietary.

ATM with AAL2 is the newest approach to VoATM. AAL2 provides a number of important improvements over AAL1 and AAL5, including support for CBR and VBR applications, dynamic bandwidth allocation, and support for multiple voice calls over a single ATM PVC. An additional and significant advantage of AAL2 is that cells carry content information. This feature enables the traffic prioritization for packets (cells) and is the key to dynamic bandwidth allocation and efficient network use.

Integrated Access at the Local Loop

Because DSL links are ready-made for voice and data, and ATM excels at carrying varied traffic, using VoATM over DSL over the local loop to the customer is a natural extension of these services. To enable the combination, equipment that supports VoATM is needed at each end of the local loop: a next-generation *integrated access device* (IAD) at the customer premises and a voice gateway at the CO as shown in Figures 12-16 and 12-17.

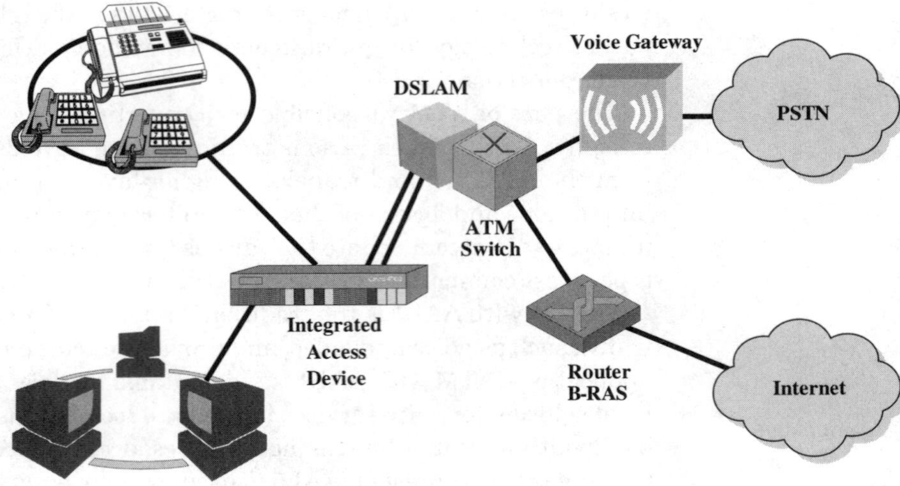

Figure 12-16 The IAD will add services for the future.

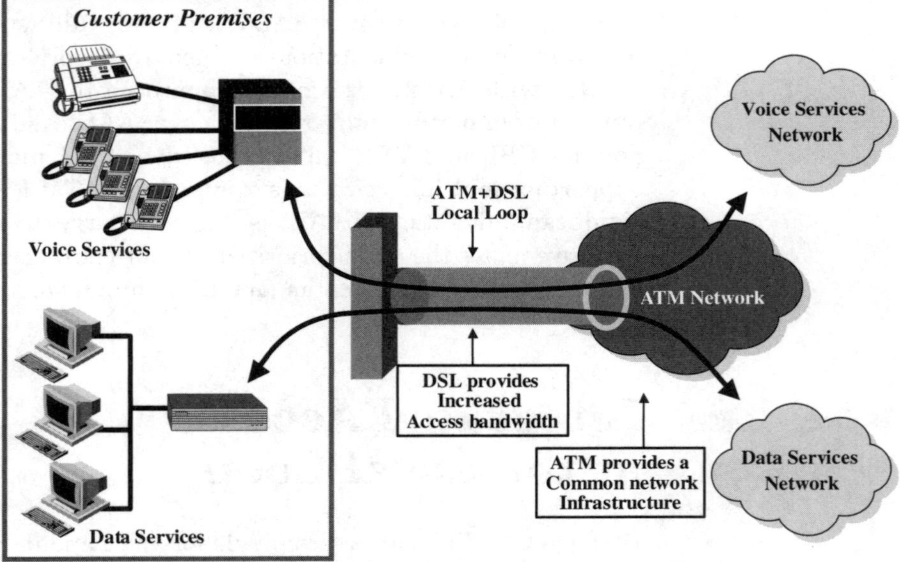

Figure 12-17 Combining the ATM and DSL at the local loop

These devices make it possible for the local exchange provider or the competitive DSL provider to use the existing facilities and still satisfy the needs of voice and data over the existing local loop. Using voice over the DSL circuit enables up to 16 simultaneous VoIP calls and Internet access to simultaneously run over the bandwidth on the single cable pair.

CHAPTER 13

ATM and Frame Relay Internetworking

Since 1992, two major developments have rolled into the carrier networks. The first was the implementation and rollout of Frame Relay networking protocols. Frame Relay met with immediate success because of its ability to handle Wide Area Networking data traffic, replacing (or complementing) older X.25 networks. The network suppliers had to upgrade their equipment to support the newer protocols in their packet switches. Although Frame Relay met with instant success, as described in Chapter 11, "Frame Relay," it was introduced as a data-only service for the WAN. The commitment to equipment and labor to upgrade the network became a very heavy burden.

As with most carriers, network equipment has traditionally been depreciated over a 20-year period. Thus, when Frame Relay began rolling into the marketplace in 1992, the depreciation window opened with an end-date of 2022 and counting. Frame Relay was deployed in every major network around the world because of its flexibility and cost advantages over the older protocols. The result was a widely deployed and well-accepted international standard for data communications.

The second major development introduced also in 1992 was ATM. ATM is a robust set of protocols that works in more than the WAN, but is designed to work across the various platforms of network from the LAN, CAN, MAN, and WAN. Because it was designed as a transport set of protocols to work at layer 2 of the OSI equivalent model, ATM both competes with and complements the use of Frame Relay. Yet, ATM goes further than just being a data transmission set of protocols. It is designed to carry voice, data, video, and high-speed multimedia traffic. As a broadband communications set of protocols, ATM is the one set of operating protocols that meets the expectations of the end user, local and long distance carrier, and the equipment manufacturers alike. But, like all new protocols that cross boundaries (between the voice and data worlds), ATM needed some added enhancements that were not ready in 1992. Consequently, ATM did not start to catch on until late 1995 and early 1996. Moreover, ATM has been specified and studied to death, slowing its acceptance. Just before 2000, ATM was accepted and standardized as a set of protocols for the network providers throughout the world.

What remains is a problem with the rollout of the equipment and other associated interfaces. Where the carriers have endorsed and embraced Frame Relay, they now have to upgrade to an ATM backbone network. This will require significant investments. Moreover, the carriers will still have the Frame Relay switches in the networks that are on the books, usable and still viable traffic handling machines. To solve the problem, several techniques were developed to enable legacy systems, new systems, Frame Relay, and ATM all work together. This concept has been deemed as Frame Relay and ATM internetworking. In reality, it is a form of interworking instead of internetworking.

ATM and Frame Relay Compared

One way to understand the interworking functionality of the two sets of protocols is to compare and contrast the capabilities of the two protocols. Table 13-1 is used as an overall summary of the two techniques in use today.

What you can see from the comparison is a summary of the characteristics of the two service offerings from the carriers. This is not the whole story, but it does give the reader a visual means of seeing where and why the two techniques have become popular.

Table 13-1

Comparing Frame Relay and ATM characteristics

Features	Frame Relay	ATM
Connection Type	Connection Oriented	
Connection Mechanism	PVC or SVC	
Switched Access (SVC arrangement)	Yes but not widely implemented	Yes
Multiplexing Arrangement	Statistical TDM	
Current Speeds Available	Typically 56 Kbps–2 Mbps, implementations up to 50 Mbps	1.544 Mbps (T1) 12.3 Mbps 25.6 Mbps 34 Mbps (E3) 45 Mbps (T3) 51 Mbps (OC1) 155 Mbps (OC3) 622 Mbps (OC12)
Area Served	WAN	LAN CAN MAN WAN
Sequencing of Data	No	No
Protocol Data Units	Variable	Fixed
Protocol Data Unit Size	≤4096 bytes	53 bytes

Table 13-1 cont.

Comparing Frame Relay and ATM characteristics

Features	Frame Relay	ATM
Flow Mechanisms	Circuit by circuit	
Traffic Congestion management	DE bit	CLP bit
Congestion Notification	FECN/BECN	Payload Type field
Bursty	Yes; defined by B^c and B^e rates	Yes, defined by PCR (maximum burst sizes)
Addressing Method	DLCI	VPI and VCI
Address Size	10 bits (normal)	24 bits
Standards-based	Joint development with Frame Relay Forum and ATM Forum, ANSI specifications, ITU standards and others	

Frame Relay Revisited

To take this a step further, Frame Relay is faster than the older networking X.25 and meets the demands of the older applications from a user's perspective. Very few applications today need a full 50 Mbps speed, but the 1.544 Mbps speeds and below are very robust and ubiquitous. This makes Frame Relay attractive also from a pricing perspective. The primary applications used with Frame Relay are SNA internetworking for mainframes and LAN- to LAN-connections across the WAN. Other services, such as remote access and internetworking between major corporations and their branch and small offices, are well suited for Frame Relay. The use of Frame services also fits well in the smaller corporate networks, but only scales to certain sizes before congestion and addressing become problematic. Although well deployed around the world, Frame Relay is primarily suited for the small- to mid-sized corporate network. When the network has to grow to tens of thousands of nodes, Frame Relay may run out of capacity and capability very quickly. Moreover, Frame Relay is already in place so it is used more heavily. Figure 13-1 is a graphic representation of the use of Frame Relay in the WAN by carriers and end-users alike. In this graphic, the Frame Relay switches are deployed in the carriers' networks, whereas

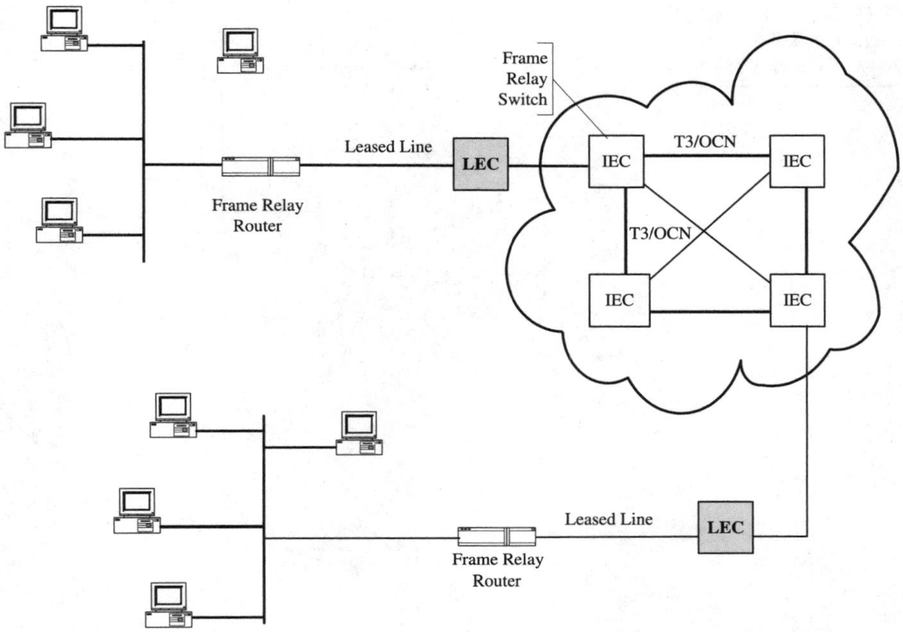

Figure 13-1 Frame Relay in various places

the routers on a LAN use the Frame Relay protocols. This allows the interconnectivity across the WAN in a straightforward and cost efficient manner. Frame Relay is also well suited for the bursty traffic typified across the LAN-to-WAN networks.

ATM Revisited

ATM, on the other hand, uses a fast relay systematic approach to data handling. By using the fixed-sized cells across the ATM switching fabric, the process of cell transmission can occur very quickly. ATM supports added functionality and capability in differing applications, such as the integration of voice and video across multiple platforms. Even though voice across ATM was slower in developing, ATM was designed to carry the various traffic types in the cells. The ATM model is shown in Figure 13-2 where ATM was designed to work. This is a quick review of the two techniques so that you will have an understanding of why they must interwork in order to achieve harmony across the networks.

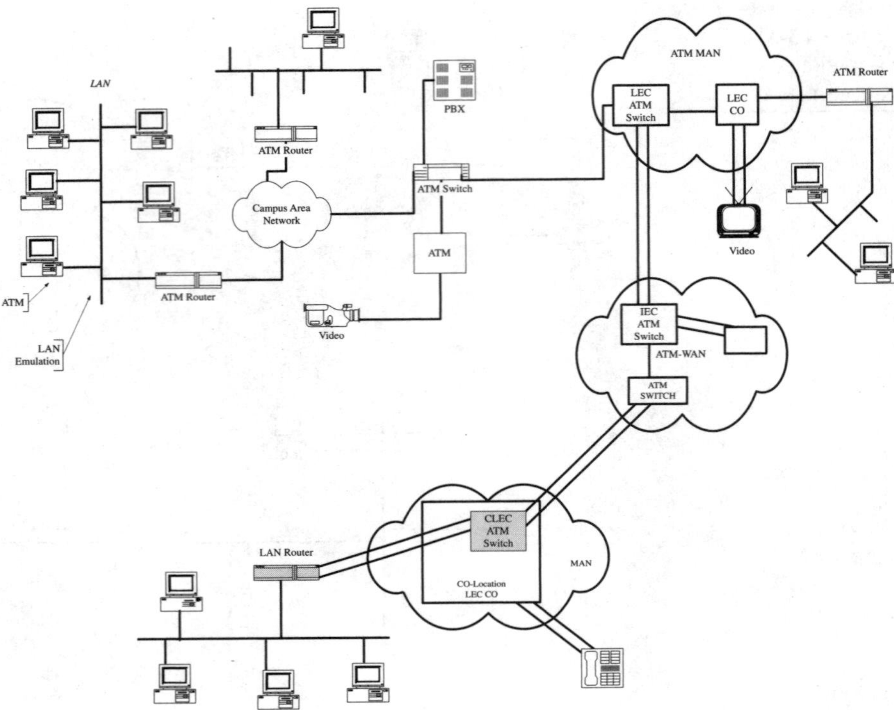

Figure 13-2 ATM in various uses

This graphic shows the integration of the various forms of information (voice, data, and video), as well as the locations in the LAN, CAN, MAN, and WAN. You can see that the piece parts installed at the customer locations may include an ATM router, an ATM switch, or an ATM PBX (not shown in the graphic). These changes require significant investments, not only by the consumer, but also by the carriers where each of the LEC, CLEC, and IEC interfaces use ATM switches in the backbone. ATM switches cost great amounts of money for a carrier to install and maintain.

The Frame and ATM Merger

Because of the dilemma between the two differing techniques, both the ATM and Frame Relay Forums took action to align the two standards more closely to allow for the interworking and internetworking between them. Both protocols have very defined characteristics that make them desirable. Both also have parameters that are easily matched up to the other. By aligning these parameters together, the interworking function is more easily accomplished. The similarities between the two technologies make them complements of each other without creating a direct conflict when properly planned. By combining both technologies, carriers and end users alike can protect their investments and preserve the infrastructure of the networks in place today. Without major renovations to the architecture, logical progression from frame-based services to cell-based services in the future will become far more acceptable. As a matter of fact, the combined technologies are rapidly being accepted and implemented throughout the industry.

Transparency Across the Network

The real goal is to provide for transparency across the network. End users running Frame Relay need not concern themselves with the fact that the carrier is likely running ATM in the backbone. This is shown in Figure 13-3 where the carrier is providing high-speed connectivity across the WAN, but the local attachment is using Frame Relay.

Still another way of looking at the network interworking is to have Frame Relay on one end and ATM at the other end through the network (see Figure 13-4).

Figure 13-3 Frame and ATM coming together

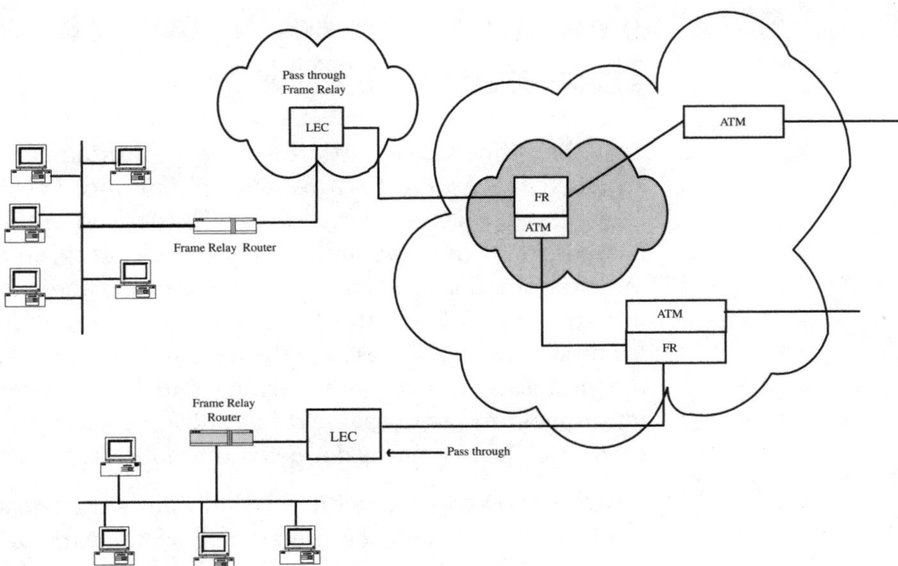

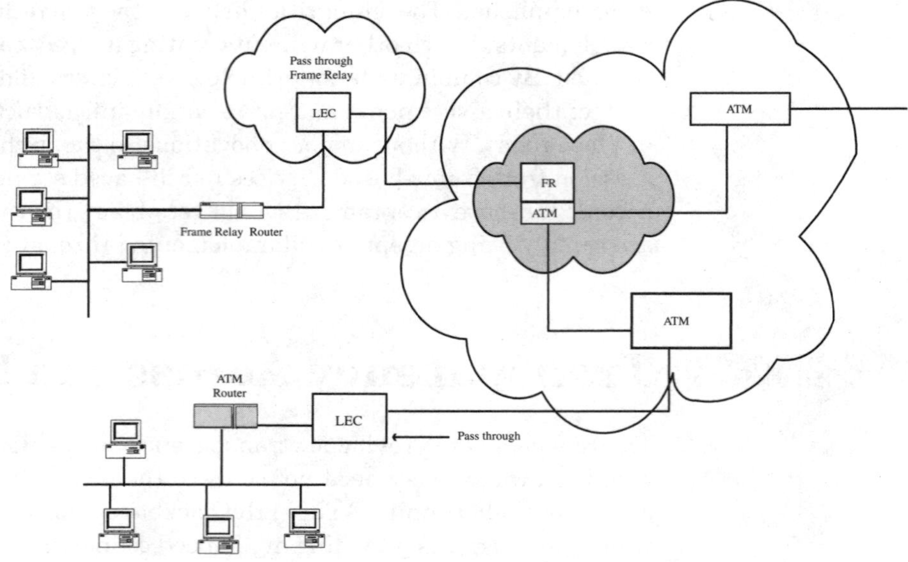

Figure 13-4 Frame Relay to ATM conversions

Either way, this is achieved with relative ease from the perspective of the carrier and the end user. Transparency is what is necessary to make this happen. The interworking functionality is a reality today.

Frame User-to-Network Interface (FUNI)

Providing the interfaces to the internetworking services, the ATM forum specified the *Frame User-to-Network Interface* (FUNI) in order to allow frame-based services to access an ATM network. FUNI allows traffic to be passed from the two different networking technologies by using an industry-standardized interface. This provides the carrier and the user with a migration path for the future, while serving today's needs. The ATM forum was careful to describe the differences in two terms before specifying the protocols needed: interworking and Internetworking. Although they may appear to say the same thing, subtle differences exist between the two concepts. The two terms defined are as follows:

- **Interworking** A technique that allows the two systems on the ends to run Frame Relay as shown earlier in Figure 13-3, yet run ATM across the backbone. Thus, the systems interwork with each other.

Frame Relay is on both sides of the network so we can encapsulate or tunnel the traffic through the ATM network, but the traffic entering and exiting the backbone is still Frame Relay on both sides.

- **Internetworking** Uses techniques to convert the traffic from one form to another, much the same as a protocol converter or a gateway. A device that sits at the edge of the network earlier performs the internetworking (or interworking) function (see Figure 13-4). The device is typically referred to as an IWF.

Taken one step further, the ATM Forum defined this interworking in two different categories:

- Service interworking
- Network interworking

Network interworking means that Frame is supported through the ATM backbone. The operations of the network nodes are performing actual actions on the data in the form of a convergence sublayering function. Network interworking supports AAL5 as a specified and optional AAL3/4 in the performance of its interface.

FUNI is capable of handling and supporting the basic functions at the UNI such as the following:

- VPI/VCI multiplexing
- Network management
- Traffic and congestion control (shaping)
- *Operation, administration, maintenance, and provisioning* (OAM & P) functions

Although these functions are supported, they may be limited in their overall performance and functionality. FUNI does not support some of the traffic types (such as AAL 1 and 2), so therefore it does not support the corresponding signaling and traffic parameters for these traffic types. In addition, the FUNI is designed to support variable bit rate (nonreal time) and unspecified bit rate class of traffic.

Data Exchange Interface (DXI)

Data Exchange Interface (DXI) is another of the interfaces used to internetwork various services across an ATM backbone. Functionally, the FUNI and DXI provide the same services with minor modifications. DXI is used with legacy routers that are not ATM equipped, but use the DXI protocols and interfaces to provide frame-based services to the *Data Service Unit* (DSU) on the circuit where the cells are then generated. The DXI uses the DSU whereas the FUNI does not. Moreover, DXI does not use fractional T1 services, where FUNI can. A comparison of the FUNI and the DXI interfaces is shown in Figure 13-5.

When comparing the DXI and the FUNI interfaces as the means of internetworking the services, it is appropriate to view the overhead comparisons of frame-based services contrasted with the cell-based services of ATM (see Figure 13-6).

The FUNI reference model as defined by the ATM forum is shown in Figure 13-7. The interworking functionality gives an idea of the mapping of services across the FUNI to an ATM network. The upper portion of the graphic represents the block flow; whereas the lower portion of the graphic shows the protocol stacks as they compare between the interworking functions.

With this protocol analysis in mind, there are some subtleties between the Frame Relay protocol and the mapping across the ATM platform through the IWF on a network. In this case, the graphic shown in Figure 13-8 represents the interworking functions as they align to the reference model shown. Once again, the upper portion of the graphic is the block diagram of the network, and the lower portion of the graphic shows the protocol stack.

Figure 13-5 DXI and FUNI interfaces compared

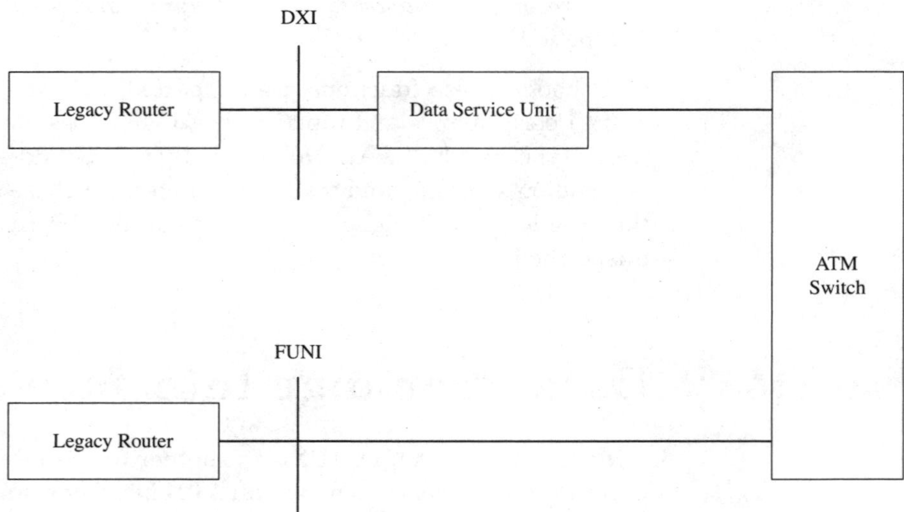

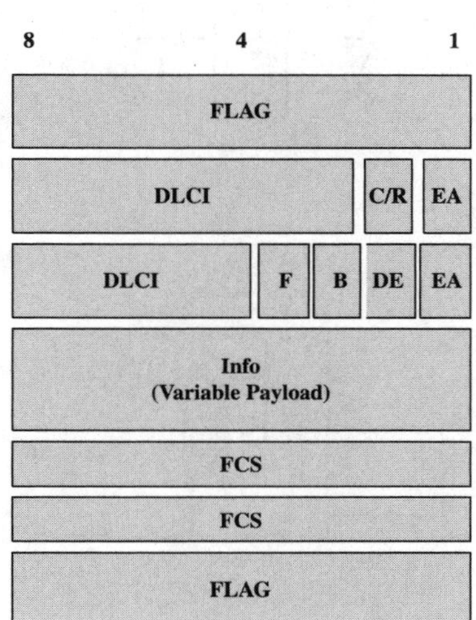

Figure 13-6 Contrast of frame- and cell-based services

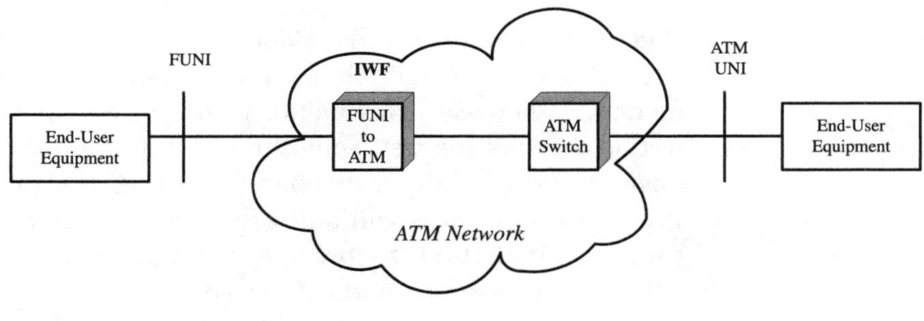

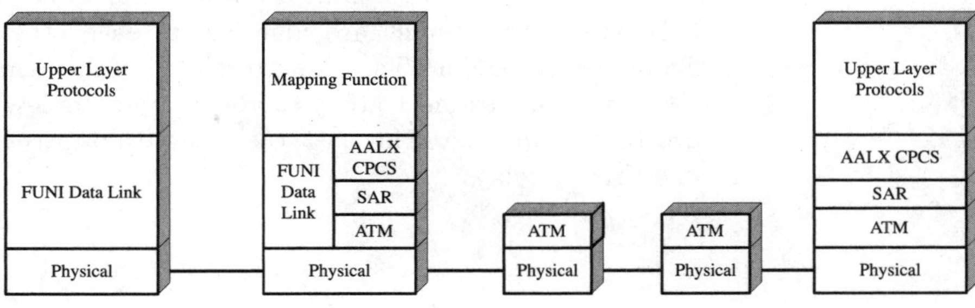

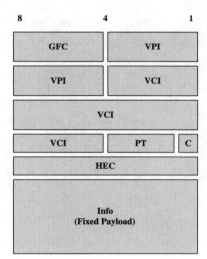

Figure 13-7 FUNI reference model

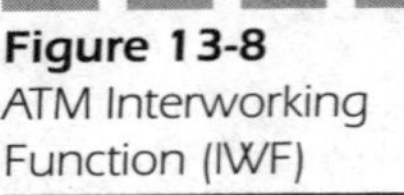

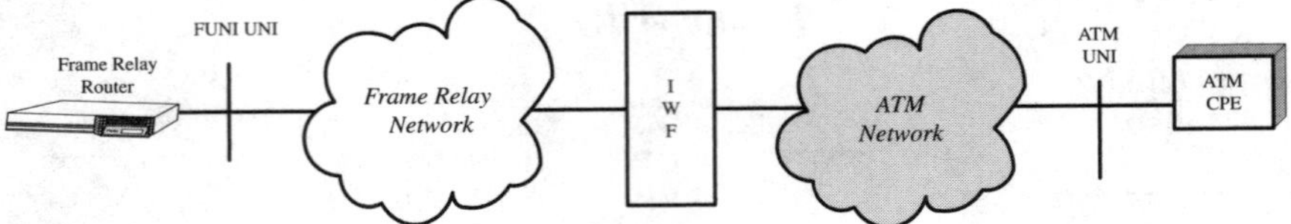

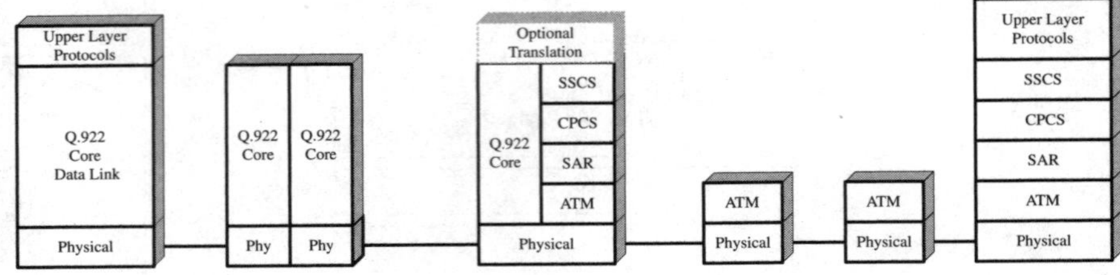

Figure 13-8 ATM Interworking Function (IWF)

What Constitutes a Frame?

The market still is unsettled. More low-end, low-cost devices are being sold every day. Users are trying to preserve their investments and hold the line on new, higher-cost investments as long as possible. With the growth projection for this market, something had to be done. Interworking is the answer. However, the frame-based services in legacy equipment and the Frame Relay devices will still exist for some time to come. The frame is therefore important to provide the interoperations and interworking between frame-based and call-based networking components. FUNI and DXI allow the frame-based access to the ATM networks, while Frame Relay allows frame-based access to a Frame Relay network. DXI, FUNI, and Frame Relay have similar frame structures. The headers of FUNI and DXI within the frames are identical to each other, but different from Frame Relay headers. The bit patterns fall in the same logical structure as a Frame Relay frame, so they can be mapped somewhat consistently. Figure 13-9 compares the bit patterns of the frame structures for FUNI, DXI, and Frame Relay.

Figure 13-9 Comparing the frames

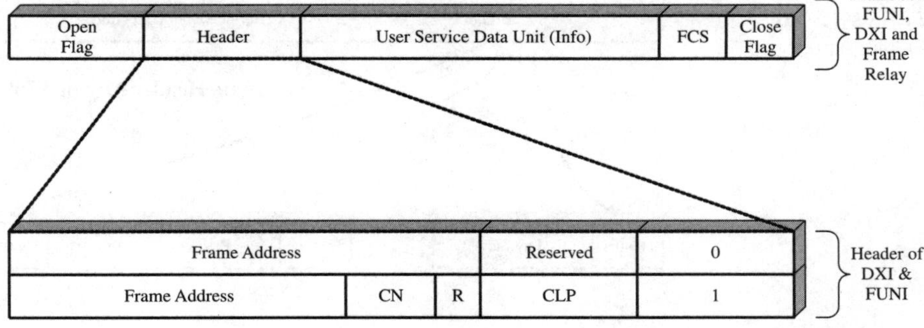

DLCI Upper			C/R	0
DLCI Lower	FECN	BECN	D/E	1

Frame Relay Header

When the frame is then segmented into cells, the DXI frame address and the FUNI frame address fields can both be mapped to a VPI/VCI in ATM by using the same process. Frame Relay interworking functions can use the same addressing map. The Frame Relay to ATM interworking functions can also use other mapping of the data formats.

The next step from the original frame format is to segment the FUNI frame into ATM cells, using a mapping of the frame address into an ATM VPI/VCI. This is shown in Figure 13-10 where the actual header information is segmented into the cells and delineated into the appropriate VPI/VCI. The *Congestion Notification* (CN) bit performs much the same function as the *Forward Explicit Congestion Notification* (FECN) bit in Frame Relay. The network will set this bit during periods of network congestion in the same direction as the traffic flow where the congestion exists. The Frame Relay *Backward Explicit Congestion Notification* (BECN) bit does not have a suitable equal in the FUNI or the DXI.

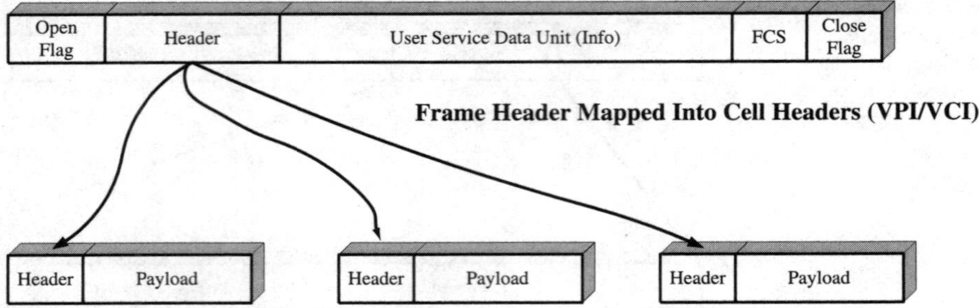

Figure 13-10 FUNI segmented into cells

FUNI Interoperability

When a FUNI device needs to establish a switched virtual circuit to some end point, it doesn't matter what the other device is (FUNI, DXI, or ATM). The use of a call set-up mechanism in signaling is the same as the ATM signaling procedure at the UNI. When FUNI traffic is delivered across the ATM networks, it doesn't matter if the data terminates at another FUNI or an ATM device. These service aspects are transparent to the network and the devices. Because FUNI is an ATM protocol, the multiprotocol encapsulation procedures at the ATM UNI are used at the FUNI, so there are no issues related to interoperability. Looking at Figure 13-11, one can see that the interoperability issues were addressed up front to prevent any downstream problems in the protocol or interoperability standards for ATM.

Network Interworking

The network interworking function provides the transport of Frame Relay user traffic transparently across the ATM link. It also handles the PVC signaling traffic over ATM. As already discussed, this is sometimes called tunneling through the network. Other times it is called encapsulating the traffic. Regardless of the name it is given, the function provides for the transparent movement of end user Frame Relay information across the ATM network. The benefit is that with the tunneling or encapsulation for-

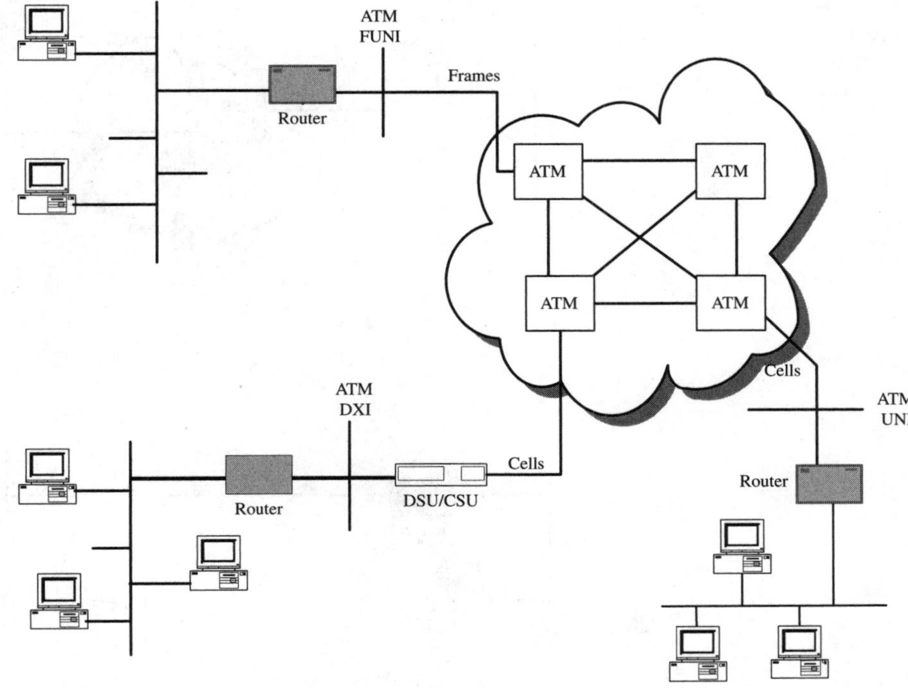

Figure 13-11 FUNI, DXI, and ATM interoperability

matting, the service is as good as though the end user has a leased line service between the two end points. The benefit of this tunneling approach is connecting two Frame Relay networks across an ATM backbone. This is shown through the use of the network-interworking unit in Figure 13-12. The interworking function is shown as a separate piece of equipment between the Frame Relay and ATM network, which in some cases it is. However, newer implementations of this architecture place the network interworking function and interfaces inside an ATM switch. Regardless of where it resides, the functionality is really important, not the location of the box. The interworking function will allow for each Frame Relay PVC connection to be mapped on a one-to-one basis over an ATM PVC. In other cases, many Frame Relay PVCs can be bundled together across a single, higher-speed ATM PVC. The one-to-one or the one-to-many services allow more flexibility.

Figure 13-12 The network interworking function

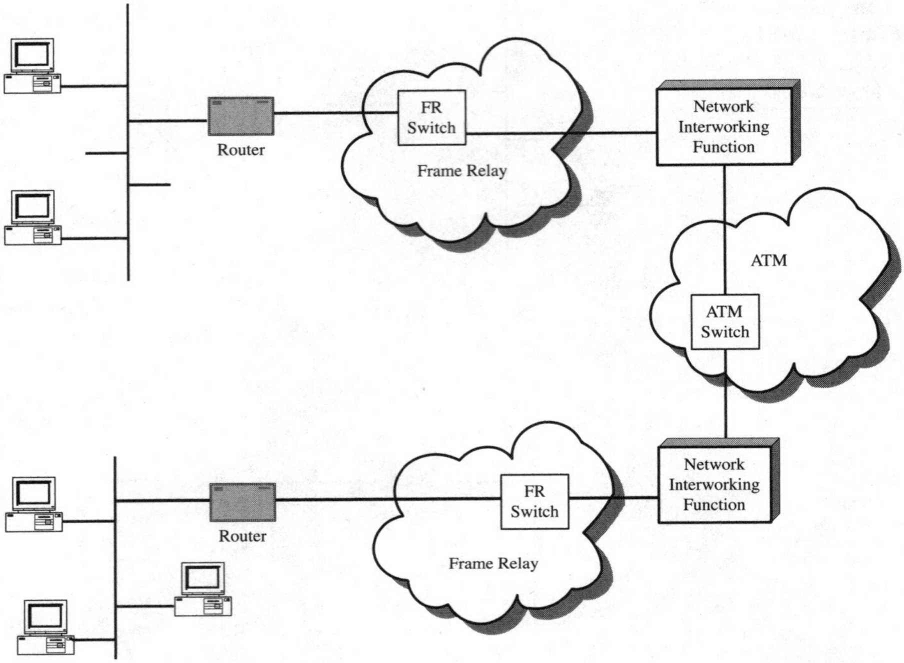

Service Interworking Functions

The use of a service interworking function takes away some of the flexibility and transparency across the network. It actually acts more like a gateway (protocol converter) to facilitate the connection and communications between different disparate pieces of equipment. Figure 13-13 is a representation of the interconnection of the service interworking devices across the network. The end user actually sends traffic out across a Frame Relay network on its own PVC, and then it gets passed through the Frame Relay network to the service interworking function where the data is then mapped to an ATM PVC. The IWF functionality provides the mapping of the DLCI to the VPI/VCI, as well as other optional features. The IWF is shown as a separate box, whereas newer implementations will have the dual mode functionality inside an ATM switch.

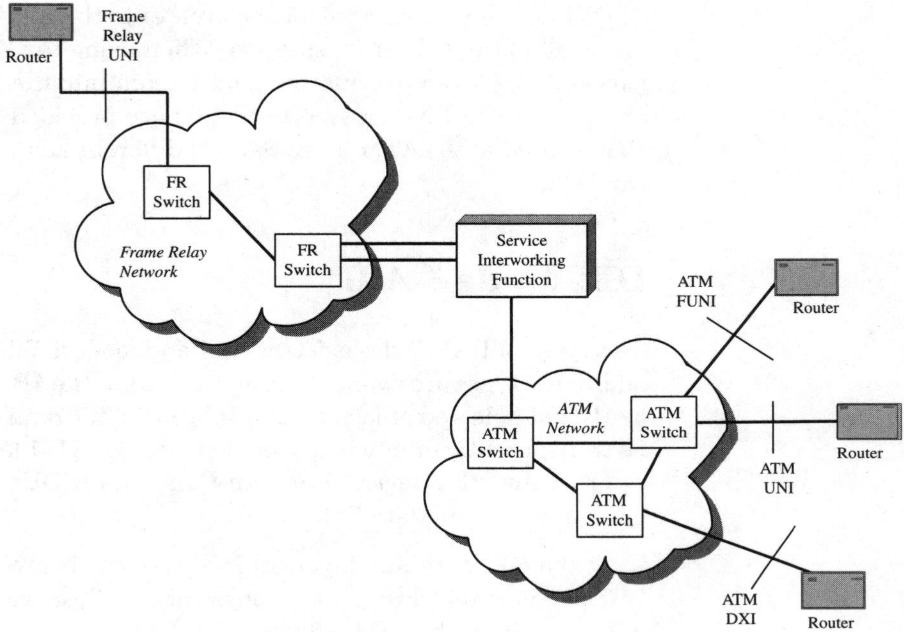

Figure 13-13 Frame/ATM service interworking function

The DXI Interface

Several times now the additional device mentioned in the frame and ATM interworking function is the DXI interface. Using a legacy Frame Relay router, for example, one can still have access to the ATM network on an internetworking basis. This prevents the old *forklift mentality* where all users have to change hardware en masse when a new service is introduced. DXI is fairly straightforward, but does deserve some mention in its workings.

Preparing information for ATM coming from a router or a brouter DXI typically deals with a link to frame the data and prepare it for a DSU. Connecting the DXI link will be a high-speed connection at either one of the following:

- V.35
- HSSI
- EIA449/422

DXI is an open interface for the brouter to the DSU where the DSU performs all of the DXI encapsulation. When using the V.35, all that is necessary is HDLC frame-formatted data stream into the DSU from the router or brouter. The DXI transport services are provided within the DSU and then passed to the ATM layer. Several different adaptation layer processes can be used.

DXI Mode 1 A/B

Two types of DXI modes exist: mode 1 and mode 2. When using mode 1, two additional types are available: either the 1A or the 1B. When using mode 1A or 1B, a simple and efficient encapsulation of the data is provided. Data can be transparently or efficiently passed into the ATM layer.

From the DTE device, a *Service Data Unit* (SDU) will be prepared and passed to the DXI data link.

- At the DXI data link layer, an HDLC frame is provided. This HDLC frame is mapped to the DXI physical interface, which is then connected to the DCE (that is, a router).
- At the DCE, the input interface from the DTE terminal devices will be at the DXI physical layer and again through a DXI data link. The DTE SDU will then be passed to the AAL5.

The AAL5 *Common Part Convergence Sublayer* (CPCS) will then prepare the data for the SAR. From the SAR, the payload is handed to the ATM layer for the 5-byte header generation and mapped to a physical layer interface using the ATM UNI. The DXI modes 1 A and B are shown in Figure 13-14.

DXI Protocol Mode 1A

The DXI protocol mode 1A is synonymous to what occurs with SMDS (802.6 protocols). The DTE service unit will be 9,233 bytes of actual data. Around the data will be the typical HDLC frame formatting. The frame starts with a 1-byte opening flag. Following the opening flag, is a 2-byte DXI address. Next comes the data (up to 9,232 bytes). The frame check sequence (or CRC), which is 2 bytes long, follows. The closing flag is 1 byte long.

From the DTE, the service data unit will then be passed to a DSU through a physical interface where the ATM conversion process takes place.

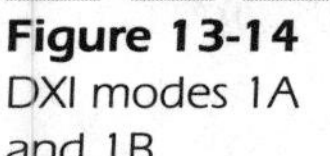

Figure 13-14 DXI modes 1A and 1B

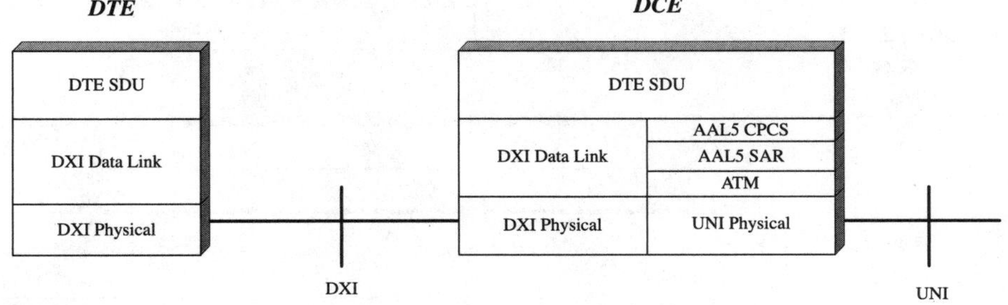

DXI mode 1 A/B simple & efficient encapsulation of SDU

- Mode 1A supports AAL5
- Mode 1B supports AAL5 plus AAL3/4

At the DSU network, translation information and the entire service data unit is passed through the AAL5 common part convergence sublayer. This transfer will become an AAL5 *Protocol Data Unit* (PDU). The AAL5 SDU is then broken down at the SAR layer into 48-byte payloads (SAR PDUs). These SAR PDUs will be mapped into the ATM layer where the header is generated. The DXI information in the original HDLC frame will be mapped both to and from the ATM VPI/VCI. This address mapping will help to keep everything in order. Using the mode 1A, up to 1,023 addressable devices can be used. The AAL5 layer is the easiest of the DXI protocols that can be accommodated (see Figure 13-15).

DXI Protocol Mode 1B

As noted earlier, the DXI mode 1A supports the AAL5. The mode 1B supports AAL3/4 and AAL5 (see Figure 13-16).

The same process takes place at the DTE. First, a DTE SDU will be created. Next, the SDU will be passed into an AAL3/4 CPSC. This layer will be framed and put across the data link layer at the DXI interface and mapped onto the physical layer.

At the DCE, the data is passed to the DXI data link layer. Then it is mapped to the router as AAL3/4 or AAL5. The data will be handed down into the SAR layer, passed to ATM, and finally processed out across the UNI to the ATM network (see Figure 13-17).

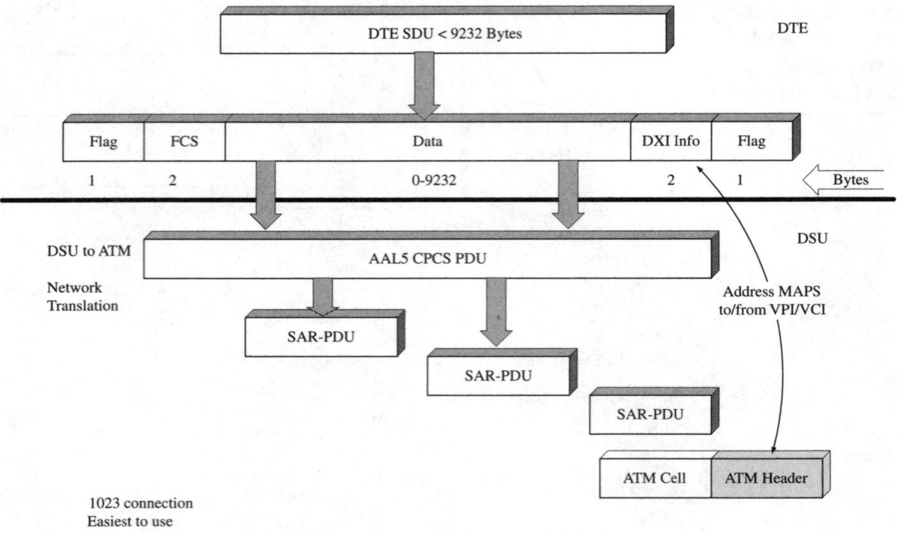

Figure 13-15 DXI protocol mode 1A

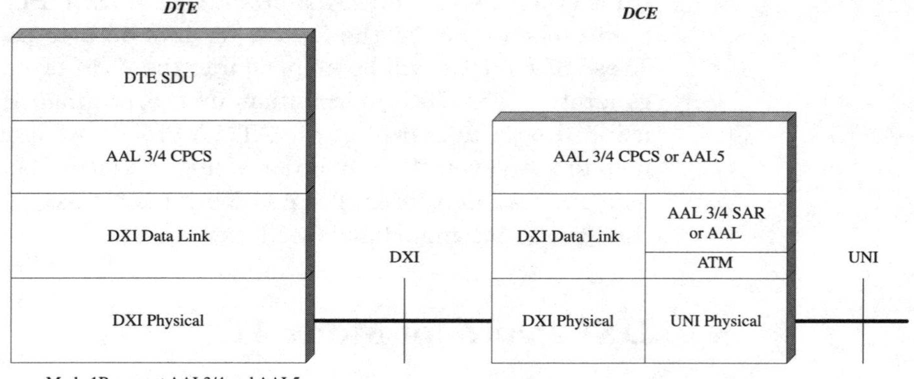

Figure 13-16 DXI mode 1B

DXI Mode 2

The second way of handling the DXI is to use the protocol mode 2. DXI is used to provide connectionless ATM network services, but is still important. The difference between mode 2 and mode 1A and B is that mode 2 can support up to 16,777,000 virtual connections either for AAL5 or AAL3/4 services. A DTE SDU of up to 64 KB long and a frame check sequence are all that are used for the SDU. The architecture of mode 2 uses the SDU from

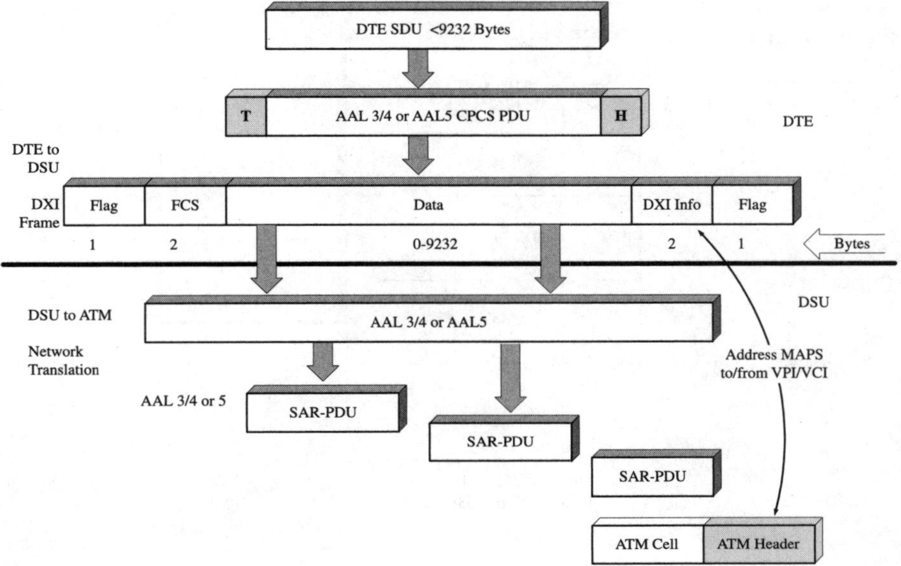

Figure 13-17 DXI protocol mode 1B

the DTE passed directly to the AAL CPCS and then framed into a DXI data link layer and passed to the DXI physical interface. This mode supports more virtual path identifiers than the normal mode 1. Entering the router, the physical interface at the DXI level maps the SDU into the DXI data link and passes to the AAL3/4 CPCS. The data is passed to the AAL3/4 SAR. Now the SAR is processed to the ATM layer where the header is generated and the data is mapped onto the UNI. The mode 2 SDU supports 64 KB compared to mode 1, which supports 9,232 bytes. Refer to Figure 13-18 for a representation of the DXI mode 2.

DXI Protocol Mode 2

The DTE processes the protocol SDU of 64 KB. The DTE will provide AAL3/4 encapsulation of the entire 64 KB, adding both header and trailer information, creating the AAL CPCS PDU. Now the PDU is mapped to the DSU as a DXI frame. The AAL3/4 CPCS PDU is then encapsulated as an HDLC frame containing the DXI information field after an opening flag. Note that the DXI information field is 4 bytes long this time. Greater addressing capabilities are possible for the virtual connections. Using a 32-bit address, we can support 16,777,000 addresses.

Figure 13-18 DXI mode 2

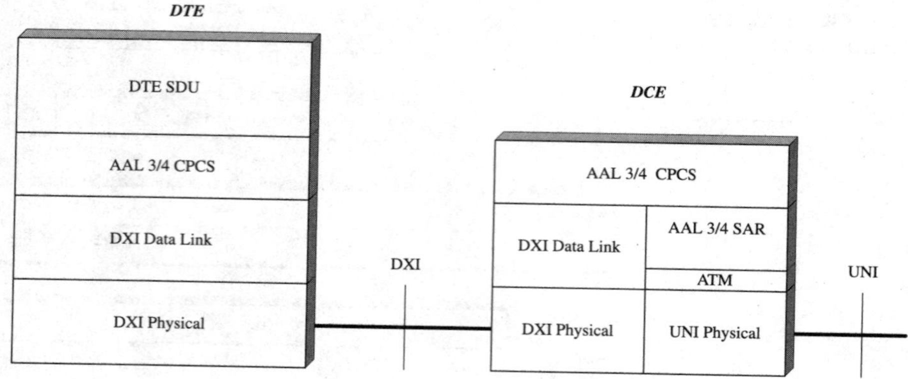

- Supports more connections (16, 777, 215 vs. 1023
- More VP (256 vs. 16)
- Larger SDU (65, 536 vs. 9232)

Figure 13-19 DXI protocol mode 2

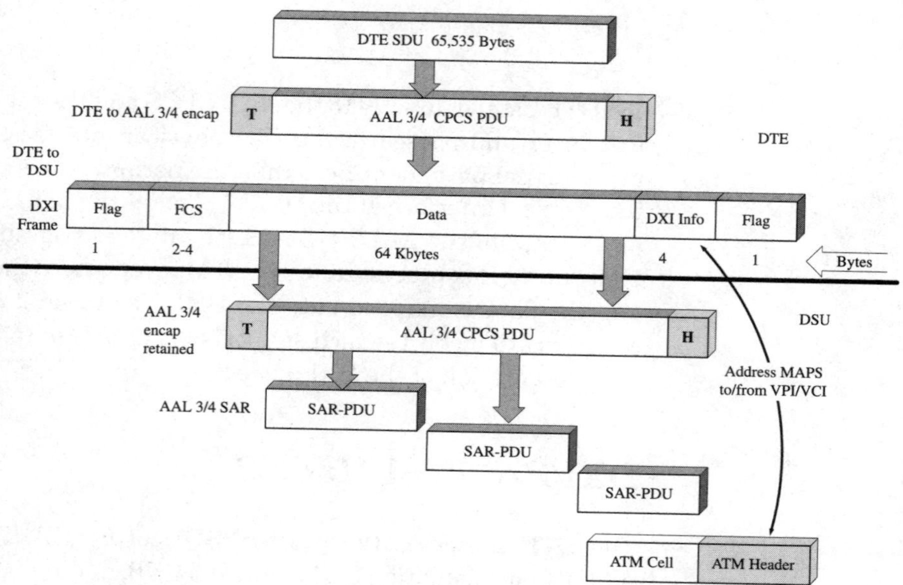

Now the data will be framed. From there the frame check sequence and the closing flag are added. At the DSU, the protocol data unit will then be mapped using an encapsulation technique. The encapsulation is retained for the PDU and then processed into SAR PDUs and onto the ATM portion of the DSU. Once again, the DXI addressing information is mapped both to and from the VPI/VCI within the ATM header for delivery across the ATM network. This is shown in Figure 13-19 as the data is prepared in mode 2.

Summary

Frame Relay and ATM can interwork and interoperate through several different techniques, which gives the carrier a sense of comfort. Because the legacy systems of the past can still be accommodated, the Frame Relay investments made in the early 1990s are still viable, and the ATM investments will be around for some time to come. Through the ATM protocols such as FUNI, end users have comfort knowing their networks are not obsolete. By using the DXI interfaces and protocols, the older Frame Relay routers, bridges, and brouters can still be used in an ever-changing network environment. One can now see why the interworking functions from the networking and service functionality are so important. Millions of dollars of investment can still be used, and newer protocols can be deployed without making the entire network obsolete. This is what internetworking is all about!

CHAPTER 14

Cable TV Systems

The television broadcast signal, regardless of the standard used, is one of the most complex signals used in commercial communications. The signal consists of a combination of amplitude, frequency, phase, and pulse modulation techniques all on a 6 MHz channel with a single sideband transmission process called *vestigial sideband* (VSB).

Cable TV appeared in the industry during the early 1960s. The initial networks installed used a basic tree architecture, in which all signals emanated from the head-end location and were distributed to individual subscribers via a series of main trunks (trees), subtrunks (branches), and feeders (twigs), as shown in Figure 14-1. This topology requires analog amplifiers to periodically boost signals to acceptable levels based on the service area being covered. However, all the benefits of solving the gain/loss problems were offset by the introduction of noise and distortion directly attributed to the amplifiers. Analog amplifiers, as noted in any communications discussion, do nothing to eliminate noise (such as cross talk, white noise, *Electromagnetic*, and *Radio Frequency Interference* [EMI/RFI]).

In early 1988, the *Community Antenna Television* (CATV) companies discovered that fiberoptic cables could be used as a means of improving the cable infrastructure both in quality and in capacity. The initial deployments

Figure 14-1 The CATV architecture

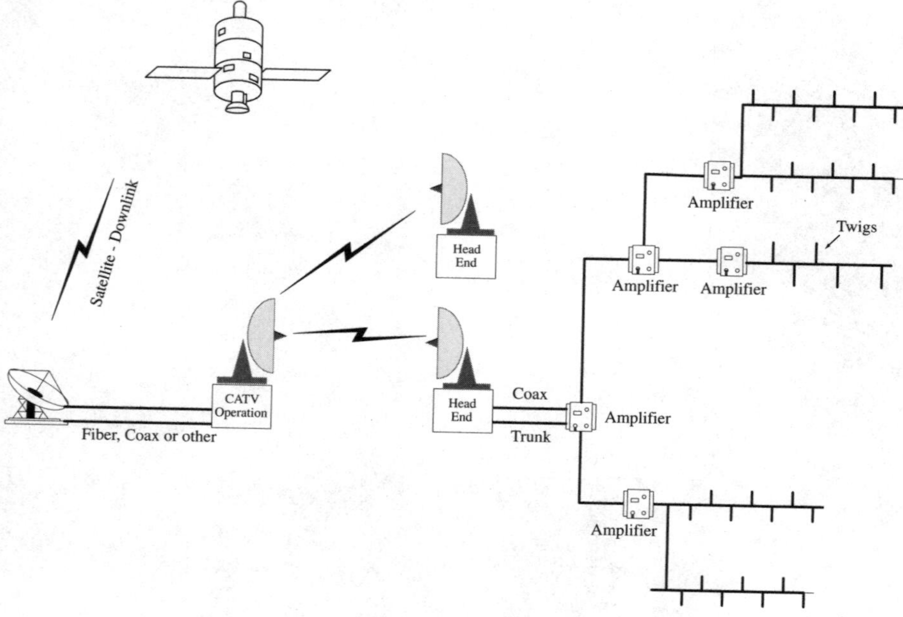

used a *Fiber-Based Backbone* (FBB) overlay placed on top of the existing tree networks to do the following

- Improve performance
- Reduce cascading amplifier problems
- Increase reliability
- Segment systems into smaller, regional areas
- Facilitate targeted programming
- Improve upstream performance

These *Hybrid Fiber Coaxial* (HFC) networks drive the fiber closer to the consumer's door. They still use the conventional tree architecture, which branches off at the *last mile* (even if it is only a few hundred feet, the reference to the last mile still prevails in the communications business) from the node to the subscriber. Unfortunately, amplifiers may still be placed inefficiently, amplifier cascades are often longer than necessary, and active counts per mile may be higher than needed.

Each of these factors increases the initial investment costs without improving the reliability of the network or reducing the ongoing operating costs. Over the past decade, CATV networks have migrated from the tree architecture, to the FBB, to the current HFC platform. Comparing the number of nodes during this period of evolution, the industry has reduced node servicing from 5,000 to 20,000 homes per node to approximately 500 homes. This 10-year migration illustrates the rapid advancements that have taken place in the technology/cost structure.

Many CATV providers believe a balance between cost and service capacity at 500 homes per node is acceptable. However, if subscription rates explode, a CATV company's platform must be easily adjustable to fewer nodes. In light of the initial investment costs of a system upgrade/rebuild, operators must continually consider a system that will improve reliability for the end user, while reducing the initial costs and ongoing operating costs.

In the past, the CATV industry has designed systems uniformly; whether design began at the head end or at the node site, the coaxial portion of the system maintained the same look and feel of a tree and branch. This usually meant optimizing the design through trial-and-error methods of cascading devices to reach all areas of the system. This conventional approach to design was based on proven design techniques, which were developed prior to the broader range of design products available today.

The advent of fiber, combined with innovative new amplifier products and the need to consider future service capabilities (such as data and Internet access) and optimal costs, is the driving force behind the development of alternative design methods.

Cable Television Transmission

Television signals that are broadcast over the air in the United States are transmitted in 6 MHz channels that are allocated to broadcasters by the *Federal Communications Commission* (FCC). The FCC regulates the location, power, and frequencies used by television stations, ensuring that stations that use the same channel are sufficiently far apart so that they do not interfere with one another in any of the areas where they may be received. But preventing interference may also require that two stations do not use adjacent channels in the same area for a more subtle reason. In the air, broadcast signal strength falls off rapidly with the distance from the transmitter, and as a result, the signal from a nearby transmitter can be several orders of magnitude stronger than that from a distant transmitter.

Because transmitters cannot contain their signal perfectly within their designated bandwidth and because receivers cannot perfectly discriminate between signals from adjacent channels, a strong nearby signal can interfere with a weak distant signal. This is known as the near-far problem, and the result is that it is not always practical to use adjacent television channels in one area.

Television signals delivered over traditional cable television networks are sent the same way as they are over the air: by dividing the cable spectrum into 6 MHz channels of bandwidth and modulating each television signal into one channel. (This is an example of *frequency-division multiplexing* [FDM].) But these cable systems can carry many more channels than broadcast television for two reasons.

First, the near-far problem of broadcast television is not a problem on cable systems because all channels can be transmitted at the same power level throughout the network, enabling adjacent channels to be used on the cable.

Second, cable systems are not limited to the bandwidth that is designated by the FCC for broadcast television; a cable can carry as many channels as the infrastructure will permit, which in modern systems can be one hundred channels or more.

Because the transmission, or encoding, of analog television signals is done in the same manner as it is for broadcast television, receiving these signals is straightforward. But a television that is not built specifically to receive cable will require an external receiver because a wider range of frequencies is used on cable systems and sometimes because a television receiver cannot properly discriminate between adjacent channels. This external receiver (a set-top box) retransmits a selected channel to one that the television can receive. Modern televisions, however, can often tune cable channels directly.

The Cable Infrastructure

The coaxial cable and broadband amplifier technologies define the essential capabilities of cable networks. But we also need to understand how real cable systems are actually built out of these and other components such as optical fiber since this introduces both technological and economic constraints on using cable to support other communication applications. Although cable systems in the United States were built independently by a variety of cable companies and equipment providers, the infrastructure is similar enough from system to system that we can talk about cable systems in general terms. Much of what we need to understand is a matter of terminology used to describe different parts of the network.

CATV has always been considered a one-way transmission system, designed to deliver TV and packaged entertainment to the residential marketplace. As the development of the architectures rolled out across the country, the CATV business, like other utility functions, became a monopoly. The industry enjoyed the freedom to offer products and services packaged to serve a local community. As a utility, the CATV companies were granted exclusivity. There were few demands placed on the cable systems because the providers delivered a sufficient number of channels to the consumer based on the coaxial technology used. The typical operator uses a 6 MHz channel capacity that is then filtered to prevent overlapping signals from interfering with other TV channels as shown in Figure 14-2. The channel filters then limit the overall capacity to approximately 4.2 to 4.5 MHz to carry the TV signal. This is sufficient to deliver CATV broadcast quality signals to the consumers' door for a fee.

In the event that the operator runs out of capacity, new techniques exist to offer as many as three or four channels of TV on a single CATV 6 MHz

Figure 14-2 Coaxial channel capacities at 6 MHz

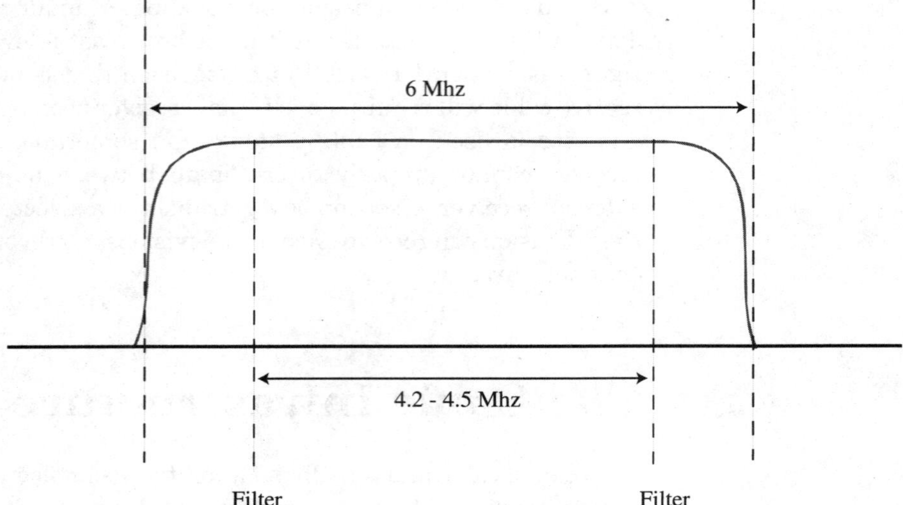

capacity through the use of *Moving Picture Experts Group* (MPEG) 2 or 3 standard compression. With a three- to four-fold increase in the channel carrying capacity on the same cable, the operators find this rewarding. This is especially true when the new MPEG standard allows the increased capacity without requiring the CATV operator to trench new fibers or coaxial cables throughout the neighborhoods.

Once the cable is channeled into the current 6 MHz techniques with the bandpass filters allowing a 4.5 MHz capacity, the carriers use the sideband capacities for other operations:

- Filter guard bands (see Figure 14-3) can prevent two channels from interfering with each other.
- The horizontal black bar at the bottom of the TV screen, the vertical interval, usually contains test signals used by the broadcast system for online performance tests that will not interfere with the regular programming.
- The audio information in a TV channel is a frequency-modulated carrier placed 4.5 MHz above the visual carrier at less than 1/8 of its power.

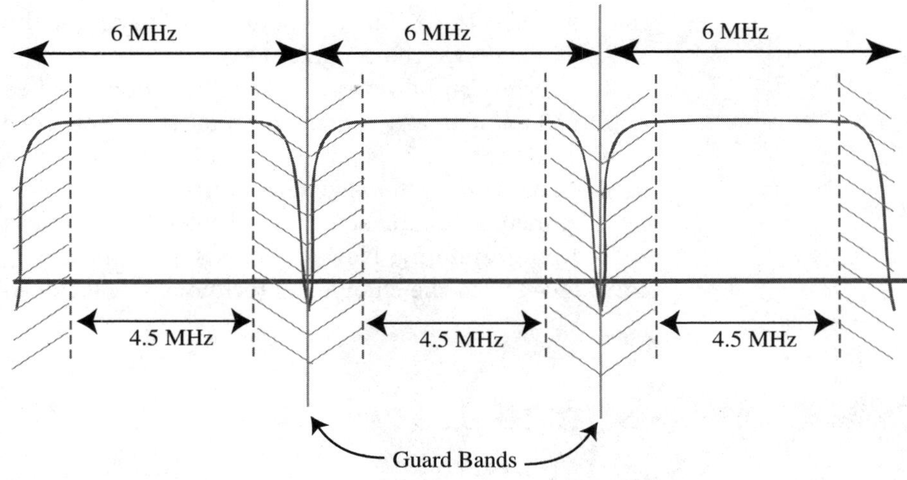

Figure 14-3 Guard bands prevent the channels from interfering with each other.

The Cable Television Distribution System

In the beginning, early television cable distribution systems were established to serve communities where residents could not receive over-the-air programming because of geographical interference. Atmospheric conditions and fading were crucial in the delivery of the airborne signals. Distance from the transmitters was also a critical factor, but may have been limited by the regulatory bodies (such as the FCC and local utility commissions). When an area could not receive the signals for whatever reason, the logical distribution of a cable to these areas made prudent business sense. Regardless of the reason, CATV was used to supplement commercial TV systems whether the consumer was

- Behind a mountain
- Out of the line of sight from the Omni directional antennae
- Beyond the reception power area of a major broadcast provider

The real issue was the monopoly that was established to provide an incentive for the cable companies to make the huge investments and attempt to service the consumer.

The term *Community Antenna Television* (CATV) has long since been extended to mean any region wired for the reception of broadcast programming, whether or not good residential antenna reception is available. Subscribers to these systems generally pay a monthly fee for the service, which usually includes increased channel selection, *pay per view*, and locally originated programming. Newer services are continually being reviewed and introduced with the changes in technology and the shift in regulation.

Signal Level

The cable system requires certain parameters in order to function properly. Originally distributed via a microwave distribution (still done in some cases) and coaxial cables, the systems are under constant scrutiny and change. Throughout the CATV system, power is distributed in the form of

- TV and FM carriers
- Test signals
- Pilot tones
- DC power
- Noise

Specified levels must be maintained at each point in the system to assure good performance. The signal levels at different frequencies are just as important. Here is how they are measured: Cable TV uses FDM onto the coaxial cable, as shown in Figure 14-4.

This frequency division allows the provider to multiplex various channels onto a single analog carrier system and deliver the various forms of entertainment available on the cable. Not all providers are the same, but many offer more than just TV and entertainment. Some of the providers offer high-quality stereo music. Others offer information such as Dow Jones or Reuters in the guard bands that travel along the cable. If the end user has a special receiver, the data are available. If not, then the data are not available to the individual set. Actually, the data are travelling along the cable, but fit in the black horizontal bar on the bottom of the screen. These value-added services are an additional revenue generator for the cable com-

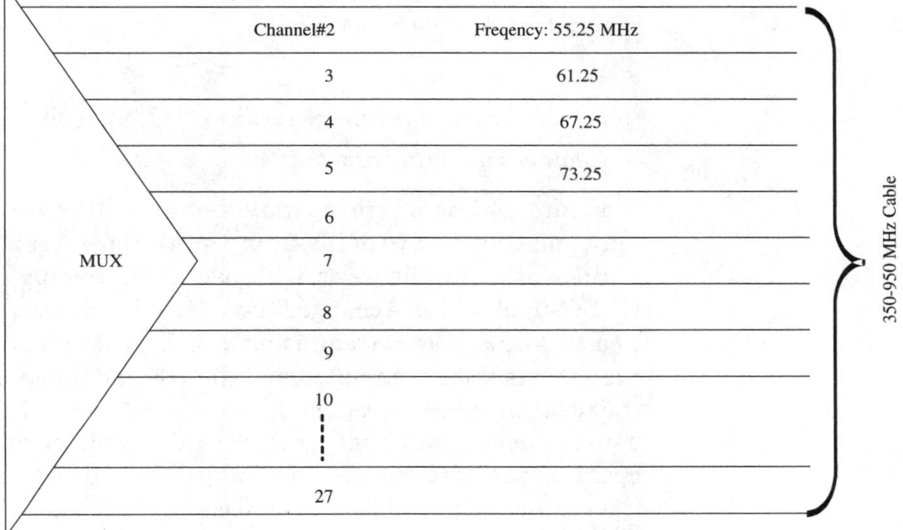

Figure 14-4 Frequency-division multiplexing (FDM) on CATV

pany. The convenience for the end user is what drives the business case. Most users who cannot receive over-the-air programming (public TV) probably have trouble receiving over-the-air radio, too!

Let's start with the cable itself. The *characteristic impedance* of distribution cable is 75 ohms (75Ω). This impedance is the amount of resistance that the cable signals "see" from the center conductor to the outer shield of the cable at the transmission frequencies. This impedance governs all the signal voltage and currents traveling the cable as covered in Ohm's Law. Most CATV system measurements involve a signal power difference that is one level relative to another. Voltage differential is an awkward measure of power differential because each time a power change is measured, the formula must *be calculated*. The *decibel* resolves difficulties in handling system power figures.

Digital Video on Cable TV Systems

Digital service is a key enabling technology that will allow cable systems to deliver a multitude of emerging services. High spectral efficiency, robust resistance to noise, and exceptional flexibility permit the installation of premium digital services, such as

- Video on demand
- *Personal communications services* (PCS) telephony
- Commercial data transport

Figure 14-5 is a representation of a multiservice cable operation that may well be the wave of the future for the operators.

Shown in this figure is the basic cable coming to the residence. The CATV suppliers have changed their basic architecture from a one-way cable to a two-way cable system using a FBB. At the local hub along the route, the cable is then terminated and the HFC equipment delivers the cable to the door interface. At the home (or office), the cable serves the telephony, entertainment, and high-speed data demands of the end user. The cable operator provides the network interface, and the ancillary equipment is then hung off the cable. Subscribers already view these new capabilities with new expectations of high value and more reliable service.

As telephone companies (the *local exchange carriers* [LECs]), long distance companies (the *interexchange carriers* [IECs]), and the cable companies compete to deliver digital services, a key differentiation will be the

Figure 14-5 CATV services of the future

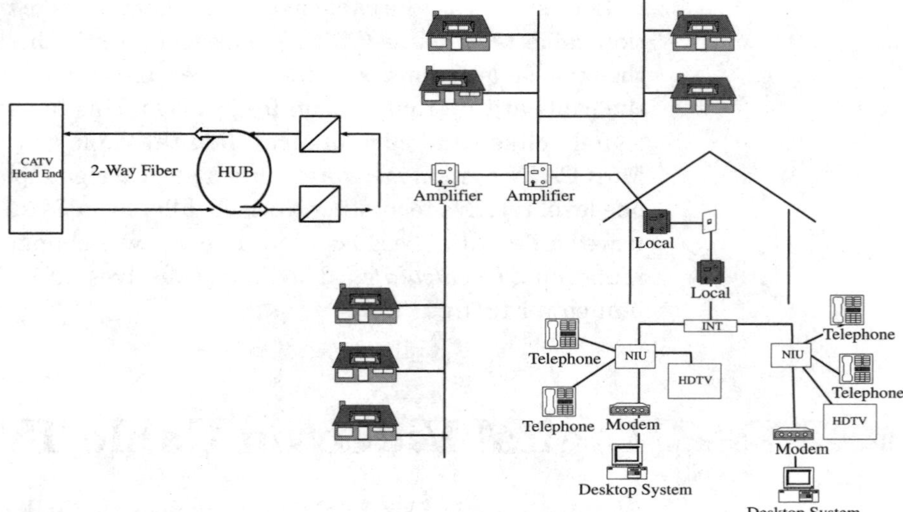

quality and reliability of service. Recent acquisition and merger activity shows the IEC and CATV merger taking off as a means of getting to the consumer's door. Ensuring *quality of service* (QoS) requires testing digitally modulated signals. Digital services will revolutionize the way consumers view their CATV suppliers. As in an analog cable TV system, power and interference measurements are essential to maintaining digital cable TV services. Although the effects are different from impairments on an analog television signal, amplifier compression and spurious interference will degrade digital signals.

Forming a Digital Video Signal

In CATV, all digital modulation formats define how the data bits correspond to both carrier phase and amplitude, and how transitions between bits are made. Digital modulation formats often associated with cable systems are 16 VSB, 64 *Quadrature Amplitude Modulation* (QAM), *quadrature partial response* (QPR), and *Offset Quadrature Phase Shift Keying* (OQPSK). However, there has been much discussion about 256 QAM for the future implementation of broadband communications services. Regardless of the technique, these formats share common characteristics in the time and frequency domains, driving common needs for measuring power and interference. Cable digital audio services are often delivered to the head end by using the OQPSK format.

Key Features of Digital Modulation

In the frequency domain, digital modulation produces a noise-like spectrum whose bandwidth depends on the symbol rate, coding, and filtering used. The spectrum of unfiltered 64 QAM is 4.167 Msymbols per second. This produces a 25 Mbps digital data stream. The bandwidth of the filtered 64 QAM channel is approximately 4.2 MHz. Note that the broadband digital video signal is susceptible to spurious interference across the entire channel bandwidth, making control of analog and digital transmission spurs critical. Proper symbol filtering avoids spilling interference from a digital channel into adjacent video channels. 64 QAM uses six bits (2^6) and produces the technique discussed here. With 4.167 Msymbols per second and six bits per symbol, then the resultant yield is (4.167×6 bits $= 25$ Mbps).

DTV Solution Introduction

November 1, 1998, marked the beginning of a new era in the broadcast industry. After 10 years of R&D and standards development, the *Digital TV* (DTV) revolution entered its implementation stage. Broadcasters are still gearing to plan as well as building up their DTV facilities. The broadcasters actually completed their implementation in the 10 largest cities three days early. Approximately 13,176 pioneers (first purchasers) had acquired DTV sets at a roaring $7,000 apiece. The broadcasters implemented the standard based on the digital HDTV systems developed by the Grand Alliance. The *Advanced Television Systems Committee* (ATSC) standardized this specification. It consists of three subsystems:

1. Source coding and compression
2. Service multiplex and transport
3. *Radio frequency* (RF) transmission

Although this has met with some success, more has to be done with it. For example, antennas for DTV surfaced to the forefront of the discussion when everyone realized that Digital TV is an all or nothing proposition. Analog TV fades and gets snowy from distortion, distance, atmospheric conditions, and other issues. DTV is either there or not! If a user has a problem with reception, then a new antenna may be needed. If not enough bits make it to the set, the picture will either freeze or disappear altogether. Anyone with a satellite transmission (for example, Direct TV or others) knows what this is like.

Cable TV providers should be able to overcome this problem with the cable itself, so long as their receivers are properly tuned. This may drive more customers to the CATV operators for their HDTV and DTV needs. The major networks will be broadcasting some of their programming by using DTV channels. The cable companies are not mandated (yet) to carry DTV broadcasts. But if they choose to do so, at issue is at what resolution they will carry and deliver DTV. DTV uses an 8-VSB signal, so the QAM used by some systems may have difficulty passing through the 8 VSB. Some incompatibilities may exist for the short term.

The source coding and compression deal with bit-rate reduction of video and audio. The compression layer transforms the raw video and audio samples into a coded bit stream that can be decoded by the receiver to recreate the picture and sound. The video compression syntax conforms to the MPEG-2 video standard, at a nominal data rate of approximately 18.9

Mbps. The Dolby AC-3 audio compression is used in the ATSC DTV standard to provide 5.1 channel surround sound at a nominal rate of 384 Kbps.

The service multiplex and transport layer based on the MPEG-2 Systems Standard provides for dynamic allocation of video, audio, and auxiliary data. It utilizes a layered architecture with headers/descriptors to provide flexible operating characteristics. The flexibility of the multiplex and transport layer provides the means for multiple *Standard Definition Television* (SDTV) services. The cable operators can send a single channel of HDTV programming or use a lower resolution SDTV and split the channel to simulcast multiple programs, including data transmissions. This all depends on which digital picture format the broadcaster uses and how the station allocates bits in the data transport rate of the channel. Currently, the data rate of a 6 MHz channel is rated at 19.4 Mbps.

The flexibility of the ATSC system specifications could allow a broadcaster to mix different streams of high and low data densities and transmit them simultaneously. An example of this may be to use 8 Mbps for a sporting event and have three lower-rate programs at 3.8 Mbps each (such as a newscast or a soap opera). At the same time, all four programs will be very unlikely to use all the allocated bandwidth, so a statistical *time-division*

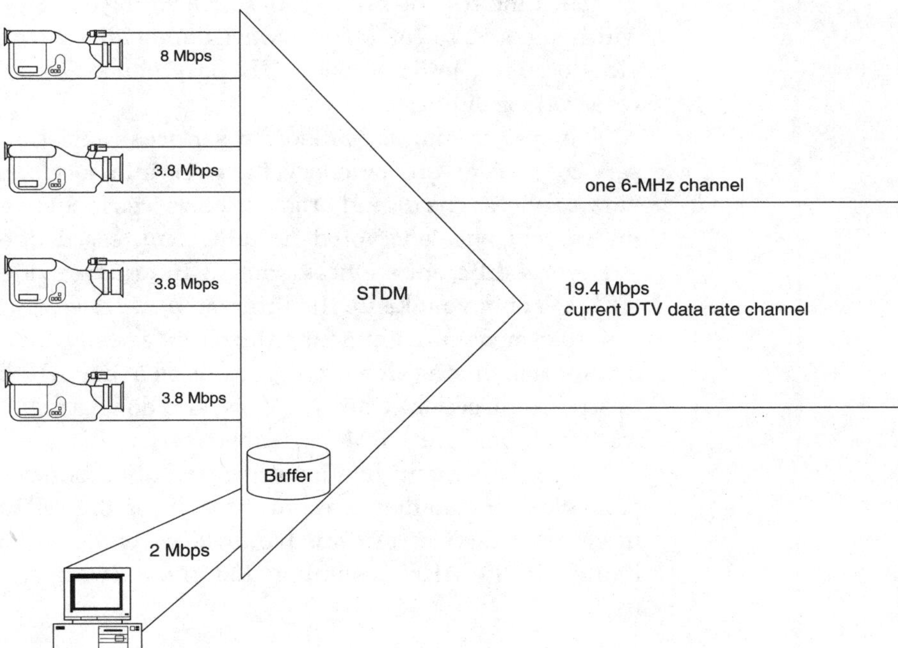

Figure 14-6 Combining multiple streams on a single DTV channel

multiplexing (TDM) technique can use the remaining bits for data transmission (see Figure 14-6).

The transmission layer modulates a serial bit stream into a signal that can be transmitted over a 6 MHz television channel. The transmission system is based on a trellis-coded, 8-level VSB modulation technique for terrestrial broadcasting.

ATSC standard-based encoding systems are one of the key elements in DTV implementation. Encoding systems will be used in the entire broadcast chain. However, not every encoder in a DTV broadcast chain has to be ATSC standard based. It should be noted that the FCC standardizes only the terrestrial-broadcasting signal. Cable operators are not required to adapt to this standard. In addition to the source coding, compression, and multiplexing, an encoding system provides ATSC standard-based systems information, program guide, data, and interactive services along with video and audio.

The ATSC DTV standard, as well as MPEG 2, describes the bit stream syntax and semantics. The standards also specify the constraints and decoder models. However, encoding parameters are not specified by the standards. Thus, encoder performance and systems implementation are left to encoder designers. Thus, standard compliance does not guarantee encoder performance.

In a standard 750 MHz system, consisting of a 5 to 40 MHz return bandwidth and a 52 to 750 MHz forward bandwidth, downstream broadcast services occupy the 52 to 550 MHz pass band. The remaining 200 MHz is reserved for digital services.

The system's analog broadcast services consist of analog nonencrypted services (basic), analog encrypted premium services, and analog pay-per-view services. The digital broadcast services include digital audio programming, along with encrypted digitally compressed video programming.

Several different services, such as interactive video, cable modems, and HFC telephony make up the interactive services.

Cable modem equipment utilized a two-way RF system with 64 QAM modulation in the downstream direction and QPSK modulation in the upstream direction. Data rates for the downstream and upstream paths were 27 Mbps and 1.7 Mbps, respectively.

Using the system bandwidth information shown earlier, the available bandwidth in the downstream direction is 698 MHz. Both broadcast and interactive services use this bandwidth. Available upstream bandwidth is limited to 35 MHz, assuming that the entire 5 to 40 MHz bandpass is useable.

Migration strategies for the network must be carefully planned. Selection of downsizable architecture is crucial in this planning process. Also, conducting detailed traffic studies about these services when they are first offered in the network is crucial.

CATV operators are currently in the limelight and will remain there for some time to come. The emphasis on access to the door, deregulation, and merger mania in this industry has propelled the CATV operators into one of the driver seats of technology. Whether they choose to adjust to their newfound popularity or be gobbled up by new players looking for access depends on the individual cable operators across the country (and the world).

CHAPTER 15

Cable Modem Systems and Technology

In the late 1970s, a major battle arose in the communications and the computer industries. Convergence of the two industries was happening as a result of the implementation of the *Local Area Networks* (LANs). In the local networking arena, users began to implement solutions to their data connectivity needs within a localized environment. Two major choices were available for their installation of wiring: baseband coaxial cable and broadband coaxial cable.

The baseband cable was based on the Ethernet developments using a 20 MHz, 50Ω coax. Designed as a half-duplex operation, Ethernet allowed the end user to transmit digital data on the cable at speeds of up to 10 Mbps. Clearly, the 10 Mbps was maximum throughput, but was attractive in comparison to the technology of twisted pair at the time (telephone wires were capable of less than 1 Mbps bursty data). Moreover, the use of the baseband technology allowed the data to be digitally applied directly onto the cable system. No analog modulation was necessary to apply the data. It was dc input placed directly onto the cable. The signal propagates to both ends of the cable before another device can transmit. This is shown as a quick review in Figure 15-1. To control the cable access, the attached devices used *Carrier Sense Multiple Access with Collision Detection* (CSMA/CD) as the access control. CSMA/CD allowed for the possibility that two devices may attempt to transmit on the cable at the same time, causing a collision and corruption of the actual data. As a result, the cable had to be very controlled, but in the late 1970s this was not a real issue.

A second alternative was to use the broadband coaxial cable, operating at a total bandwidth of approximately 350 MHz in an analog FDM technique

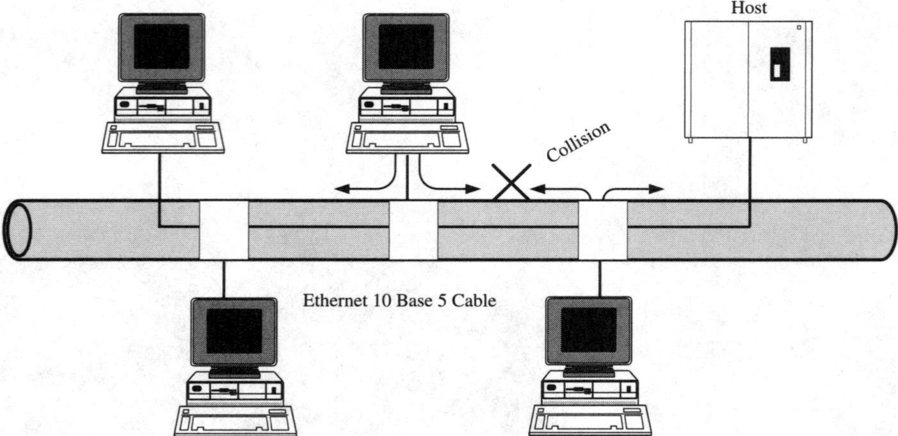

Figure 15-1 CSMA/CD cable networks are collision domains.

on a 75Ω cable. As a technology, the broadband systems were well known because they were the same as *Community Antenna Television* (CATV), which had surfaced in the early 1960s. As a result, the technology was well deployed and commodity priced. Moreover, the 350 MHz of capacity was attractive to the computer industry and the communications industry partisans. The issues began to surface quickly regarding the benefits and losses of using each technique (see Figure 15-2).

What the issue really boiled down to was one of analog versus digital and the baseband versus broadband implementations to achieve this goal. One world-renowned consultant and research house in the New England area of the United States even said that people who installed a broadband cable to provide their LAN services did not know what they were doing and were wasting their company's money! This was a hot issue throughout both industries. In reality, the issue of using a broadband cable was under the turf of the voice communications departments, whereas the baseband cables were under the primary control of the data processing/data communications departments within corporations. If one technology was chosen over another, the lines in the sand would be washed away, and the convergence of voice and data would force the convergence of the two groups.

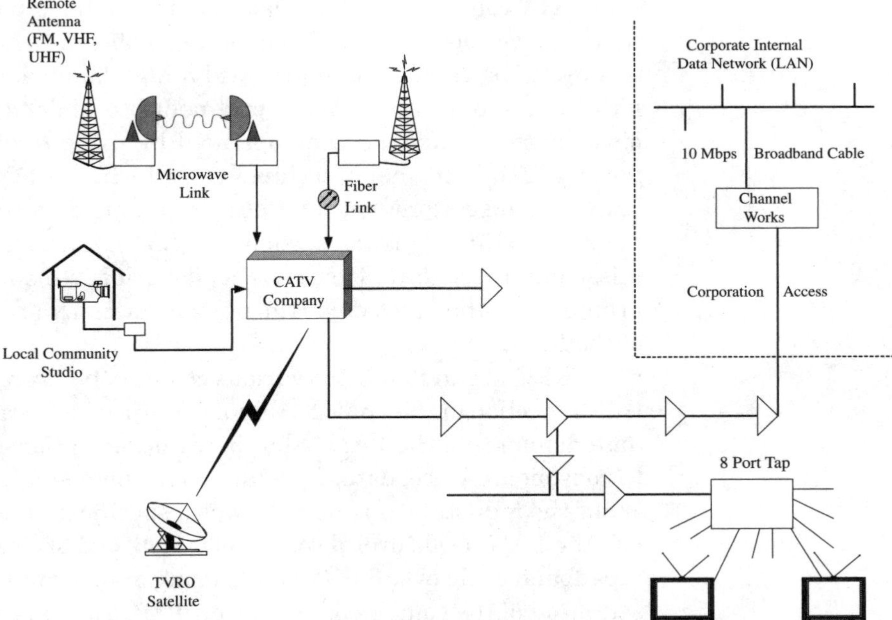

Figure 15-2 Broadband coaxial cable system from the beginning

The issue was therefore not whether to use a cable, but which type of cable to be used so that the LAN would fall under the correct jurisdictional authority within the organization. Unfortunately, control is not the goal of organizations, but access and profitability are! As an industry, too much time was wasted over semantics. However, what ultimately rolled out of the bandwidth argument was that the baseband cable systems were better for the LAN. This was the decision of the 1980s, when all traffic on the LAN was geared to data only at speeds of 10 Mbps and less.

Cable TV Technology

As discussed in Chapter 14, CATV has been around since the early 1960s. It is a proven technology. In the early days of Ethernet, *Digital Equipment Corporation* (DEC) rolled out many of their systems using baseband (Ethernet) cables. However, recognizing that some organizations needed more than just data on a large localized network, they worked with two major providers at the time to develop the interfaces for the broadband cable systems to attach an Ethernet to the CATV cable.

DEC developed several working arrangements with various suppliers to provide a *Frequency Agile Modem* (FAM) to work on the cable TV systems. The CATV companies did not necessarily own the broadband cable. Instead, this cable was locally owned in a high-rise office or a campus complex by the end user. The cable system provided a high bandwidth, but was very complex for the data and LAN departments to understand. The reason was obvious: the broadband coax operated by using *frequency division multiplexing* (FDM, or analog techniques), which was beyond the scope of the LAN administrators and the Data Processing departments. The voice people knew of analog transmission, but had a hard time with digital transmission in those days. There was a silent department in the crux of all the arguments—the video departments within many organizations stayed out of the fight.

As DEC began to roll out various choices, the average user had to justify the connection of the analog technologies (used as a carrier) with the digital data demands of the LAN. What many organizations did on a campus was to consolidate voice, data, LAN, and video on a single cable infrastructure. What the industry came up with was a specification for 10 Broad 36 to satisfy the LAN needs over a coax cable. 10 Broad 36 stands for 10 Mbps on a broadband cable over 3,600 meters using analog amplifiers. A classic representation of the combined services on 10 Broad 36 is shown in Figure 15-3.

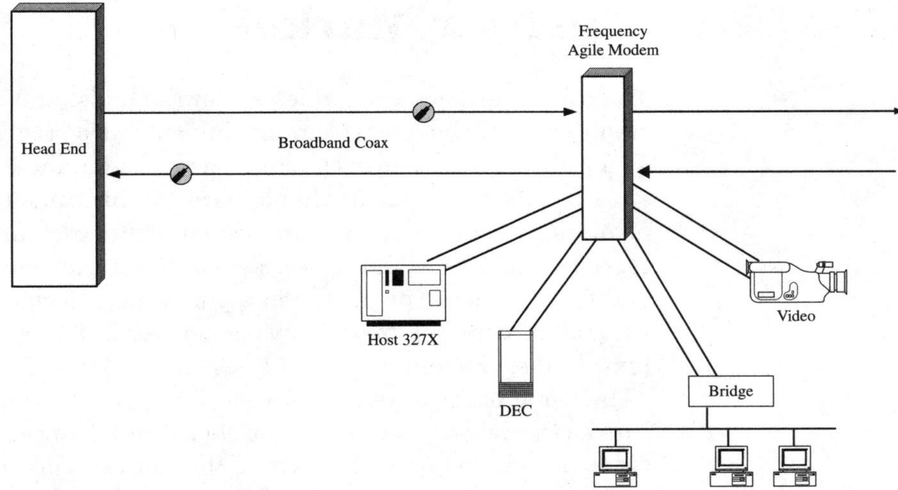

Figure 15-3 Mixing services on a 10 Broad 36 cable

The data industry was distraught because this encouraged the use of an analog carrier system to move the digital data. Over the years, however, this has been revisited several times. Wang Computer Company developed a proprietary cable system for connecting WangNet systems together by using a dual broadband coax cable. This met with only limited success because of the proprietary nature of their interfaces and the pricing model they used. Technologically, the system was sound.

Later in the evolution of this service, the term broadband LANs became popularized. Ten Mbps Ethernet grew to 100 Mbps, and then on to Gb Ethernet and now the introduction of 10 Gb Ethernet in the LAN and *metropolitan area network* (MAN). Justifying and enabling this high-speed communications service met with some resistance until the use of the various fiber and coaxial systems emerged. By taking a quantum leap in the industry, the data and voice departments saw the benefit and need of converging the two services to the desktop in order to offer voice and video over the LAN. The 10 Mbps Ethernet and coaxial cables could not handle this offering. Enter the new technology! Hybrids of coax and fiber were introduced to the desktop. Moreover, with the access to the Internet under constant scrutiny and pressure to add speed and capacity (voice and video on the Internet), the industry began to seek a new method of bypassing the local copper loop provided by the telephone companies. The technology already at the door, of course, was the CATV. So a new idea emerged: use the CATV to support the high-speed Internet access and bypass the local loop from telephone companies. Hence, cable modem technology changed the way we will do business for the future.

The New Market

The cable television companies are in the midst of a transition from their traditional core business of entertainment video programming to a position as a full-service provider of video, voice, and data telecommunications services. Among the elements that have made this transition possible are technologies such as the cable data modem. These companies have historically carried a number of data services. These services have ranged from news and weather feeds, presented in alphanumeric form on single channels or as scrolling captions, to one-way transmission of data over classic cable systems as discussed in Chapter 14, "Cable TV Systems."

Information providers are targeting upgraded cable network architecture as the delivery mechanism of choice for advanced high-speed data services. These changes stem from the commercial and residential data communications markets. The PC and LAN explosion in the early 1980s was rapidly followed by leaps in computer networking technology. More people now work from home, depending on connectivity from commercial online services (such as AOL, CompuServe, and Prodigy) to the global Internet.

Increased awareness has led to increasing demand for data service, and for higher speeds and enhanced levels of service. Cable is in a unique position to meet these demands. The same highly evolved platform that enables cable to provide telephony and advanced video services also supports high-speed data services. There appear to be no serious barriers to the cable deployment of high-speed data transmission.

System Upgrades

The cable platform is steadily evolving into a hybrid digital and analog transmission system. Cable television systems were originally designed to optimize the one-way, analog transmission of television programming to the home. The underlying coaxial cable, however, has enough bandwidth to support the two-way transport of signals. The hybrid network is shown in Figure 15-4.

Growth in demand for Internet access and other two-way services has dovetailed with the trend within the industry to enhance existing cable systems with fiber optic technology.

Many cable companies are in the midst of the upgrade to HFC plants to improve the existing cable services and support data and other new ser-

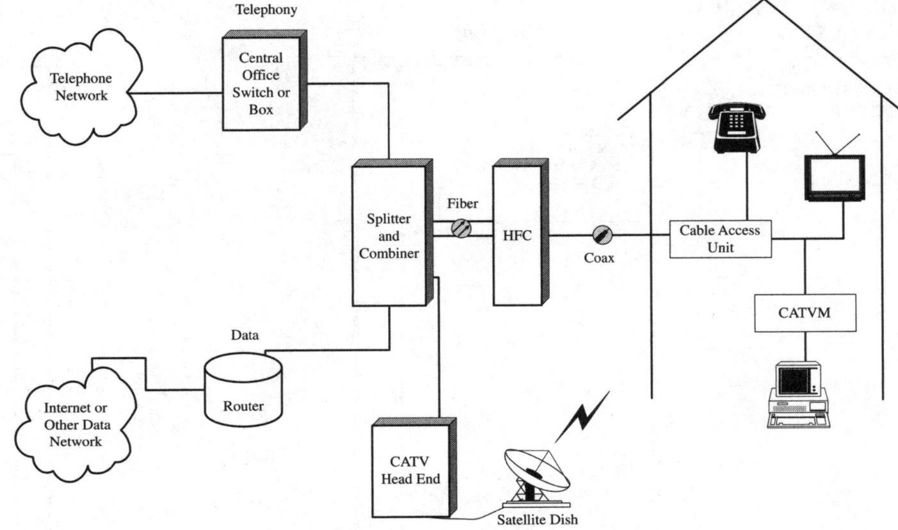

Figure 15-4 The new hybrid data network

vices. Companies are taking different approaches to online service access. For some applications, customers may be accessing information stored locally at or near the cable head-end or regional hub, such as the @Home®[1] services being offered in many cities. This may be temporary until wide area cable interconnections and expanded Internet backbone networks are in place to allow information access from any remote site.

Cable Modems

Digital data signals are carried over *radio frequency* (RF) carrier signals on a cable system. Digital data utilizes cable modems, devices that convert digital information into a modulated RF signal and convert RF signals back to digital information. The conversion is performed by a modem at the subscriber premises, and again by head-end equipment handling multiple subscribers. Look at Figure 15-5 for a block diagram of the cable modem.

[1]The reference here to @Home was the registered trademark of @Home Corporation. In 2001 @Home filed bankruptcy and the company prepared to cease operations after February 28, 2002.

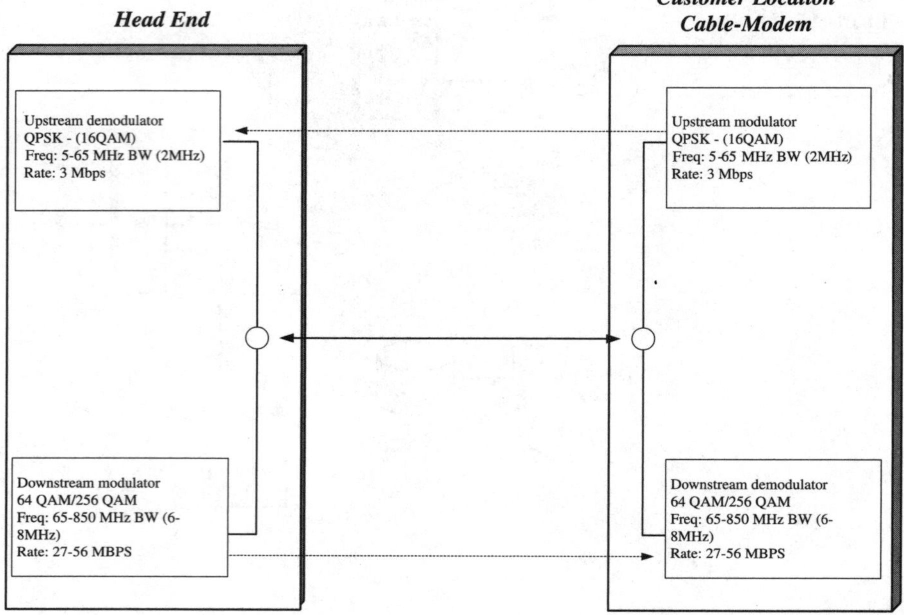

Figure 15-5 Block diagram of the cable modem

A single 6 MHz channel can support multiple data streams or multiple users through the use of shared LAN protocols such as Ethernet, commonly used in business office LANs today. This is where the industry began in the late 1970s when Ethernet networks were applied to the broadband coaxial networks.

Different modulation techniques are being tried to maximize the data speed that can be transmitted through a 6 MHz channel. Modulation techniques include *Quadrature Phase Shift Keying* (QPSK), *Quadrature Amplitude Modulation* (QAM), and *Vestigial Side Band* (VSB) amplitude modulation. Comparing the data traffic rates for different types of modems shows why the cable modem is so popular under today's environment. Table 15-1 is a comparison of a file download of 500 KB using different techniques.

Careful traffic engineering is being performed on cable systems so that data speeds are maximized as customers are added. Just as office LANs are routinely subdivided to provide faster service for each individual user, so too can cable data networks be custom tailored within each fiber node to meet customer demand. Multiple 6 MHz channels can be allocated to expand capacity as well.

Some manufacturers have designed modems that provide asymmetrical capabilities, using less bandwidth for outgoing signals from the subscriber. Cable systems in some locations may not have completed system upgrades, so manufacturers have built migration strategies into such modems to allow for the eventual transmission of broadband return signals when the systems are ready to provide such service and customers demand it. A representative sample of the way data speeds are provided on cable modems is shown in Table 15-2.

Table 15-1

Comparison of transmission speeds

Time to Transmit a Single 500 KB Image		
Telephone Modem	28.8 Kbps	6–8 minutes
ISDN	64 Kbps	1–1.5 minutes
Cable Modem	10 Mbps	Approximately 1 second

Source: CableLabs

Table 15-2

Representative asymmetrical data cable modem speeds

Sample Cable Modem Speeds	Upstream	Downstream
General Instrument	1.5 Mbps	30 Mbps
Hybrid/Intel	96 Kbps	30 Mbps
LANcity	10 Mbps	10 Mbps
Motorola	768 Kbps	30 Mbps
Zenith	4 Mbps	4 Mbps

Standards

Modems are available today from a variety of vendors, all with their own unique technical approach. These modems are making it possible for cable companies to enter the data communications market now. In the longer term, modem costs must drop and greater interoperability is desirable. Customers who buy modems that work in their current cable system need assurance that the modem will work if they move to a different geographic location served by a different cable company. Furthermore, agreement on a standard set of specifications will allow the market to enjoy economies of scale and drive down the price of each individual modem. Ultimately, those modems will be available as standard peripheral devices offered as an option to customers buying new personal computers at retail stores. The cable companies and manufacturers came together formally in December 1995 to begin working toward an open standard.

Leading U.S. and Canadian cable companies were involved in this development toward an open cable modem standard. Specifications were to be developed in three phases, and then be presented to standards-setting bodies for approval as standards. Individual vendors were free to offer their own implementations with a variety of additional competitive features and future improvements. A data interoperability specification will comprise a number of interfaces. The resultant specification is called *Data Over Cable Service Interface Specification* (DOCSIS), which architecturally is shown in Figure 15-6 as it relates to the TCP/IP protocol stack. Note that there are several sublayers added into the DOCSIS specification at the bottom layers (for example, layer 1 and 2) of the protocol stack. This is to simplify the connection and add the dimension of security into the DOCSIS specifications.

Some interfaces reside within the cable network. Several of these system-level interfaces also will be specified in order to ensure interoperability of such important functions as authentication for login/logout, ease of installation of cable modems for reliable customer activation, and spectrum management over the cable network's hybrid fiber/coaxial plant.

Return Path

The portion of bandwidth reserved for return signals (from the customer to the cable network) is usually in the 5 to 40 MHz portion of the spectrum. This portion of the spectrum can be subject to ingress and other types of

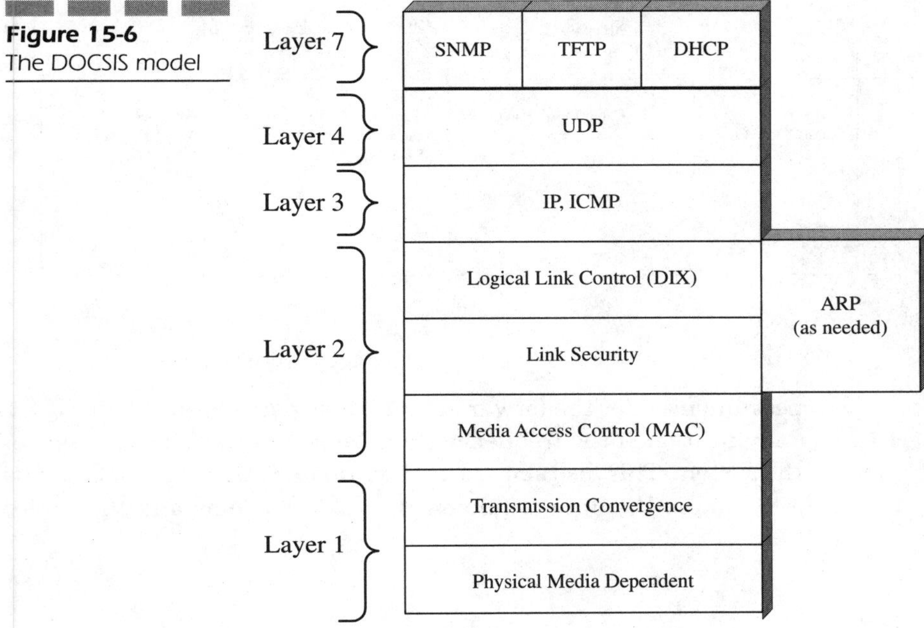

Figure 15-6 The DOCSIS model

interference, and so cable systems offering two-way data services have been designed to operate in this environment.

Industry engineers have assembled a set of alternative strategies for return path operations. Dynamic frequency agility (shifting data from one channel to another when needed) may be designed into modems so that data signals may avoid unwanted interference as it arises. Other approaches utilize a *gate* that keeps the return path from an individual subscriber closed except for those times when the subscriber actually sends a return signal. Demarcation filters, different return laser types, and reduced node sizes are among the other approaches, each involving tradeoffs between capital cost and maintenance effort and cost.

Return path transmission issues have already been the subject of two years of lab and field testing and product development. The full two-way capability of the coaxial cable already used in most U.S. homes is now being utilized in many areas, and will be available in most cable systems soon. Full activation of the return path in any given location will depend on individual cable company circumstances, ranging from market analysis to capital availability.

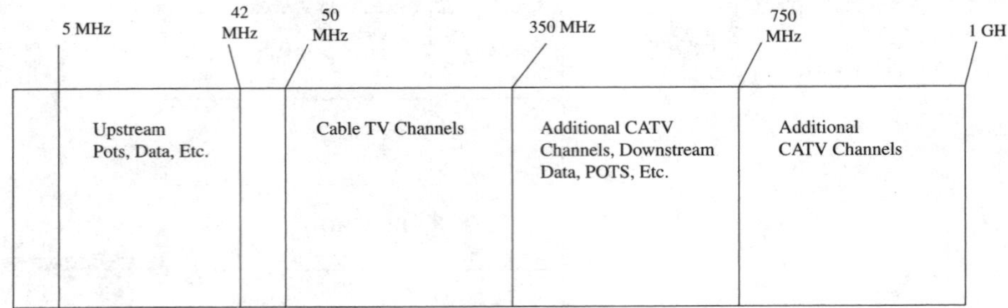

Figure 15-7 Frequency spectrum allocated to the cable modems

The spectrum used for the forward and reverse paths is shown in Figure 15-7 as an indication of the frequencies available and the overall management of the system. This also shows that additional 6 MHz channels can be set aside to handle the data traffic on the cable modems and the cables themselves.

Applications

Cable modems open the door for customers to enjoy a range of high-speed data services, all at speeds hundreds of times faster than telephone modem calls. Subscribers can be fully connected, 24 hours a day, to services without interfering with cable television service or phone service. Among these services are

- **Information services** Access to shopping, weather maps, household bill paying, and so on
- **Internet access** E-mail, discussion groups, and the World Wide Web
- **Business applications** Interconnecting LANs or supporting collaborative work
- **Cablecommuting** Enabling the already popular notion of working from home
- **Education** Allowing students to continue to access educational resources from home

The promises of advanced telecommunications networks, once more hype than fact, are now within reach. Cable modems and other technology are being deployed to make it happen. Regardless of the technology selected, the main goal is to get the high-speed data communications on the cable adjacent to the TV and entertainment. This gives the CATV companies the leverage to act in an arbitrage situation, competing with the local telephone companies who have dragged their feet in moving high-speed services to the consumer's door. As shown in Figure 15-8, there are several up-and-down speed capabilities that can be shared to deliver asymmetrical speeds to the consumer's door. In the particular figure, the download speed is up to 30 Mbps, whereas the upstream operates at 1.5 Mbps. For many, this is sufficient based on their applications.

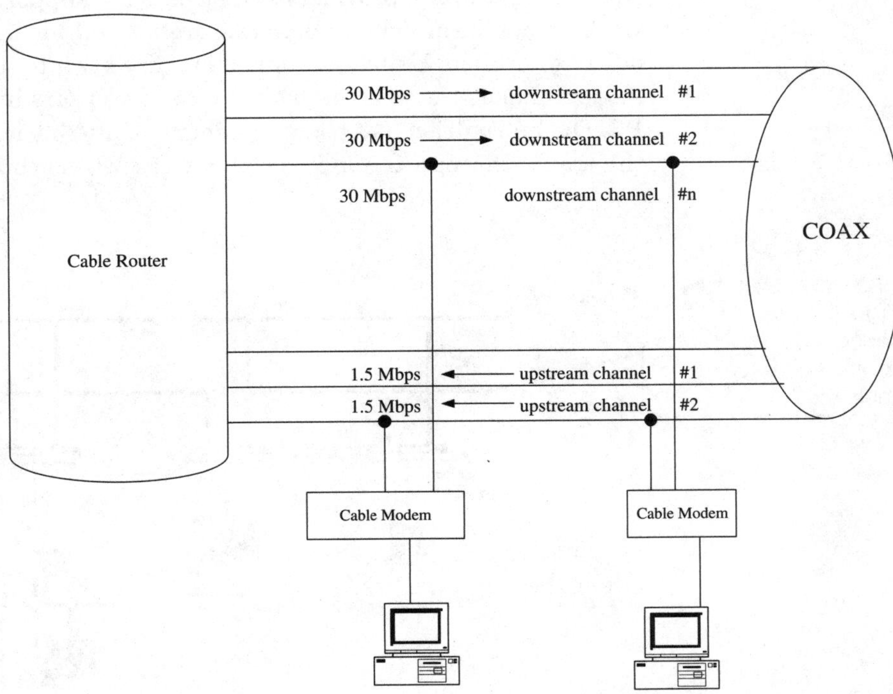

Figure 15-8 Different speeds on the up-and-down stream flows

The Combined Corporate and End User Networking Strategies

The use of a single PC on a cable system is fine for the telecommuter (or cable commuter now), but what of the small office or home office where more than a single PC is connected? Figure 15-9 is an example of various ways the CATV connection can be accomplished. This figure uses an example of local home networking with two PCs connected to a single cable modem. Most of the providers have instructions on how to accomplish this and require a home user (or small user) to download additional software to accommodate the dual connection on a single modem. The second alternative to this is to have a router connected to the cable modem, such as a *branch office router*. The network is attached to the router, and the router is responsible to handle the dispersing of the traffic onto the cable system. This can be a very effective use of the link. Next in the figure is a connection to a hub, such as a 10 or 100 Base T connection into a LAN hub. Although the CATV providers state this is not supported and will likely not work, it has been done and works pretty well for a small office or home office connection. Using the connection directly into the hub from the cable modem makes the modem available to more users instead of just a single PC. The hub will act as a bridging function onto the modem and concentrate the traffic through the individual devices. These configurations all work,

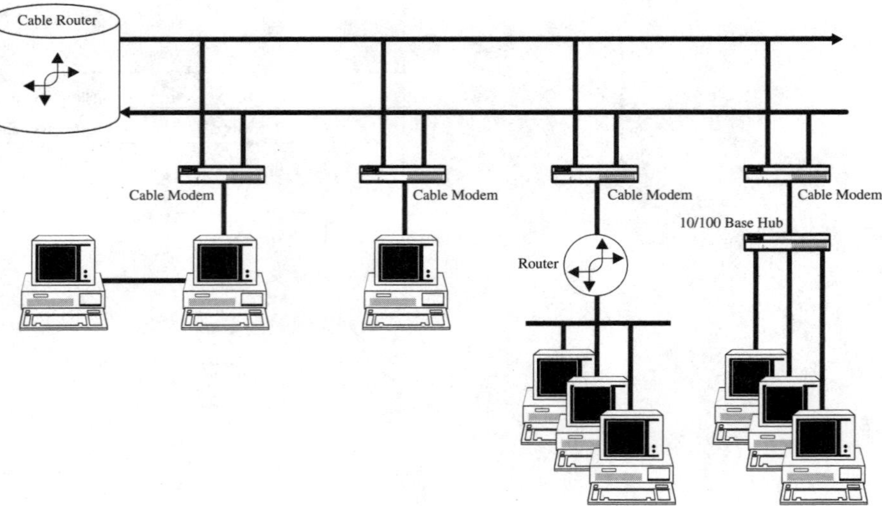

Figure 15-9 Multiple ways of connecting to the cable modem

but the providers do not support problems if they arise. You are on your own if it does not work.

A Final Thought

A final thought concerning the use of this technology is to understand the security concerns associated with the cable modem. When the CATV systems are used, they are shared, high-speed Ethernet backbone access to the Internet or other connections. One must be aware that on a shared cable, the PC is a peer to all others on the same cable, even though they are in physically different locations. In a community of 500 connections, there will be many people who acquire the service from the CATV suppliers. The CATV company installs according to the appropriate technology, not according to security parameters. This is okay because they are merely providing the bandwidth to gain access. It is the end user's responsibility to turn off all the leaks in the local system (the PC). By default, when you run a Microsoft windows environment and the appropriate networking software, the *shares* on the PC are turned on. The end user must therefore go in and turn them off. This means that if the shared services are not turned off, a user down the street, or across the town, can double-click their *network neighborhood* icon and see all the other PCs connected to the cable. Not only can they see the devices, but they can double-click the PC and see the resources available on that PC. From there when a remote device has double-clicked your PC, they can open your drives and see your files. Unless some provisions have been taken to block this access, the intruder (used as a method of entry only) can read, write, edit, or delete your files. Worse yet, while the intruder is on your system, you do not even know he is there.

Many users who have cable modem service from the cable companies are not aware of the risks. Worse, the installation personnel on these systems do not totally understand or they forget to point out the risks. Therefore, users leave the PC on 100 percent of the time (day and night), making access to their computers totally available. The cable modem is available 100 percent of the time, making the computer a target for hackers and mischievous users, without the permission or the knowledge of the penetrated computer owner. Be aware that the risk is there and find out how to shut these open doors before leaving your computer on the network. Do not assume that just because you shut your system off when you're not using it that you are protected, because when you do log on, you are exposed, and the perpetrator can get on your system while you are on it, too! Technology is a wonderful thing, but it must be controlled.

CHAPTER 16

xDSL

One of the major problems facing the *incumbent local exchange carriers* (ILEC) is the ability to maintain and preserve their installed base. Ever since the Telecommunications Act of 1996, there has been mounting pressure on the ILECs to provide faster and more correct Internet access. In order to provide the higher-speed communications abilities, these carriers have continually looked for new means of providing the service.

However, the ILECs have an installed base of unshielded twisted pair in the local loop that cannot be ignored or abandoned. Therefore, a new form of communications was needed to work over the existing copper cable plant. One of the technologies selected was the use of xDSL. The DSL family includes several variations of what is known as digital subscriber line. The lower case x in front of the DSL stands for the many variations. These will include

- *Asymmetrical digital subscriber line* (ADSL)
- *ISDN* (like) *digital subscriber lin*e (IDSL)
- *High bit-rate Digital Subscriber Line* (HDSL)
- *Consumer Digital Subscriber Line* (CDSL)
- *Single High Speed DSL* (SHDSL)
- *Rate-adaptive digital subscriber line* (RADSL)
- *Very high-bit rate digital subscriber line* (VDSL)
- *Single or symmetric digital subscriber line* (SDSL)

One can see that the variations are many. Each DSL capability carries with it differences in speed, throughput, and facilities used. The most popular of this family under today's technology is the use of ADSL.

ADSL is a technology being provided primarily by the ILECs because the existing cable plant can be supported, and the speed throughput can vary, depending on the quality of the copper. However, the most important and critical factor in dealing with ADSL technology is the capability to support speeds between 1.5 Mbps up to 8.192 Mbps. At the same time, the ILEC can also support *Plain Old Telephone Service* (POTS) for voice or fax communications on the same line. What this means is that the ILEC does not have to install all new cabling to support high-speed communications access to the Internet.

ADSL Defined

ADSL is the new modem technology to converge the existing twisted pair telephone lines into the high-speed communications access capability for various services. Most people consider ADSL as a transmission system instead of a modification to the existing transmission facilities. In reality, ADSL is a modem technology used to transmit speeds of between 1.5 Mbps and 6 Mbps under current technology. It is stated that in the future ADSL will support speeds of about 8.192 Mbps. This definition of the higher range of ADSL speeds is one that is yet to be proven; however, with changes in today's technology one can only imagine that the speeds will be achievable.

Some of the many capabilities that are being considered through the use of the DSL family are the services for converging voice, data, multimedia, video, and Internet streaming protocols services. It is through these demands that the carriers see their future rollout of products and services to the general consuming public. In Table 16-1, various theoretical speeds and distances of ADSL technologies are shown.

One should remember that the speeds and distances shown here are the theoretical limits based on good copper. If the copper has been damaged or impaired in any way, then the speed and distances will change accordingly (downward). The theoretical limits are what most engineers and providers have been touting that their technologies can support. Reality is another thing, and the actual distances and speeds very likely will be less than those shown here. What is most important is to assume that these speeds can be established and maintained on the installed base of *unshielded twisted pairs* (UTP) of wire. As long as the ILEC can approximate these speeds today, the consumer will most likely not have much to complain about.

Table 16-1

Data rates for ADSL, based on installed wiring at varying gauges

Current Data Rate	Wire Gauge	Distance in K Feet	Distance in Kilometers
1.5 to 2.048 Mbps	24	18	5.5
1.5 to 2.048 Mbps	26	15	4.6
6.3 Mbps	24	12	3.7
6.3 Mbps	26	9	2.7

Modem Technologies

Before proceeding too far in this discussion, a review of modem technology is probably in line. Modems, or modulator/demodulator, were designed to provide for data communications across the voice dial-up communications network. Through the use of modem technology introduced back in the 1960s, users were able to transmit data across the voice networks at speeds varying between 300 bps to 33,600 bps. Although this may seem like high-speed communication, our demands and needs for faster communications quickly outstripped the capabilities of our current modem services, making the demand for newer services more evident. Higher-speed modems could be produced, but the economics and variations on the wiring system prove this to be somewhat impractical. Instead, the providers looked for a better way to provide data communications that mimic the digital transmission speeds we are accustomed to.

Using the telephone companies' voice services, the end user installed a modem on the local loop. This modem acts as the *Data Circuit terminating Equipment* (DCE) for the link. (DCE can also stand for data communications equipment.) As shown in Figure 16-1, a modem is used on the ILEC's wires to communicate across the *Wide Area Networks* (WANs) such as the

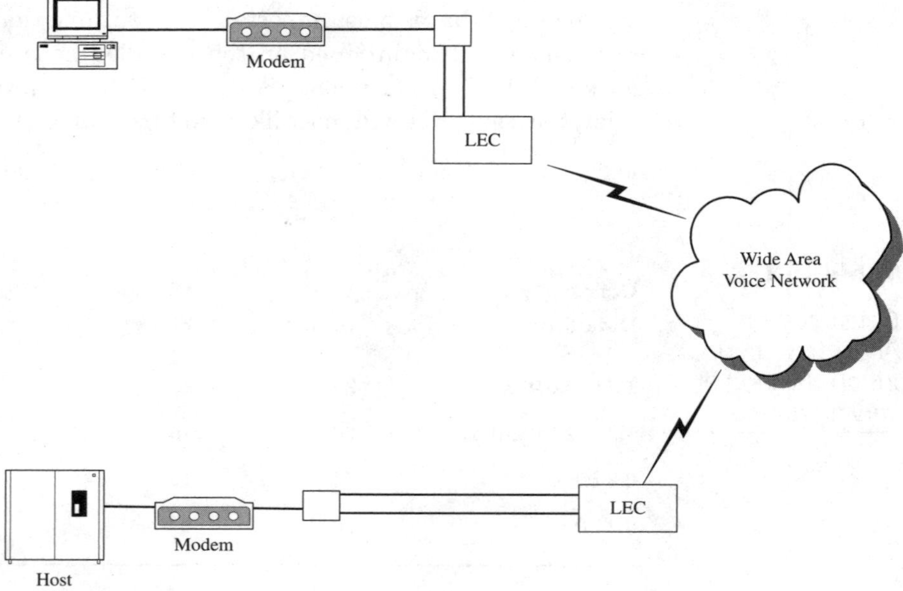

Figure 16-1 Modems are installed at the customer's location and use the existing telephone wires to transmit data across the voice network.

long-distance voice networks. This figure shows that the modem is the interface to the telephone network limited to the quality of the local loop. The ILEC installs a voice-grade line on the local copper cable plant and enables the end user to connect the modem. The modem then converts the data from a computer terminal into a voice-equivalent analog signal. There is no real magic in modem communications today, but in the early days of data communications, this was considered *voodoo science*. The miracle of data compression and other multimodulation techniques quickly expanded the data rates from 300 bps to today's 33.6 Kbps. Newer modems are touted to handle data at speeds of up to 56 Kbps, but there are few who actually get data across the network at these rates. So, the reality of all the pieces combined still has the consumer operating at approximately 33.6 Kbps. Newer technologies will produce much higher compressed speeds of up to 230 to 300 Kbps on a modem, but these are now in their infancy. They are not a major factor in our communications networks yet.

The Analog Modem History

In the early days of modem communications, the Bell telephone companies (or the independent telephone companies) provided all services across North America. A customer desiring to transmit data needed only to call the local supplier who would then install the dial-up telephone line, the modem, and all associated services to accommodate the desired data rates available. Leased lines were used when specific speeds or volumes were anticipated, but not guaranteed by the dial-up services. Regardless of the modem and lines used, the main provider was the key ingredient. The local providers supported only what they knew they could meet, so speeds were often kept very low from a guaranteed standpoint.

If the customer had a leased line and needed better or faster data, special equipment was installed on the line to reach these goals but at a higher monthly fee. Moreover, the technological advancement of modem technology was not a priority for the local providers because they owned the installed base.

In 1968, things began to change! With court decisions allowing the introduction of competitive devices and the connection of these devices on the regulated carrier's network, demands began to escalate. Restrictions on power output and energy levels were in place to prevent any interference from the modems on the voice network. Also, the customer-provided modems were interconnected through a data coupler (called a *Data Access*

Arrangement [DAA]) provided by the local regulated carriers. This, of course, involved a fee for the connection through the telephone company that provided protection equipment.

Later, the *Federal Communications Commission* (FCC) in the United States and the *Communications Radio and Television Commission* (CRTC) in Canada allowed changes in the way the interconnection was handled. Modem manufacturers were allowed to produce their products according to a set of specifications and registrations, eliminating the need for the telephone company protection equipment and the fee associated with the monthly rentals.

Soon the market began to swell with modem products that could take advantage of the voice network to transmit data. However, limitations still existed on the speeds and services enabled by these newer devices. Most of the communications limitations came from the intent of the voice network. The telephone network was designed to carry a voice call with reasonable and reproducible voice characteristics and quality. Limitations were placed on the overall throughput of the physical wires using special filters on the wires. However, the competition spurred the development of modem technologies over approximately 18 years from the old 300 bps speeds to the current speeds we now accept (28.8 to 33.6 Kbps). In 1997, the introduction of the 56 Kbps modem was going to revolutionize the market and speed up data transmission to meet the demands of the consumers. However, even at 56 Kbps, users were looking for more. The modems just did not satisfy the demands for higher-speed Internet access and video demands. Hence, the movement to newer techniques to provide faster data communications across the local voice telephone networks. Enter the DSL modem to meet the need.

IDSL

DSL refers to a pair of modems that are installed on the local loop (also called *the last mile*) to facilitate higher speeds for data transmission. Network providers do not provide a line; they use the existing lines in place and add the DSL modems to increase the throughput. DSL modems offer duplex operations—transmission in both directions at the same time. The speed of a DSL modem may be 160 Kbps on copper at distances up to 18K using the twisted pair wires. The bandwidth used is from 0 to 80 kHz, as opposed to the arbitrarily limited 0 to 3300 Hz on a voice line. This is the IDSL using the 144 Kbps full duplex, which gives us what is known as the *Basic Rate Interface* (BRI). As shown in Figure 16-2, the IDSL technique is all digital

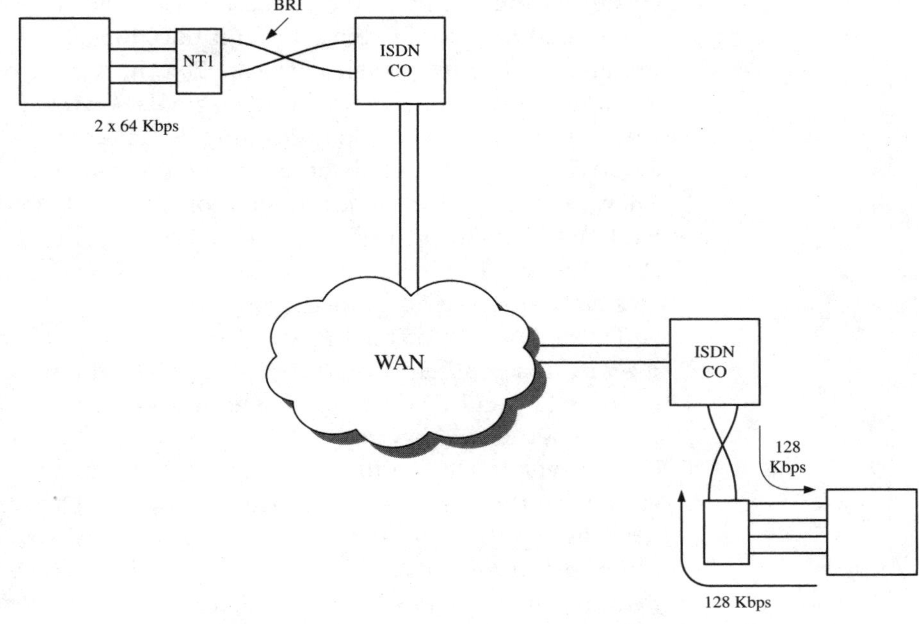

Figure 16-2 The IDSL line connection enables 128 Kbps in total simultaneously.

operating at two channels of 64 Kbps for voice or nonvoice operation and a 16 Kbps data channel for signaling, control, and data packets. ISDN was very slow to catch on, but the movement to the Internet created a whole new set of demands for the carriers to deal with. In fact, the carriers were caught off-guard when user demand, which was moderate, escalated so quickly. Now more telephone companies (ILECs) and the newer competitive LECs (CLECs) offer ISDN services for data at a very reasonable fee. The term IDSL is new, but the gist is the same. A DSL is used to deliver ISDN services. As the deployment of IDSL was speeding up on the local loop, the providers developed a new twist, called "*always on, ISDN*" mimicking a leased set of channels that are always connected. By bonding the channels together, Internet users can surf the Net at speeds of 128 Kbps in each direction. Note this is asymmetrical DSL.

HDSL

In 1958, the Bell Laboratories developed a voice multiplexing system that used a 64 Kbps voice modulation technique called *pulse coded modulation*

(PCM). Using the PCM techniques, voice calls were sampled 8,000 times per second and coded using an 8-bit encoding. These samples were then organized into a framed format using 24 time slots to bundle and multiplex 24 simultaneous conversations onto a single, 4-wire circuit. Each frame carries 24 sample of 8 bits, plus 1 framing bit (making the frame 193 bits long), 8,000 times a second. This produces a data rate of 1.544 Mbps or what we know as a T1 (for further discussions on T1, see Chapter 28, "The T Carrier Systems [T1, T2, and T3]"). We now refer to this as a *Digital Signal Level 1* (DS-1) at the framed data rate. This rate of data transfer is used in the United States, Canada, and Japan.

Throughout the rest of the world, standards were set to operate using an E1 with a signaling rate of 2.048 Mbps. The differences between the two services (T1 and E1) are significant enough to prevent the seamless integration of the two services.

However, in the digital arena, T1 required that the provider install the circuits to the customer's premises on copper. (Other technologies can be used, but the UTP is easiest because it is already there.) The local provider could install the circuit by using a 4-wire circuit with repeaters spaced at 3K from the Central Office and 3K from the customer's entrance point. In between these two points, repeaters are used every 5 to 6K. Moreover, when installing the T1 on the copper local loop, limitations of the delivery mechanism get in the way. T1 (and E1) uses *Alternate Mark Inversion* (AMI), which demands all of the bandwidth and corrupts the cable spectrum quickly. As a result, the providers can only use a single T1 in a 50-pair cable and could not install another in adjacent cables because of the corruption. Figure 16-3 is a representation of this cable layout. This is inefficient use of the wiring to the door, making it impractical to install T1s to *small office/home office* (SOHO) and residential locations. Further limitations required the providers to remove bridge taps, clean up splices, and remove load coils from the wires to get the T1 to work.

To circumvent these cabling problems, HDSL was developed as a more efficient way of transmitting T1 (and E1) over the existing copper wires. HDSL does not require the repeaters on a local loop of up to 12K. Bridge taps will not bother the service, and the splices are left in place. This means that the provider can offer HDSL as a more efficient delivery of 1.544 Mbps. The modulation rate on the HDSL service is more advanced. Sending 768 Kbps on one pair and another 768 Kbps on the second pair of wires splits the T1. This is shown in Figure 16-4.

As already mentioned, HDSL runs at 1.544 Mbps (T1 speeds) in North America and at 2.048 Mbps (E1 speeds) in other parts of the world. Both speeds are symmetric (simultaneous in both directions). Originally, HDSL

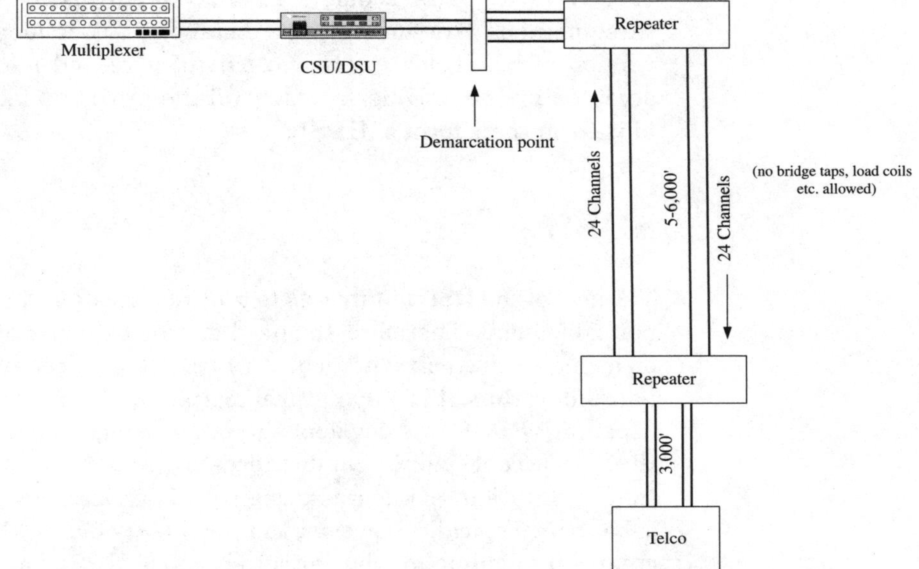

Figure 16-3 The typical layout of the T1

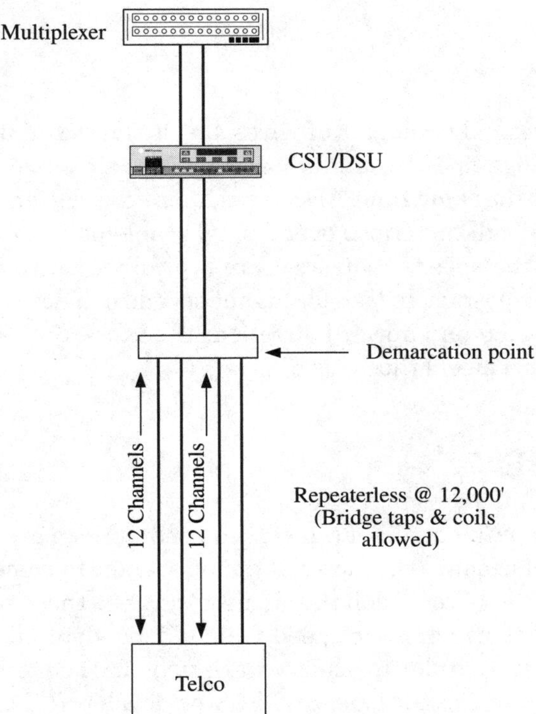

Figure 16-4 HDSL is impervious to the bridge and splices. The T1 is split onto two pairs.

used two wire pairs at distances of up to 15K. HDSL at 2.048 Mbps uses three pairs of wire for the same distances, but no longer. The most recent version of HDSL uses only one pair of wire and is expected to be more accepted by the providers. Nearly all the providers today deliver T1 capabilities on some form of HDSL.

SDSL

The goal of the DSL family was to continue to support and use the local copper cable plant. Therefore, the need to provide high-speed communications on a single cable pair emerged. Most local loops already employ single cable pair today; thus, it is only natural to assume the providers would want this capability. SDSL was developed to provide high-speed communications on that single cable pair but at distances no greater than 10K. Despite this distance limitation, SDSL was designed to deliver 1.544 Mbps on the single cable pair. Typically, however, the providers offer SDSL at 768 Kbps. This creates a dilemma for the carriers because HDSL can do the same things as SDSL.

ADSL

SDSL uses only one pair of wires, but is limited in its distance to provide duplex, high-speed communications. Not all users require symmetrical speeds at the same time. ADSL was, therefore, designed to support differing speeds in both directions over a single cable pair at distances of up to 18K. Because the speeds requested are typically for access to the Internet (or intranet), most users look for higher speeds in a download direction and the lower speed for an upward direction. Therefore, the asymmetrical nature of this service meets those needs.

RADSL

Typically with equipment, installed assumptions are made based on minimum performance characteristics and speeds. In some cases, special equipment is used to condition the circuit to achieve those speeds. However, if the line conditions vary, the speed will be dependent on the sensitivity of the equipment. In order to achieve variations in the throughput and be sensitive to the line conditions, RADSL was developed. This gives the flexibility to adapt to the changing conditions and adjust the speeds in each direction to potentially maximize the throughput on each line. Additionally, as line conditions change, you can see the speeds changing in each direction during the transmission. Many of the ILECs have installed RADSL as their choice, given the local loop conditions. Speeds of up to 768 Kbps are the preferred rates offered by the incumbent providers.

CDSL

Consumers are not all looking for symmetrical high-speed communication in order to achieve access to the Internet. Furthermore, the speeds of ADSL technology are more than the average consumer may be looking for. As a result, the lower-speed communications capability was developed by using CDSL as the model. With other forms of DSL (such as ADSL and RADSL), splitters are used on the line to separate the voice and the data communications. CDSL does not use, nor need, a splitter on the line. Moreover, speeds of up to 1 Mbps in the download direction and 160 Kbps in the upward direction are provided. It is expected that the speeds and DSL will meet the needs of the average consumer for some time to come. As a result, a universal ADSL working group developed what is called *ADSL-lite*. This was ratified in late 1998, using the specifications from this working group for delivery to the average consumer. Because of the changes in speeds with this technique, the telephone companies are in a position to support a lower-speed DSL strictly through the use of the modems without the concern for local loop. An example of this DSL-lite (G.Lite) service is provided with the Nortel 1 Mb modem.

SHDSL

SHDSL conforms to the International Telecommunications Union G.991.2 recommendations, leveraging capabilities of older DSL and other transport technologies, such as SDSL, HDSL and HDSL2, IDSL, ISDN, T-1, and E-1. One of the most significant improvements SHDSL brings to the business market is increased reach—at least 30 percent greater than any earlier symmetric DSL technology. Furthermore, SHDSL supports repeaters, which further increase the reach capability of this technology.

Another critical advantage of SHDSL is its increase in symmetric bandwidth. In a typical installation, up to 2.3 Mbps will be available on a single copper pair. For greater bandwidth needs in the future, a 4-wire model that can provide up to 4.6 Mbps is also supported by the new standard. SHDSL is also rate adaptive, enabling flexible revenue-generation models and enabling service providers to offer service-level agreements that ensure businesses get the service they want, when they want it.

G.SHDSL stands for *Symmetric High Bit Rate Digital Subscriber Loop* defined by the new ITU Global Standard G991.2 as of February 2001. This service delivers voice and data services based on highly innovative communication technologies and will thus be able to replace older communication technologies such as T1, E1, HDSL, HDSL2, SDSL, ISDN, and IDSL in the future.

SHDSL provides high symmetric data rates with guaranteed bandwidth and low interference with other services. By supporting equal upstream and downstream data rates, G.SHDSL better fits the needs of

- Remote LAN access
- Web-hosting
- Application sharing
- Video conferencing

G.SHDSL targets the small business market. Multiple telephone and data channels, video conferencing, remote LAN access, and leased lines with customer-specific data rates are among its many exciting characteristics. Spectrally friendly with other DSLs, it supports symmetric data rates varying from 192 Kbps to 2.320 Mbps across greater distances than other technologies.

In an ATM-based network on the customer side, an *Integrated Access Device* (IAD) is installed to convert voice and data into ATM cells. An IAD can also contain some routing functionality. Data is converted using AAL5 (ATM Adaptation Layer), while voice requires AAL1 (without compression) or AAL2 (with compression and micro cells). These cells are mapped together in the SHDSL frame and recovered later on in the DSLAM. An ATM switch routes the cells either to an *Internet service provider* (ISP) or to a voice gateway that translates the voice cells back into the TDM world.

The voice part of the SHDSL frame will be treated in a similar fashion to normal ISDN or POTS services. However, the data needs to be converted into ATM. This can be done either in an IAD, resulting in a mix of TDM and ATM on the SHDSL line, or at the central office side. In the second case, it is necessary to protect the data on the line. This can be easily done by an HDLC protocol. The division between voice and data should be done in the loop carrier so that the ATM cells can be sent directly to the ATM backbone in order not to congest the PSTN network. This approach has the advantage of being more bandwidth efficient because the HDLC overhead is smaller than the ATM overhead. Additionally, the *Segmentation And Reassembly* (SAR) functionality can be centralized in the DLC. However, because an IAD normally uses Ethernet to connect to a LAN, some intelligence is required at the subscriber side to process the Ethernet MAC and also have SAR functionality.

VDSL

Clearly, changes will always occur as we demand faster and more reliable communications capabilities. It was only a matter of time until some users demanded higher-speed communications than was offered by the current DSL technologies. As a result, VDSL was introduced to achieve the higher speeds. If, in fact, speeds of up to 50 Mbps are demanded, then the distance limitations of the local cable plant will be a factor. In order to achieve the speeds, you can expect that a fiber feed will be used to deliver VDSL. This technique will most likely carry ATM traffic (cells) as its primary payload. The pilot program of Qwest Communications in Arizona leaves plans to deliver fiber to the door, *Fiber to the Home* (FTTH), to provide voice, data, video, and multimedia communications to the consumer. Although this pilot is still emerging, a lot of excitement has been generated over the possibilities.

Table 16-2 summarizes the speeds and characteristics of the DSL technologies discussed. These are the typical installation and operational characteristics; others will certainly exist in variations of installation and implementation.

The Hype of DSL Technologies

Why all the hype? Well the local providers are extremely excited if they can install higher-speed communications and preserve their local cable plant. No one wants to abandon the local copper loop, but getting more data reliably across the local loop is imperative. Therefore, the ability to breathe new life into the cable plant is an extension of the facilities in place.

The consumer is looking for higher-speed access (primarily to access the Internet) for whatever the application. Yet, at the same time, consumers are looking for a bargain. They do not want to spend a lot of money on their communications services.

Table 16-2

Summary of DSL speeds and operations using current methods

Service	Explanation	Download	Upload	Mode of Operation
ADSL	Asymmetric DSL	1.5 to 8.192 Mbps	16 to 640 Kbps	Different up and down speeds, one pair wire.
RADSL	Rate Adaptive DSL	64 Kbps to 8.192 Mbps	16 to 768 Kbps speeds	Different up and down. Many common operations use 768 Kbps. One pair wire.
CDSL	Consumer DSL	1 Mbps	16 to 160 Kbps	Now ratified as DSL-lite (G.lite). No splitters. One pair wire.
HDSL	High-data rate DSL	1.544 Mbps in North America, 2.048 Mbps in rest of world	1.544 Mbps 2.048 Mbps	Symmetrical services. Two pairs of wire.
IDSL	ISDN DSL	144 Kbps (64+64+16) as BRI	144 Kbps (64+64+16) as BRI	Symmetrical operation. One pair of wire. ISDN BRI.
SDSL	Single DSL	1.544 Mbps, 2.048 Mbps	1.544 Mbps, 2.048 Mbps	Uses only 1 pair but typically provisioned at 768 Kbps. One pair wire.
VDSL	Very High data rate DSL	13 to 52 ± Mbps	1.5 to 6.0 Mbps	Fiber needed and ATM probably used.
SHDSL (G.SHDSL)	Single High-speed DSL	192 Kbps to 2.360 Mbps or 384 Kbps to 4.720 Mbps	192 Kbps to 2.360 Mbps or 384 Kbps to 4.720 Mbps	Using 1 pair. Using 2 pair.

The providers are trying to bump up their revenues without major new investments. They would like to launch as many new service offerings on their existing cable plant and increase the costs to the end user. This is a business decision, not a means of trying to rake the consumer over the coals. Yet there has to be a happy medium of providing services and generating revenues with limits on expenses. To do this, the xDSL family offers the opportunity to meet the demands while holding down investment costs. The key ingredient for success is to minimize costs and satisfy the consumer. Make no mistake, if the local provider does not offer the high-speed services, someone else will.

xDSL Coding Techniques

Many approaches were developed as a means of encoding the data onto the xDSL circuits. The most common are *Carrierless Amplitude Phase Modulation* (CAP) and *discreet multitone* (DMT) modulation. *Quadrature with Phase Modulation* (QAM) has also been used, but the important part is the standardization. The industry, as a rule, selected DMT, but several developers and providers have used CAP. It is, therefore, appropriate to summarize both of these techniques. The SHDSL technology uses a *trellis-coded pulse amplitude modulation* (TCPAM) technique to gain the benefits of the single-pair services or two-pair service.

Discreet Multitone

DMT uses multiple narrowband carriers, all transmitting simultaneously in a parallel transmission mode. Each of these carriers carries a portion of the information being transmitted. These multiple discrete bands, or, in the world of frequency division multiplexing, subchannels, are modulated independently of each other using a carrier frequency located in the center of the frequency being used. These carriers are then processed in parallel form.

In order to process the multicarrier frequencies at the same time, a lot of digital processing is required. In the past, this was not economically feasible, but integrated circuitry has made this more realistic.

The *American National Standards Institute* (ANSI) selected DMT with the use of 256 subcarriers, each with the standard 4.3125 kHz bandwidth. These subcarriers can be independently modulated with a maximum of 15 bits/second/Hz. This enables up to 60 Kbps per tone used. Figure 16-5 shows the use of the frequency spectrum for the combination of voice and two-way data transmission. In this representation, voice is used in the normal 0 to 4 kHz band on the lower end of the spectrum (although the lower 20 kHz is provided). Separation is enabled between the voice channel and the upstream data communications, which operates between 20 kHz and 130 kHz. Then a separation is enabled between the upstream and the downstream channels. The downstream flow uses between 140 kHz and 1 MHz. As shown in this figure, the separation allows for the simultaneous up- and downstreams and the concurrent voice channel. It is on this spectrum that the data rates are sustained. Each of the subchannels operates at approximately 4.3125 kHz, and a separation of 4.3125 kHz between channels is allocated.

Figure 16-5 The ANSI DMT specification

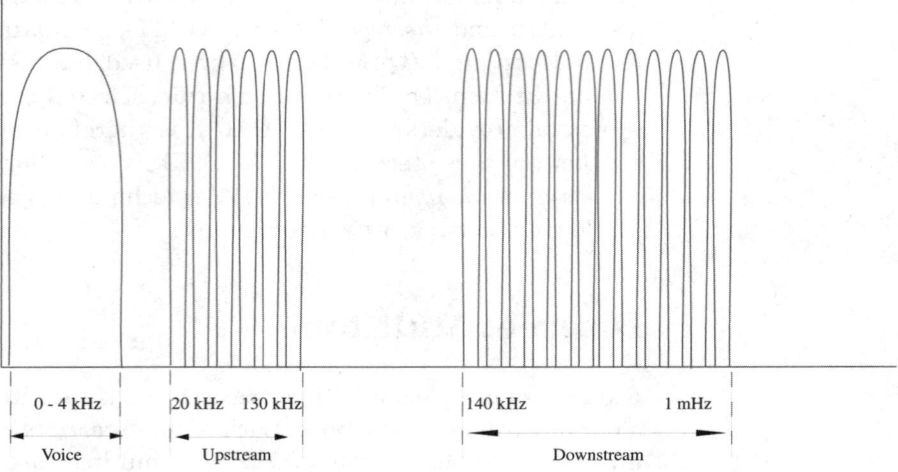

Using DMT for the Universal ADSL Service (G.Lite)

Provisions for the high-speed data rates of full ADSL are good, but not every consumer is looking for the high data rates afforded on ADSL. Therefore, the Universal ADSL Working Group decided to reevaluate the need for the end user. What they determined is that many consumers need download speeds of up to 1.5 Mbps and upload speeds between 9.6 to 640 Kbps. As a result, the ADSL Lite specification was designed to accommodate these speeds, as a logical steppingstone to the higher-speed needs for the future. Initially introduced in early 1998, the specification was ratified in late 1998 to facilitate the lower throughput needs of the average consumer. DMT is the preferred method of delivering the G.Lite specification and service, as it is now known. This involves a slightly different method of delivering the service, but does accommodate the providers with a less expensive solution to provide full-rate ADSL.

There is no way to know if the network providers can support hundreds of multimegabit ADSL up- and download speeds on their existing infrastructure. But using the G.Lite specification can support lower-demand users more efficiently. Similar to the DMT used in the ANSI specification, the carriers are divided as shown in Figure 16-6. Note that in this case, the high end of the frequency spectrum tops out at approximately 550 kHz instead of the 1 MHz range with ADSL.

To Split or Not to Split

Another issue of using ADSL is the use of splitters on the line. In normal ADSL and RADSL, the local provider uses a splitter on the line. ADSL modems usually include a POTS splitter, which enables the simultaneous access to telephony applications and high-speed data access. Some vendors provision the service with an active POTS splitter device that allows the simultaneous telephone and data access. Unfortunately, with an active device, if the power or the modem fails, then the telephone also fails. This is problematic because we are accustomed to having lifeline services with our telephone systems that are always available, even if the power fails. The splitter is shown in Figure 16-7 as it is installed.

Figure 16-6
The ANSI and UAWG G.Lite spectrum

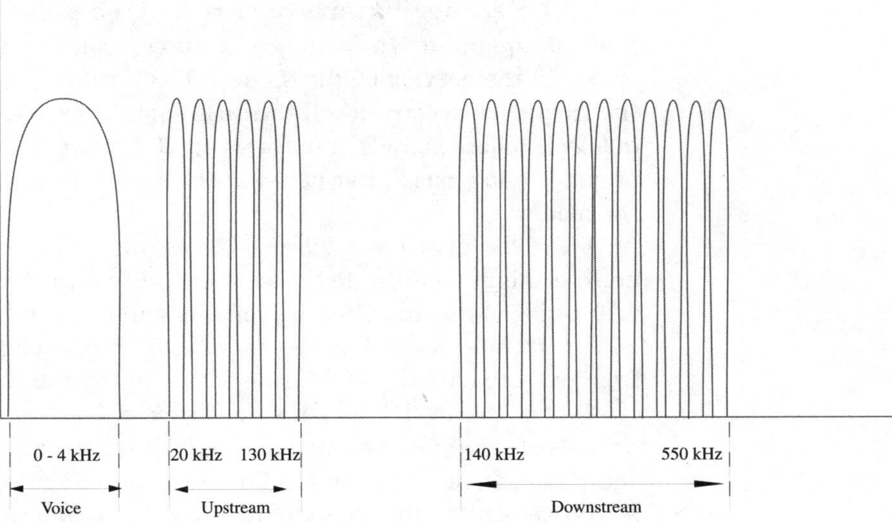

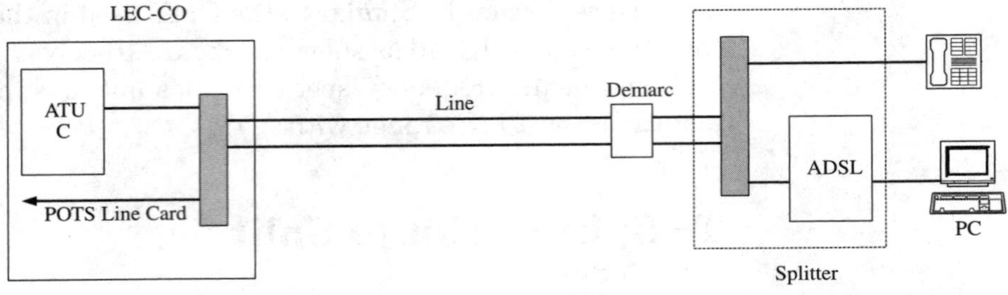

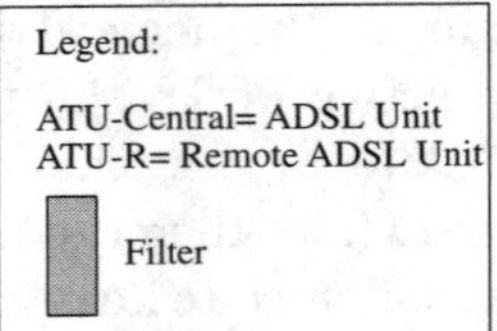

Figure 16-7 The splitter-enabled ADSL service

A passive POTS splitter maintains the lifeline service of telephony even if the modem fails. This is important because the telephone line is powered from the line instead of an external power source. Telephony service will be available as much as we have always expected from the normal service we have always received.

A POTS splitter is a three-pronged device, enabling telephony and simultaneous up-and-down loading data access on the same copper loop. As shown in the previous figures, the POTS service operates at the low end of the frequency spectrum. All the data signals are located in the higher frequencies. These signals start between 20 to 25 kHz and above. The splitter provides a low pass filter between the copper line and the ADSL point on the modem.

One of the primary concerns is to enable a filter to block any transient noise coming from the POTS side of the line from crossing over into the ADSL side of the line. Ringing voltages can cause significant impulse noise across the line, destroying the data. ADSL modem signals are also blocked from passing onto the POTS side of the line by the filters.

Most of the POTS splitters are passive. Passive filters provide higher degrees of reliability because they don't require power, they can isolate the equipment from surges on the line, and they can arrest lightning that may be coupled on the line. Active filters require power and are less adapted to suppress the surges. In many countries around the world, active filters are a requirement.

The G.Lite specification uses a splitterless device to facilitate the installation of low-speed DSL services. This is shown in Figure 16-8 for implementation. Without the splitter, there is less cost for the provider and the issues of active versus passive splitters are eliminated. This provides a different approach.

CAP

CAP is closely aligned to QAM. QAM as a technique is widely understood in the industry and well deployed in our older modems. Both CAP and QAM are a single-carrier signal technique. The data rate is divided into two and modulated onto two different orthogonal carriers before being combined and transmitted. The main difference between CAP and QAM is in the way they are implemented. QAM generates two signals with a sine/cosine mixer and combines them onto the analog domain.

CAP, on the other hand, generates its two orthogonal signals and executes them digitally. Using two digital transversal bandpass filters with equal amplitude characteristics and a p/2 difference in phase response, the signals are combined and fed into a digital-to-analog converter. Then the data is transmitted. The advantage of CAP over QAM is that CAP is done in silicon, which is more efficient and less expensive.

CAP was one of the original proposals for use with ADSL technology. Unfortunately, this was a proprietary solution offered by a single vendor, which turned heads away from acceptance. CAP is shown in Figure 16-9 in its use of the frequency spectrum of the line. Most industry vendors agree that CAP has some benefits over DMT but also that DMT has more benefits over CAP. The point here is that two differing technologies were initially

Figure 16-8
The splitterless G.Lite installation

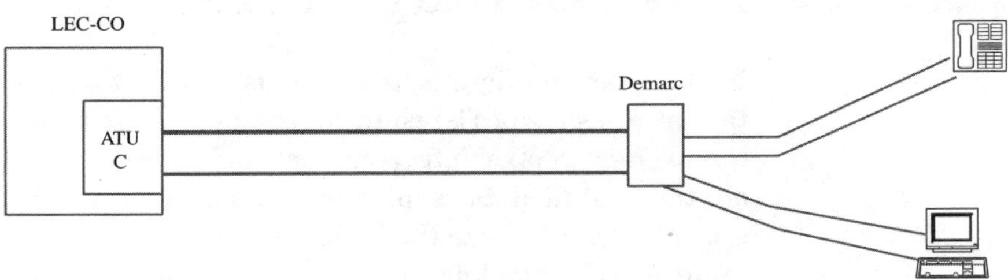

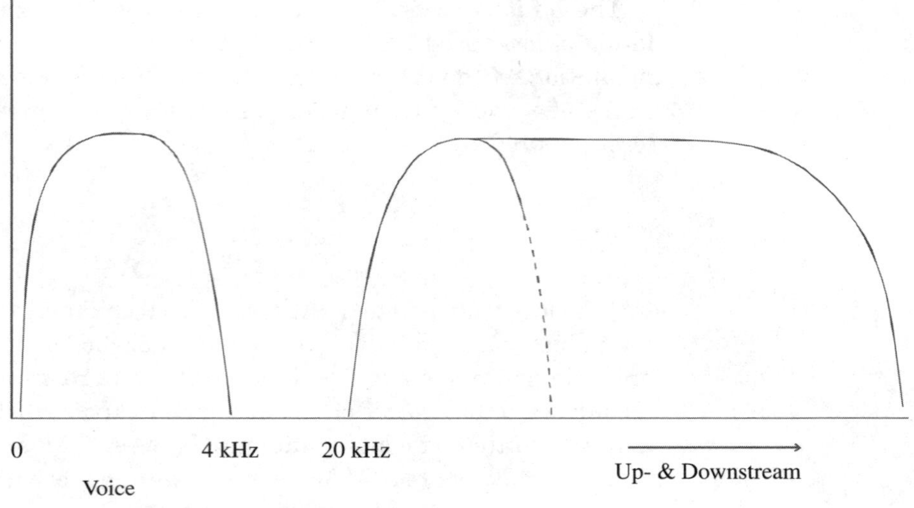

Figure 16-9 The spectral use of CAP

rolled out for ADSL (and the other family members), which contradict each other in their implementation.

CAP uses the entire loop bandwidth (excluding the 4 kHz baseband analog voice channel) to send the bits all at once. There are no subchannels, as found in the DMT technique. The lack of subchannels removes the concern about the individual channel transmission and problems. To achieve the simultaneous send and receive capability, frequency division multiplexing is used, as is echo cancellation. Many of the *Regional Bell Operating Companies* (RBOCs) have used or tried CAP in their installations, but have moved away from the CAP to a uniform use of DMT.

Provisioning xDSL

In the following figures, the various architectures of the xDSL implementations are shown. The point to remember here is the goal of xDSL is to use the existing copper infrastructure and improve the speed and throughput on the installed base of wires. Consequently, the installation process attempts to minimize the added equipment (particularly at the customer's premises) and the labor required to get the equipment installed.

In Figure 16-10, the design of an ADSL model and the model components are shown. The intent of the model is to show the infrastructure of the network from the customer premises to the network provider. This model also shows the splitters in place to facilitate the ADSL model.

Figure 16-11 demonstrates the connection from the service provider to the rest of the world. In many cases, ADSL access to the local network access provider (the ILEC or other local loop provider) is then passed on to the ISP. This is designed to run over an ATM backbone but not a firm requirement. Therefore, the NAP will assign a *DSL Access Multiplexer*

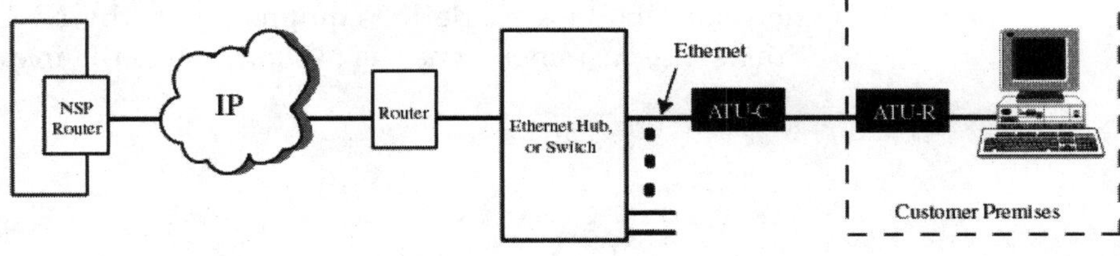

Figure 16-10 The ADSL model as it is laid out from the customer premises to the service provider

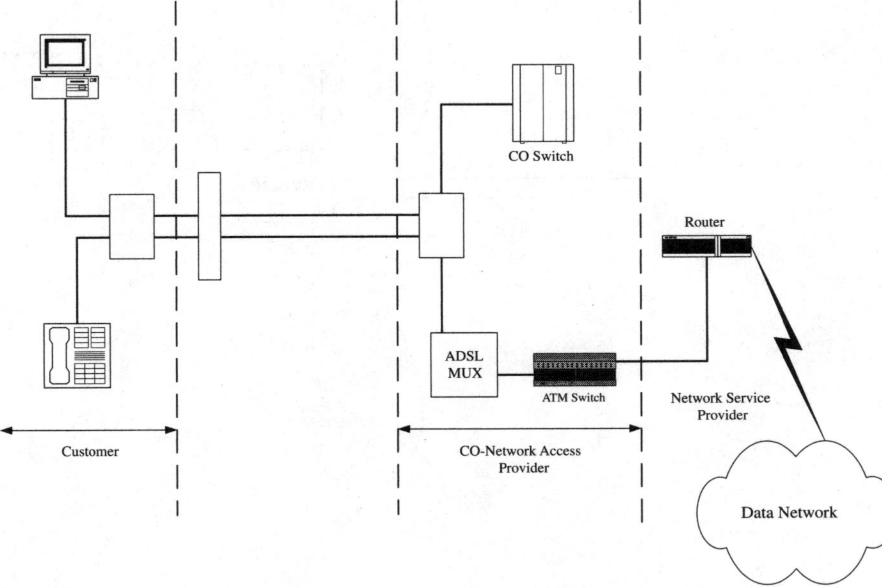

Figure 16-11 Access from the NAP to the NSP

(DSLAM) card and assign an ATM VPI and VCI as a default to carry the data into the ISP or other *Network Service Provider* (NSP).

The application most commonly used is to gain high-speed access to the Internet. Many of the local service providers install the ADSL service into a single PC at the end-user location, as shown in Figure 16-12. The local providers offer the customer a packaged deal with the following components:

- LAN NIC card operating at 10 Mbps
- DSL modem
- Splitter
- Management cables

The local provider will normally advise the customer that the termination must be to a single PC equipped with the NIC card. In the United States, the customer owns the package when the installation is completed

Figure 16-12 The typical local installation

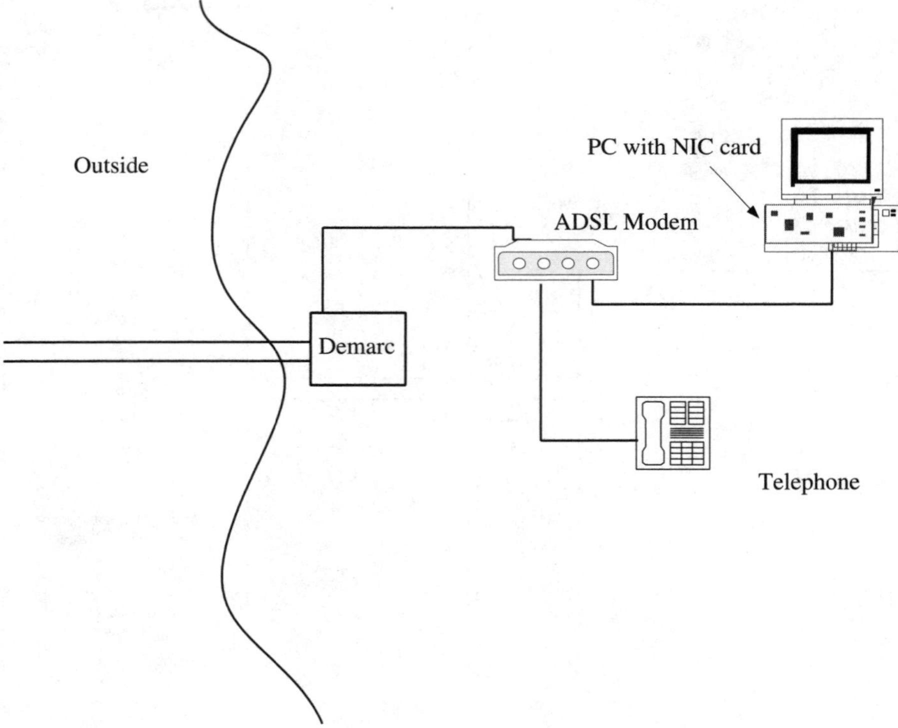

due to some of the regulatory constraints and the Public Utility Commission rulings. This places the burden of maintenance and diagnostics on the end user rather than the local service provider. In the case of a LAN attachment described previously, the ADSL modem is set to bridge from the LAN to the ATM network interfaces rather than route.

In other cases when a LAN is present at the customer location and the end user wants to connect all the LAN devices to the high-speed outside network for Internet access or private network access, the local carrier may suggest that a proxy server is a requirement. The proxy server (PC dedicated to act in this function) will then act as the gateway to the outside world for all devices attached to the LAN (see Figure 16-13).

An alternative to this approach is the direct connection to a LAN hub, such as that found in the telephone closet. Keeping the connection active, the carrier will normally assign an IP address, using DHCP for a contracted period of time. Normally, this is a lease period of four hours, and then the

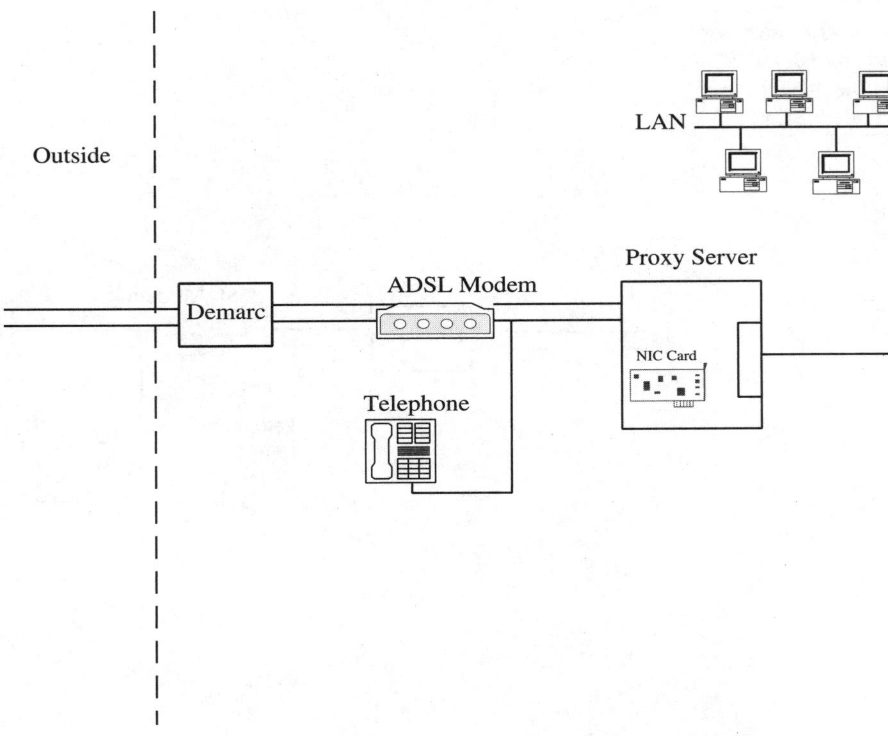

Figure 16-13 The proxy server in lieu of a single attached PC

network server (outside) will renegotiate and assign a new IP address for the end user. This protects the end-user network from becoming visible on the Internet and helps to prevent some of the normal security risks associated from a hard connection to the Net (see Figure 16-14).

Many of the hubs located in customer locations are 10 Base T, or 10 Mbps Ethernet hubs. Occasionally, a customer may have a 10/100 auto-sensing hub or a 100 Base T hub. The local providers have been known to tell the customer that this arrangement will not work. Specific networks are already attached with a direct attachment of 10/100 and 100 Base hubs with no impact, as shown in Figure 16-15. The connection allows for a specific number of simultaneous connections onto the ADSL service. The local providers will always try to configure the network connections in ways they can guarantee will work and with a standard way to troubleshoot problems. By working with the previous variations, the local providers still need to come up to speed on the way the data networks actually perform.

Figure 16-14
Connecting the ADSL service to a hub

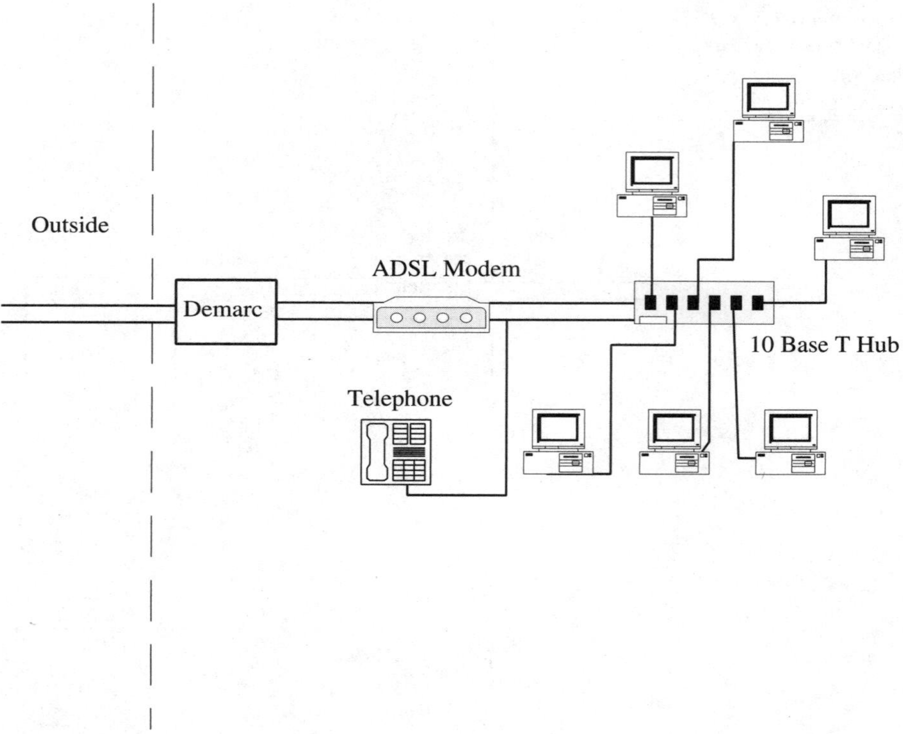

Figure 16-15 Connecting the ADSL modem directly through a 10/100 or a 100 Base T hub

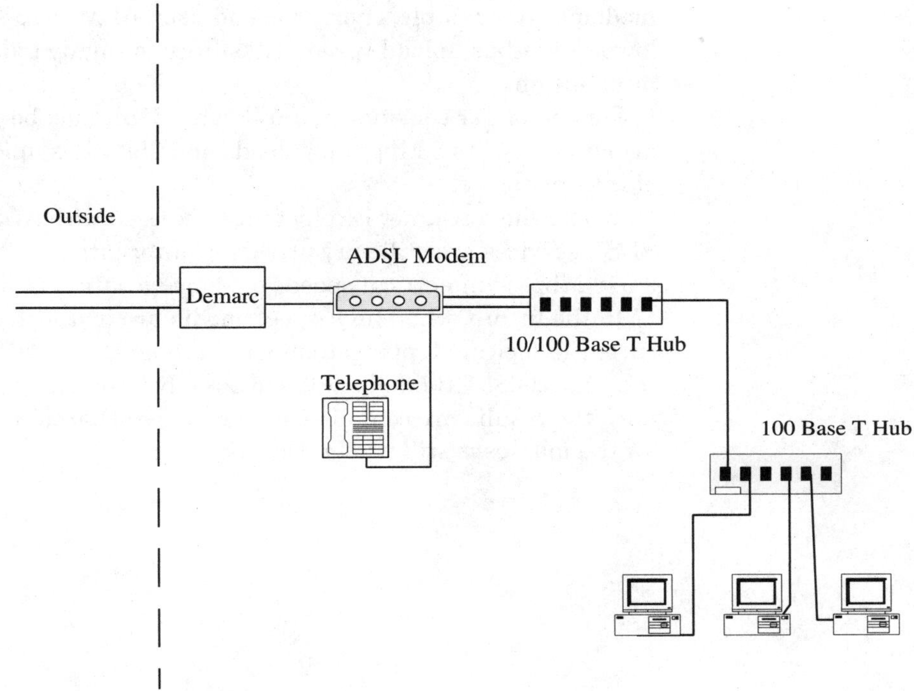

Final Comment on Deployment

The use of ADSL service is catching on. However, the local providers (ILECs and CLECs) are dragging their feet. As of late 1998, there were only about 15 to 20,000 total ADSL modem pairs installed in the United States. In contrast, there were over 300,000 cable modems installed in residences and businesses across the country. The local owners of the copper loop have to take a more aggressive approach to delivering the high-speed services, or the consumer will go somewhere else.

As the market continues to mature and standards continue to develop, the local providers must preserve their infrastructure. Consumers (small and large alike) are demanding the higher-speed services. As a stepping stone for residential- and home-based businesses, the acceptance and standardization of the G.Lite specification will provide suitable transmission rates until the carriers can complete their data strategies. 1 Mbps modems, for example, giving the end user a 1 Mbps download speed and a lower 160 Kbps upload speed, will suffice for many today and into the next millennium.

For the larger consumers, a full-rate ADSL may be just what is needed, bringing 1.544 to 8 Mbps downloads and 768 Kbps uploads to the forefront of Internet access.

Where the consumer is reluctant to proceed with ADSL, the HDSL or the SDSL services are still very attractive alternatives, offering up to 1.544 to 2.048 Mbps symmetrical speeds or some variation as already discussed.

In the future, when high-speed media are installed to the door or to the curb, the logical steppingstone will become the VDSL service, perhaps in the year 2003. Although trials are already underway, too much time passes until the results are compiled and analyzed. Therefore, the reality of VDSL for the masses is still a long way off.

CHAPTER 17

Microwave- and Radio-Based Systems

No one ever pays much attention to the microwave radio dishes mounted on towers, on the sides of buildings, or any other place. This technology has been taken for granted over the years. However, this nondescript industry has quietly grown into a $4.6 billion global business annually with expectations that the market will reach approximately $10 billion by 2006. Table 17-1 is a possible breakdown of the distribution of wireless microwave services by category between 2002 and 2006.

Four major suppliers provide one-half of all the radio-based systems globally. Microwave has also become a vital link in the overall backbone networks over the years. Now, it has achieved new acclaim in the wireless revolution, relaying thousands of telephone conversations from place to place, bypassing the local landlines.

Microwaves (the actual radio waves) are between 1 mm and 30 cm long, and operate in a frequency range from 300 MHz to 300 GHz. Microwaves were first used in the 1930s, when British scientists discovered the application in a new technology called radar.

In the 1950s, microwave radio was used extensively for long-distance telephone transmission. With the need to communicate over thousands of miles, the cost of stringing wires across the country was prohibitive. However, the equipment was both heavy and expensive. The radio equipment used vacuum tubes that were bulky as well as highly sensitive to heat. All of that changed dramatically when integrated circuits and transistors were used in the equipment. Now the equipment is not only lightweight, but also far more economical and easy to operate. In 1950, the typical microwave radio used 2,100 watts to generate three groups of radio channels (each group consists of 12 channels), yielding 36-voice-grade-channel capacity. Each voice grade channel operated at the standard 4 kHz. Today, equipment from many manufacturers (and Harris/Farinon, specifically) requires only 22 watts of output to generate 2,016 voice channels. Although there have been two orders of magnitude improvements in the quality of the voice

Table 17-1 Possible market share for microwave products

Service	2002	2006
Point-to-point microwave radios services	$2.5 billion	$4 billion
Point-to-multipoint microwave	$700 million	$2 billion
Wireless LAN microwave products	$1.4 billion	$4 billion
Total	$4.6 billion	$10 billion

transmission, the per-channel cost has plummeted from just over $1,000 to just under $37. This makes the transmission systems very attractive from a carrier's perspective. However, the use of private microwave radio has also blossomed over the years because of the cost and performance improvements. This is shown in the graph reflected as Figure 17-1, which details why the use of microwave has become so well accepted in the industry.

Today's microwave radios can be installed quickly and relocated easily. The major time delays are usually in getting through the regulatory process in a governmentally controlled environment. Several installations have taken over a year to be approved, only to have the radio system installed and running within a day or two. In many situations, microwave systems provide more reliable service than landlines, which are vulnerable to everything including flooding, rodent damage, backhoe cuts, and vandalism. Using a radio system, a developing country without a wired communications infrastructure can install a leading-edge telecommunications system within a matter of months. For these reasons, regions with rugged terrain or without any copper landline backbone in place find it easier to leap into the wireless age and provide the infrastructure at a fraction of the cost of installing wires.

Throughout the world, government-owned and -controlled monopolies are being eliminated. Brazil opened its doors to international Telecom competition, allowing microwave radio systems and a mobile telephone system supplied by North American firms. In Russia, one of the leading systems

Figure 17-1 Comparison of cost per channel over the years

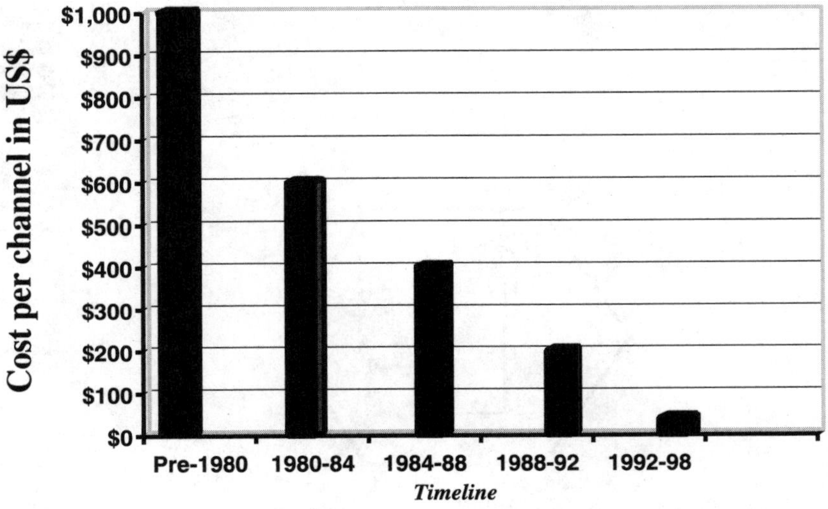

manufacturers installed an integrated network along a 3,600-mile gas pipeline. Using microwave radios and digital telephone switches, this link sends data and voice from Siberia to Russia's southern border. A similar system is used for the railroad system, using a trunked radio system, microwave radio-relay, and digital telephone switches. The telephone service may serve more than the railroad, incorporating hundreds of thousands of people residing alongside the railway line.

The cellular and *Personal Communications Service* (PCS) industries invested heavily in microwave radios to interconnect the components of their networks. This is shown in Figure 17-2 where the interconnection is used in the cellular world. In addition, a new use of microwave radio, called micro/millimeter wave radio, is bringing transmission directly into buildings through a new generation of tiny receiver dishes.

WinStar Communications, a *Competitive Local Exchange Carrier* (CLEC), pioneered the use of micro/millimeter wave radio communications in the 30+ GHz frequency range (actually 28 to 38 GHz). This allowed the CLEC to deliver broadband communications to the consumer's door without the use of telephone company wires. Unfortunately, the results were mixed as WinStar Communications developed financial trouble and filed for bankruptcy protection in 2001 and was acquired by IDT Corporation in December 2001.

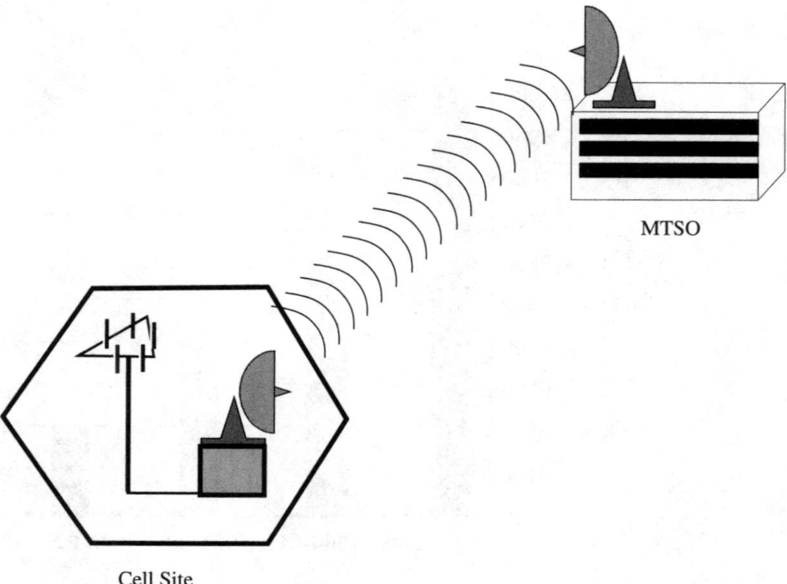

Figure 17-2 Cellular interconnection of microwave radio

The PCS industry chose microwave radio technology for the interconnection and backhaul transport on its expanding network. The PCS suppliers and the cellular suppliers do not want to pay the local telephone company for monthly T1 access lines from the cell sites to the mobile switching sites. Therefore, to eliminate the monthly recurring charges, they have installed microwave radio systems in the 18 to 23 GHz frequency range. Tens of thousands of new cell sites and PCS sites have been constructed and will continue to be constructed over the next few years, further expanding the use of microwave radio systems in each of these sites. As third-generation, handheld devices make their way into the industry, more wireless interconnectivity will be used.

Microwave also played a very crucial part of the PCS industry as the PCS systems use the 1.9 to 2.3 GHz frequency band. Fixed systems operators such as police, fire, electric utilities, and some municipal organization occupied these frequencies. To accommodate the move of these users from the 2 GHz frequency band, microwave was used to relocate the users to a new band, as mandated by the FCC. One study indicated that the PCS industry would spend over $3 billion in microwave equipment and services by 2005.

Another large demand for microwave emerged in the *Competitive Access Providers'* (CAPs) market. CAPs offer long-distance access to customers at lower prices than the local telephone companies and the newer competitors. The CAPs normally install their own fiber-optic wires. However, they recognize the benefit of expanding coverage to consumer building entrances, using a wireless, high-speed connection. The CAPs are supplementing their fiber-based networks with *Wireless CAPs* (WCAPs). WCAPs use microwave transmission to deliver the telecom service without the need for a costly, wire-based infrastructure (see Figure 17-3).

The newer micro/millimeter-wave radios, which are smaller and usually less expensive than other microwaves, are also popular with these CAPs and PCS suppliers. They are used in urban areas to extend the fiber networks. These radio units use the high-frequency (or millimeter) bandwidth that hadn't been used before. Now, they are seen as a solution to increasing congestion in the lower frequency bands. An advantage of these systems is the small antennas that can be hidden on rooftops without interfering with zoning ordinances or creating aesthetic controversy.

Microwave is heavily used in radio and television systems. Satellite TV relies on microwave repeaters on the satellite to retransmit TV signals to a receiving station. Microwave communication via satellite provides a more reliable signal than longer, land-based radio waves. It also improves the reception of the picture.

Some TV stations have been using microwaves to facilitate wireless communications from field cameras since 1992. What we continually hear

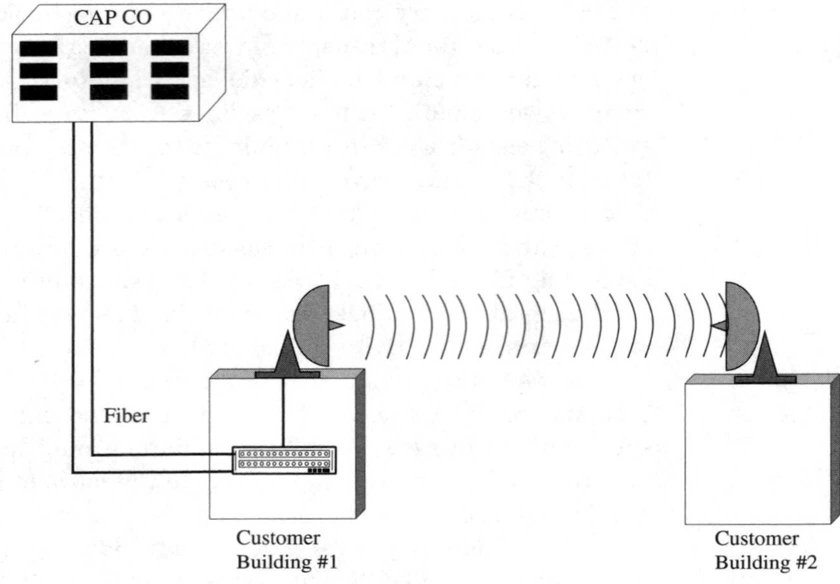

Figure 17-3 Wireless interconnection of fiber and CAPs

about the "action cams" is a portable microwave system connected to a camera for real-time broadcast. Instead of being constrained to a fixed location, a news van can be driven and hooked up instantly, as shown in Figure 17-4. The systems hook up with a field camera with microwave units the size of a deck of cards. These can go anywhere and can operate from locations up to two miles from the van. Action and news is transmitted back to the van where it is relayed via microwave to the TV station. We have all experienced the "news as it was happening" on TV from local events to worldwide events.

Other Applications

A laptop computer with a credit card-sized PRISM radio chip set can now convert incoming microwave messages into binary code for computer processing and then convert them back into microwaves for transmission (see Figure 17-5). Similarly, microwave transmission is used in LANs, on corporate or college campuses, in airports, and elsewhere. Whether it is collecting data, relaying conversations, or beaming messages from space, microwave makes the wireless revolution possible.

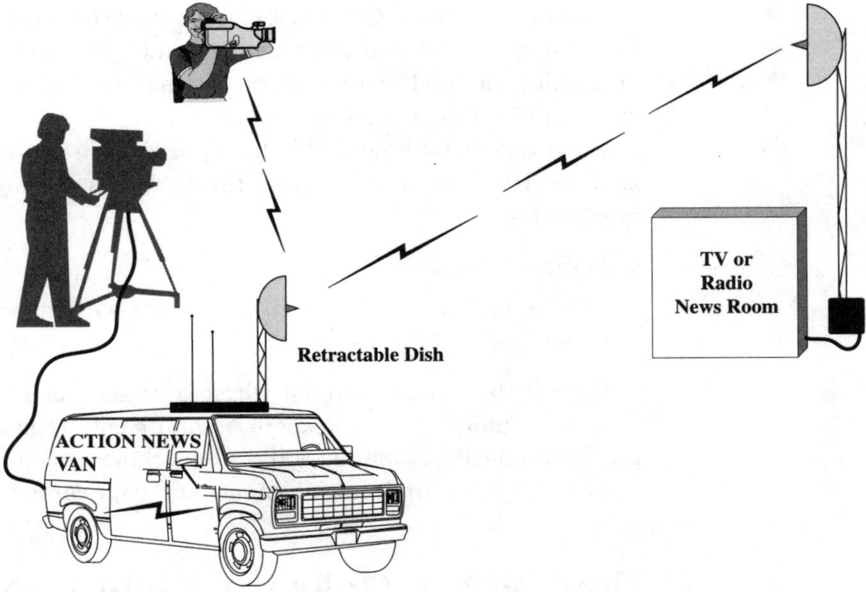

Figure 17-4 Action camera and microwave systems working together

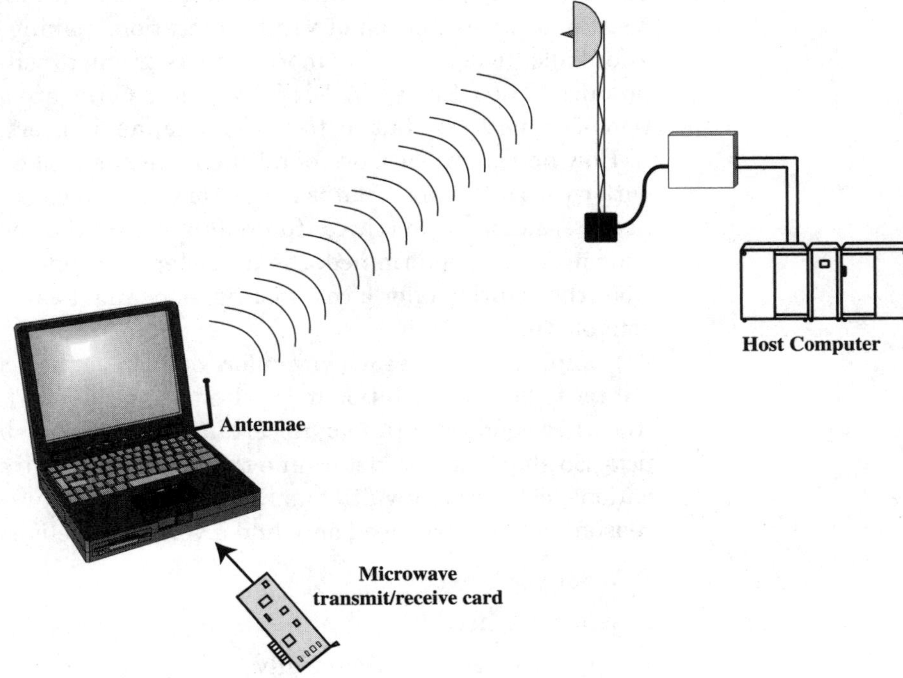

Figure 17-5 Laptop computers can now send and receive microwave radio transmissions.

No one can escape the wireless hype these days. The challenge is in wading through all the confusion and misleading statements to decide whether an application fits the need. If you can make sense of it all, you may find the solution to your connectivity needs.

First, one distinction will help to narrow the playing field. In this explosion of wireless technologies, there are two major categories worth mentioning:

- Personal wireless devices
- Wireless devices that are used between buildings (for voice, data, and video)

Buying the wrong personal wireless device, such as a pager or cell telephone, is annoying, but inexpensive and easily replaceable. Yet, if you're a telecommunications manager, the wrong choice to connect your sites together can have significant financial impact and your career can be shortened.

How Do You Make the Right Choices?

Telecommunications and information managers have many options to connect remote sites. The options have expanded with all the wireless excitement and the expansion of VLSI integration, making the devices far more affordable. Today, there are more vendors trying to sell the end user on their products. The CLECs, CAPs, and Wireless CAPs are all vying for this portion of connectivity, but so, too, are the manufacturers.

How do you make sense of all the hype and make a decision consistent with corporate goals? That is not as complicated as it sounds. Conducting a needs assessment is critical to understanding the connectivity goal. After your needs are determined, the next step is to find a solution that works. Does the solution offer a cost justification, and it can be delivered in a reasonable time?

Complications arise when vendors don't have the actual solution needed, but try to force the solution to fit with their product. They further disparage the other vendors with fear, uncertainty, and doubt about competing products. So step one is to determine the technical requirements. Most organizations look for bandwidth and reliability. If these two requirements can be reasonably met, one need only find a vendor to deliver the following:

- What you need
- When you need it
- What is reasonable financially

What About Bandwidth?

Bandwidth is always a touchy subject. It can become a "never satisfied drain" on the corporate bottom line if due diligence is not practiced. There is a direct relationship to cost and total bandwidth. The more bandwidth needed, the greater the cost. Everyone would like as much bandwidth as possible, and at the same time wants it to be affordable. Many people make the mistake of buying more than they need, anticipating future growth. In this industry, prices keep falling as competition increases. If an organization needs an OC-3 (155 Mbps) today, then laying fiber is probably the most affordable solution. However, 155 Mbps microwave systems are available and the prices are constantly dropping, giving short-haul fiber a run for the money.

Conversely, if 10 Mbps Ethernet is the current rate of transmission, then this demand can be immediately met. Additional bandwidth can be bought later. In two to three years, the costs will plummet so that the new requirements can be met with incremental or marginal costs.

It's wiser to buy bandwidth as you need it and not before (there will be a small amount of incremental add-on, but limited). In the future, there will be the following:

- More choices
- Increased providers
- Greater availability
- Lower costs

What should be done in the interim to satisfy the need? The answer is the following:

- Lease (dark) fiber instead of paying the cost of installation
- Lease services from the *Incumbent Local Exchange Carrier* (ILEC) or CLEC if sufficient bandwidth is available
- Buy a wireless connection such as point-to-point microwave

With a leased line (or a dark fiber) solution, the costs can be predictable, based on demand and agreed-upon bandwidth. Assuming physical facilities are available, take what is needed for the interim and order more only when necessary. This is a good intermediate step to get the bandwidth so long as the recurring charges are not exorbitant.

The alternative to leasing physical circuits from the ILEC or CLEC is wireless acquisition. For a one-time fee and limited recurring maintenance charges, bandwidth can be purchased for the immediate and future needs. If the wireless product delivers the bandwidth and reliability desired and the payback is reasonable, then wireless may be the best choice. Consider, however, the financial life and return on the investment.

It doesn't make sense to order an OC-3 (155 Mbps) connection when the need is only for 10 Mbps. Yet there should be an upgrade path. As mentioned, leasing copper lines or fiber from the ILEC allows a migration path. Growth can be accommodated as needed. Conversely, some wireless products limit this growth option. Some products only handle growth to a T1 or a 10 Mbps channel. Consider the expandability before buying a wireless product.

How Much Is Enough?

The risks associated with buying bandwidth fall into the two categories pointed out earlier:

- Buying too little bandwidth will increase incremental growth costs that can add up to more than buying a larger quantity at the onset would.
- Buying more bandwidth than immediately needed means paying for bandwidth that may not be required for some time, or that will be less expensive in the future.

If a T1 line slows the voice and data access to the point that users are frustrated or unproductive, then T1 is not the solution! One year of the unproductive environment costs a fortune in lost productivity.

If a couple of Ethernet channels are needed (at 10 Mbps each), and the organization invests in a fiber optic connection, justifying the expense makes an interesting paradox.

Consider approximately how much bandwidth is needed. The answer is as much as it takes to keep the data moving, voice calls coming, and users productive enough to sell (or whatever else the mission is). There is no requirement for more, yet there must be at least that much. Meeting this equation is the one that keeps the industry guessing, including the ILECs and the CLECs, as they design their networks.

What About Reliability?

Having too much bandwidth is possible. Having too much reliability is just the opposite. Organizations lose significant amounts of money when the network connection is too slow, but far more when the link is down completely. One hour of network downtime can cost more than the profits and productivity achieved from a year of uptime. In this scenario, automatic backup is an absolute must. Buy the appropriate amount of bandwidth and make sure that the reliability is built in. Plan for the worst-case scenario! Consider an alternate backup plan. Use circuit-switched or packet-switched (frame-switched) alternative connections in case of an outage.

The Choices Are Leased Lines, Fiber, or Microwave

Leased lines, fiber, and microwave each have benefits and drawbacks in terms of bandwidth, reliability, price, and delivery. The tendency is usually very application specific. Every case is different in terms of terrain, line of sight, right of way, location of a Bell central office, and so on. A 1 to 5 mile fiber choice can range in cost between $20,000 and several million dollars, depending on the terrain requiring traversal. Similarly, a high-speed leased line can cost $600 a month or $20,000 a month if it crosses *Local Access and Transport Area* (LATA) or other rate boundaries. No one solution fits all possibilities for connectivity.

Using a wireless connection, the first, full-speed Ethernet speed solution was a major milestone. From this innovation, users had viable options, bridging the extremes between leased lines and fiber. Many users still limit their choices to the bottleneck of a T1 leased line or overpaying for a T3, which is too much bandwidth for the need.

Instead, you should find a solution that provides the needed bandwidth for a justifiable cost. All three solutions offer high reliability for the most critical connections, although a mission-critical path must be backed up.

Keep in mind also that the three choices are not mutually exclusive. They frequently work well together. Microwave handles "last-mile connectivity" to a fiber backbone or serves as a lower-cost, automatic backup as insurance against "backhoe fade."[1] You should weigh each choice, based on which offers the best cost-to-benefit ratio. Cost includes installation, ongoing charges, upkeep, losses due to downtime, and organizational productivity.

Microwave and the Other Wireless Solutions

Prior to the 1970s, microwave was the most widely used wireless communications medium in the world. Microwave usage is making a comeback now with end users. Many user organizations were reluctant to experiment with microwave radio transmission due to misconceptions surrounding the technology as well as confusion between the "wireless" products. It is important to recognize that the difference between one wireless device and another can be as different as fiber and copper wire. Both fiber and copper are "wired," but that is where the commonality ceases. The same is true between microwave and laser, spread spectrum, or cellular service.

There are even differences between one type of microwave and another. The differences are due primarily to their respective operating frequencies. Some frequencies are good for distances of 30 or 40 miles and others can barely get you across an office park. Some can only support a couple of T1s or a single video channel and others go to 10 to 45 Mb.

In Table 17-2, a comparison of distances and frequencies is used for representative purposes. Many times this is the best-case scenario.

Microwave Radio Solutions

Private-user microwave systems are essentially the same as what the telephone company, FM radio stations, broadcasters, and fixed-site utility companies relied on before the implementation of fiber. For example, the corkscrew-type antennas on news vans are shooting microwave signals

[1]As the author was preparing this update in Phoenix, Arizona, a major cable cut occurred in the Chandler area, severing multiple fiber lines and disrupting local and long-distance service for more than six hours.

back to their TV and radio stations as shown earlier. FM radio stations still rely mainly on microwave. In fact, most microwave radio uses *frequency modulation* (FM) radio technology.

What can microwave offer an organization? Primarily, microwave combines huge bandwidth and reliability that is better than other wireless devices. In fact, microwave is typically far more reliable than the leased-line specification (99.985 percent) for distances across the street to 20-plus miles away. Microwave can deliver bandwidth up to 45 Mbps (and most demanded speeds in between). A properly configured system will sustain operation except in the most severe rainstorm where power and telephone line outages would be expected.

The myths run rampant with radio-based systems. Despite the rumors about the various risks and perils for the radio signal, microwave usually operates 99.99-plus percent of the time. Microwave is normally impervious to the following:

- Snow
- Sleet
- Fog
- Birds
- Pollution
- Sandstorms
- Sunspot activity

The real risk is water fade (water absorption) and multipath fade across bodies of water. These can be accommodated for the most part in design of the radio path.

A microwave link can transmit Gb of data without dropping a single bit (or packet when a data transmission uses packetized information). On

Table 17-2

Comparison of frequency bands and distances (line of sight)

Frequency	Distance
2–6 GHz	30 miles
10–12 GHz	20 miles
18 GHz	7 miles
23 GHz	5 miles
28–30 GHz	1–2 miles

copper wire, noise is always present. Thermal noise causes a continuous hum, white noise, and the like. A microwave path can be so clear that if no one is talking or sending data, the line is perfectly silent. This is difficult for the average layperson to understand.

Private User Microwave

Having proven that the bandwidth and reliability are readily available on a microwave system, most people look for the negative side of private microwave. The unaware consumer assumes that there must be some "gotcha" lurking in the background. One technology cannot be the most robust and reliable and yet not be the most favored. There has to be a major drawback to using microwave. The two largest drawbacks have always been the availability and the sticker price. Microwave *appears* to be the most expensive wireless option (typically, between $20,000 to $100,000 for an end-to-end link). However, if greater bandwidth and reliability are the goals, then microwave is not as expensive as it appears superficially. The initial upfront cost avoids the bandwidth and reliability trade-off with less expensive or less robust technologies.

If a critical connection requires high bandwidth and high reliability to prevent catastrophic losses from downtime, then more money is required initially. The old saying "pay me now, or pay me later" applies here.

Another perceived drawback (handling over a Mbps) is the requirement to have line-of-sight between locations. Line-of-sight issues are rarely show-stoppers.

First, only a small part of the remote site needs to be visible. Even if the other site is not visible, solutions exist. Actually, this is quite common. Look for high points that can be used to get visibility between both sites. A passive or active repeater site can be implemented. Setting up a repeater is not difficult, particularly with passive repeaters. The antennas are small and lightweight enough to be placed almost anywhere: on a water or radio tower, utility pole, other rooftops, and so on. An alternative is to bounce the signal off a physical obstacle (such as a mountain) and use obstacle gains to get the signal through (see Figure 17-6).

If a repeater is needed, rental space is often available on other towers. This is true in long haul and also in localized communications. Renting space on an existing tower or a rooftop is not very dramatic. Leased space typically costs $200 per month per dish. Repeaters are shown in Figure 17-7.

Figure 17-6 Obstacle gain uses a bounced signal off a natural obstacle, such as a mountain.

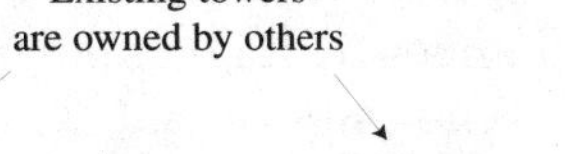

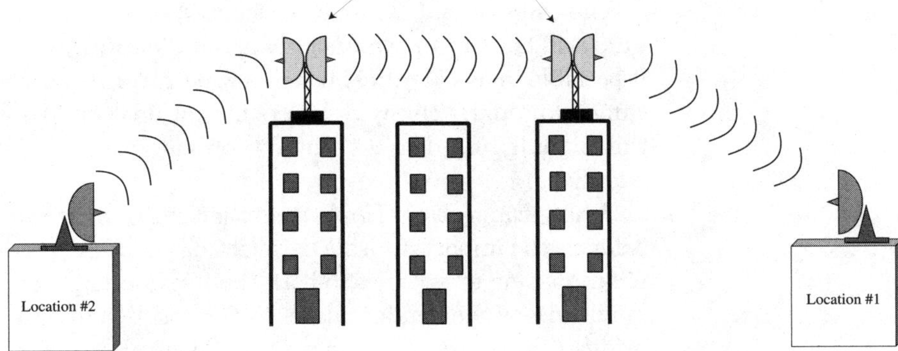

Figure 17-7 Repeater space can be rented from other suppliers.

This assuages the fact that line-of-sight is a major drawback. Anyone who has ever negotiated right-of-way for fiber knows how tedious that can be. With fiber, the right-of-way includes the entire physical length of the cable or conduit required for the connection. With microwave, the concern is limited to a few physical points where the dish placement will occur and a few legal issues.

Another perceived drawback of microwave, aside from price and line-of-sight, has to do with getting and maintaining the license to use. Licensing is a protection for the end user. It is a tedious process that provides structure and prevents interference and overlays in the frequency spectrum. Most organizations see this as a step for governmental control, but if a conflict arises, it can be the salvation for the licensed user. License gives you the right to use a good, clear transmission path, and that is definitely positive. The FCC is quite efficient in approving licenses for private users, and spectrum is readily available for point-to-point applications.

Licensing involves contracting someone to do a frequency search and filing a FCC license application. The frequency search is to find unused, available frequencies. Those frequencies are then reserved and filed, along with other pertinent data, on the appropriate FCC form. The process is normally completed within a few days, and it can cost $2,000 to have your vendor handle it for you. An experienced vendor can usually assure that the FCC will grant your license by avoiding amateurish mistakes.

The 23 GHz frequency band, for example, is a very common frequency band for short-haul, private-user microwave systems. Most people confuse the band with the actual operating frequency. The 23 GHz band actually consists of 24 pairs of frequencies, ranging between 21.200 and 23.600 GHz. The number can be doubled to 48 pairs of frequencies with minor antenna changes (changing polarization from vertical to horizontal).

The radio signal is narrowly focused by the antennas at each end of the link, and transmit power is only about 60 milliwatts. These variables make it possible to use identical frequency pairs for two links originating from the same rooftop! By changing the polarization of the antennas and separating the signals, 10 degrees should provide the necessary separation and isolation.

When connecting LANs together with microwave radio systems or bridges, the important issue is selecting the right vendor. The vendor's qualifications and experience in both the LAN and microwave systems are paramount. Does the vendor possess the expertise to differentiate a network or bridge failure from a radio problem? Can the vendor provide turnkey services and assume total system responsibility? The correct vendor will ensure the successful implementation of a LAN microwave link.

CHAPTER 18

MMDS and LMDS

Multichannel Multipoint Distribution Service (MMDS), also known wireless cable, is another wireless broadband technology for Internet access. MMDS has been around since the 1970s and is a well-tested wireless technology, which has been used for TV signal transmission for more than 30 years. The service is delivered using terrestrial-based radio transmitters located at the highest location in a metropolitan area. Each subscriber receives the MMDS signal with an exclusive, small, digital receiver placed at your location with line of sight to the transmitters. MMDS channels come in 6 MHz chunks and run on licensed and unlicensed channels. Each channel can reach transfer rates as high as 27 Mbps (over unlicensed channels: 99 MHz, 2.4 GHz, and 5.7 to 5.8 GHz) or 1 Gbps (over licensed channels). Typically, a block of 200 MHz is allocated to a licensed carrier in an area.

MMDS is a broadcasting and communications service that operates in the *ultra-high-frequency* (UHF) portion of the radio spectrum between 2.1 and 2.7 GHz. MMDS is also known as *wireless cable*. It was conceived as a substitute for conventional cable television (TV). However, it also has applications in telephone/fax and data communications.

In MMDS, a medium-power transmitter is located with an omnidirectional broadcast antenna at or near the highest topographical point in the intended coverage area. The workable radius can reach up to 70 miles in flat terrain (significantly less in hilly or mountainous areas). There is a monthly fee, similar to that for satellite TV service.

MMDS frequencies provide precise, clear, and wide-ranging signal coverage. Customers are protected from interference from other users when the provider uses the licensed frequencies. Rain, snow, and fog do not interfere with signal performance as we saw in the microwave radio chapter (see Chapter 17, “Microwave- and Radio-Based Systems”). Many of the carriers use a super-cell concept with a service area spanning a 35-mile radius from each of its MMDS transmitters.

The MMDS wireless spectrum originally consisted of 33 analog video channels, which were 6 MHz wide. The evolution of video technology into digital capacities enables the carriers to convert these 33 analog MMDS channels into 99 digital, 10 Mbps data streams, enabling full Ethernet connectivity. Therefore, a carrier with a normal operation can have as much as 1 Gbps of capacity at a single transmitter providing adequate capacities for most applications. This capacity is also readily expandable by using a sector cell concept (see the analog cellular chapter to get a handle on sectors), which reuses the same frequency many times. The combination of super cells and sectors enable the carrier to reuse the same frequency many times by building multiple cell sites. When enough customers sign on and as their bandwidth demands grow, the growth in traffic can be handled expeditiously through a new cell or a new sector.

Limited Frequency Spectrum

The limited number of channels available in the lower *radio frequency* (RF) bands characterizes MMDS networks. Only 200 MHz of spectrum (between 2.5 GHz and 2.7 GHz) is allocated for MMDS use. This constraint reduces the effective number of channels in a single MMDS system. For TV signals using 6 MHz of bandwidth, 33 channels can be fit into the spectrum. The FCC allowed for digital transmission utilizing *Code Division Multiple Access* (CDMA), *quadrature phase shift keying* (QPSK), *vestigial sideband* (VSB), and *Quadrature Amplitude Modulation* (QAM) schemes yielding up to five bits per hertz (one gigabit per second total raw capacity for the band), and return transmission from multiple sites within a 35 mile radius protected service area.

A new frequency band dedicated to digital MMDS services has been proposed, but this may be impractical in the lower microwave frequencies due to the political and business pressures from alternative video service providers. Moreover, higher transmitter power and antenna gain are required for broadcasting in this frequency range, which will require higher system costs. Higher frequency bands are not chosen for MMDS due to higher free space or path attenuation. However, the FCC has actually allocated the 27.5 to 29.5 GHz band in the United States to *Local Multipoint Distribution Service* (LMDS). It is presently intended to operate FM-based TV services, with each service occupying a 20 MHz bandwidth. Due to its limited range of transmission (3 to 5 miles radius), LMDS is not a good choice to provide wide area coverage of digital television service.

System Configuration

The typical configuration of an MMDS system is shown in Figure 18-1. The wireless system consists of head-end equipment (satellite signal reception equipment, radio transmitter, other broadcast equipment, and transmission antenna) and reception equipment at each subscriber location (antenna, frequency conversion device, and set-top device).

Signals for MMDS broadcast at the transmitter site originate from a variety of sources, just like at cable head-ends. Satellite, terrestrial, and cable delivered programs, in addition to local baseband services, comprise the material to be delivered over MMDS. All satellite-delivered baseband formats are remodulated and subsequently up-converted to microwave frequencies. Terrestrially delivered signals are usually passed through a

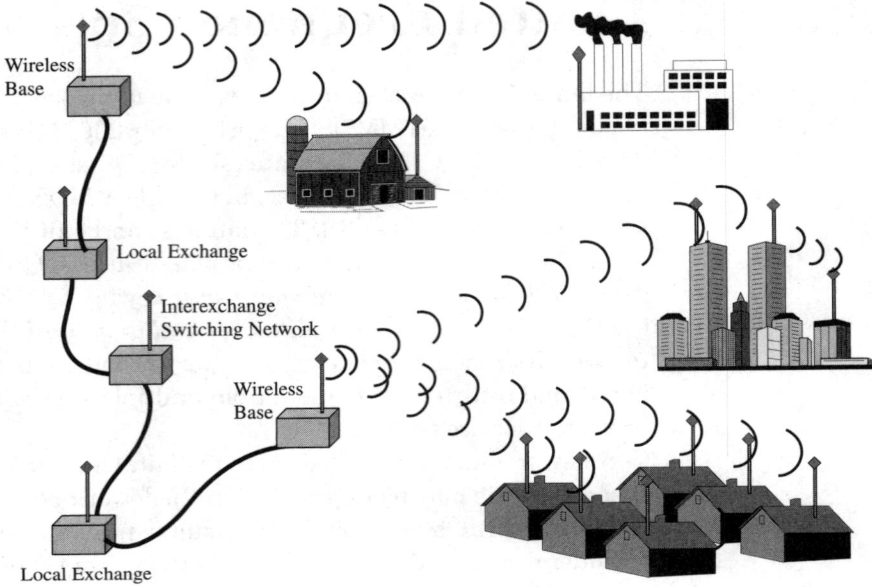

Figure 18-1
A typical MMDS arrangement
Source: AMD

heterodyne processor prior to up-conversion to the desired MMDS frequencies. Repeater stations can be used to direct MMDS signals to blocked areas. The typical range of a transmitting antenna can reach up to 35 miles, depending on the broadcast power. Transmitters usually operate in the 1 to 100 watt range. This is substantially lower than the transmission power requirements of VHF and UHF terrestrial broadcasting stations. MMDS is a line-of-sight service, so it does not work well around mountains, but it will work in rural areas, where copper lines are not available.

The antenna is designed to receive vertically polarized or horizontally polarized signals, or both, at the subscriber's location. The microwave signals are passed to a down-converter that converts the received signal to standard cable VHF or UHF channel frequencies. TV signals are then sent directly to a TV set or a set-top converter box.

Here's how a wireless cable system works:

1. The cable studio, along with the head-end, receives programming from a variety of sources (see the following section). Each source is assigned a channel number, processed to improve quality, encoded, and then sent to a transmitter. The signal is broadcast in the *super-high-frequency* (SHF) range. Using an omni-directional transmit pattern, the signal reaches subscribers located up to 50 KM from the antenna, depending on the terrain and transmit power.

2. Wireless cable signals are received by the subscriber's small rooftop antenna, decoded (pay TV), and down-converted to standard TV channels on the subscriber's TV set.
3. One of the two systems are normally used for multiple-dwellings (condo, apartment, and so on) to receive and distribute wireless TV.
   a. The building management pays for all units to receive the programming from a single communal antenna. This agreed fee is usually based on the number of potential viewers.
   b. In other buildings, a single community antenna is installed with each tenant subscribing separately and billed separately by the cable company.
4. In all cases, deposits are paid by subscribers that cover receiver system costs, much like cable subscribers.

Wireless Cable Sources

Programming can be provided from a variety of sources including

- Reception of broadcasts from local TV stations
- Playback of video tapes
- Direct "live" feeds from various locations
- Multiple satellite dishes receiving TV signals from around the world

Unfortunately, the wireless cable industry has been riddled with failure. The smaller operators were unable to generate a profitable business using the frequencies for the transmission of analog video. Several *regional Bell operating companies* (RBOCs) announced that MMDS would be their means of effectively competing with the cable TV operators. Later they sold their MMDS properties off and circled the wagons around their telephony infrastructure. BellSouth remained a significant provider of MMDS video service alongside its landline cable service (though several RBOCs like SBC and Qwest built semi-successful landline cable TV services). The U.S. markets for residential video are crowded by broadcast TV, *direct broadcast satellite* (DBS) and cable, and the limited channel capacity of analog MMDS simply could not compete.

However, in 1999, the floodgates were opened as many providers began to revisit the opportunities to use the MMDS frequencies. What caused this resurgence of interest in this portion of the spectrum was MMDS. It is seen as a viable broadband service delivery option. The Internet has changed everything. MMDS providers created Internet-focused subsidiaries. They upgraded their networks with digital compression capabilities and rapidly installed a return channel to create interactive capability. Unlike their counterparts operating in the LMDS band who mainly target businesses in metro areas, the MMDS providers mostly want to tap the pent-up demand for broadband digital data and TV directly into the home.

Advantages of Using MMDS

The following list includes some advantages of using MMDS:

- It has chunks of under-utilized spectrum that will become increasingly valuable and flexible.
- System implementation, which is little more than putting an installed transmitter on a high tower and a small receiving antenna on the customer's balcony or roof, is quick and inexpensive.
- Because MMDS services have been around for 30 years, there is a wealth of experience regarding the use and distribution of the services.

A single tower can provide coverage to a very large, densely populated area at a very reasonable cost to the service provider. Because a large number of users may share the same radio channels, data throughputs will be lower than they are for other broadband wireless options. The net result is practical data throughput of 500 Kbps to 1 Mbps, which is ideal for small and midsize business customers as well as residential consumers. Sprint's "Pizza Box"[1] service is relatively inexpensive. In Phoenix, a 2 to 3 Mbps download and 256 Kbps upload capability typically costs between $29 and 39 per month.

Internet Access

The hottest application for MMDS is Internet access; this differs from MMDS' original application of one-way "wireless cable" service to deliver

[1]The reference to the pizza box is the diamond shape and dimensions (13.5 × 13.5") of the antenna looks like a pizza box on its side.

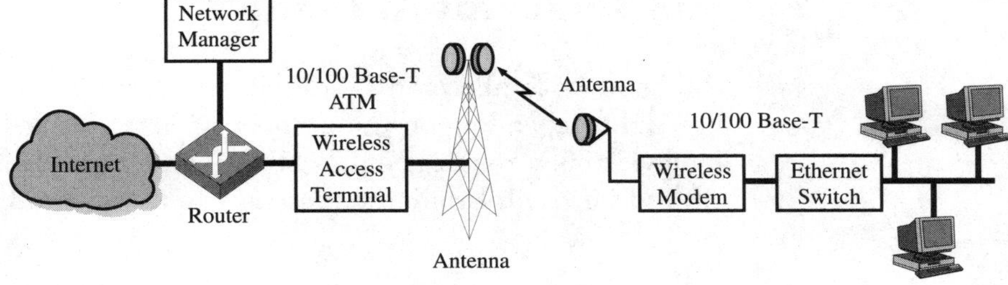

Figure 18-2 The MMDS architecture key elements

television programming. This application never proved popular, and most license holders are now concentrating on data service. An MMDS connection is just like any other ISP connection: normally a router port with a connection for the external ISP network as shown in Figure 18-2. This is an Ethernet connection to a wireless modem. Alternatively, some vendors provide a wireless modem card for their routers. A cable runs from the modem to a radio, which connects to the antenna. The radio and antenna can be combined in one compact unit. This antenna is mounted directly on your building or on a pole and points at the service provider's tower. Future versions of the technology will omit the line-of-sight requirement.

Key Elements

The key elements of an MMDS system consist of the following pieces.

The Head-End

The key elements in optimizing transmitted signal levels are the selection of the head-end site and the transmitting antenna, transmission feeders, channel combiners, channel diplexers, and transmitters. The head-end's task is to distribute the signal to as many subscribers as possible. Choosing a site with good elevation and a clear line of sight to the service area provides real dividends. This is how the CATV companies do it with their community antenna, which then delivers the signal over coax cables.

The Transmit Antenna

The bandwidth allocated to MMDS operators can vary from 200 to over 300 MHz, depending on the number of channels and their spacing. Wide bandwidth is a requirement of MMDS antennas together with downward tilt and horizontal radiation patterns to concentrate on the signal in the service area.

The Transmission Line

This is another critical component that can have a substantial effect on system losses. Major head-end sites typically use 50 or 100 watt transmitters, yet often only 50 percent of this power reaches the antenna after passing through channel combiners and transmission feeders. Waveguides from the antenna to the radio equipment vary to reduce loss and add gain.

Channel Combiners

MMDS sites normally transmit a number of channels. Special filters (channel combiners) are used to combine the outputs of the transmitters to the transmission feeder and antenna. The design of these combiners is critical to ensure they are stable with temperature, have low return loss, and provide low pass band loss.

Local Multipoint Distribution Service (LMDS)

Whenever the concept of the competitive environment enters a discussion, two other discussions ensue: the WLL and the use of LMDS. This chapter will look at some of the movement in this area to understand how and why the last mile has become so critical in meeting the demands for higher-speed broadband communications. Moreover, when looking at the *incumbent local exchange carriers'* (ILEC) copper-based plant, one can only marvel at the lack of foresight in fending off the competition. The LECs have always been in control of the last mile and invested heavily in the copper-based plant they use. Given the unshielded, twisted-pair wiring scheme and the band-limited channel capacity they deliver, one would expect them to write

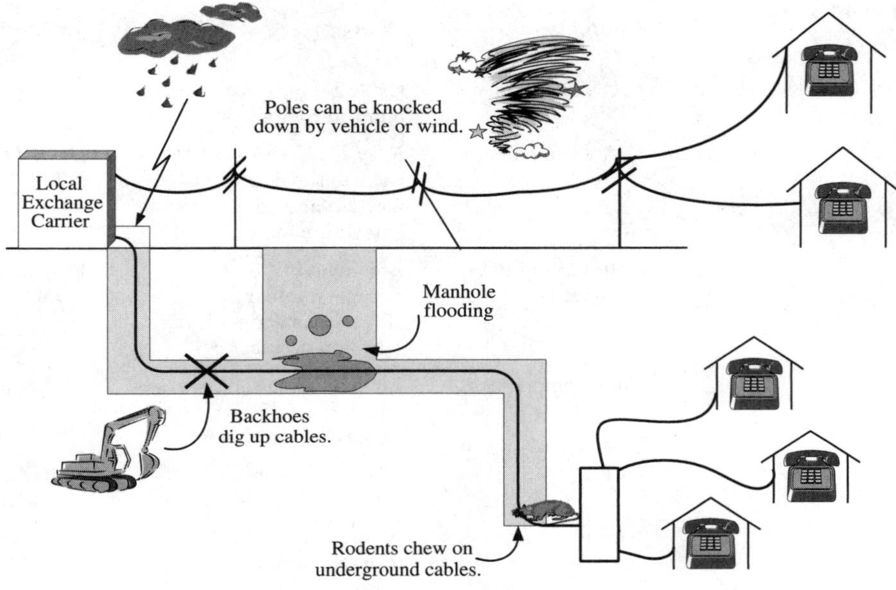

Figure 18-3
The local loop is prone to problems.

the cabling systems down (depreciate) as quickly as possible. Yet, to keep their costs competitively low, they have chosen the opposite route. Instead of fast depreciation, the LECs use a 20 to 30 year depreciation schedule on their cables. What this boils down to now is a cable plant that has not kept pace with the demands of the user and is still on the books for the provider. This is a problem for the carriers because they can ill afford to walk away from the cables in place, yet they have to breathe new life into an infrastructure that is limited in capacity. Moreover, the existing cable plant is prone to noise and disruption as a matter of course. A graphic representation of the local loop and some trouble spots is shown in Figure 18-3.

Enter the Competitive Discussion

As one discusses the possibility of the LEC having an inherent problem, there are several areas under attack at this last mile. Already discussed in earlier chapters was the use of the local cable TV operation's cable to carry voice and data communications. However, there are many areas that are being considered as competitive local loop concepts, as seen in Table 18-1.

Table 18-1

Multiple areas of competition at the local loop

Competitor	Concept	Technology
CATV companies	Cable TV for voice and Internet access	Cable modem technology
CATV companies	*Fiber to the Curb* (FTTC), Coax to the door for data and voice plus entertainment	*Hybrid Fiber Coaxial* (HFC), (FTTC)
Cellular and PCS suppliers	Broadband PCS as a single number for all wireless voice, paging, and data access	PCS on TDMA and CDMA or GSM
Local competitors	Broadband voice and data services on wireless local loop	LMDS and MMDS
CLECs	Fiber or copper to the door	SONET and local drops on copper
New wireless providers	Wireless access through various methods	*Wireless Local Loop* (WLL)

Although many of these discussions center on new technology, the two that have gained the most momentum are the concept of the WLL and the LMDS. The following information will consider these concepts more closely.

WLL

The industry in general has placed a lot of emphasis on the WLL and predicts that millions of subscribers will enjoy the benefits of untethered communications before the turn of the century. This may or may not be aggressive, but it signals the point that the competitive machine is in full swing at the last mile. Much of the growth being discussed will occur in areas where an infrastructure does not exist, such as third world countries installing the initial communications systems to the residential and business user for the first time. Many countries across the globe still do not have basic *Plain Old Telephone Service* (POTS), so it makes sense to consider a wireless connection. In some cases, the use of a *Radio in the Loop* (RITL) concept or a *Fixed Wireless Radio Access* (FWRA) concept is what the countries have dubbed the services. Countries like Brazil and China will reap many benefits from using a WLL concept, both financially and in the speed of installation. The cost of installation on a per user basis is much more favorable. One set of statistics shows the difference of the installed cable versus a wireless local access method as seen in Table 18-2.

However, the emerging underdeveloped countries are not the only places where WLL technology will be used. Instead, the developed countries around the world may also take advantage of the economies of scale and the financial benefits of installing the wireless local access. As a result, as many as 50 million access lines may be deployed worldwide shortly after the turn of the century and rapid growth may follow the initial installations. The day of installing copper to the door has ceased; instead, the wireless technologies may be the mode of choice for the future. No longer can the carriers afford the cost of installation and maintenance for the copper local loop.

Figure 18-4 is a representation of the overall concept of the WLL concept, without specific technology used but as a model for the carriers considering the use of wireless technology.

Table 18-2
Cost comparison for wired versus WLL

Technology Used	Cost per User (in U.S. dollars)
Copper local loop	$ 5,500.00
WLL	$500.00 to $800.00[1]

[1]This figure will drop rapidly to approximately $200 to $300 per user as deployment continues and economies of scale are achieved.

Figure 18-4
WLL conceptual model

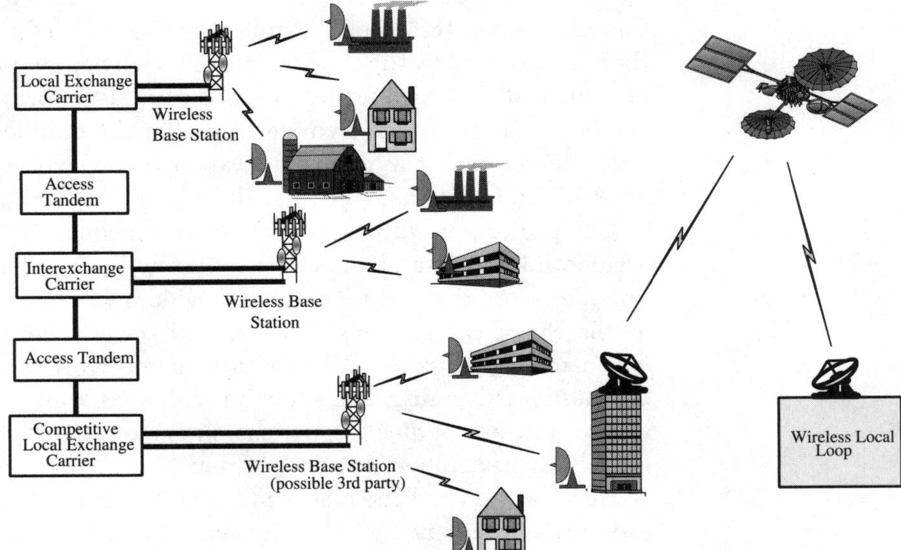

Not for Everyone

The WLL will encourage many new opportunists to jump into the market, but few will survive. Either the providers will be underfunded and will not survive the competition, or the larger providers looking for market share in an area of operation will gobble up the smaller local providers. In either case, the number of providers will change, and the operators will continually be looking for new and competitive approaches to attract customers. Full service providers will offer the list of services as shown in Table 18-3. Others may offer pieces of these services. The point is that the end user is looking for a one-stop shopping approach and the leverage that comes with bundled services.

Too many providers will jeopardize the success of many, causing some form of shakeout, but one must consider that the end user is willing to use one or more of the providers listed in the previous table. What will likely occur is a merger or a joint offering with partnering providers to get to the consumer's door. When one looks at the offerings and the carriers shown previously, it is obvious that the services are disjointed. Some providers offer all the services, whereas others are just planning the possible services they will offer. However, the infrastructure they choose to install may have an impact on their ability to service future demands. No one answer or solution jumps out right now, but changes will occur rather quickly in this business. The economics of getting the consumer (both residential and business) to buy into more than one offering will set the stage for future services. One can add the numbers and see where the providers want to take this. Table 18-4 shows a summary of service offerings (on average) for the services used by the consumer. In this particular scenario, the consumer is a home office-based user or a residential user whose needs include various bundled services. If a carrier can offer the bundled services for a moderate decrease in the monthly costs, one can expect 65 percent of those approached to churn.

Using these bundled, one-stop pricing models, one can expect that the residential and small business customer will be tempted to use the service provider. Note that not all service providers will offer the equipment (such as the PC or the modems), whereas others may. Many of the *competitive local exchange carriers* (CLECs) are toying with this idea for their total business provisioning. The smart provider will consider this bundled offering. As a means of meeting the customer's communications needs on a single bill, the consumer is a ready target.

However, some of the pieces may not be required. For example, many of the WLL providers include the analog cellular suppliers and the PCS

Table 18-3

Summary of service offerings and providers today

	Services Offered by Carriers Today and Future Offerings						
Provider	**Voice**	**Data Low Speed**	**Point to Point Data**	**CATV**	**Video Conferencing**	**Internet Access (High Speed)**	**Multimedia Services**
Cable companies	Yes	Not avail, but possible	Not avail, but possible	Yes	Not avail, but possible	Yes, 10 Mbps	Not avail.
LECs	Yes	Yes	Yes	No, but planned	Yes	Yes, 1.5 Mbps	Limited
CLECs	Yes	Yes	Yes	No, but planned	Yes/Limit	Yes, 1.5 Mbps	Limited
IECs	Yes	Yes	Yes, but not local loop	No, merge with CATV	Yes, Limited	Yes, 1.5 Mbps	Limited
WLL	Yes	Yes	Limited	No, but possible	Limited	Yes, 10 Mbps	Possible
Cellular providers	Yes	Yes	No	No	No	No	No
PCS providers	Yes	Yes	No	No	No	Limited	No

Table 18-4

Bundled versus individual services plans

Service Offering	Average Monthly Price	Bundled Price
CATV (basic cable)	$8.00	$100.00
Extended channel services[1]	$23.00	
Basic Internet service provider access	$20.00	
Data access for Internet at dial-up rates	$25.00	
Dial tone for voice	$25.00	
High speed Internet access 6 1+ Mbps	$50.00	$35.00
Long distance services (typical customer)	$25.00	$15.00
Equipment costs amortized (PC, modem, telephone, and so on) monthly	$30.00	$30.00
Cellular phone basic plan	$50.00	$30.00
Total monthly fees	$256.00	$210.00[2]

[1] Excludes premium channel services (HBO, Showtime, and so on) and pay-per-view services.

[2] The goal of the providers with one-stop shopping is to offer all the services at approximately $200.00 per month.

suppliers whose recent advertisements state that consumers can remove their wired telephone and use the cellular or PCS service for their home and business needs. This is possible and has some merit. Thus, if the consumer takes this provider up on this advertisement, the carrier loses the bundling of a $25.00 monthly dial tone service. However, the cost of the cellular plan will increase in the number of minutes used, driving that plan cost up higher. Effectively, this may become an even trade.

Another point here is the cost of the infrastructure. Once the CATV companies have delivered the basic cable services, for example, the cost of any added usage or shared bandwidth on their infrastructure is usually marginal. Thus, the profitability and mark-up is that much higher. The wired carriers understand the benefit of one-stop shopping; now the WLL carriers are learning very quickly.

What About the Bandwidth?

The bandwidth necessary for each of these services listed previously changes the rules considerably. In many of the WLL providers' backbone, there is not enough bandwidth to support the number of users and the higher-speed services. For this reason, the marriage of the providers may occur sooner than expected. If a cellular provider joins forces with a WLL microwave supplier, then the bandwidth for the fixed needs at the door is assured while the cellular provider handles the demands of the roaming user. These combinations and permutations can be very complicated as the number of providers expands and the services they offer shift in any direction. The interesting point will be to see how the total market plays out with an expectation that approximately five to seven providers will dominate, and the rest will be absorbed or fail.

Enter LMDS

LMDS, as its name implies, is a broadband wireless technology that is used to deliver the multiple service offerings in a localized area. The services possible with LMDS include the following:

- Voice dial-up services
- Data
- Internet access
- Video

Just as the network providers were getting used to the battlegrounds between the ILECs and the new providers, RF spectrum was freed up around the world to support access and bypass services. Typically, the services operate in the RF spectrum above 25 GHz, depending on the licenses and spectrum controlled by the regulatory bodies. This offering operates as a point-to-point, broadband wireless access method, which can provide two-way services. Because LMDS operates in the higher frequencies, the radio signals are limited to approximately five miles of point-to-point service. This makes it somewhat like a cellular operation in the way the carriers lay out their operations and cells. An architectural concept for the LMDS operation is shown in Figure 18-5 from the perspective of the supplier to the user. This figure uses some of the premises that the service is constrained to a localized area. (Occasionally in uncongested and unpopulated areas,

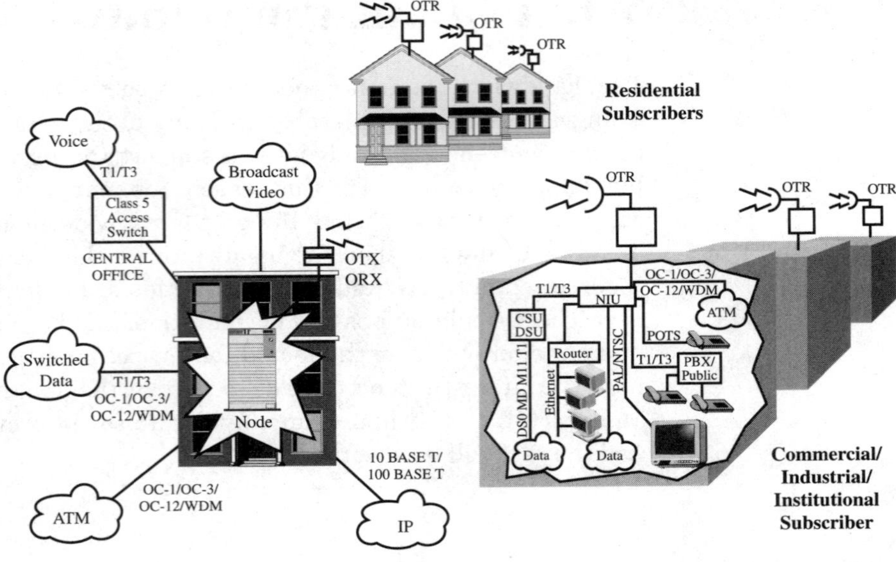

Figure 18-5 Typical LMDS service areas Source: LMDS Org.

the signals are transmitted in much wider areas of coverage, similar to other wireless technologies. This reference to the five miles is within populated areas and the obstacles that will be encountered within the areas.)

The Reasoning Behind LMDS

Point-to-point fixed microwave radio has been in use for decades in the local loop environment. Many organizations (individual businesses, utility companies, and so on) required dedicated access to their own private network facilities or to a carrier's *point of presence* (POP). As they would approach the ILEC, the cost to run high-speed services to the business consumer's door was typically prohibitive. The monopoly owning the embedded infrastructure could literally demand any price that seemed appropriate. This met with objections from the user, but as long as the monopoly existed, there were few choices. The businesses, therefore, demanded frequency spectrum to install their own infrastructure at the last mile. The connection was typically in a special, set-aside frequency band as shown in Figure 18-6, using distance-sensitive frequencies as listed in Table 18-5.

The problems of the fixed access methods using microwave in the past included the following:

- Local ordinances were unfavorable to the use of the technology.
- Local regulatory bodies had several restrictions.

Figure 18-6 Typical microwave point-to-point services of the past

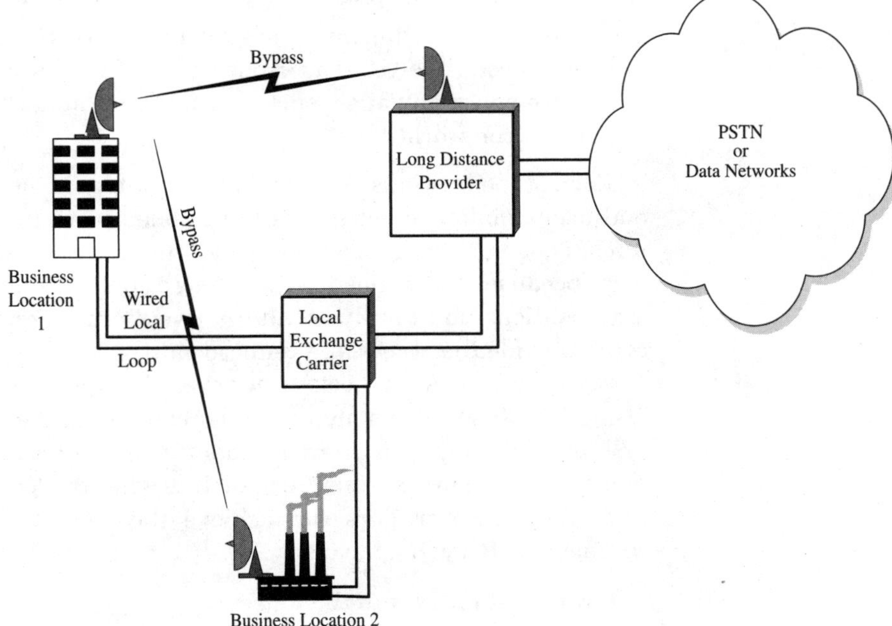

Table 18-5 Typical microwave distances, bands, and operations

Frequency Band	Distances	Use
2 to 6 GHz	30 miles	Commercial, utility, fixed operation, TV
10 to 12 GHz	20 miles	Commercial, utility, fixed operation, TV, DBS
18 GHz	7 miles	Business, limited fixed operation
23 GHz	5 miles	Business, limited fixed operation
25 GHz and above	3 to 5 miles	Business, bypass operation

- Federal authorities limited the use.
- Cost of building the tower.
- Cost of security for the site.
- Local power and other utilities might not be readily available.
- Cost of the equipment was very high.
- Maintenance costs were unnecessarily high.
- Line-of-sight frequencies might not be readily available.
- FCC-licensed technicians were required to do the maintenance (Reference to the U.S.-based operations. In other countries, similar requirements prevailed, such as CRTC in Canada and PTT in other parts of the world)

Each of the previous areas was somewhat limiting to the demands of the end user's ability to get access to the fixed point-to-point microwave systems. The largest organizations could financially justify the use of this service because their needs were more demanding. However, smaller organizations had to rely on alternative methods or service bureaus that could provide the access at a reduced rate.

From a carrier's perspective, however, the equation changes very quickly. Using LMDS services, a new provider can install the systems more readily due to the competitive environment being introduced worldwide. The monopolies no longer mandate or dictate what the local connectivity will be like. The new providers can achieve the benefits of the LMDS world through the following means:

- Lower cost entry into the market.
- Costs are deferred to later when services are needed. This moves the pricing model from fixed to variable costs associated with demand, as opposed to fixed size increments.
- *Return on investments* (ROI) are achieved more quickly, encouraging the provider to enter the market.
- Less risk of customer churn, leaving the carrier stuck with large investments.
- Ease of installation and licensing makes the implementation faster.
- Standards-based services and equipment, minimizing obsolescence and proprietary solutions.

The carriers seem to have found a nirvana of technology and financial benefit in a single solution. The real issues then begin to work around the need, demand, and the method of delivery. Not all systems are implemented the exact same way, so the carrier still has some choices, enabling even greater flexibility in delivering the bandwidth to the door.

Network Architectures Available to the Carriers

As already stated, various means of installing leave the carrier choices. The bulk of the carriers will likely standardize on a straight, point-to-point connectivity solution for their customers. Point-to-point TV distribution can also be provided with the LMDS offering. This increases the attractiveness of the LMDS supplier when the other services desired by the end user, such as voice, data (IP), and multimedia applications, are added to their TV distribution capability. The architecture of the LMDS will lend itself to these point-to-point services nicely. The primary pieces constituting the LMDS system are as follows:

- The *network operation center* (NOC) contains all the management functions that manage all the components of a much larger infrastructure.
- The cabled infrastructure is usually fiber-based to connect the components of the LMDS to the public-switched and private networks. The cabling will consist of T1/T3 or OC-1, OC-3 or OC-12 connecting to the ATM and Internet backbones.
- The *base station* (BS) is where the fiber-to-radio frequency conversion takes place; the modulation of the signal across the airwaves occurs here also.
- The customer equipment, which can vary from user to user and by vendor, has to satisfy the demands of the consumer.

The architecture also varies in the modulation of the signals onto the RF (airwaves), based on the chosen strategy of the vendor. Two predominant methods of using the technology are to use an analog interface such as *Frequency Division Multiple Access* (FDMA) or a digital interface, using *Time Division Multiple Access* (TDMA). The choice will vary, depending on the density of the sectors being served, the available financial choices and the overall quality desired. The more common implementation is to use a FDMA technique to serve the customer. The use of FDMA enables better coverage and density of the applications being served, and it uses a higher modulation technique to satisfy the system demands.

Modulation and Access Techniques

As already discussed, the modulation and access method falls into two primary categories, FDMA and TDMA. Each of these techniques differs but also creates other submodulation capabilities. For the broadband LMDS services, the system is usually separated into both phase and amplitude modulation of the RF. *Phase-shift keying* (PSK) and amplitude modulation combinations have been successfully used to achieve high rates of multiplexing and carrying capacities. The options of using FDMA and TDMA are similar for the RF spectrum and will be discussed generically here.

The ultimate goal is to multiplex the most services and modulate the least amount of RF spectrum to achieve the same throughput. Table 18-6 is an example of the desired results of modulating less amount of RF to get the same amount of effective throughput (in this case a 2 Mbps data rate). The table highlights the various methods used by different vendors to achieve the data rates such as the techniques used to modulate the signal.

We can see the capacities and differences available to modulate the signal over an FDMA technique. These variations are what will separate the supplier's ability to satisfy future demands to sustain a 2 Mbps data rate with the least amount of bandwidth.

Table 18-6

Summary of modulation techniques available for LMDS in FDMA

Technique Used	Modulation Method	Bandwidth Required to Sustain 2 Mbps Data Rate	Number of Bits per Modulation Technique
BPSK	Binary Phase Shift Keying	2.8 MHz	1:1
DQPSK	Differential Quaternary Phase Shift Keying	1.4 MHz	2:1
QPSK	Quaternary Phase Shift Keying	1.4 MHz	2:1
8PSK	Octal Phase Shift Keying	0.8 MHz	4:1
4-QAM	Quadrature Amplitude Modulation, 4 state	1.4 MHz	2:1
16-QAM	Quadrature Amplitude Modulation, 16 states	0.6 MHz	4:1
64-QAM	Quadrature Amplitude Modulation, 64 states	0.4 MHz	6:1

In the TDMA alternative, time slots are more efficiently used to deliver the capacity to the end user, but the same techniques as listed previously are used, with the exception of the 64-QAM. The problem with a TDMA method is that the time slotting (fixed in most cases) uses more of the spectrum in overhead and therefore produces less efficiency in the RF side of the business. Each carrier must be aware that the extra overhead associated with this time slot usage can detract from the overall system performance.

Two-Way Service

The TDMA and FDMA modulation techniques on the LMDS network allow for the bidirectional flow between the carrier and the end user. In many cases, a different upstream is required than the downstream. The ability to modulate differently enables this to be compensated for. The more important factor is that the service will offer two-way communication. Many of the past services enabled only a one-way downstream with a dial-up upstream. This sounds incomprehensible for high-speed data, but for Internet access this method was used initially on an MMDS service and then on LMDS in the initial rollout. Later evolution of the network architectures enabled the carriers to change direction and satisfy both directions for their data needs. Moreover, for two-way voice the two-way simultaneous transmission is a must.

When lower speed users are using the system, TDMA is an effective tool for their two-way voice and data (dial up) needs. For simultaneous up and downstream services, using approximately 250 MHz in each direction, the average number of TDMA users per 5 MHz of spectrum handling use dial-up service at the DS0 rate 80. This means that a sector (or cell) using five separate streams of 5 MHz each, can achieve up to 4,000 simultaneous dial-up users. This is a reasonable use of the bandwidth. The overall network design on a wired world uses a 10:1 ratio of trunks to users. Using a slightly lower ratio for a wireless network connection to accommodate for fax and long-hold time traffic (that is, Internet surfing) of approximately 8:1, the network supplier can achieve a service level of 32,000 possible customers with normal demands in a sector (cell) with 250 MHz of RF spectrum. This is a reasonable amount of traffic capacity based on standard traffic engineering design. Better ratios have been used in some networks, but the issues are coming with long hold times that exceed expectations. Using a conservative ratio of 5:1, the average network supplier can achieve service for 20,000 DS0 level users in a sector. Keep in mind that when higher-speed demands are the requirement, an FDMA arrangement will enable more flexibility. This all depends on the overall demands of the network users.

Propagation Issues

Like any radio-based system, the issue of propagation is always a concern. Like the analog cellular networks of the past, there are several factors that contribute to the quality of the signal. Many operators have to consider that at the higher frequencies (over 25 GHz), rain fade will be a critical factor. The higher the frequency, the more susceptible to rain fade than lower frequencies. One CLEC chose to use all 31 GHz radio equipment in their infrastructure to get to the customer's door. Other issues have a bearing on the design and layout of the system such as the following:

- Distance
- *RF Interference* (RFI)
- *Electromagnetic Interference* (EMI)
- Multipath fade
- Frequency reuse

In each case, the individual carrier will have to assess the overall system design specifications to meet the needs of the consumer, either residential or business. No one solution is going to satisfy all systems providers or consumers. The constant shift in network architecture will be required in a fine-tuning approach to provide the necessary quality.

Because these systems mimic the cellular network of the early 1980s, they have similar concerns. None of the concerns are insurmountable. The issue is that the carriers have a more fixed target, rather than a moving target in the wireless mobile networks. Each case, however, offers the capability to service a wide-range of needs dictated by the customer, rather than the network itself. Progress comes in many ways.

CHAPTER 19

Specialized Mobile Radio (SMR)

The *Specialized Mobile Radio* (SMR) service was first established by the Wireless Radio Commission (part of the *Federal Communications Commission* [FCC]) in 1979 to provide land/mobile communications on a commercial basis. An SMR system consists of the following:

- One or more base station transmitters
- One or more antennas
- End user radio equipment that usually consists of a mobile radio unit, either provided by the end user or obtained from the SMR operator for a fee

SMR users operate in either an *interconnected* mode or a *dispatch* mode:

- **Interconnected mode** Links the mobile radio with the *Public Switched Telephone Network* (PSTN). An end user transmits a message via the mobile radio unit to the SMR base station. The call is then routed to the local dial-up telephone network, which enables the mobile radio unit to function as a mobile telephone.
- **Dispatch mode** Allows two-way, over-the-air, voice communications between two or more mobile units or mobile units and fixed locations.

SMR customers using dispatch communications include construction companies with several trucks at different jobs or an on-the-road trucking company with a dispatch operation in a central office.

SMR systems consist of two distinct types: conventional and trunked systems. A *conventional* system allows the use of only one channel. If someone else is already using that end user's assigned channel, the end user must wait until the channel is available, as shown in Figure 19-1. In contrast, a *trunked* system combines channels and contains microprocessor capabilities that automatically search for an open channel. This search capability enables more users to be served at any one time. A majority of the current SMR systems are trunked systems.

Although SMR is primarily used for voice communications, systems are also being developed for data and facsimile services. Additionally, the development of a digital SMR marketplace allows for the features shown in Table 19-1.

SMR growth is attributed to these developments and features. For example, at the end of 1998, approximately 8.8 million vehicles and portable units were served by SMR systems. Several radio-based systems have been introduced and used throughout the years. Many of these operated on a very specific frequency band and in specific geographic areas of the country in which they were approved. The purpose of a radio system is to provide the wireless technology to satisfy various needs as shown in Table 19-2.

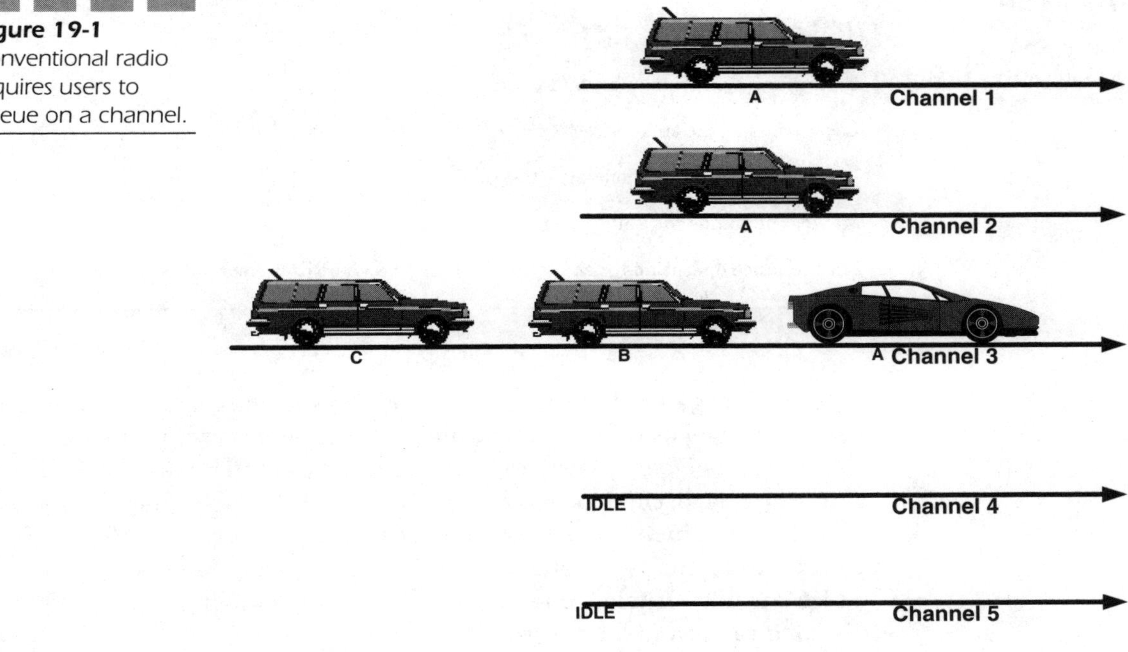

Figure 19-1 Conventional radio requires users to queue on a channel.

Table 19-1 Features Possible with Digital SMR

Features of a Digital SMR Network
Two-way acknowledgment paging
Inventory tracking
Credit card authorization
Automatic vehicle location
Fleet management
Remote database access
Voicemail

Although these services are old hat, they still demand the connectivity necessary to be reachable at a moment's notice. Over the years, the demands have changed and matured. The FCC and *Canadian Radio-television and Telecommunications Commission* (CRTC) have issued several standards and set aside frequency bands to accommodate these services. One of the primary goals was to employ the use of SMR to help meet the need.

Table 19-2

Radio Services Meeting the Need

Needs Addressed by Radio-Based Systems
Emergency operations (fire and police)
Local transportation services
Long haul communications over-the-road
Medical demands for ambulance services
Maintenance operations such as the Dept. of Transportation for road systems

As SMR began rolling out, the obvious players who commanded the market share developed the products and services necessary to support the use of SMR. The bigger developers and manufacturers (for example, Motorola, Ericsson, Sharp, Uniden, and others) developed products and accessories to work in the industry and meet the demand of these users. SMR also incorporated different types of service into the major systems and operation, which included a telephone, pager and short messaging service, two-way dispatch radio, and data transmission.[1]

Motorola's iDEN™ technology is a classification of SMR that is based on a variety of proven *radio frequency* (RF) technologies. The technology offers increased spectral efficiency and full-service integration, two of the main benefits of digital communications. The iDEN is also the basis for the SMR system operated by NEXTEL, a competitor in the wireless business who now touts that they have better services and coverage than most of the *personal communications services* (PCS) suppliers.

Improved Spectral Efficiency

The capacity to accommodate crowded markets and worldwide growth is a critical component of iDEN. The development of this spectrally efficient technology allows multiple communications to occur over a single analog channel. This expansion of the network gives users greater access to the network and provides space for new and expanded services to be added without rebuilding the infrastructure.

[1]The data transmission service of the SMR system is limited to slow speed data, but improvements will enhance this and increase the data rate to over 33 Kbps.

iDEN represents a significant step toward the integration of wireless business communications systems that meet today's demands for a one-stop process. Motorola used a combination of technologies to create the increased capacities and the combination of services. Much of the enhancements and increased capacities come from Motorola's *Vector Sum Excited Linear Predictor* (VSELP)[2] vocoding technique and *Quadrature Amplitude Modulation* (QAM) modulation process, as well as the *Time Division Multiple Access* (TDMA) channel splitting process.

Motorola's VSELP-Coding Signals for Efficient Transmission

The key to the expanded capacity is the reduced transmission rate needed to send information. Motorola has developed a vocoder technology that handles the process as shown in Table 19-3.

This vocoder, known as VSELP, compresses the voice signals to reduce the transmission rate needed to send information. Moreover, VSELP provides for clear voice transmission by digitizing the voice and providing high-quality audio under conditions that normally will result in a distorted analog voice. Using speech extrapolation, the VSELP decoder can "repair" the loss of a speech segment over the radio channel. The result is less distortion and interference (for example, break-up, static, and fading) as users move toward the periphery of the coverage area, enhancing the clarity and quality of voice communications at the outskirts of a cell.

Table 19-3

The VSELP Coding Process is Straightforward

The VSELP Coding Process
Compresses voice signals
Creates digital packets of information (voice)
Assigns the packets a time slot
Transmits
Receives the information on the iDEN network

[2]VSELP is a trademark of Motorola.

QAM Modulation

While the VSELP compresses the signal and reduces the transmission rate, QAM increases the density of the information. QAM modulation technology was specifically designed to support the digital requirements of the iDEN network. Motorola's unique QAM technology transmits information at a 64 Kbps rate. No other existing modulation technology transmits as much information in a narrow-band channel.

Multiplied Channel Capacity

Another essential element is the TDMA. TDMA is a technique for dividing the wireless radio channel into multiple communication pathways. In the iDEN system, each 25 kHz radio channel is divided into six time slots. During transmission, voice and data are divided into packets. Each packet is assigned a time slot and transmitted over the network. At the receiving end, the packets are reassembled according to their time assignments into the original information sequence.

The Advantage of Integration

More than ever before, users are demanding multifunction devices that are simple to use. With the iDEN network, users need only one telephone to access voice dispatch, two-way telephony, short message service, and future data transmission. This integration provides business users with flexible communications that enable users to access information in the most efficient and convenient way, no matter where they are in the system.

The SMR systems are part of a larger family of products that are classified as trunked radio systems. Trunked radio involves a combination of wired and wireless communication typically found in the emergency service operation, such as fire, police, and road-maintenance operations. What is trunked radio?

A Short Overview of Trunked Radio

Historically, these systems have provided one-to-many and many-to-one voice communications service—also known as mobile dispatch services.

These systems are operated by commercial entities, otherwise known as service providers that are in the business to resell their services to other entities for a profit. For over 100 years, the lines between telephone exchanges have been shared between customers. In early telephone exchanges, operators would patch calls through when a customer was located there. The operator just selected the next available circuit. Now the allocation is automatic, but the result is the same.

Radio channels were shared since the early days of radio. The operators had to listen on a frequency to determine if it was in use. Mobile telephone systems also required a customer to find an inactive channel manually. These systems were upgraded by hardware that could find a vacant mobile-telephone channel automatically and by two-way radios with subaudible (*Compatible Time-Sharing System* [CTSS]) tone equipment. This equipment was available in the 1950s.

In the 1980s, microcomputers brought a revolution in controls. A computer could be installed inside a two-way radio. The result was the development of cellular telephones and trunked radios. Both systems have a central computer managing the system. The main computer communicates with the mobile radios via an inaudible data signal.

When a trunked radio user wants to talk with someone on the same logical channel, he/she presses the microphone (Push to Talk) button. The radio sends a data signal to the controller requesting a channel. The controller responds with a physical channel number. The requesting radio switches back to receive long enough to hear this information. At the same time, all radios in the system hear the same data. These radio systems use repeaters. The base station transmits from a tall tower or building on the base frequency. The mobile units listen on that frequency, but transmit on a paired frequency (the mobile frequency).

A trunked radio system gets its name from the trunk used in commercial telephone systems. A *trunk* is a communications path between two or more points, typically between the telephone company's central offices. Several different users share the telephone trunk, but each user is not aware of this sharing of lines. The caller places a call to another party, and the call is completed.

Radio communication over a trunked system is similar to the telephone system. The transmitting and receiving radio units can be thought of as the calling and called parties. Instead of telephone lines, the radio system uses radio channels to place calls. As with the telephone system, the radio users are not aware of which trunk or radio channel they are communicating over. Trunking a multichannel radio system increases the efficiency of the radio and the radio channels.

The concept behind trunking is very simple. Trunked radio is the pooling of several radio channels so that all the users in a given area have automatic access to any free channel. The result is a system that can provide private, wide-area, wireless communication to many different users without interference or interruption.

Trunked radio is very different from the old conventional radio where you had to contend with interference and congestion. Trunked radio, the least expensive wireless service, allows both dispatch communication and the ability to place and receive phone calls, similar to cellular.

A trunked radio system is always comprised of several radio channels. One channel acts as the *control channel* (CC), while the others carry traffic. The CC is used for registrations, the transmission of status messages, and for call requests. This is not unlike the cellular, radio-based system using the paging and CCs in a cellular operation. The difference is that in a cellular radio network, there are several (19 to 21) channels set aside, whereas in trunked radio only one channel is needed (see Figure 19-2).

Upon requesting a call, a talk path is allocated on an exclusive basis to the subscriber from a pool of radio channels. The call is processed on this channel. If the trunked radio system receives additional call requests, a different channel is allocated to the calling party from the pool. As soon as all channels are in use, new call requests are stored in a queue. When a channel becomes available, the requested call is switched to the first available channel on a first-in, first-out basis.

This method means that a call request need only be sent once. If the call cannot be set up immediately, the system stores the call request and processes it later.

The Control Channel (CC)

Each radio cell consists of a Trunked Radio Exchange and a *Radio Base Station* (RBS). The Trunked Radio Exchange can be used as a *Master System Controller* (MSC) or as a *Trunking System Controll*er (TSC). The Trunked Radio Exchange manages the radio channels of the RBSs. One of these channels is used as the CC. The SMR base is shown in Figure 19-3.

When a mobile set is powered on, it automatically registers using the CC. Once the subscriber receives a positive acknowledgement from the network, the mobile is registered on the trunked radio system and can be used.

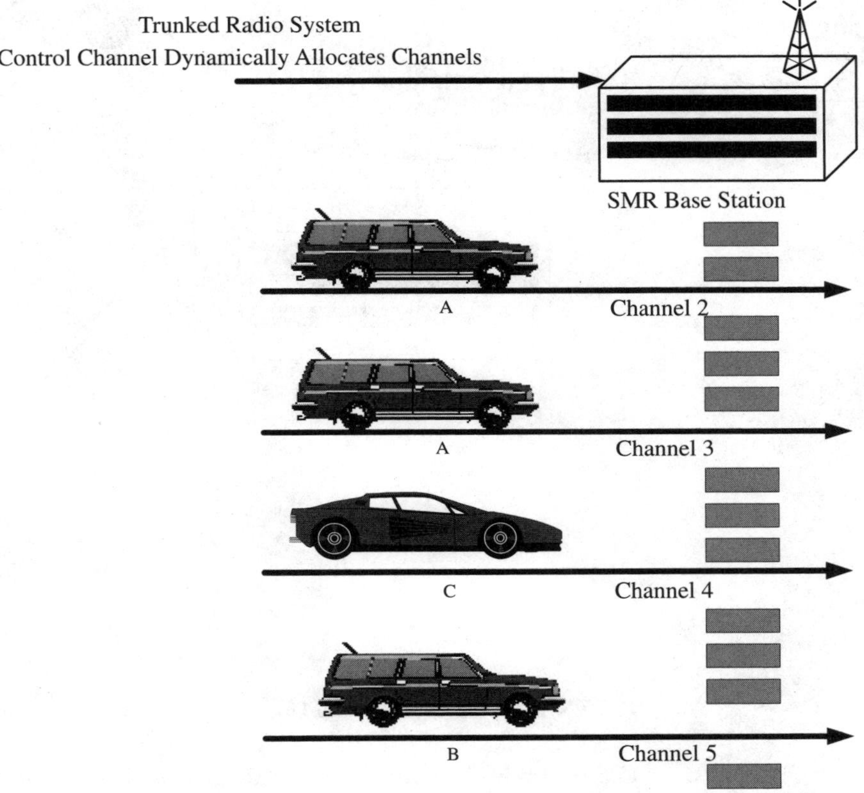

Figure 19-2 Control and talk channels in a trunked radio system

The mobile is constantly in contact with the CC. If there is a call request, the trunked radio system checks whether the addressed subscriber is available. If he is not available, not registered, or engaged, this information is given to the caller. If the requested subscriber is available, the call is set up by the Trunked Radio Exchange, using a free traffic channel. Status messages and short data are submitted on the CC.

Trunked radio systems are those which share a small number of radio channels among a larger number of users. The physical channels are allocated as needed to the users who are assigned logical channels. The users only hear units on the same logical channel. This uses the available resources more efficiently since most users do not need the channel 100 percent of the time.

Figure 19-3
SMR base station and radio service

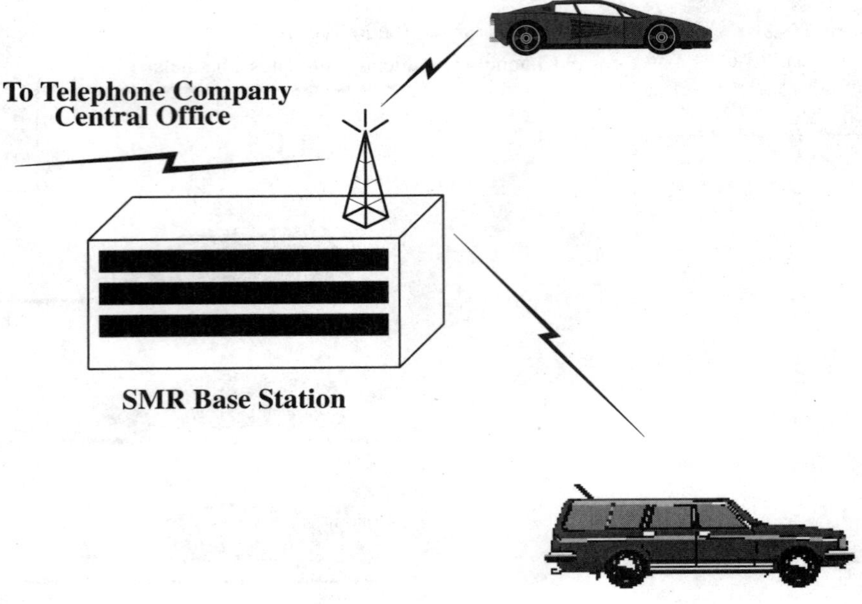

Service Areas and Licensing Blocks

Older 400 MHz channels are still in operation, as well as TV channel operation. Two sets of frequency bands are available for SMR operation: 800 MHz and 900 MHz. Approximately 19 MHz of spectrum is available for use by SMR operators (14 MHz in the 800 MHz band and 5 MHz in the 900 MHz band). The 800 MHz SMR systems operate on two 25 kHz channels paired, while the 900 MHz systems operate on two 12.5 kHz channels paired. Due to the different sizes of the channel bandwidths allocated for 800 MHz and 900 MHz systems, the radio equipment used for 800 MHz SMR is not compatible with the equipment used for 900 MHz SMR.

The 900 MHz SMR service was first established in 1986 and initially employed a two-phase licensing process. In Phase I, licenses were assigned in 46 *Designated Filing Areas* (DFAs), comprised of the top 50 markets. Following Phase I, the FCC envisioned licensing facilities in areas outside these markets in Phase II. Meanwhile, licensing outside the DFA was frozen while the commission completed the Phase I process. The freeze on licensing outside DFAs continued until 1993, when Congress reclassified most SMR licensees as *Commercial Radio Service* (CMRS) providers and established the authority to use competitive bidding to select from among mutually exclusive applicants for certain licensed services. During the freeze, however, some DFA licensees elected to become licensed for secondary sites (for example, facilities that may not cause interference to primary licensees and must accept interference from primary licenses) outside their DFA to accommodate system expansion.

In response to Congress' reclassification of the SMR service in 1993, the commission revised its Phase II proposals and established a broad outline for the completion of licensing in the 900 MHz SMR band. The 200 channel pairs in the 900 MHz service have been allocated in the 896–901 MHz and 935–940 MHz bands. Each MTA license gives the licensee the right to operate throughout the MTA on the designated channels, except where a cochannel incumbent licensee is operating already.

Frequencies have standard separation between the base and mobile pairs. Table 19-4, shows the operating bands for the base and mobile radio and the separation between the channels.

Innovation and Integration

Motorola's integrated digital-enhanced network technology and protocols combine dispatch radio, full-duplex telephone interconnect, short message service, and data transmission into a single integrated business communications solution. The digital technology was the result of studies indicating that a high percentage of dispatch users carried cellular telephones and 30 percent of cellular users carried pagers, along with an increase in demand for data communications. For network design efficiency, iDEN uses a standard seven-cell, three-sector reuse pattern.

Table 19-4
Frequency Pairing for SMR

Band	Base Station	Separation	Mobile Device
800 MHz	851–869 MHz	45 MHz lower	806–824 MHz
900 MHz	935–940 MHz	39 MHz lower	896–901 MHz
450–470 MHz	450–455 MHz	5 MHz higher	455–460 MHz
450–470 MHz	460–465 MHz	5 MHz higher	465–470 MHz
TV band	470–512 MHz	3 MHz higher	6 MHz TV channel

The technology is designed to work around many SMR spectrum limitations as well. You can take individual channels and group them together to work as a single capacity. In cellular communications, the spectrum must be contiguous. Enhanced voice places iDEN-based services more on par with TDMA, GSM, and *Code Division Multiple Access* (CDMA) vocoders.

Although dispatch mode is simplex and not full duplex, connections are quick. It's very efficient and very fast. A typical cellular call with speed dial would take 7 to 10 seconds for a path to be established. With Motorola's product it takes about 1 second. Add in 140-character alphanumeric displays (for short message capability) and direct, circuit-switched data support, and you get innovation and integration in one neat little package.

Spectral Efficiency with Frequency Hopping

Geotek targets the same wireless niche with its digital frequency hopping, multiple access-based technology, and networks. Geotek holds a near unique position as manufacturer and spectrum owner/service operator, offering an integrated suite of mobile office solutions for dispatch, fleet management, and mobile businesses. Service offerings include dispatch, telephony, two-way messaging, automatic vehicle location, and packet data.

Frequency Hopping Multiple Access (FHMA) technology lies at the core of Geotek's networks. This TDMA and spread-spectrum derivative, originally developed by the research and development arm of the Israeli military, employs frequency hopping to achieve substantial benefits in flexibility and spectral efficiency.

FHMA achieves 25 to 30 times the capacity of existing analog technologies using a macrocell approach. Macrocells typically cover areas up to 70 miles in diameter with up to 10 radial slices or sectors, incorporating from 5 to 20 microsites.

Within sectors, FHMA implements synchronized hops from one discrete frequency to another, in a predetermined manner at both the transmitter and receiver—you stay on one piece of spectrum for only a short time.

FHMA slices information packets into pieces, shuffles, and then transmits them. Packet losses result in minimal degradation since sequential losses don't occur. In addition, the system uses two-branch diversity, incorporating two separate antennas and receivers on both ends. This space diversity ensures that the best incoming signals are chosen.

The system is built on TCP/IP, and every system user has an IP address. With its inherent data integration, the Internet logically becomes a larger part in Geotek's service strategy. Geotek's automatic vehicle location capability is illustrative of things to come. It has sophisticated business data applications that will drive the mobile business markets.

Digital Transition

Digital technology is an integral part of Ericsson's SMR and private radio systems (*enhanced digital access communications system* [EDACS]), but not as an all-or-nothing requirement. Its systems can migrate to all-digital configurations, as customer needs dictate. Ericsson's EDACS technology package employs standard trunked radio.

Ericsson's Aegis system, which is an all-digital system, uses an adaptive multiband encoding vocoder to transform analog voice signals to digital. After the vocoding process, error protection codes are added to the digitized audio stream. This process is further augmented with synthetic audio regeneration, which replaces certain portions of the voice signal that are corrupted by noise with usable segments of speech.

EDACS embeds its control information on the channel as well. This includes unit identification or push-to-talk identification, priority scan information, and talk group segmentation. With error correction and detection added, as well as channel signaling and synchronization, the combined signal transmits at 9,600 bits per second.

The combined voice and data encoding is an EDACS feature not all systems can match. Many others require separate channels for voice, data, and control signaling. EDACS combines all three, automatically compensating for periods of high demand and with no sacrifice in capacity or reliability.

Ericsson plans system enhancement and upgrades, such as TDMA. To date, its upgrade strategy has been to protect the customers' embedded investment in private or SMR technology and provide a painless and cost-effective move to digital.

Is There Still a Benefit from Two-Way Radio?

Frankly, yes! Any business with personnel in the field or any business where efficient use of resources can reduce operating costs and increase profits will benefit from two-way radio. Now that doesn't leave out too many businesses, does it? In fact, hundreds of businesses, of all sizes and types, have mobile radios in use, and the list grows with each year. Police departments, delivery services, realtors, school systems, industrial plants, farmers, repair services, construction companies, contractors, vending companies, and many more organizations save time and money through the continued use of two-way radios. Clearly, some Cellular operators offer unlimited local calling services or local-to-local calling plans. These offer some very attractive benefits, but they are cellular calls requiring a call set up and tear down, different from the instant-on two-way mobile radio operation. Moreover, the cellular caller may experience congestion or busy tones that would not be the case with the two-way radio option.

What Kind of Savings Can Your Business Expect?

How much does it cost you to operate just one of your vehicles per hour, including labor? How many hours of vehicle time could be saved every day with better scheduling and more control of your people? When you multiply these two numbers together, you'll quickly conclude that what a two-way radio system costs doesn't even come close to what you can save.

By reducing operating costs and increasing productivity with your current workload, you may find yourself able to expand your business dramatically without investing a great deal more.

When Will You Need a Radio Service Provider?

Two-way radio as a concept is simple: wireless communication between two or more points. However, because radio waves follow a line-of-sight path, your required coverage area may require a radio service provider to satisfy the need and the demand for coverage areas greater than a local private system can meet.

A two-way radio system generally consists of three types of units: a base station at a central dispatching location, mobile stations used in vehicles, and hand-held portables utilizing battery power. In today's business, conventional radio typically will not provide the wide-area communications that most businesses require. The cost for a business to construct its own radio system consisting of multiple tower sites can be too expensive. That's why an SMR system provider is used. Customers pay a small monthly fee to use the service, similar to cellular service. The difference is that you only pay a flat monthly fee, which allows for unlimited communication without high monthly cellular bills! There's no per-minute charge!

CHAPTER 20

Cellular Communications

A lot has happened in the cellular world since its original introduction in 1984. In 1984 when cellular communications became the hot button in the industry, all systems used analog radio transmissions. Many reasons were used to justify the cellular networks. These included very limited service areas, where you just could not get service where you wanted or needed it. Poor transmission haunted the operators because of the nature of the radio systems at the time. Users experienced excessive call setup delays. Heavy demand and limited channels were some of the most common problems in an operating area.

Analog cellular radio systems used *Frequency Division Multiple Access* (FDMA), which is an analog technique designed to support multiple users in an area with a limited number of frequencies. Analog radio systems use analog input, such as voice communications. Because these systems were designed around voice applications, no one had any thought of the future transmission of data, fax, packet data, and so on from a vehicle.

Back then, no one was sure what the acceptance rate would be. Currently, there are over 100 million cellular users in the United States. Approximately 100 to 150,000 new users sign up every month, yet 2001–2002 saw some significant slowdowns in the overall new subscriptions. Estimates are that four of five new telephones sold today are wireless telephones. Therefore, acceptance has become a nonissue. The new problem is not one of acceptance, but of retaining users. The churn ratio has been as high as 15 to 20 percent and it costs the carriers approximately $300 to $400 to acquire and set up a new subscriber. This means that if the subscriber contributes $30.00 per month to the carrier's payback revenue, it takes a minimum of 10 months for the carrier to break even.

The answer to the retention problem lies in packaging the service with the handset. By offering service plans with a usage fee of $0.10 to 0.15/minute, the acceptance rate has skyrocketed. The cellular industry is still primarily an analog backbone. Estimates are that 90 to 95 percent of the United States has analog coverage, whereas the digital counterpart to the cellular networks only covers between 70 and 75 percent. Dual-mode telephones have become the salvation for cellular providers because they would not be able to sustain their customer base without this offering.

Coverage Areas

The cellular operators build out their networks to provide coverage in certain geographically bounded areas. This poses the following dilemmas for the providers:

- The carriers need users (more) to generate higher revenues to pay off their investment.
- They must continue the evolution from analog to digital systems, allowing more efficient bandwidth use.
- Security and protection against theft is putting pressure on the carriers and users alike (analog cellular carriers lose close to $500 million in fraud). Today, cellular fraud takes many forms:
  - **Access fraud** Illegally modifying phones to gain access to a cellular carrier's network.
  - **Counterfeiting (cloning)** Duplicating a valid *Electronic Serial Number* (ESN)/*Mobile Identification Number* (MIN) combination for use on a cloned phone.
  - **Tumbling** Randomly changing the ESN and/or MIN after each phone call.
  - **Subscription fraud** Applying for cellular phone service with fraudulently obtained customer information or false identification, and with no intention of paying for it.
  - **Call sell** Reselling fraudulently obtained service, creating a cash-per-call scam.
  - **Cellular theft** Using a stolen phone until it is reported stolen.

Analog Cellular Systems

Analog systems do nothing for these needs. Using amplitude or frequency modulation techniques to transmit voice on the radio signal uses all of the available bandwidth. This means that the cellular carriers can support a single call today on a single frequency. The limitations of the systems include limited channel availability.

The analog system was designed for quick communication while on the road. Because this service could meet the needs of users on the go, the thought process regarding heavy penetration was only minimally addressed. However, as the major *Metropolitan Service Areas* (MSA) began expanding, the carriers realized that the analog systems were going to be too limiting. With only a single user on a frequency, congestion in the MSA became a tremendous problem. For example, a cellular channel uses 30 kHz of bandwidth for a single telephone call!

Cellular was designed to overcome the limitations of the conventional mobile telephone. Areas of coverage are divided into honeycomb-type cells and a hexagonal design of smaller sizes, as shown in Figure 20-1. The cells overlap each other at the outer boundaries. Frequencies can be divided into bands or cells to prevent interference and jamming of the neighboring cell's frequencies. The cellular system uses much less power output for transmitting. The vehicular transmitter uses 3 watts of power, while the hand-held sets use only 3/10 watts. Frequencies can be reused much more often and are closer to each other. The average cell design is approximately 3 to 5 miles across. The more users who subscribe to a network, the closer the transmitters are placed to each other. In rural areas, the cells are much farther apart.

For normal operation, 3 to 5 miles may separate cell sites, but as more users complain of *no service* due to congestion, cell splitting occurs. A cell can be subdivided into smaller cells, reallocating frequencies for continued use. The smaller the cell, the more equipment and other components are necessary. This places an added financial burden on the carriers as they attempt to match customer needs with returns on investments.

Figure 20-1
The cell patterns

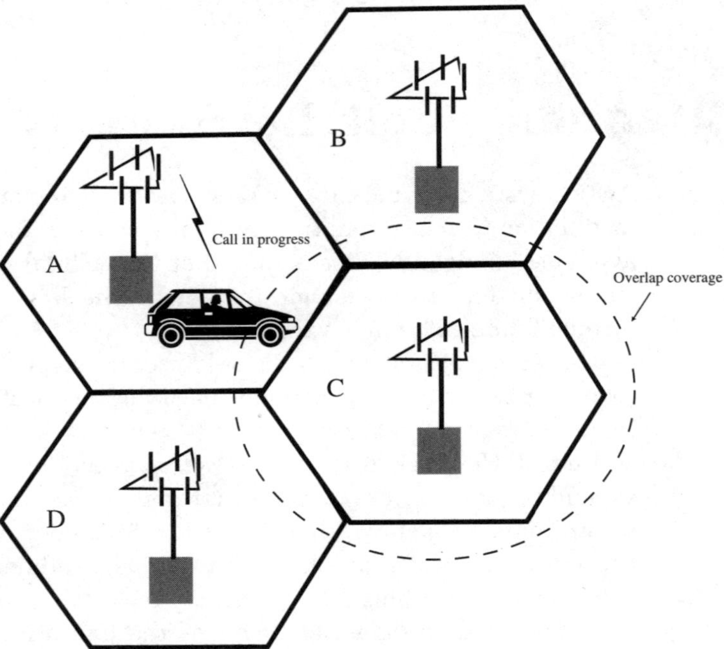

Log On

When the vehicle telephone powers on, it immediately logs onto the network. First, the telephone set sends a message to the *Mobile Telephone Switching Office* (MTSO). The MTSO is the equivalent of a Class 5 Central Office. It provides all the line-in-trunk interface capabilities, much the same as the CO will do.

The information sent to the MTSO includes the electronic serial number and telephone number from the handset. These two pieces of information combined will identify the individual device.

The telephone set will use an information channel to transmit the information. Several channels are set aside specifically for the purposes of logon capabilities (see Figure 20-2).

Monitoring Control Channels

Once a telephone set has logged on, it will then scan the 21 channels set aside as control channels. Upon scanning, the telephone will lock in on the channel that it receives the strongest. It will then go into monitoring mode.

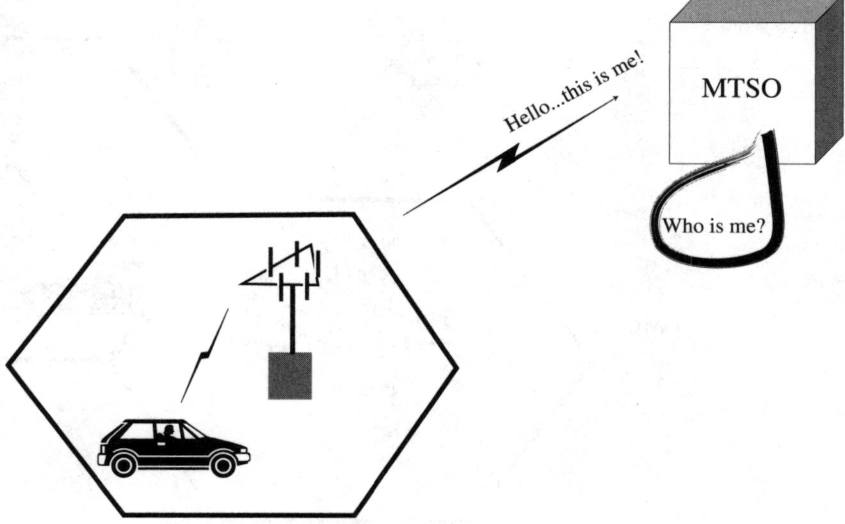
Hello...this is me!
MTSO
Who is me?

Figure 20-2 The logon process uses specific channels.

Although the set has nothing to send, it will continue to listen to the monitored channel in the event the mobile telephone switching office has an incoming call for it. The telephone user has actually done nothing. Upon power-up, the set immediately will log onto the network, identify itself, and go into the monitoring mode.

Failing Signal

One can assume in vehicular communications that the vehicle will be in motion. As the vehicle moves from cell to cell, it will move out of range from the first site, but come into range of the second site. The received signal on the monitoring channel will begin to fail (get too weak to hear), as shown in Figure 20-3. Immediately the telephone will rescan all the monitoring channels and select a new one. After the vehicle finds a new channel, it will continue to monitor that channel until such time as it rolls out of range again. This concept of rolling from site to site allows the vehicle to be in constant touch with the mobile telephone switching office so long as the set is on.

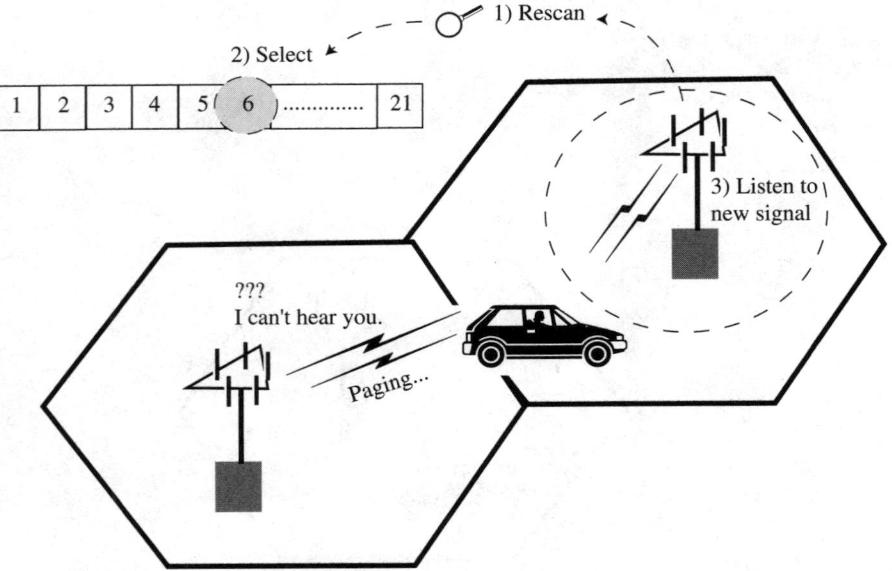

Figure 20-3
The failing signal procedure

Setup of a Call

When the user wants to place a call, the steps are straightforward. The process is very similar to making a wired call:

1. Pick up the handset and dial the digits.
2. After entering the digits, press the send key.
3. The information is a dialogue between the MTSO and the handset.
4. The MTSO receives the information and begins the call setup to a trunk connection.
5. The mobile office scans the available channels in the cell and selects one.
6. The mobile office sends a message to the handset, telling it which channel to use.
7. The handset then tunes its frequency to the assigned channel.
8. The mobile office connects that channel to the trunk used to set up the call.
9. The call is connected, and the user has a conversational path in both directions.

Setup of an Incoming Call

When a call is coming in from the network, things are again similar to the wireline network:

1. The mobile office receives signaling information from the network that a call is coming in.
2. The mobile office must first find the set, so it sends out a page through its network.
3. The page is sent out over the control channels.
4. Upon hearing the page, the set will respond.
5. The mobile office hears the response and assigns a channel.
6. The mobile office sends a message telling the set that it has a call and to use channel X.
7. The set immediately tunes to the channel it was assigned for the incoming call.
8. The phone rings, and the user answers.

Handoff

While the user is on the telephone, several things might happen. The first is that the vehicle is moving away from the center of the cell site. Therefore, the base station must play an active role in the process of handling the calls:

1. As the user gets closer to the boundary, the signal will get weaker.
2. The base station will recognize the loss of signal strength and send a message to the mobile office.
3. The mobile office will go into recovery mode.
4. The MTSO must determine what cell will be receiving the user.
5. The MTSO sends a message to all base stations advising them to conduct a quality of signal measurement on the channel in question.
6. Each base station determines the quality of the received signal.
7. They will advise the MTSO if the signal is strong or weak.
8. The MTSO decides which base station will host the call. The handset is a passive player in this role; the MTSO is in control of the hand off.

Setting Up the Handoff

After the MTSO has determined which base station will be the new host for the call, it will then select a channel and direct the new base station to set up a talk path for the call. This is all done in the background. An idle channel is set up in parallel between the base station and MTSO.

The Handoff Occurs

1. The original base station is still serving the call.
2. The new base station will host the caller.
3. The parallel channel has been set up.
4. MTSO has notified the cells to set the parallel channel in motion.
5. The MTSO sends a directive to the telephone to retune its frequency to the new one reserved for it.

6. The telephone set moves from one frequency to the new one.
7. The call is handed off from one cell to another.
8. The caller continues to converse and never knew what happened.

This procedure is shown in Figure 20-4.

Completion of the Handoff

After the telephone has moved from one base station to the other, and one channel to another, the handoff is complete. However, the original channel is now idle, but in parallel to the original call. Therefore, the base station notifies the MTSO that the channel is now idle. The MTSO is always in control of the call. It manages the channels and the handoff mechanisms. The MTSO commands the base station to set the channel to idle and makes it available for the next call.

Figure 20-4
The handoff process

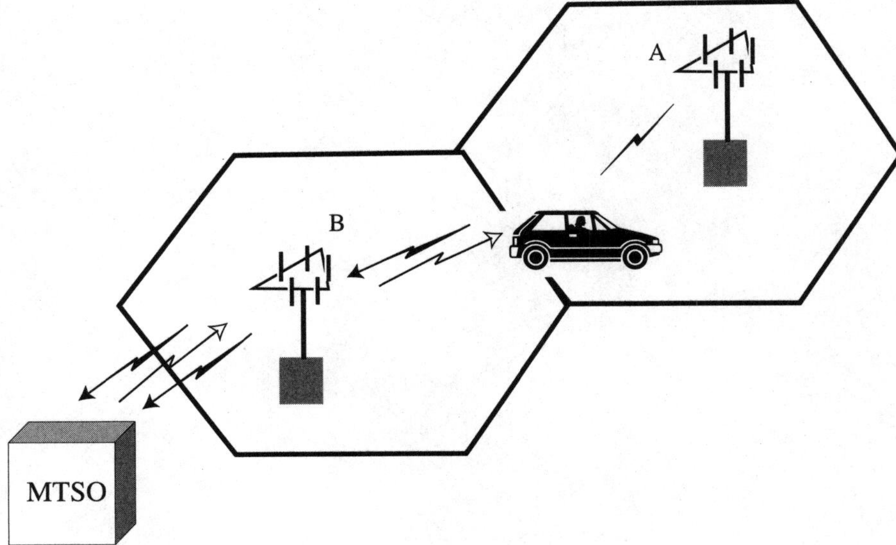

The Cell Site (Base Station)

The preceding discussion centered on the process of the call and referred to the base station quite a bit. The cell is comprised of a 3- to 5-mile radius. The base station is comprised of all the transmission and reception equipment between the base station and MTSO and the base station to the telephone. The cell has a tower with multiple antennae mounted on the top. Each cell has enough radio equipment to service approximately 45 calls simultaneously as well as to monitor all the channels in each of the adjacent cells to it (see Figure 20-5). The equipment varies with the manufacturer and the operator, but typically, an operator will have 35 to 70 cells in a major location.

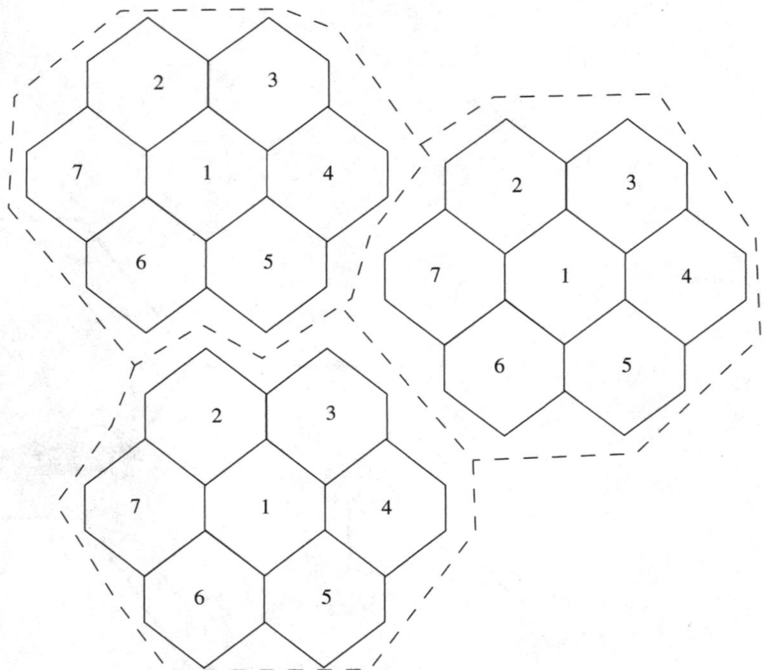

Figure 20-5
Seven-cell pattern

The Mobile Telephone Switching Office (MTSO)

The MTSO is a Class 5 *Central Office* (CO) equivalent. It provides the trunks and signaling interfaces to the wireline carriers. It has a full-line switching component and the necessary logic to manage thousands of calls simultaneously. Like the CO infrastructure, the MTSO uses digital trunks between the MTSO and the wireline carriers (*Incumbent Local Exchange Carrier* [ILEC], *Competitive LEC* [CLEC], or *Interexchange Carrier* [IEC]) either on copper, fiber, or microwave radio systems.

At the MTSO, there is a separate trunk/line interface between the MTSO and the base station. This is the line side of the switch, and it is used for the controlling call setup. Normally, the MTSO connects to the base station via a T1 operating line at 32 Kbps *Adaptive Differential Pulse Code Modulation* (ADPCM). This T1 will be on copper or microwave. A MTSO is a major investment, ranging from $2 to $6 million, depending on the size and the area being served.

Frequency Reuse Plans and Cell Patterns

Frequency reuse is what started the cellular movement. Planning permits the efficient allocation of limited radio frequency spectrum for systems that use frequency-based channels (*Advanced Mobile Phone System* [AMPS], *Digital AMPS* [DAMPS], and *Global System for Mobile Communications* [GSM]). Frequency reuse enables increased capacity and avoids interference between sites that are sharing the frequency sets. Frequency plans exist that specify the division of channels among 3, 4, 7, and 12 cells. They define the organization of available channels into groups that maximize service and minimize interference.

As a mobile unit moves through the network, it is assigned a frequency during transit through each cell. Because each cell pattern has one low-power transmitter, air interface signals are limited to the parameters of each cell. Air interface signals from nonadjacent cells do not interfere with each other. Therefore, a group of nonadjacent cells can reuse the same frequencies.

CDMA systems do not require frequency management plans because every cell operates on the same frequency. Site resources are differentiated by their PN offset (phase offset of the *Pseudorandom Noise* reference). Mobile channels are identified by a code that is used to spread across the baseband signal, and each can be reused in any cell. Using an N=7 frequency reuse pattern, all available channels are assigned to their appropriate cells. It is not necessary to deploy all radios at once, but their use has been planned ahead of time to minimize interference in the future (see Figure 20-5).

Overlapping Coverage

Each cell has its own radio equipment with an overlap into adjoining cells. This allows for the monitoring of the adjacent cells to ensure complete coverage. The cells can sense the signal strength of the mobile and hand-held units in their own areas and in the overlap areas of each adjoining cell. This is what makes the handoff and coverage areas work together. The graphic in Figure 20-6 shows the overlap coverage in the 7-cell pattern described previously.

Figure 20-6 Overlap coverage between the cells

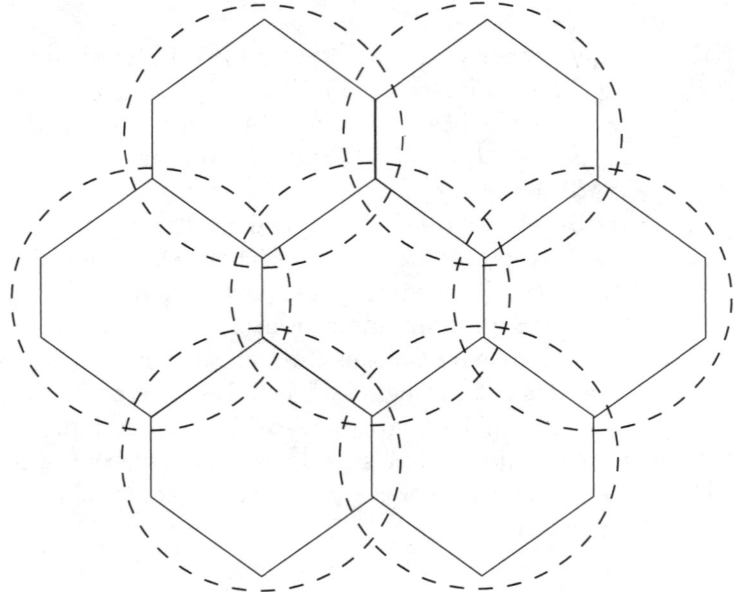

Cell Site Configurations

The type of antenna used to support the air interface between the cell site and the mobile phones determines the mode of operation of a cell site. When an omnidirectional antenna is used, the site serves a single 360-degree area around itself (see Figure 20-7).

A single antenna supports the cell sites using the omnidirectional antenna for both send and receive operations. These devices cover the full 360-degree site independently. One transmit antenna is used for each radio frame at the site (one frequency group per radio frame). Two receive antennae distribute the receive signal to every radio, providing diversity reception for every receiver at the site.

Due to the site's capability to receive signals from all directions, transmissions from neighboring sites may interfere with the site's reception. When interference reaches unacceptable levels, the site is usually sectorized to eliminate its capability to receive interfering information. Sectoring may also come into play when the site becomes so congested that the omnidirectional antennae cannot support the operations.

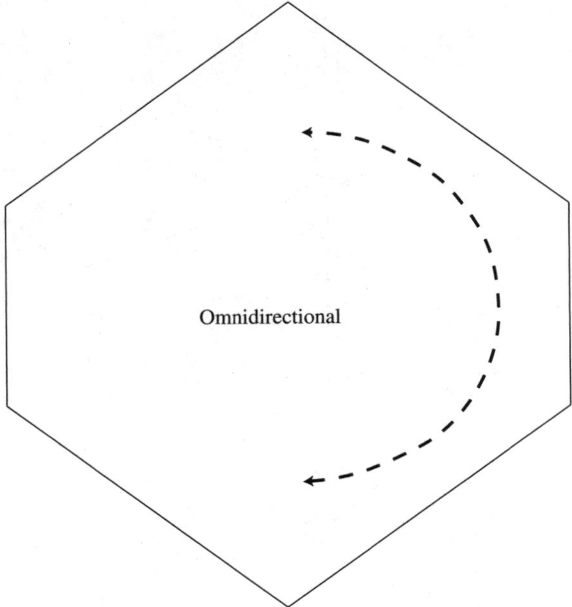

Figure 20-7 The omnidirectional antenna

Sectorized Cell Coverage

Directional (sectorized) sites use reflectors positioned behind the antenna to focus the coverage area into a portion of a cell. Coverage areas can be customized to the needs of each site, as shown in Figure 20-8, but the typical areas of coverage are as follows:

- Two sectors using 180-degree angles
- Three sectors using 120-degree angles
- Six sectors using 60-degree angles

At least one transmit antenna is used in each sector (one per radio frame) and two receive antennae provide space diversity for each sector of a two- or three-sectored site. One receive antenna is used in each of the 60-degree sectors, with neighboring sectors providing the sector diversity. This is an economic issue because of the number of antennae required.

Figure 20-8
Sectorized coverage

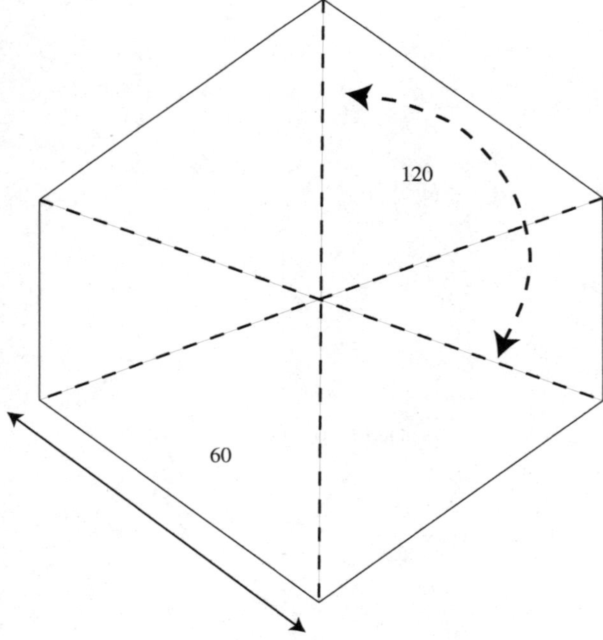

Tiered Sites

This configuration places a low-power site in the same location with a high-power site. Mobiles change channels as they move across the boundary between the two in order to relieve congestion in the center, as shown in Figure 20-9. This configuration is used in all GSM and *Code Division Multiple Access* (CDMA) applications today. It is not supported on the older AMPS and DAMPS configurations, but may be used in newer implementations of AMPS and DAMPS. Each sector requires its own access/paging control channel to manage call setup functions. Voice traffic in each sector is supported by radios connected to antennae supporting that sector.

Reuse of Frequencies

Frequency reuse enables a particular radio channel to carry conversations in multiple locations, increasing the overall capacity of the communications systems. Within a cluster, each cell uses different frequencies; however, these frequencies can be reused in cells of another cluster.

One centralized radio site with 300 channels can have 300 calls in progress at any one time. The 300 channels can be divided into four groups

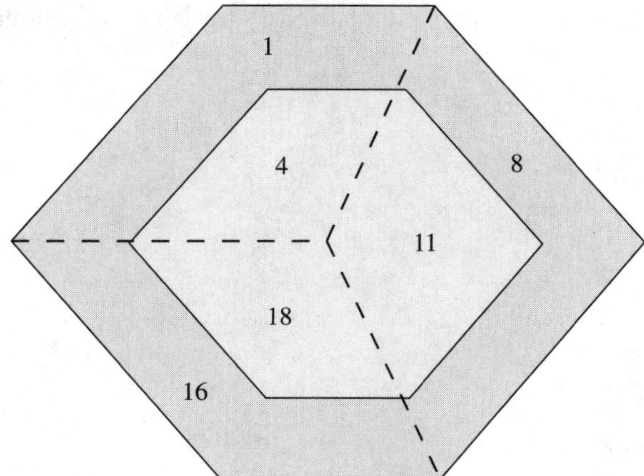

Figure 20-9
Tiered cell coverage

of 75 channels and still provide 300 calls at once. Dividing the service area into 16 sections called *cells* allows each cell to use one of the four groups of channels, increasing the call-carrying capacity of the system by a value of 4 (1,200 calls at one time).

The service area can be continually divided into smaller and smaller cells to obtain greater call-carrying capacity, increasing the number of calls by a factor of four with each division. The limit on how many cells can be used is determined by this information:

- The cost infrastructure at each cell
- The processing power of the switch that controls the system
- The minimum power output at each site

Allocation of Frequencies

The allocation of frequencies based on the first cellular arrangement of AMPS was designed around 666 duplex channels. The frequency ranges were allocated in the 825–845 MHz and 870–890 MHz frequency bands. In each band, the channels use a 30 kHz separation, and 21 channels are allocated to control channels. Figure 20-10 is a representation of the channel allocation.

The FCC approved licenses for two operators of the cellular service; the wireline carrier (usually the telephone company in the area) and the non-wireline carrier (a competitor to the local telephone company). The fre-

Figure 20-10 Frequency allocation for cellular

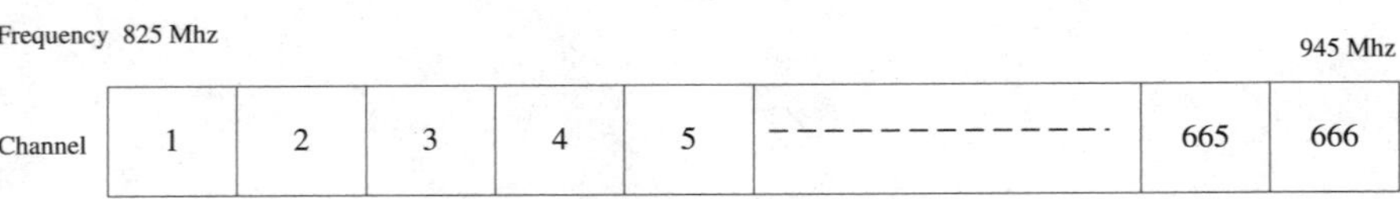

Frequency 870 Mhz
890 Mhz
Channel
1
2
3
4
5
665
666
Receive frequencies

quencies were equally split between the wireline and nonwireline operators. This meant that only half the channels were available to each carrier and two sets of control channels were required.

Four signaling paths are used in the cellular network to provide for signaling and control, as well as voice conversation. These can be broken into two basic function groups:

- Call setup and breakdown
- Call management and conversation

Establishing a Call from a Landline to a Mobile

From a wired telephone, the local exchange office pulses out the cellular number called to the MTSO over a special trunk connecting the telephone company to the MTSO. The MTSO then analyzes the number called and sends a data link message to all paging cell sites to locate the unit called. When the cellular unit recognizes the page, it sends a message to the nearest cell site. This cell site then sends a data link message back to the MTSO to alert the MTSO that the unit has been found. This message further notifies the MTSO which cell site will handle the call.

The MTSO next selects a cell site trunk connected to that cell and sets up a network path between the cell site and the originating CO trunk carrying the call (see Figure 20-11).

The MTSO is now also called the *Mobile Switching Center* (MSC). It is the controlling element for the entire system. The MSC is responsible for the following information:

- All switching of calls to and from the cells
- Blocking calls when congestion occurs
- Providing necessary backup to the network
- Monitoring the overall network elements
- Handling all the test and diagnostic capabilities for the system

This is the workhorse of the cellular system. The MSC relies on two different databases within the system to keep track of the mobile stations in its area.

The first of the two databases is called the *Home Location Register* (HLR). The HLR is a database of all system devices registered on the

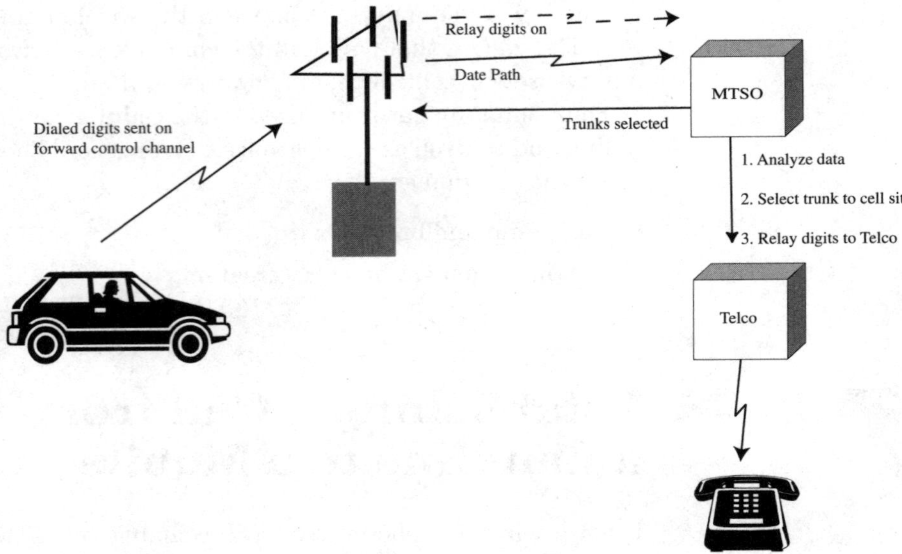

Figure 20-11 Call establishment

system and owned by the operator of the MSC. These are the local devices connected to the network. The HLR keeps track of the individual device's location and stores all the necessary information about the subscriber. This includes the name, telephone number, features and functions, fiscal responsibilities, and the like.

The second database is called the *Visiting Location Register* (VLR), which is a temporary database built on roaming devices as they come into a particular MSC's area. The VLR keeps track of temporary devices while they are in an area, including the swapping of location information with the subscriber's HLR. When a subscriber logs onto a network, and it is not home to that subscriber, the VLR builds a data entry on the subscriber and tracks activity and feature usage to be consistent for the user.

Another set of databases is used by the MSC in some networks. These again are two separate and distinct functions called the *Equipment Inventory Register* (EIR) and the *Authentication Center* (AuC). These are very similar to the databases used in the GSM networks and keep track of manufacturer equipment types for consistency. The authentication center is used to authenticate the user to prevent fraudulent use of the network by a cloned device.

CHAPTER 21

Global Services Mobile Communications (GSM)

History of Cellular Mobile Radio and GSM

The idea of cell-based mobile radio systems appeared at Bell Laboratories in the early 1970s. However, the commercial introduction of cellular systems did not occur until the 1980s. Because of the pent-up demand and newness, analog cellular telephone systems grew rapidly in Europe and North America. Today, cellular systems still represent one of the fastest growing telecommunications services. Recent studies indicate that three of four new phones are mobile phones. Unfortunately, when cellular systems were first being deployed, each country developed its own system, which was problematic because

- The equipment only worked within the boundaries of each country.
- The market for mobile equipment manufacturers was limited by the operating system.

Three different services had emerged in the world at the time. They were

- *Advanced Mobile Phone Services* (AMPS) in North America
- *Total Access Communications System* (TACS) in the United Kingdom
- *Nordic Mobile Telephone* (NMT) in Nordic countries

To solve this problem, in 1982, the Conference of *European Posts and Telecommunications* (CEPT) formed the *Groupe Spécial Mobile* (GSM) to develop a pan-European mobile cellular radio system (the acronym later became *Global System for Mobile* communications). The goal of the GSM study group was to standardize systems to provide

- Improved spectrum efficiency
- International roaming
- Low-cost mobile sets and *base stations* (BSs)
- High-quality speech
- Compatibility with *Integrated Services Digital Network* (ISDN) and other telephone company services
- Support for new services

The existing cellular systems were developed on analog technology. However, GSM was developed using digital technology.

Benchmarks in GSM

Table 21-1 shows many of the important events in the rollout of the GSM system; other events were introduced, but had less significant impact on the overall systems.

Commercial service was introduced in mid-1991. By 1993, 36 GSM networks were already operating in 22 countries. Today, you can be instantly reached on your mobile phone in over 171 countries worldwide and on 400 networks (operators). As of May 2001, over 550 million people were subscribers to mobile telecommunications. GSM truly stands for Global System for Mobile telecommunications. Roaming is the ability to use your GSM phone number in another GSM network. You can roam to another region or

Table 21-1

Major events in GSM

Year	Events
1982	CEPT establishes a GSM group in order to develop the standards for a pan-European cellular mobile system.
1985	A list of recommendations to be generated by the group is accepted.
1986	Field tests are performed to test the different radio techniques proposed for the air interface.
1987	*Time Division Multiple Access* (TDMA) is chosen as the access method (with *Frequency Division Multiple Access* [FDMA]). The initial *Memorandum of Understanding* (MoU) is signed by telecommunication operators representing 12 countries.
1988	GSM system is validated.
1989	The responsibility of the GSM specifications is passed to the *European Telecommunications Standards Institute* (ETSI).
1990	Phase 1 of the GSM specifications is delivered.
1991	Commercial launch of the GSM service occurs.
1992	The addition of the countries that signed the GSM Memorandum of Understanding takes place. Coverage spreads to larger cities and airports.
1993	Coverage of main roads' GSM services starts outside Europe.
1995	Phase 2 of the GSM specifications occurs. Coverage is extended to rural areas.

country and use the services of any network operator in that region that has a roaming agreement with the GSM network operator in your home region/country. A roaming agreement is a business agreement between two network operators to transfer items, such as call charges and subscription information, back and forth as their subscribers roam into each other's areas.

GSM Metrics

The GSM standard is the most widely accepted standard and is implemented globally, owning a market share of 69 percent of the world's digital cellular subscribers. *Time Division Multiple Access* (TDMA), with a market share close to 10 percent, is available mainly in North America and South America. GSM, which uses a TDMA access, and North American TDMA are two of the world's leading digital network standards. Unfortunately, it is currently technically impossible for users of either standard to make or receive calls in areas where only the other standard is available. Once interoperability is in place, users of GSM and TDMA handsets will be able to roam on the other network type—subject to the agreements between mobile operators. This will make roaming possible across much of the world because GSM and TDMA networks cover large sections of the global population and together account for 79 percent of all mobile subscribers, as shown in Figure 21-1.

Figure 21-1 Market penetrations of GSM and TDMA

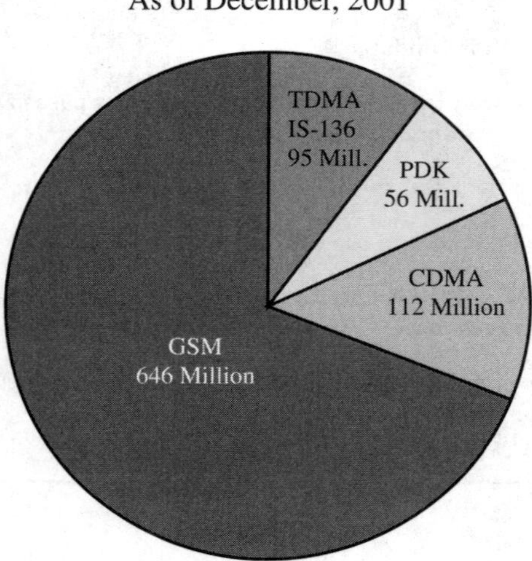

Cell Structure

In a cellular system, the coverage area of an operator is divided into cells. A cell is the area that one transmitter or a small collection of transmitters can cover. The size of a cell is determined by the transmitter's power. The concept of cellular systems is the use of low-power transmitters in order to enable the efficient reuse of the frequencies. The maximum size of a cell is approximately 35 km (radius), providing a round-trip communications path from the mobile to the cell site and back. If the transmitters are very powerful, the frequencies cannot be reused for hundreds of kilometers, as they are limited to the coverage area of the transmitter. In the past when a mobile communications system was installed, the coverage blocked the reuse beyond the 25-mile coverage area and created a corridor of interference of an additional 75 miles. This is shown in Figure 21-2.

The frequency band allocated to a cellular mobile radio system is distributed over a group of cells, and this distribution is repeated in all of an

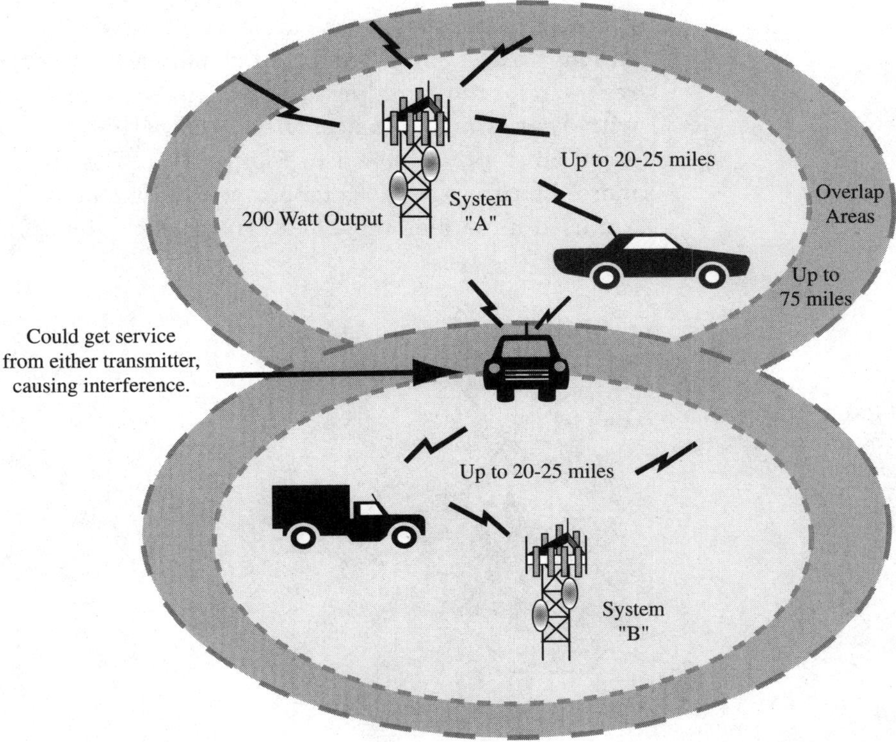

Figure 21-2
The older way of handling mobile communications

operator's coverage area. The entire number of radio channels available can then be used in each group of cells that form the operator's coverage area. Frequencies used in a cell will be reused several cells away. The distance between the cells using the same frequency must be sufficient to avoid interference. The frequency reuse will increase the capacity in the number of users considerably. The patterns can be a four-cell pattern or other choices. The typical clusters contain 4, 7, 12, or 21 cells.

In order to work properly, a cellular system must verify the following two main conditions:

- The power level of a transmitter within a single cell must be limited in order to reduce the interference with the transmitters of neighboring cells. The interference will not produce any damage to the system if a distance of about 2.5 to 3 times the diameter of a cell is reserved between transmitters. The receiver filters must also conform.
- Neighboring cells cannot share the same channels. In order to reduce the interference, the frequencies must be reused only within a certain pattern. The pattern may also be a seven-cell pattern similar to the AMPS networks, which is shown in Figure 21-3.

In order to exchange the information needed to maintain the communication links within the cellular network, several radio channels are reserved for the signaling information. Sometimes we use a 12-cell pattern with a repeating sequence. The 12-cell pattern is really a grouping of three 4-cell clusters, as shown in Figure 21-4. The larger the cell pattern, the more the coverage areas tend to work. In general, the larger cell patterns are used in various reuse patterns to get the most out of the scarce radio

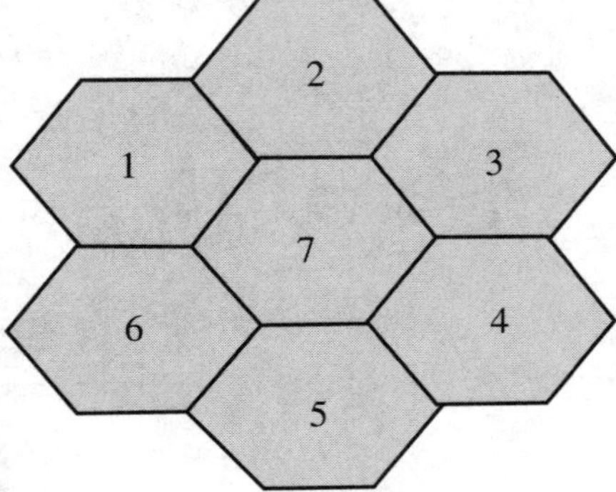

Figure 21-3
The seven-cell pattern
Source: ETSI

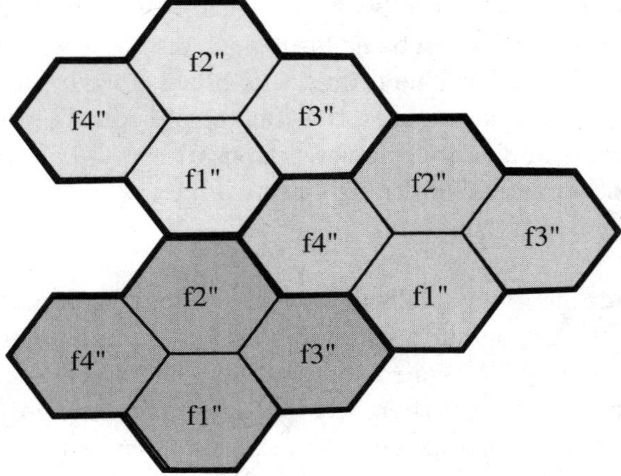

Figure 21-4
The 12-cell pattern
Source: ETSI

resources as possible. The 21-cell pattern is by far the largest repeating pattern in use today. The cells are grouped into clusters. The number of cells in a cluster determines whether the cluster can be repeated continuously within the coverage area.

The number of cells in each cluster is very important. The smaller the number of cells per cluster, the greater the number of channels per cell. Therefore, the capacity of each cell will be increased. However, a balance must be found in order to avoid the interference that could occur between neighboring clusters. This interference is produced by the small size of the clusters (the size of the cluster is defined by the number of cells per cluster). The total number of channels per cell depends on the number of available channels and the type of cluster used.

Types of Cells

The density of population in a country is so varied that different types of cells are used:

- Macrocells
- Microcells
- Selective or sectorized cells
- Umbrella cells
- Nanocells
- Picocells

Macrocells Macrocells are large cells for remote and sparsely populated areas. These cells can be as large as 3 to 35 km from the center to the edge of the cell (radius). The larger cells place more frequencies in the core, but because the area is rural, the macrocell typically has limited frequencies (channels) and higher-power transmitters. This is a limitation that prevents other sites from being closely adjacent to this cell. Figure 21-5 shows the macrocell.

Microcells These cells are used for densely populated areas. By splitting the existing areas into smaller cells, the number of channels available and the capacity of the cells are increased. The power level of the transmitters used in these cells is then decreased, reducing the possibility of interference between neighboring cells. Some of the microcells may be as small as .1 to 1 km, depending on the need. Oftentimes the cell splitting will use the reduced power and the greater coverage to satisfy hot spots or dead spots in the network.

Another need may well be a below-the-rooftop cell that satisfies a very close-knit group of people or varied users. The picocell will be in a building and is typically a smaller version of a microcell. The distances covered with a picocell are approximately .01 to 1 km. These are used in office buildings for "close in" calls, part of a *private branch exchange* (PBX) or a wireless *local area network* (LAN) application today. A small group of users will share this cell because of the close proximity to each other and larger cells around. Nanocells also fall into the below-the-rooftop domain where the

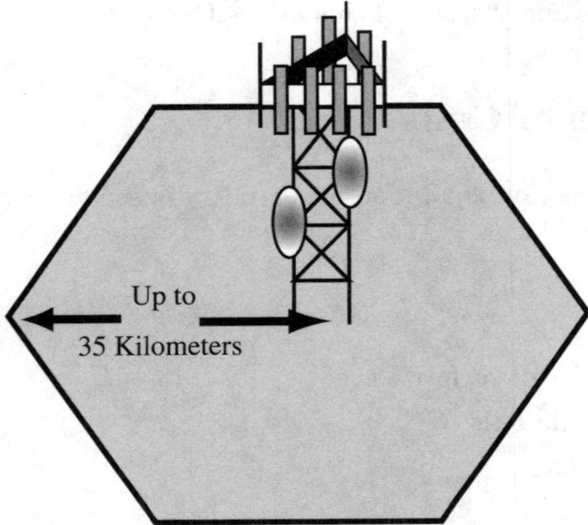

Figure 21-5
The macrocell

distances for this type of cell are from .01 to .001 km. These are just smaller and smaller segments that are built within a building as an example. Figure 21-6 shows a combination of a microcell and a picocell.

Selective Cells or Sectorized Cells It is not always useful to define a cell with a full coverage of 360 degrees. In some cases, cells with a particular shape and coverage are needed. These cells are called *selective cells*. Selective cells are typically the cells that may be located at the entrances of tunnels where 360-degree coverage is not needed. In this case, a selective cell with coverage of 120 degrees is used. This selective cell is shown in Figure 21-7.

Tiered Cells A tiered cell is one where an overlay of radio equipment operates in two different frequencies and uses different sectors. The tiered cell is also a form of a selective cell.

Umbrella Cells Alongside a high-speed freeway, crossing very small cells produces an overabundance of handovers among the different small neighboring cells. To solve this problem, the concept of umbrella cells was introduced. An umbrella cell covers several microcells, as shown in Figure 21-8. The power level inside an umbrella cell is increased compared to the power levels used in the microcells that form the umbrella cell.

Figure 21-6 The microcell and picocell

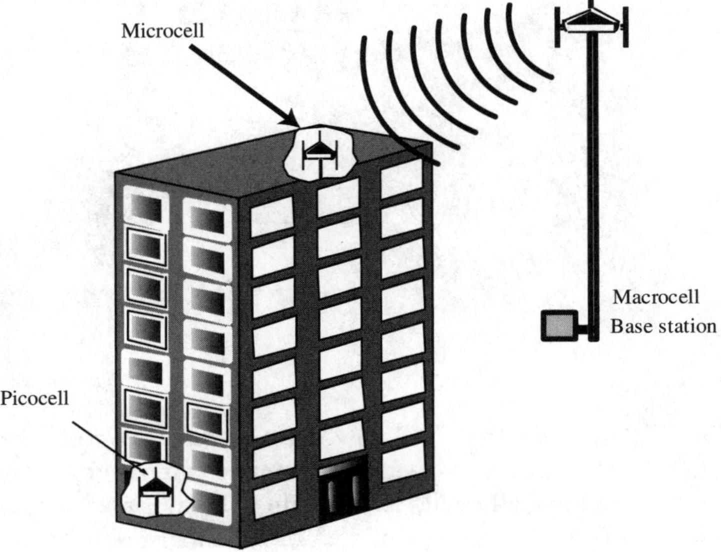

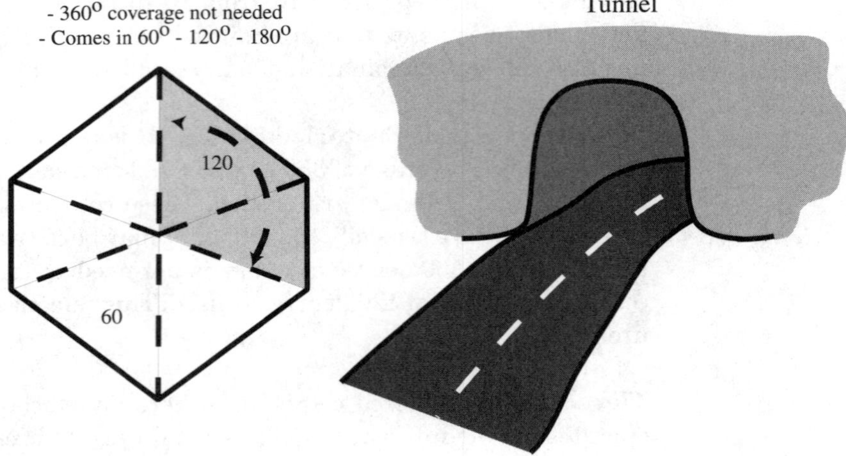
- 360° coverage not needed
- Comes in 60° - 120° - 180°
120
60
Tunnel

Figure 21-7
The selective cell

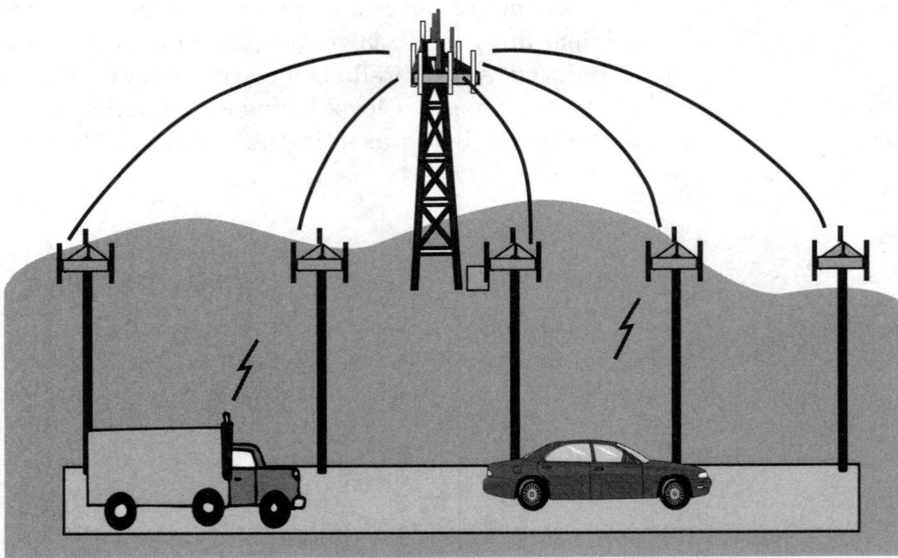

Figure 21-8
The umbrella cell

How does the cell know when to shift from a microcell to an umbrella cell? When the speed of the mobile is too high, the mobile is handed off to the umbrella cell. The mobile will then stay longer in the same cell (in this case, the umbrella cell). This will reduce the number of handovers and the work of the network. A large number of handover demands and the propagation characteristics of a mobile can help to detect its high speed. The radio equipment is no longer forced to constantly change hands from cell to cell when using this umbrella. This meets the goal of GSM in that the efficient use of the *radio frequency* (RF) spectrum is what is being achieved.

Analog to Digital Movement

In the 1980s, most mobile cellular systems were based on analog systems, including AMPS, TACS, and NMT. In fact, 95 percent of the United States has coverage from AMPS services, whereas only 70 percent is covered with digital service. The roaming agreements used between cellular carriers in North America use AMPS for roaming. In many cases, the analog networks are starting to wind down in the major metropolitan areas; however, in rural communities, AMPS is still predominant. The GSM system was the first digital cellular system created from the onset. Different reasons explain the transition from analog to digital technology. Cellular systems experienced phenomenal growth. Analog systems were not able to cope with this increasing demand. To overcome this problem, new frequency bands and new technologies were suggested. Many countries rejected the possibility of using new frequency bands because of the restricted spectrum (even though later on, other frequency bands were allocated for the development of mobile cellular radio). New analog technologies were able to overcome some of the problems, but were too expensive. The digital radio was the best option (but not the perfect one) to handle the capacity needs in a cost-efficient manner.

The decision to adopt digital technology for GSM was made in the course of developing the standard. During the development of GSM, the telecommunications industry converted to digital networking standards. ISDN is an example of this evolution. In order to make GSM compatible with the services offered by ISDN, it was decided that digital radio technology was the best option available.

Quality of service (QoS) can also be improved dramatically by using digital rather than analog technology. From the beginning, the planners of GSM wanted ISDN compatibility in the services offered and to control the signaling used. The radio link imposed some limitations because the standard ISDN bit rate of 64 Kbps could not be practically achieved.

Using the *International Telecommunication Union-Telecommunication Standardization* (ITU-T) definitions, telecommunication services can be divided into the following categories:

- Teleservices
- Bearer services
- Supplementary services

Teleservices

The most basic teleservice supported by GSM is telephony, the transmission of speech. It has an added emergency service, where the nearest emergency service provider is notified by dialing three digits. The emergency number 112 is used like 911 in North America. Group 3 fax, an analog method described in ITU-T recommendation T.30, is also supported with the use of an appropriate fax adapter.

A unique feature of GSM compared to older analog systems is the *Short Message Service* (SMS). SMS is a bidirectional service for sending short alphanumeric (up to 160 bytes) messages in a store-and-forward fashion. For point-to-point SMS, a message can be sent to another subscriber to the service, and an acknowledgement of receipt is provided to the sender. SMS can also be used in a cell broadcast mode for sending messages such as traffic updates or news updates. Messages can be stored in the *Subscriber Identity Module* (SIM) card for later retrieval. The SMS service has been very well accepted with over one billion SMS messages being sent monthly.

As things progressed, Phase II of GSM introduced enhancements. For example, in the teleservices, half-rate voice coding was introduced. In the first phase, full-rate voice coding was used at a rate of 13 Kbps for a voice conversation. Later the 6.5 Kbps vocoders were introduced for use at a network operator's choice. This enables the network operator to offer good speech quality to twice as many users without any additional radio resources. Essentially, we split the channel in half because people actually carry traffic on the channel only 25 to 30 percent of the time.

Enhancements also included better SMS informational flow for point-to-point communications and the use of point-to-multipoint communications. The 160-character SMS message was finally documented and became fully store-and-forward.

Bearer Services

The digital nature of GSM enables data, both synchronous and asynchronous, to be transported as a bearer service to or from an ISDN terminal. Data can use either the transparent service, with a fixed delay but no guarantee of data integrity, or a nontransparent service, which guarantees data integrity through an *Automatic Repeat Request* (ARQ) mechanism, which unfortunately introduces a variable delay. The data rates supported by GSM are 300 bps, 600 bps, 1,200 bps, 2,400 bps, and 9,600 bps. One can imagine in this new millennium that these data speeds are intolerable for the mainstay of data transmission. In fact, if someone were to offer us Internet access at speeds of up to 9,600 bps, we would probably become very disinterested in the service. Yet, from a mobile perspective, these speeds were considered quite fast.

Enhancements from Phase II also included better throughput for data transmission using a synchronous dedicated packet data access operating at 2.4 to 9.6 Kbps. Phase I only accepted asynchronous access to a dedicated *packet assembler/dissembler* (PAD). The access of data through a dedicated PAD at the entrance of an X.25 network enables access to a higher degree of reliable data transport, helping to overcome the link layer problems on the radio.

Data is now available over the GSM Phase II at both send and receive speeds of up to 9.6 Kbps. In the earlier releases, slower data was more prevalent. The use of the GSM network enables the integration of various network platforms such as

- *Plain old telephone service* (POTS)
- ISDN access and emulation
- Packet data network access (X.25 and IP are the most common)
- Circuit-switched data transfers across and X.25, X.31, and X.32 standards

Because the data is being sent across a digital air interface, no modem is required at the *mobile station* (MS) end.

Supplementary Services

Supplementary services (which are really the added features of the cellular networks) are provided on top of teleservices or bearer services and include features such as

- *Caller identification*
- *Call forwarding* The subscriber can forward incoming calls to another number if the called mobile is busy (CFB), unreachable (CFNRc), or if no reply (CFNRy) occurs. Call forwarding can also be applied unconditionally (CFU).
- *Call waiting*
- *Multiparty conversations*
- *Barring of outgoing (international) calls* Different types of call barring services are available:
  - *Barring of All Outgoing Calls* (BAOC)
  - *Barring of Outgoing International Calls* (BOIC)
  - *Barring of Outgoing International Calls except those directed toward the Home PLMN Country* (BOIC-exHC)
  - *Barring of All Incoming Calls* (BAIC)
  - *Barring of incoming calls when roaming* (BICR)

Phase II enhancements to the supplementary services include the following:

- *Calling/Connected Line Identification Presentation (CLIP)* This supplies the called user with the ISDN of the calling user.
- *Calling/Connected Line Identification Restriction (CLIR)* This enables the calling user to restrict the presentation.
- *Connected Line identification Presentation (CoLP)* This supplies the calling user with the directory number he or she receives if his or her call is forwarded.
- *Connected Line identification Restriction (CoLR)* This enables the called user to restrict the presentation.
- *Call Waiting (CW)* This informs the user, during a conversation, about another incoming call. The user can answer, reject, or ignore this incoming call.

These are added supplementary services finishing off the list:

- *Call hold* This puts an active call on hold.
- *Advice of Charge (AoC)* This provides the user with online charge information.
- *Multiparty service* This creates the possibility of establishing a multiparty conversation.
- *Closed User Group (CUG)* This corresponds to a group of users with limited possibilities of calling (only the people of the group and certain numbers).
- *Operator-determined barring* This provides restrictions of different services and call types by the operator.

GSM Architecture

A GSM network consists of several functional entities whose functions and interfaces are defined. Figure 21-9 shows the layout of a generic GSM network. The GSM network can be divided into three broad parts. The subscriber carries the MS, the *Base Station Subsystem* (BSS) controls the radio link with the MS, and the Network Subsystem, the main part of which is the *Mobile Switching Center* (MSC), performs the switching of calls between the mobile and other fixed or mobile network users, as well as the management of mobile services such as authentication. The Operations and Maintenance Center, which oversees the proper operation and setup of the network, is not shown in the figure. The MS and the BSS communicate

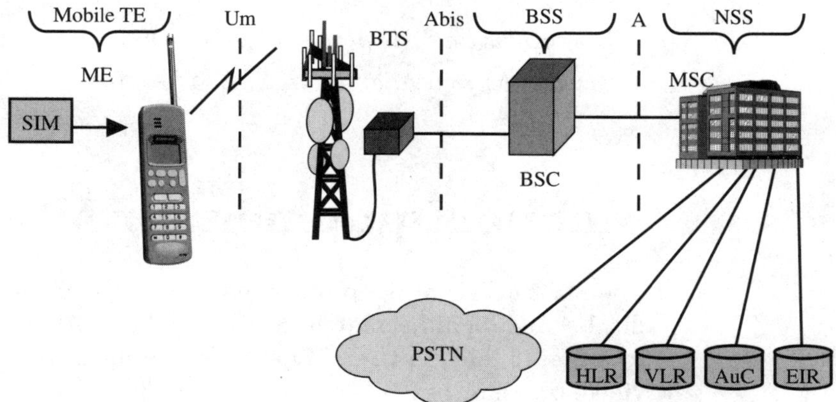

Figure 21-9 The GSM architecture

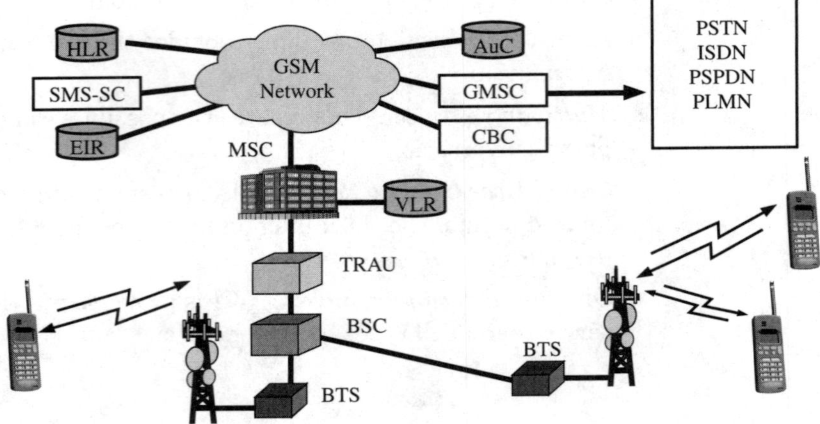

Figure 21-10 The added components of a GSM network

across the Um interface, also known as the air interface or radio link. The BSS communicates with the network service switching center across the A interface.

The added components of the GSM architecture (see Figure 21-10) include the functions of the databases and messaging systems:

- *Home Location Register* (HLR)
- *Visitor Location Register* (VLR)
- *Equipment Identity Register* (EIR)
- *Authentication Center* (AuC)
- *SMS Serving Center* (SMS SC)
- *Gateway MSC* (GMSC)
- *Chargeback Center* (CBC)
- *Operations and Support Subsystem* (OSS)
- *Transcoder and Adaptation Unit* (TRAU)

Mobile Equipment or MS

The MS consists of the physical equipment, such as the radio transceiver, display and digital signal processors, and the SIM card. It provides the air interface to the user in GSM networks. As such, other services are also provided, which include

- Voice teleservices
- Data bearer services
- The features' supplementary services

SIM

The SIM provides personal mobility so that the user can have access to all subscribed services irrespective of both the location of the terminal and the use of a specific terminal. By inserting the SIM card into another GSM cellular phone, as shown in Figure 21-11, the user is able to receive calls at that phone, make calls from that phone, or receive other subscribed services. The *International Mobile Equipment Identity* (IMEI) uniquely identifies the mobile equipment. The SIM card contains the *International Mobile Subscriber Identity* (IMSI), identifying the subscriber, a secret key for authentication, and other user information. The IMEI and the IMSI are independent, thereby providing personal mobility. A password or personal identity number may protect the SIM card against unauthorized use.

Figure 21-11
The SIM

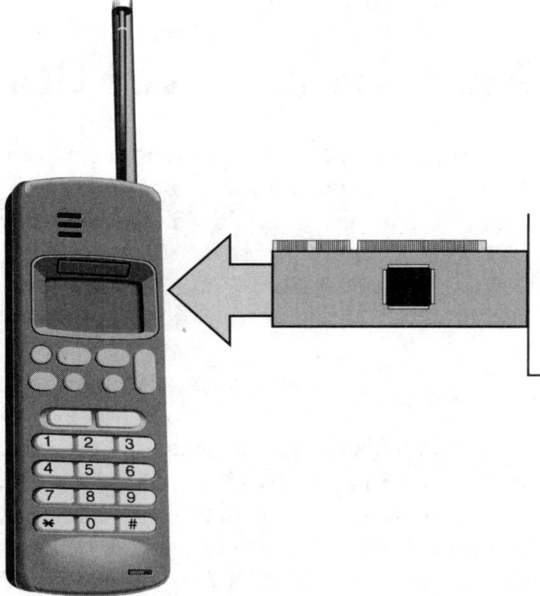

The MS Function

Different types of terminals are available that are distinguished principally by their power and application:

- The fixed terminals are terminals installed in cars.
- The GSM portable terminals can also be installed in vehicles.
- The hand-held terminals have experienced the biggest success thanks to their weight and volume, which are continuously decreasing.

The MS also provides the receptor for SMS messages, enabling the user to toggle between the voice and data use. Moreover, the mobile facilitates access to voice-messaging systems. The MS also provides access to the various data services available in a GSM network. These data services include

- X.25 packet switching through a synchronous or asynchronous dialup connection to the PAD at speeds typically at 9.6 Kbps
- *General Packet Radio Services* (GPRSs) using either an X.25- or IP-based data transfer method at speeds up to 115 Kbps
- High-speed, circuit-switched data at speeds up to 64 Kbps

The data speeds will vary by application and other conditions, such as air interfaces across a hostile link.

The Base Transceiver Station (BTS)

The BTS (see Figure 21-12) houses the radio transceivers that define a cell and handles the radio link protocols with the MS. In a large urban area, a large number of BTSs may be deployed. The requirements for a BTS are

- Ruggedness
- Reliability
- Portability
- Minimum cost

The BTS corresponds to the transceivers and antennas used in each cell of the network. A BTS is usually placed in the center of a cell. Its transmitting power defines the size of a cell. Each BTS has between 1 and 16 transceivers, depending on the density of users in the cell. Each BTS serves a single cell. It also includes the following functions:

Figure 21-12
The BTS

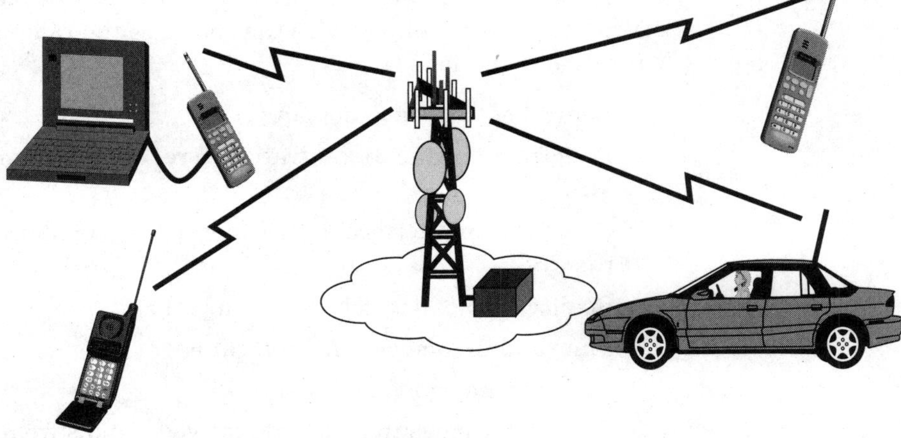

- Encoding, encrypting, multiplexing, modulating, and feeding the RF signals to the antenna
- Transcoding and rate adaptation
- Time and frequency synchronizing
- Voice through full- or half-rate services
- Decoding, decrypting, and equalizing received signals
- Random access detection
- Timing advances
- Uplink channel measurements

The Base Station Controller (BSC)

The BSC manages the radio resources for one or more BTSs. It handles radio channel setup, frequency hopping, and handovers. The BSC is the connection between the mobile and the MSC. The BSC also translates the 13 Kbps voice channel used over the radio link to the standard 64 Kbps channel used by the *Public Switched Telephone Network* (PSDN) or ISDN. The BSC is between the BTS and the MSC and provides radio resource management for the cells under its control. It assigns and releases frequencies and time slots for the MS. The BSC also handles intercell handover. It controls the power transmission of the BSS and MS in its area. The function of the BSC is to allocate the necessary time slots between the BTS and the MSC. It is a switching device that handles the radio resources. Additional functions include

- Control of frequency hopping
- Performing traffic concentration to reduce the number of lines from the MSC
- Providing an interface to the Operations and Maintenance Center for the BSS
- Reallocation of frequencies among BTSs
- Time and frequency synchronization
- Power management
- Time-delay measurements of received signals from the MS

BSS

The BSS is composed of two parts: the BTS and the BSC. These communicate across the specified Abis interface, enabling (as in the rest of the system) operations between components that are made by different suppliers. The radio components of a BSS may consist of four to seven or nine cells. A BSS may have one or more base stations. The BSS uses the Abis interface between the BTS and the BSC. A separate high-speed line (T1 or E1) is then connected from the BSS to the Mobile *central office* (CO), as shown in the architecture in Figure 21-13.

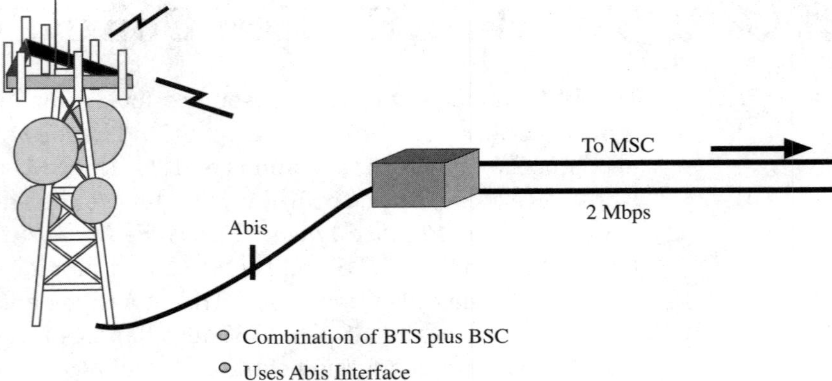

Figure 21-13 The BSS

The TRAU

Depending on the costs of transmission facilities from a cellular operator, it may be cost efficient to have the transcoder either at the BTS, BSC, or MSC. If the transcoder is located at the MSC, it is functionally still a part of the BSS. This creates maximum flexibility of the overall network operation. The transcoder takes the 13 Kbps speech or data (at 300, 600, and 1,200 bps) multiplexes 4 of them, and places them on a standard 64 Kbps digital PCM channel. First, the 13 Kbps voice is brought up to a 16 Kbps level by inserting additional synchronizing data to make up the difference of the lower data rate. Then, four 16 Kbps channels are multiplexed onto a DS0 (64 Kbps) channel.

Locating the TRAU

If the transcoder/rate adapter is outside the BTS, the Abis interface can only operate at 16 Kbps within the BSS. The TRAU output data rate is 64 Kbps standard digital channel capacity. Next, 30 of the 64 Kbps channels are multiplexed onto a 2.048 Mbps E1 service if the CEPT standards are used. The E1 can carry up to 120 traffic and control signals. The locations can be between the BTS and the BSC, whereby a 16 Kbps subchannel is used between the BTS and the TRAU, and 64 Kbps channels between the TRAU and the BSC. Alternatively, the TRAU can be located between the BSC and the MSC, as shown in Figure 21-14, using 16 Kbps between the BTS and the BSC and 64 Kbps between the BSC and the TRAU.

MSC

The central component of the Network Subsystem is the MSC, which is shown in Figure 21-15. It acts like a normal Class 5 CO in the PSTN or ISDN, and in addition provides all the functionality needed to handle a mobile subscriber, such as registration, authentication, location updating, handovers, and call routing to a roaming subscriber. The primary functions of the MSC include

- Paging
- Coordination of call setup for all MSs in its operating area
- Dynamic allocation of resources

Figure 21-14
The TRAU

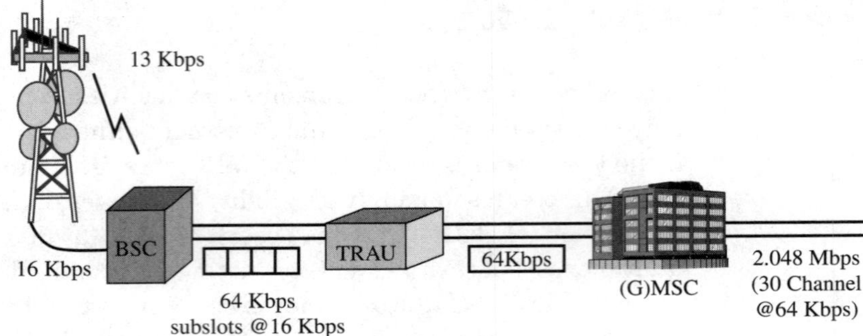

- In between BSC and MSC
- Converts GSM coding into PSTN data

13 Kbps → 64 Kbps 64 Kbps → 13 Kbps

Figure 21-15
The MSC

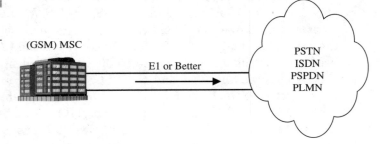

(GSM) MSC

E1 or Better

PSTN
ISDN
PSPDN
PLMN

- Location registration
- Interworking functions
- Handover management
- Billing
- Reallocation of frequencies to BTSs
- Encryption
- Echo cancellation
- Signaling exchange
- Synchronizing the BSS
- Gateway to SMS

As a CO function, it uses the digital trunks in the form of E1 (or larger) to the other network interfaces such as

- PSTN
- ISDN
- PSPDN
- *Public Land Mobile Network* (PLMN)

These services are provided in conjunction with several functional entities, which together form the Network Subsystem. The MSC provides the connection to the public-fixed network (PSTN or ISDN), and signaling between functional entities uses *Signaling System Number 7* (SS7), which is used in ISDN and is widely used in current public networks.

The *Gateway Mobile Services Switching Center* (GMSC) is used in the PLMN. A gateway is a node interconnecting two networks. The GMSC is the interface between the mobile cellular network and the PSTN. It is in charge of routing calls from the fixed network towards a GSM user. The GMSC is often implemented in the same machines as the MSC. A PLMN may have many MSCs, but it has only one gateway access to the wireline network to accommodate the network operator. The gateway then is the high-speed trunking machine connected via E1 or *Synchronous Digital Hierarchy* (SDH) to the outside world.

The Registers Completing the Network Switching Systems (NSSs)

The *Home Location Register* (HLR) and *Visitor Location Register* (VLR), together with the MSC, provide the call-routing and roaming capabilities of GSM, called the NSS. The HLR contains all the administrative information of each subscriber registered in the corresponding GSM network, along with the current location of the mobile. The current location of the mobile is in the form of a *Mobile Station Roaming Number* (MSRN), which is a regular ISDN number used to route a call to the MSC where the mobile is currently located. One HLR exists logically per GSM network, although it may be implemented as a distributed database. Figure 21-16 shows the HLR.

The VLR contains selected administrative information from the HLR, which is necessary for *call control* (CC) and provision of the subscribed services, for each mobile currently located in the geographical area controlled by the VLR. Although each functional entity can be implemented as an

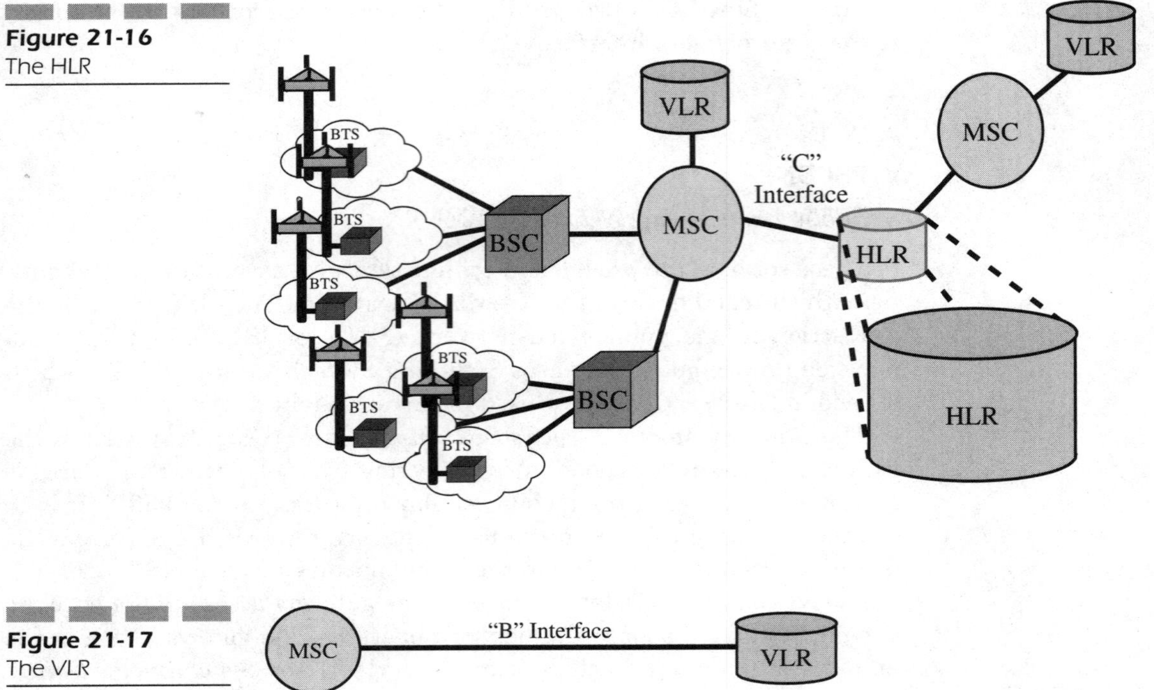

Figure 21-16 The HLR

Figure 21-17 The VLR

independent unit, most manufacturers of switching equipment implement one VLR together with one MSC (see Figure 21-17) so that the geographical area controlled by the MSC corresponds to that controlled by the VLR, simplifying the signaling required. Note that the MSC contains no information about particular MSs—this information is stored in the location registers.

The other two registers are used for authentication and security purposes. The *Equipment Identity Register* (EIR) is a database that contains a list of all valid mobile equipment on the network, where its *International Mobile Equipment Identity* (IMEI) identifies each MS. An IMEI is marked as invalid if it has been reported stolen or is not type approved. The Authentication Center is a protected database that stores a copy of the secret key stored in each subscriber's SIM card, which is used for authentication and ciphering of the radio channel.

The Cell

As has already been explained, a cell, identified by its *Cell Global Identity* (CGI) number, corresponds to the radio coverage of a base transceiver station. In a macrocell environment, the radius distance is between 3 to 35 km. The distances are calculated on the basis of a round-trip between the BTS and the mobile to provide a sufficient *bit error rate* (BER) and power to satisfy quality speech.

Location Area

A *location area* (LA), identified by its *location area identity* (LAI) number, is a group of cells served by a single MSC/VLR. One MSC/VLR combination has several LAs. The LA is part of the MSC/VLR service area in which a MS may move freely without any updating of location messaging to the MSC/VLR controlling the LA.

MSC/VLR Service Area

A group of LAs under the control of the same MSC/VLR defines the MSC/VLR area. A single PLMN can have several MSC/VLR service areas. MSC/VLR is a sole controller of calls within its area of jurisdiction. To route a call to a MS, the path through the network links to the MSC in the MSC area where the subscriber is currently located. The mobile location can be uniquely identified because the MS is registered in a VLR, which is associated with an MSC.

PLMN A PLMN is the area served by one network operator, as shown in Figure 21-18. One country can have several PLMNs, based on its size. The links between a GSM/PLMN network and other PSTN, ISDN, or PLMNs will be at the level of national or international transit. All incoming calls for a GSM/PLMN will be routed to the GMSC. A GMSC works as an incoming transit exchange for the GSM/PLMN. All *mobile-terminated* (MT) calls will be routed to the GMSC. Call connections between PLMNs or fixed networks must be routed through certain designated MSCs.

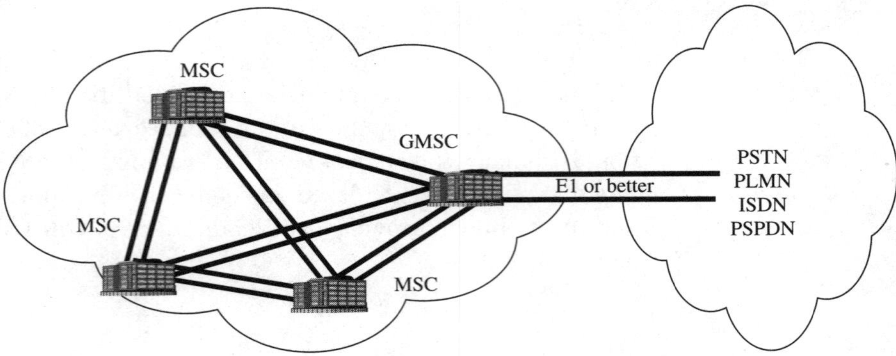

Figure 21-18
The PLMN

OSI Model—How GSM Signaling Functions in the OSI Model

The *Open Standards Interface* (OSI) is a guideline of how systems communicate transparently. SS7 is used for signaling between the outside world and the GSM architectures. Moreover, SS7 is used to communicate between the MSC and the HLR. To satisfy other functions in GSM architecture, the model is applied for other services from the MS outward. In reality, the model works at the bottom three layers of the OSI model for the bulk of the transmissions that take place in call setup and teardown, registration, and authentication, and so on. Thus, Layers 3, 2, and 1 of the OSI model are most applicable.

OSI defines a communications subsystem consisting of functions that enable distributed application processes, resident on computers, to exchange information via an underlying data network. The communications subsystem can be divided into two sublayers:

- An application-dependent sublayer providing functions that are application-dependent but network-independent
- A network-dependent sublayer providing functions that are dependent on the underlying data network but are application-independent

Ensuring the transmission of voice or data of a given quality over the radio link is only part of the function of a cellular mobile network. A GSM mobile can seamlessly roam nationally and internationally, which requires that registration, authentication, call routing, and location-updating functions exist and are standardized in GSM networks. In addition, the fact that the geographical area covered by the network is divided into cells necessitates the implementation of a handover mechanism. The Network Subsystem performs these functions using the *Mobile Application Part* (MAP) built on top of the SS7 protocol, as shown in Figure 21-19.

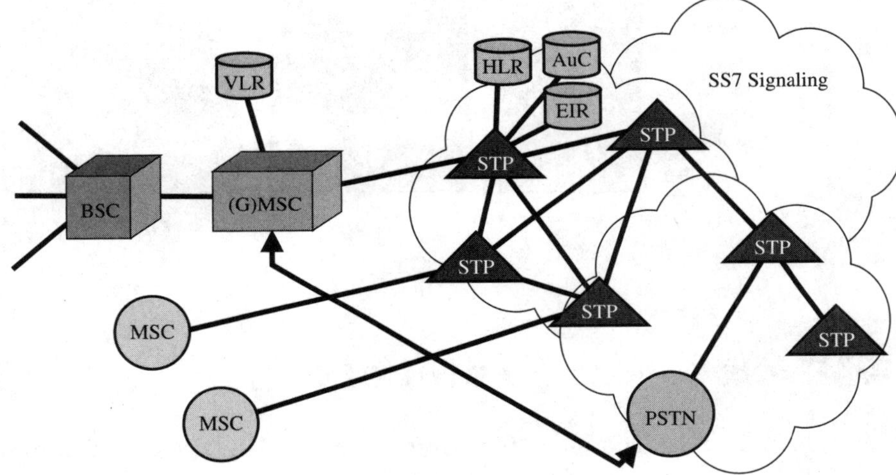

Figure 21-19 SS7 and GSM working together

Layer Functionality

In the GSM architecture, the layered model integrates and links the peer-to-peer communications between two different systems. If we look across the platform, the underlying layers satisfy the services of the upper-layer protocols. For example, at Layer 3, the *Service Access Point* (SAP) between Layer 3 and 2 addresses the services being served. *Service Access Point Identifiers* (SAPIs) describe the services that are provided by the various services from the upper and lower layers. Notifications are passed from layer to layer to ensure that the information has been properly formatted, transmitted, and received. These primitives make the process complete. Several discussions center on the chart of protocols, as shown in Figure 21-20. Refer to this chart for the next block of protocol stack discussions.

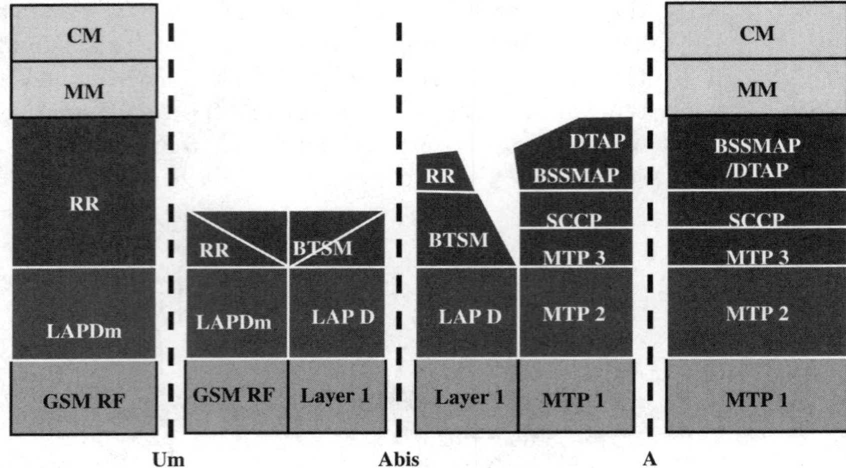

Figure 21-20 The protocol stacks

MS Protocols

The signaling protocol in GSM is structured into three general layers, depending on the interface. Layer 1 is the physical layer, which uses the channel structures discussed previously over the air interface. Layer 2 is the data-link layer. Across the Um interface, the data-link layer is a modified version of the *Link access protocol for the D channel* (LAP-D) protocol used in ISDN, called *Link access protocol on the Dm channel* (LAP-Dm). Across the A interface, the Message Transfer Part, Layer 2, of SS7 is used. Layer 3 of the GSM signaling protocol is divided into three sublayers: *radio resource management* (RR), *mobility management* (MM), and *Connection Management* (CM).

The MS to BTS Protocols

The RR layer oversees the establishment of a link, both radio and fixed, between the MS and the MSC. The main functional components involved are the MS, the BSS, and the MSC. The RR layer is concerned with the management of an RR-session, which is the time that a mobile is in dedicated mode, as well as the configuration of radio channels, including the allocation of dedicated channels.

The MM layer is built on top of the RR layer and handles the functions that arise from the mobility of the subscriber, as well as the authentication and security aspects. Location management is concerned with the procedures that enable the system to know the current location of a powered-on MS so that incoming call routing can be completed.

Location updating is when a powered-on mobile is informed of an incoming call by a paging message sent over the PAGCH channel of a cell. One extreme would be to page every cell in the network for each call, which is obviously a waste of radio bandwidth. The other extreme would be for the mobile to notify the system, via location-updating messages, of its current location at the individual cell level. This would require paging messages to be sent to exactly one cell, but would be very wasteful due to the large number of location-updating messages. A compromised solution used in GSM is to group cells into LAs. Updating messages are required when moving between LAs, and MSs are paged in the cells of their current LA.

The CM layer is responsible for CC, supplementary service management, and *Short Message Service* (SMS) management. Each of these may be considered as a separate sublayer within the CM layer. CC attempts to follow the ISDN procedures specified in Q.931, although routing to a roaming mobile subscriber is obviously unique to GSM. Other functions of the CC sublayer include call establishment, selection of the type of service (including alternating between services during a call), and call release.

BSC Protocols

After the information is passed from the BTS to the BSC, a different set of interfaces is used. The Abis interface is used between the BTS and BSC. At this level, the radio resources at the lower portion of Layer 3 are changed from the RR to the *Base Transceiver Station Management* (BTSM). The BTS management layer is a relay function at the BTS to the BSC. The RR protocols are responsible for the allocation and reallocation of traffic channels between the MS and the BTS. These services include controlling the initial access to the system, paging for MT calls, the handover of calls between cell sites, power control, and call termination. The RR protocols provide the procedures for the use, allocation, reallocation, and release of the GSM channels. The BSC still has some radio resource management in place for the frequency coordination, frequency allocation, and the management of the overall network layer for the Layer 2 interfaces.

From the BSC, the relay is using SS7 protocols so the MTP 1-3 is used as the underlying architecture, and the BSS mobile application part or the direct application part is used to communicate from the BSC to the MSC.

MSC Protocols

At the MSC, the information is mapped across the A interface to the MTP Layers 1 through 3 from the BSC. Here the equivalent set of radio resources is called the BSS MAP. The BSS MAP/DTAP and the MM and CM are at the upper layers of Layer 3 protocols. This completes the relay process. Through the control-signaling network, the MSCs interact to locate and connect to users throughout the network. Location registers are included in the MSC databases to assist in the role of determining how and whether connections are to be made to roaming users. Each user of a GSM MS is assigned a HLR that is used to contain the user's location and subscribed services. A separate register, the VLR, is used to track the location of a user. As the users roam out of the area covered by the HLR, the MS notifies a new VLR of its whereabouts. The VLR in turn uses the control network (which happens to be based on SS7) to signal the HLR of the MS's new location. Through this information, MT calls can be routed to the user by the location information contained in the user's HLR.

Defining the Channels

As we look at the radio operation, a channel can be defined in different ways. Oftentimes we hear a channel defined in RF. Other times we hear the physical channel being described thinking that it is radio frequency. Alas, the different definitions get in the way. For the definitions used, channels are defined by looking at the matrix:

1. The radio channel is defined by the frequency used.
2. Physical channels are indicative of the time slot that they occupy.
3. Logical channels are defined by the function that they provide or serve.

Frequencies Allocated

In reality, GSM systems can be implemented in any frequency band. However, several bands exist where GSM terminals are available. Furthermore, GSM terminals may incorporate one or more of the GSM frequency bands listed in the following section to facilitate roaming on a global basis.

Two frequency bands, 25 MHz in each one, have been allocated by ETSI for the GSM system:

- The band 890 to 915 MHz has been allocated for the uplink direction (transmitting from the MS to the BS).
- The band 935 to 960 MHz has been allocated for the downlink direction (transmitting from the BS to the MS).

However, not all countries can use all of the GSM frequency bands. This is due primarily to military reasons and to the existence of previous analog systems using part of the two 25 MHz frequency bands. Figure 21-21 shows the frequencies.

Primary GSM

When transmitting in a GSM network, the MS to the BS uses an uplink. The reverse channel direction is the downlink from the BS to the MS. GSM

Figure 21-21 The uplink and downlink frequencies

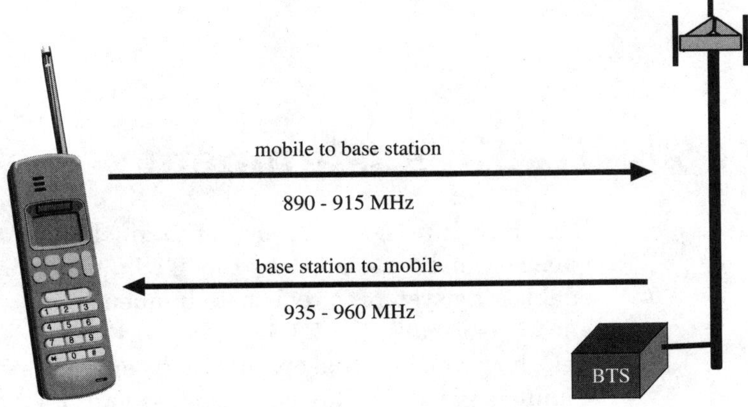

GSM frequencies initially set with 25 MHz (transmit and receive) spaced apart by 45 MHz.

uses the circa 900 MHz band. The frequency band used is 890 to 915 MHz (mobile transmit) and 935 to 960 MHz (base transmit). The duplex channel enables the two-way communications in a GSM network. Because telephony was the primary service, a full-duplex channel is assigned with the 2 separate frequencies in a 45 MHz separation.

To give the maximum number of users access, each band is subdivided into 125 carrier frequencies spaced 200 kHz apart, using FDMA techniques. The spectrum assignment is shown in Figure 21-22. Only 124 channels are used, where channel 0 is reserved and held as a guard band against interference from the lower channels. Each of these carrier frequencies is further subdivided into time slots using TDMA. The frequency bands are usually split between two or more providers who then build their networks. The channels are set at 200 kHz each. The ITU, which manages the international allocation of radio spectrum (among other functions), allocated the bands for mobile networks in Europe.

Figure 21-22 Spectrum bands for primary GSM

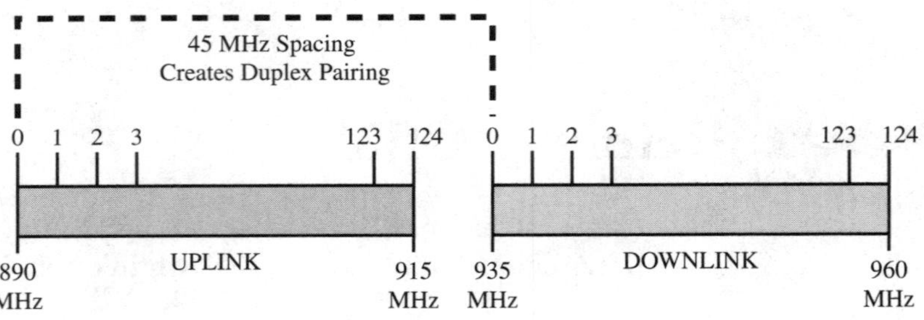

Radio Assignment

Each BTS is assigned a group of channels with which to operate. Any frequency can be assigned to the BTS, as they are frequency agile. This enables the system to reallocate frequencies as needed to handle load balancing. Normally, the BTS can support upwards of 31 channels (frequencies); however, in actual operation, the operators usually assign from 1 to 16 channels per BTS. This is a business and practicality issue. The *Absolute Radio Frequency Channel Number* (ARFCN) is used in the channel assignment at each of the frequencies.

Frequency Pairing

The pairing is shown as the way of handling the 45 MHz separations. Remember that channel 0 was not used. It was reserved as a guard band from the lower frequencies to prevent interference.

Extended GSM Radio Frequencies

After the ETSI assigned the initial block of frequencies, a later innovation was to assign an additional block of 10 MHz on the bottom of the original block. The reasoning was that future demands would require this capacity. This meant that the frequencies from 880 to 890 MHz for the uplink and 915 to 925 MHz were added. This created an additional 50 carriers. The carriers were numbered 974 to 1,023 so that the channel assignments would not be confused with the initial GSM standard. Once the added channels were implemented, the additional channels were still paired at 45 MHz separations:

- Channel 974 became the guard band for the lower frequencies below 880 MHz and 925 MHz.
- The initial channel 0 in the primary GSM band is now used because of this shift, as shown in Figure 21-23.

Figure 21-23 Extended GSM

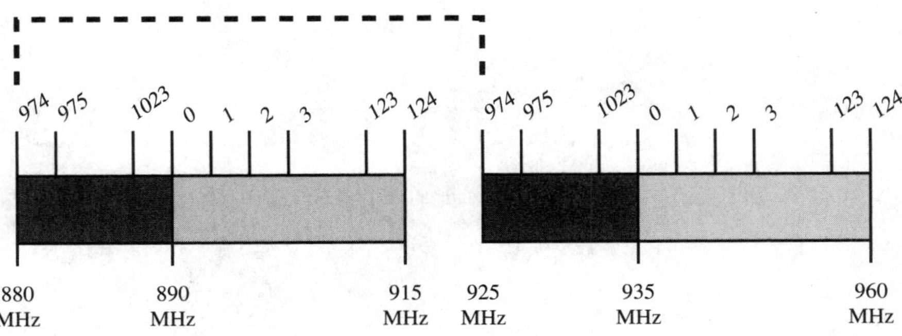

Modulation

In order to convey speech on the RF, either in analog or digital form, the transmitted information must be propagated on the radio link. It must be placed on the carrier. A carrier in this respect is a single radio frequency. The process of combining the audio and the radio signals is known as *modulation*. The resultant waveform is known as a *modulated waveform*. Modulation is a form of change process where we change the input information into a suitable format for the transmission medium. We also changed the information by demodulating the signal at the receiving end.

Three normal forms of modulation are used:

- Amplitude
- Frequency
- Phase

Amplitude Shift Keying (ASK)

In ASK (see Figure 21-24), the radio wave is modulated by shifting on the amplitude. The frequency is left constant, but the amplitude is shifted high if the data is a 0 and low if the data is a 1. Normally, we see two amplitude shifts represent a single bit.

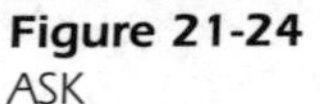

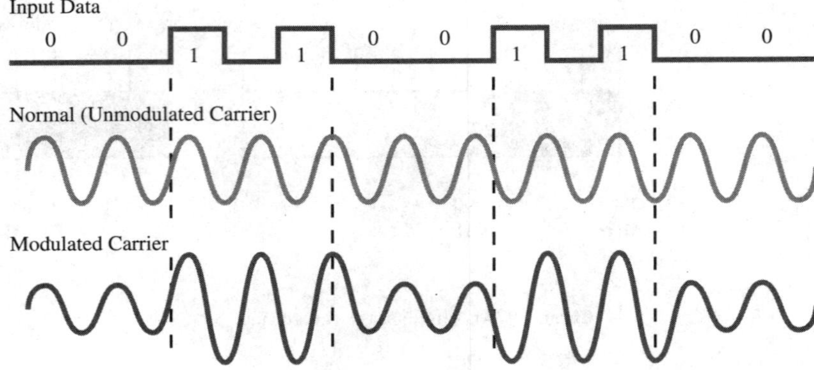

Figure 21-24
ASK

Frequency Shift Keying (FSK)

The alternative to ASK is FSK. In the case of FSK (see Figure 21-25), applying the data onto the radio wave modulates the carrier by changing the frequency. The amplitude is kept constant, and the frequency is changed. Normally, a single frequency shift represents a bit of data.

Phase Shift Keying (PSK)

In PSK, both the amplitude and the frequency are kept constant, so the changes are represented by a shift in the phase, as shown in Figure 21-26. The benefit of phase shifts is that multiple phases can be used to represent more than one modulated bit. Under normal PSK, a shift in the phase represents a single bit; however, multiphase modulation enables multiple bits to be represented. The *quadrature phase shifts* (QPSK) will enable up to 2 bits per shift, whereas a quadrature and amplitude shift will enable 4 bits per phase shift.

Gaussian Minimum Shift Keying (GMSK)

GSM modulation works differently, as shown in Figure 21-27. Using *Gaussian minimum shift keying* (GMSK), the nature of the data moved from the MS is digital. For a digital transmission in GSM, the chosen modulation

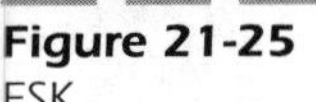

Figure 21-25
FSK

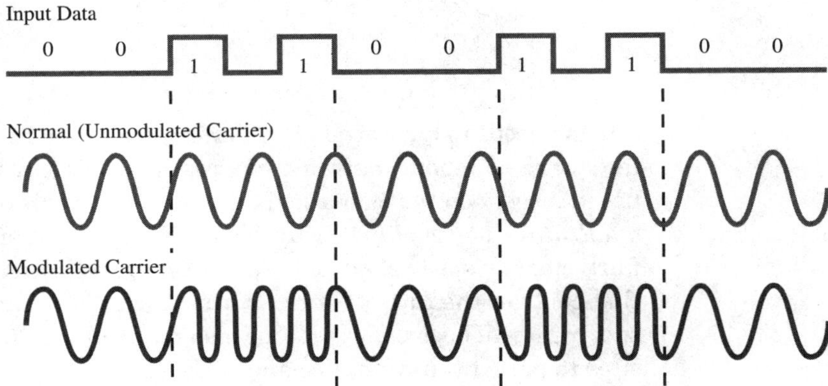

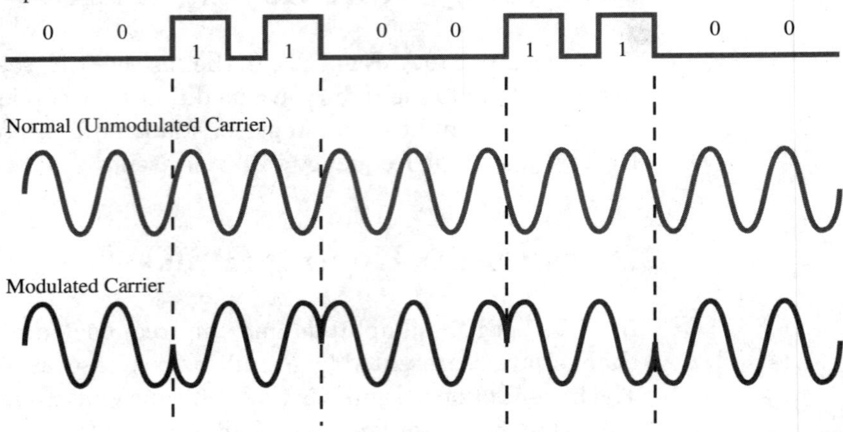

Figure 21-26
PSK

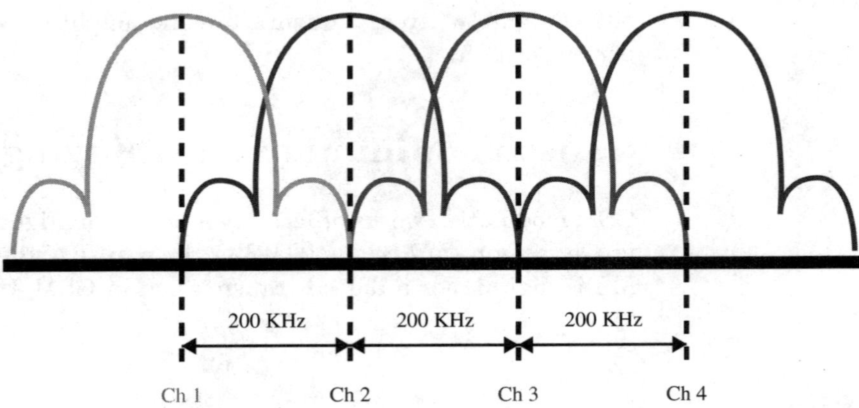

Figure 21-27
GMSK results

scheme needs to have good error performance in light of the noise and interference in a mobile network environment. GMSK is a complex scheme based largely on mathematical functions. The basis of this scheme is *offset quadrature phase shift keying* (OQPSK), which offers the advantage of a fairly narrow spectral output. This is combined with a minimum technique that controls the rate of change of the phase of the carrier and the radiated spectrum will be even lower. This also requires very careful planning at the sites to prevent interference and produces only 1 bit per symbol. The combined functions of the baseband filter, the OQPSK, and GMSK modulation work to produce a compact transmission spectrum. This is important if adequate adjacent channel interference figures are to be met. The total symbol rate for GSM at 1 bit per symbol in GMSK produces 270.833 K symbols/second. The gross transmission rate of the time slot is 22.8 Kbps.

Access Methods

Because radio spectrum is a limited resource shared by all users, a method must be devised to divide up the bandwidth among as many users as possible. The choices are

- TDMA
- FDMA
- *Code Division Multiple Access* (CDMA)

GSM chose a combination of TDMA/FDMA as its method. The FDMA part involves the division by frequency of the total 25 MHz bandwidth into 124 carrier frequencies of 200 kHz bandwidth. One or more carrier frequencies are then assigned to each BS. Each of these carrier frequencies is then divided in time, using a TDMA scheme, into eight time slots. One time slot is used for transmission by the mobile and one for reception. They are separated in time so that the mobile unit does not receive and transmit at the same time, a fact that simplifies the electronics.

FDMA

The FDMA part involves the division by frequency of the total 25 MHz bandwidth into 124 carrier frequencies of 200 kHz bandwidth. One or more carrier frequencies are then assigned to each BS. Using FDMA, a frequency is assigned to a user, as shown in Figure 21-28. Therefore, the larger the number of users in an FDMA system, the larger the number of available frequencies must be. The limited available radio spectrum and the fact that a user will not free its assigned frequency until he or she does not need it anymore explains why the number of users in an FDMA system can be quickly limited.

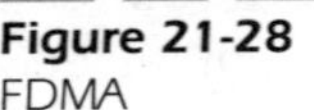

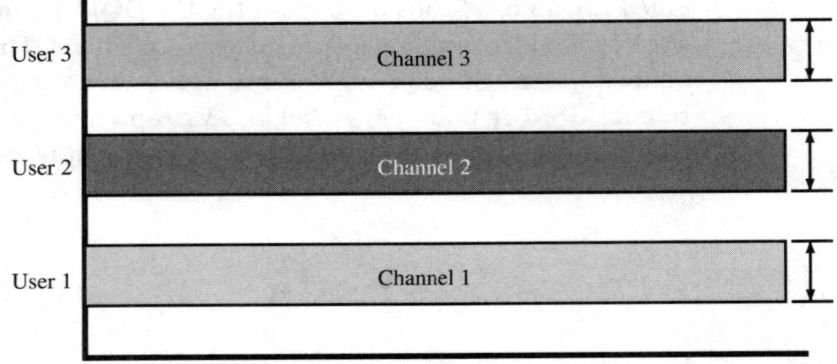

Figure 21-28
FDMA

TDMA

TDMA is digital transmission technology that enables a number of users to access a single RF channel without interference by allocating unique time slots to each user within each channel. Each of the carrier frequencies is divided in time, using a TDMA scheme, into eight time slots, as shown in Figure 21-29. One time slot is used for transmission by the mobile and one for reception. They are separated in time so that the mobile unit does not receive and transmit at the same time, a fact that simplifies the electronics.

TDMA enables several users to share the same channel. Each of the users, sharing the common channel, is assigned his or her own burst within a group of bursts called a frame. Usually, TDMA is used with an FDMA structure. In addition to increasing the efficiency of transmission, TDMA offers a number of other advantages over standard cellular technologies. First and foremost, it can be easily adapted to the transmission of data as well as voice communication. TDMA offers the capability to carry data rates of 64 Kbps to 120 Mbps (expandable in multiples of 64 Kbps). This enables operators to offer personal-communication-like services including fax, voiceband data, and SMSs, as well as bandwidth-intensive applications such as multimedia and video conferencing. Unlike spread-spectrum techniques that can suffer from interference among the users, all of whom are on the same frequency band and transmitting at the same time, TDMA's technology, which separates users in time, ensures that they will not experience interference from other simultaneous transmissions.

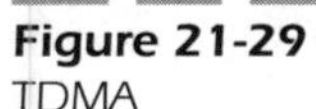

Figure 21-29
TDMA

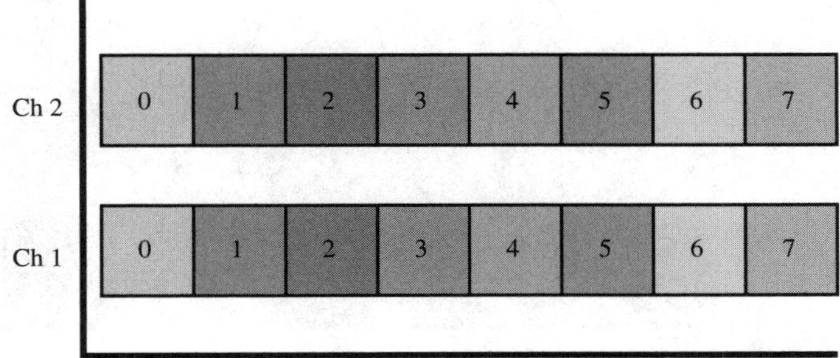

CDMA

CDMA is characterized by high capacity and a small cell radius, which employs spread-spectrum technology and a special coding scheme. CDMA is the dynamic allocation of bandwidth. To understand this, it's important to realize that in the context of CDMA, "bandwidth" refers to the capability of any phone to get data from one end to the other. It doesn't refer to the amount of spectrum used by the phone, because in CDMA, every phone uses the entire spectrum of its carrier whenever it is transmitting or receiving, as shown in Figure 21-30. One of the terms you'll hear in conjunction with CDMA is "soft handoff." A handoff occurs in *any* cellular system when your call switches from one cell site to another as you travel. In all other technologies, this handoff occurs when the network informs your phone of the new channel to which it must switch. The phone then stops receiving and transmitting on the old channel and commences transmitting and receiving on the new channel. It goes without saying that this is known as a "hard handoff."

In CDMA, however, every phone and every site are on the same frequency. In order to begin listening to a new site, the phone only needs to change the pseudorandom sequence it uses to decode the desired data from the jumble of bits sent for everyone else. While a call is in progress, the network chooses two or more alternate sites that it feels are handoff candidates. It simultaneously broadcasts a copy of your call on each of these sites. Your phone can then pick and choose between the different sources for your call and move between them whenever it feels like it. It can even combine the data received from two different sites to ease the transition from one to the other. CDMA is more efficient about that kind of thing. In both TDMA and CDMA, the outgoing voice traffic is digitized and compressed. However,

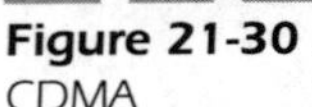

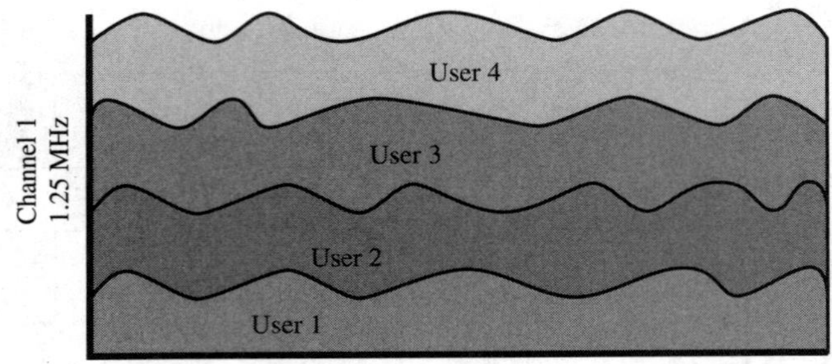

Figure 21-30
CDMA

the CDMA codec can realize when the particular packet is noticeably simpler (for example, silence or a sustained tone with little change in modulation) and will compress the packet far more. Thus, the packet may involve fewer bits, and the phone will take less time to transmit it. That's where this odd idea of what bandwidth means in CDMA comes in. In a real sense, bandwidth in CDMA equates to receive power at the cell. CDMA systems constantly adjust power to make sure as little is used as necessary and compensate for this by using coding gain through the use of forward error correction and other approaches that are much too complicated to go into. The chip rate is constant, and if more actual data is carried by the constant chip rate, then less coding gain will occur. Therefore, it's necessary to use more power instead.

TDMA Frames

In GSM, a 25 MHz frequency band is divided, using an FDMA scheme, into 124 carrier frequencies spaced 1 place from each other by a 200 kHz frequency band. Normally, a 25 MHz frequency band can provide 125 carrier frequencies, but the first carrier frequency is used as a guard band between GSM and other services working on lower frequencies. Each carrier frequency is then divided in time using a TDMA scheme. This scheme splits the radio channel, with a width of 200 kHz, into 8 bursts, as shown in Figure 21-31. A burst is the unit of time in a TDMA system, and it lasts approximately 0.577 ms. A TDMA frame is formed with 8 bursts and lasts, consequently, 4.615 ms. Each of the eight bursts that form a TDMA frame are then assigned to a single user.

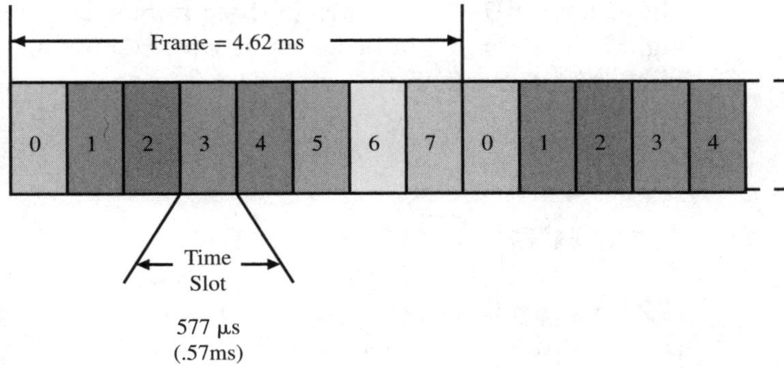

Figure 21-31 TDMA framing and time slots

Time Slot Use

One time slot is used for transmission by the mobile and one for reception. They are separated in time so that the mobile unit does not receive and transmit at the same time, a fact that simplifies the electronics. A separation is used with a three-time slot offset so that the mobile will not have to send and receive at the same time.

GSM FDMA/TDMA Combination

To enable multiple access, GSM utilizes a blending of FDMA and TDMA. This combination is used to overcome the problems introduced in each individual scheme. In the case of FDMA, frequencies are divided up into smaller ranges of frequency slots and each of these slots is assigned to a user during a call. Although this method will result in an increase of the number of users, it is not efficient in the case of high user demand. On the other hand, TDMA assigns a time slot for each user for utilizing the entire frequency. Similarly, this will become easily overloaded when encountering high user demand. Hence, GSM uses a two-dimensional access scheme. GSM uses the combined FDMA and TDMA architecture to provide the most efficient operation within the scope of price and reasonable data. The physical channels are TDMA time slots, and the radio channels are frequencies. This scheme divides the entire frequency bandwidth into several smaller pieces as in FDMA and each of these frequency slots is to be divided into eight time slots in a full-rate configuration. Similarly, 16 time slots will be in a half-rate configuration.

Logical Channels

GSM distinguishes between physical channels (the time slot) and logical channels (the information carried by the physical channels). Several recurring time slots on a carrier constitute a physical channel, which is used by different logical channels to transfer information—both user data and signaling. A channel corresponds to the recurrence of one burst every frame. It is defined by its frequency and the position of its corresponding burst within a TDMA frame. GSM has two types of channels:

- The traffic channels used to transport speech and data information
- The control channels used for network management messages and some channel maintenance tasks

The Physical Layer

Each physical channel supports a number of logical channels used for user traffic and signaling. The physical layer (or Layer 1) supports the functions required for the transmission of bit streams on the air interface. Layer 1 also provides access capabilities to upper layers. The physical layer is described in the GSM Recommendation 05 series (part of the ETSI documentation for GSM). At the physical level, most signaling messages carried on the radio path are in 23-octet blocks. The data-link layer functions are multiplexing, error detection and correction, flow control, and segmentation to enable long messages on the upper layers.

The radio interface uses the LAP-Dm. This protocol is based on the principles of the ISDN LAPD protocol. Layer 2 is described in GSM Recommendations 04.05 and 04.06. The following logical channel types are supported:

- *Speech traffic channels* (TCH)
  - *Full-rate TCH* (TCH/F)
  - *Half-rate TCH* (TCH/H)

- *Broadcast channels* (BCCH)
  - *Frequency correction channel* (FCCH)
  - *Synchronization channel* (SCH)
  - *Broadcast control channel* (BCCH)
- *Common control channels* (CCCH)
  - *Paging channel* (PCH)
  - *Random access channel* (RACH)
  - *Access grant channel* (AGCH)
- *Cell broadcast channel* (CBCH) (the CBCH uses the same physical channel as the DCCH)
- *Dedicated control channels* (DCCH)
  - *Slow associated control channel* (SACCH)
  - *Stand-alone dedicated control channel* (SDCCH)
  - *Fast associated control channel* (FACCH)

Speech Coding on the Radio Link

The transmission of speech is, at the moment, the most important service of a mobile cellular system. The GSM speech codec (coder and decoder), which will transform the analog signal (voice) into a digital representation, has to meet the following criteria:

- It must have good speech quality, at least as good as the quality obtained with previous cellular systems.
- Reduce the redundancy in the sounds of the voice. This reduction is essential due to the limited capacity of transmission of a radio channel.
- The speech codec must not be very complex because complexity is equivalent to high costs.

The final choice for the GSM speech codec is a codec named *Regular Pulse Excitation Long-Term Prediction* (RPE-LTP). This codec uses the information from previous samples (this information does not change very quickly) in order to predict the current sample. The speech signal is divided into blocks of 20 ms; these blocks are then passed to the speech codec, which has a rate of 13 Kbps, in order to obtain blocks of 260 bits.

Channel Coding

Channel coding adds redundancy bits to the original information in order to detect and correct, if possible, errors that occurred during the transmission. The channel coding is performed using two codes: a block code and a convolutional code.

- The block code corresponds to the block code defined in the GSM Recommendations 05.03. The block code receives an input block of 240 bits and adds 4 zero tail bits at the end of the input block. The output of the block code is consequently a block of 244 bits.
- A convolutional code adds redundancy bits in order to protect the information. A convolutional encoder contains memory. This property differentiates a convolutional code from a block code. A convolutional code can be defined by three variables: n, k, and K. The value n corresponds to the number of bits at the output of the encoder, k to the number of bits at the input of the block, and K to the memory of the encoder.

Convolutional Coding

Before applying the channel coding, the 260 bits of a GSM speech frame are divided in three different classes according to their function and importance. The most important class is the class Ia containing 50 bits. The class Ib is next in importance, which contains 132 bits. The class II is the least important, which contains the remaining 78 bits. The different classes are coded differently. First of all, the class Ia bits are block coded. Three parity bits, used for error detection, are added to the 50 class Ia bits. The resultant 53 bits are added to the class Ib bits. Four zero bits are added to this block of 185 bits (50 + 3 + 132). A convolutional code, with $r = 1/2$ and $K = 5$, is then applied, obtaining an output block of 378 bits. The class II bits are added, without any protection, to the output block of the convolutional coder. An output block of 456 bits is finally obtained.

CHAPTER 22

Personal Communications Services

There can be no doubt about the changes that have taken place in the industry since the inception of the *personal communications services* (PCS) in the mid-1990s. The entire industry has changed, making communications an integral component of our everyday life. What was once considered as frivolous (or for the affluent only) in the past, has become commonplace today. Look around and see where the personal communications systems have penetrated. What was once a business service is now an everyday component of students, housewives, business, and everyday people. The personal communicator, the simple wireless telephone, has become an indispensable tool for just about every occupation. The days of having a simple pager are fast becoming obsolete as the personal communicator becomes more affordable and more competitively priced. Not that long ago, the industry was taken aback by the prospect of everyone having a personal telephone set. Today, that concept is closer to reality than ever before.

The communicator of today is not a broadband device. However, it is the precursor for the future devices. The newer ones will bring more bandwidth and more sophistication to our everyday lives. We have to merely walk before we run!

The PCSs have evolved from the wireless cellular and *Global System for Mobile Communications* (GSM) networks, so there is not a lot of excitement there. However, where the original wireless networks were built on an analog-networking standard, the PCS architectures are built on digital transmission systems. Therefore, several different approaches are used to deliver the capacities at the pricing models today. We shall discuss the evolution of these wireless systems and their ability to propel us into the new millennium and the communicator of the future.

Digital Systems

Because *Frequency Division Multiple Access* (FDMA) uses the full 30 kHz channels for one telephone call at a time, it is obviously wasteful (see Figure 22-1). FDMA, an analog technique, can be improved a little by using the same frequency in a *Time Division Duplex* (TDD) mechanism. Here, one channel is used, and time slots are created. The conversation flows from A to B and then from B to A. The use of this channel is slightly more efficient. However, when nothing is being sent, the channel remains idle. Because digital transmission introduces better multiplexing schemes, the carriers wanted to get more users on an already strained *radio frequency* (RF) spectrum.

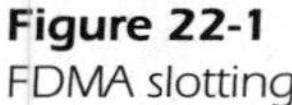

Figure 22-1
FDMA slotting

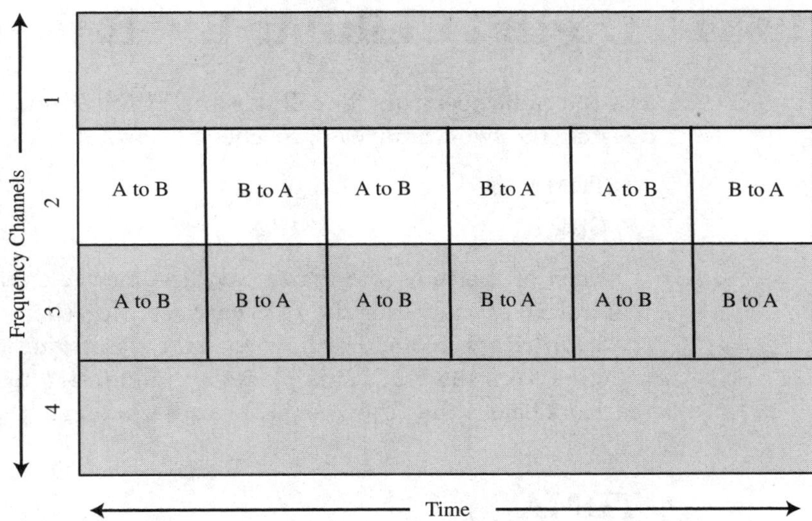

Additional possibilities for enhancing security and reducing fraud can be addressed with digital cellular and PCS. Again, this appears to be a win-win situation for the carrier because it features

- Fewer costs.
- More users.
- Better security.
- Less fraud.
- Obviously, the carriers welcomed some of these discussions.
- Once the decision was made to consider digital transmission, the major problem was how, what flavor to use, and how to seamlessly migrate the existing customer base to digital.
- The digital techniques available to the carriers are the following:
  - *Time Division Multiple Access* (TDMA)
  - *Extended Time Division Multiple Access* (ETDMA)
  - *Code Division Multiple Access* (CDMA)
  - GSM (using a slightly different form of TDMA)
  - *Narrowband Advanced Mobile Phone Service* (N-AMPS)

Carriers and manufacturers are using each of these systems. The various means of implementing these systems has brought about several discussions regarding the benefits and losses of each choice.

Digital Cellular Evolution

As the radio spectrum for cellular and PCS continues to become more congested, the two primary approaches in North America use derivatives of

- TDMA
- CDMA

Each of the moves to digital requires newer equipment, which means capital investments for the PCS carriers. Moreover, as these carriers compete for RF spectrum, they have to make significant investments during an auction from the FCC. This places an immense financial burden on these carriers before they even begin the construction of their networks.

TDMA

North American TDMA uses a *time-division multiplexing* (TDM) scheme, where time slices are allocated to multiple conversations. Multiple users share a single RF without interfering with each other because they are kept separate by using fixed time slots. The current standard for North American TDMA divides a single channel into six time slots. Then three different conversations use the time slots by allocating two slots per conversation. This provides a three-fold increase in the number of users on the same RF spectrum. Although TDMA deals typically with an analog-to-digital conversion using a typical *pulse code modulation* (PCM) technique, it performs differently in a radio transmission. PCM is translated into a quadrature phase-shift keying technique, thereby producing a four-phased shift, and doubling the data rate for data transmission. The typical time slotting mechanism for TDMA is shown in Figure 22-2.

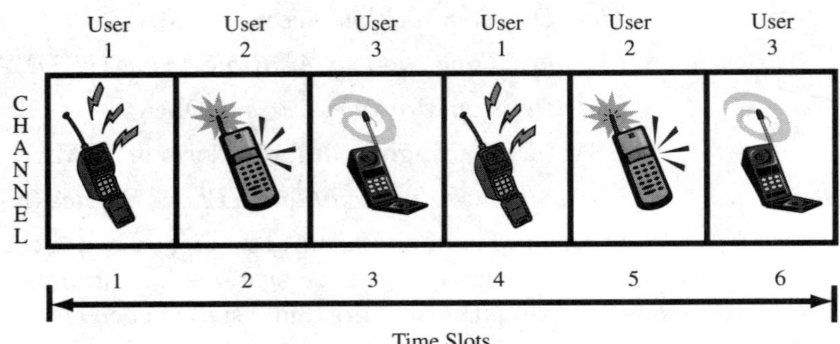

Figure 22-2 North American TDMA for PCS

The wireless industry began to deploy the use of TDMA back in the early 1990s when the scare RF spectrum problem became most noticeable. The intent was to improve the quality of the radio transmission as well as the efficiency usage of the limited spectrum available. TDMA correctly can increase the number of users on the RF by three-fold. However, as more enhanced PCS techniques, such as micro and picocells, are used, the numbers can grow to as much as a 40-fold increase in the number of users on the same RF spectrum. One can see why it is so popular.

TDMA has another advantage over the older analog (FDMA) techniques. Analog transmission across the 800 MHz frequency band (in North America, it is 800 MHz; in much of the world, it is 900 MHz) supports one primary service—voice. The TDMA architecture uses a PCM input to the RF spectrum. Therefore, TDMA can also support digital services for data in increments of 64 Kbps. The data rate can support from 64 Kbps to the hundreds of megabits/second (120 Mbps).

The carriers like TDMA because of its ability to add data and voice across the RF spectrum and the cost associated with the migration from analog to digital. A TDMA *base station* (BS) costs approximately $50 to $80,000 to migrate to digital from FDMA. This is an attractive opportunity for the carriers developing PCS and digital cellular using the industry standards. The two standards in use today include IS-54, which is the first evolution to TDMA from FDMA. The second is IS-136, the latest and greatest technology for 800 to 900 and 1900 MHz TDMA service. When the IS-136 service was introduced, the addition of data and other services was introduced. These services include the *Short Message Service* (SMS), caller id display, data transmission, and other service levels. Using a TDMA approach, the carriers feel that they can meet the needs for voice, data, and video integration for the future. TDMA is the basis of the architecture for the GSM standard in Europe and the Japanese standard for *personal digital communications* (PDC).

Advantages of TDMA Unlike spread spectrum that can experience interference from other users in the same frequency, TDMA keeps all users' conversations separate. Because the conversation is limited to a timing slot, the battery on a telephone set should be extended significantly because the conversation is only taking up one-third of the time. Moreover, TDMA uses the same hierarchical structure as traditional cellular networks so it can be increased and upgraded easily.

Enhanced TDMA Standards bodies and manufacturers all came up with variations of the use of the frequency spectrum. Nevertheless, manufacturers have been developing an ETDMA, which will enable efficient use of the system. TDMA will derive a three- to five-fold increase in spectrum use, whereas ETDMA could produce 10- to 15-fold increases. The concept is to use a *Digital Speech Interpolation* (DSI) technique that reallocates the quiet times in normal speech, thereby assigning more conversations to fewer channels, gaining up to 15 times over an analog channel. This is a form of statistical TDM. When a device has something to send, it places a bit in a buffer. As the sampling device sees the data in the buffer, it allocates a data channel to that device. When a device has nothing to send, then nothing is placed into the buffer. Then the sampling passes over a device with an empty buffer. Time slots are dynamically allocated based on need rather than on a fixed time slot architecture. An example of the ETDMA technique is shown in Figure 22-3.

CDMA

CDMA is a radical shift from the original FDMA and TDMA wireless techniques. This system has been gaining widespread acceptance across the

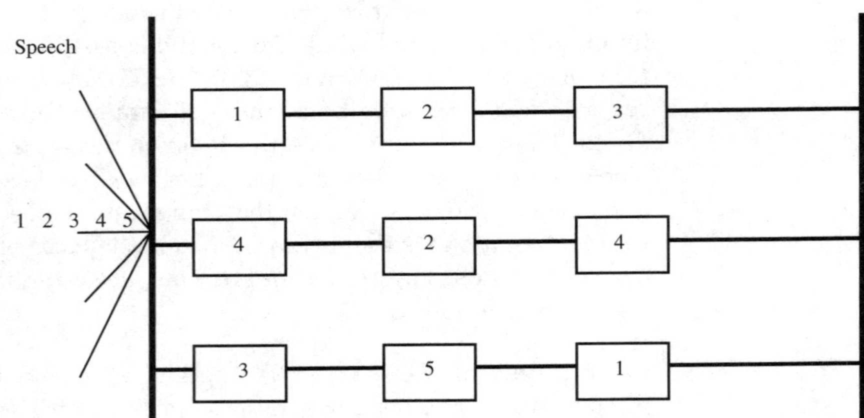

Figure 22-3
ETDMA uses a form of statistical TDM.

world in the PCS industry. The cellular providers see CDMA as an upgrade opportunity for their capacity and quality. CDMA is a form of spread spectrum, a family of digital communications techniques. The core principle behind CDMA is the use of the noise-floor to carry radio signals.

As its name implies, bandwidth greater and wider than normal constrained FDMA and TDMA channels is used. Point-to-point communication is effective on the bandwidth that uses the noise waves to carry the signal spread across a significantly wider radio carrier. Spread spectrum, which was employed back in the 1920s, has evolved from military security applications. It uses a technique of organizing the RF energy over a range of frequencies, rather than a modulation technique. The system uses frequency hopping with TDM. At one minute the transmitter is operating at one frequency; at the next instant it is on another. The receiver is synchronized to switch frequencies in the same pattern. This is effective in preventing detection (interception) and jamming; thus, additional security is derived. These techniques should produce increased capacities of 10 to 20 fold over existing analog systems. Architecturally, the model for the system is shown in Figure 22-4.

Originally conceived for commercial application in the 1940s, it was an additional 40 years before this technique became commercially feasible. The main factors holding back the use of CDMA were cost and complexity of operation. Today, the use of low-cost, high-density digital *integrated circuits* (ICs) that reduce the size and weight of the radio equipment makes the use of CDMA far more feasible. Another area is an educational one, whereby the carriers needed to understand that the use of optimal

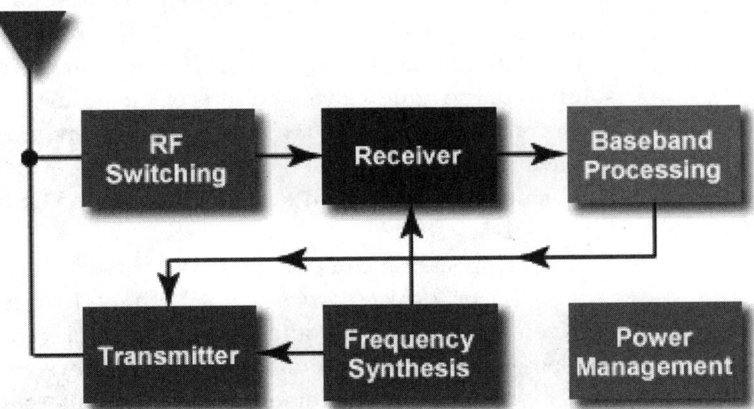

Figure 22-4
The model for the CDMA systems
Source: CDMA Org

communications requires that the station equipment must regulate the power to the lowest possible levels to achieve the maximum performance. This, of course, flies in the face of the normal operations that the carriers received their training.

Spread Spectrum Services

As the use of RF spectrum continued to put pressure on this limited resource, the systems manufacturers and regulators were searching for some way to share spectrum among multiple users. Further, the sharing is compounded by the need to secure information while on airwaves. These pressures have led to the use of spread spectrum radio. The spreading portion of these systems using a chip set coded for your specific transmitter-to-receiver system. It can use multiple frequencies (called *hopping*), or it can create a coded chip set. Regardless of the operation chosen, both are designed to spread energy over a broader range of frequencies to allow for less airtime on a specific bandwidth and to ensure the integrity of the information being sent.

Spread spectrum uses a technique of organizing the RF energy over a range of frequencies rather than a fixed *frequency modulation* (FM) technique. The system uses frequency hopping with TDM. At one minute the transmitter is operating at one frequency; at the next instant it is on another. The receiver is synchronized to switch frequencies in the same pattern as the transmitter. This is effective in preventing detection (interception) and jamming. Thus, additional security is derived. These techniques should produce increased capacities in 10- to 20-fold over existing analog systems.

In 1989, spread spectrum CDMA was commercially introduced as the solution to the bandwidth demands of the industry. By using a spread spectrum frequency-hopping technique, developers announced that they could achieve the desired frequency reuse patterns everyone wanted. The model for the network components mimics that of the GSM architecture shown in Figure 22-5.

Spread spectrum can use one of two different techniques: *frequency hopping* (FH) or *direct sequence* (DS). In both cases, the synchronization between the transmitter and receiver is crucial. Both forms use a pseudorandom carrier; they just do it in different ways.

FH is not usually implemented in the commercial versions of CDMA. DS is used by commercially available CDMA. It is accomplished as a multiple

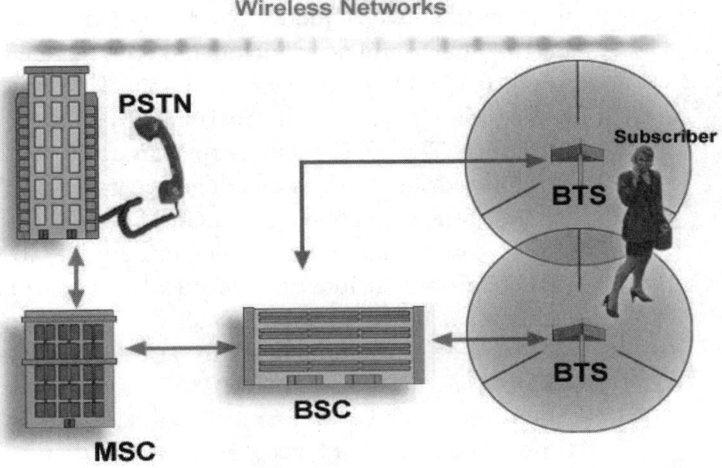

Figure 22-5 The model of the CDMA network mimics the GSM architecture.

of the more conventional waveform by using a pseudonoise binary sequence at the transmitter. Noise and interference across the waveform are uncorrelated with the pseudonoise sequence, and thus become like noise. This increases the bandwidth when they reach the detectors. The *signal-to-noise ratio* (SNR) can be enhanced by using filters that reject the interference.

The technique for the spread spectrum service is called CDMA. Many of the PCS carriers have chosen CDMA as their coding of choice. Since spread spectrum has been introduced at the commercial level, the FCC allocated spectrum in the 1.9 GHz range. This group of frequencies is heavily used by microwave users operating in the 2 GHz range. The FCC feels that the spread spectrum won't seriously impair the microwave users because it will appear only as random noise to any other frequency system operating in the same frequency ranges. However, as more users of spread spectrum are inserted into a specific frequency range, the possibility of congestion and interference exists. If you attempt to use this range of frequencies and congestion builds, then the decision could be a wrong one. Despite the benefits of FH, the benefits may be overshadowed with distance, utilization, and power constraints.

Consequently, when the FCC decided to allocate frequency spectrum to the PCS carriers, the fixed microwave users were told they had to find a new home. This means that the operators who had radio systems in operation had to find new frequencies to use. Moreover, to facilitate the frequency change, many of the fixed operators had to upgrade their equipment to work in the new frequency band they found acceptable. Part of the auctions for the PCS frequencies included money set aside to help the fixed operators upgrade and move to a new frequency. This put added pressure on the wireless carriers to pay for the auctioned frequencies before building their infrastructure. The FCC raised approximately $13 billion in the auctions, but that placed the carriers in a negative position. They spent a lot of their capital up front to get the frequencies, before they could buy any equipment.

The concepts of spread spectrum and of CDMA seem to contradict normal intuition. In most communications systems, we try to maximize the amount of useful signals we can fit into a minimal bandwidth. In spread spectrum, we try to artificially spread a signal over a bandwidth much wider than necessary. In CDMA, we transmit multiple signals over the same frequency band, using the same modulation techniques at the same time. There are very good reasons for doing this. In a spread spectrum system, we use some artificial technique to broaden the amount of bandwidth used.

Capacity Gain

When you use the Shannon-Hartly law for the capacity of a bandwidth-limited channel, it is easy to see that for a given signal powers the wider the bandwidth used, the greater the channel capacity. So if we broaden the spectrum of a given signal, we get an increase in channel capacity and/or an improvement in the SNR. This is true and easy to demonstrate for some systems but not for others. Ordinary FM systems spread the signal above the minimum theoretically needed, and they get a demonstrable increase in capacity. Some techniques for spreading the spectrum achieve a significant capacity gain, but others do not.

CDMA alters the way we communicate by

- Improving the telephone capacity of cellular operators
- Improving quality of the voice communications and eliminating audible impairments of multipath fade
- Reducing the incidence of dropped calls, especially during handoff
- Providing reliability for data communications, that is, fax and Internet traffic
- Reducing the number of sites to support a specific volume of traffic

- Simplifying the site selection process
- Reducing the average transmitter power output requirements
- Reducing or eliminating interference with other electronic devices in the area
- Limiting potential health risks
- Reducing the operating costs because fewer cell sites are needed

The CDMA Cellular Standard

With CDMA, unique digital codes, rather than separate radio frequencies are used to differentiate subscribers. The codes are shared by both the *mobile station* (MS) and the BS and are called *pseudorandom code sequences*. All users share the same range of radio spectrum. For cellular telephony, CDMA is a digital multiple access technique specified by the TIA as IS-95. In 1993, the TIA gave its approval of the CDMA IS-95 standard. IS-95 systems divide the radio spectrum into carriers that are 1.25 MHz wide as shown in Figure 22-6. One of the unique aspects of CDMA is that although the number of phone calls that a carrier can handle is certainly limited, it is not a fixed number. Rather, the capacity of the system will be dependent on a number of different factors.

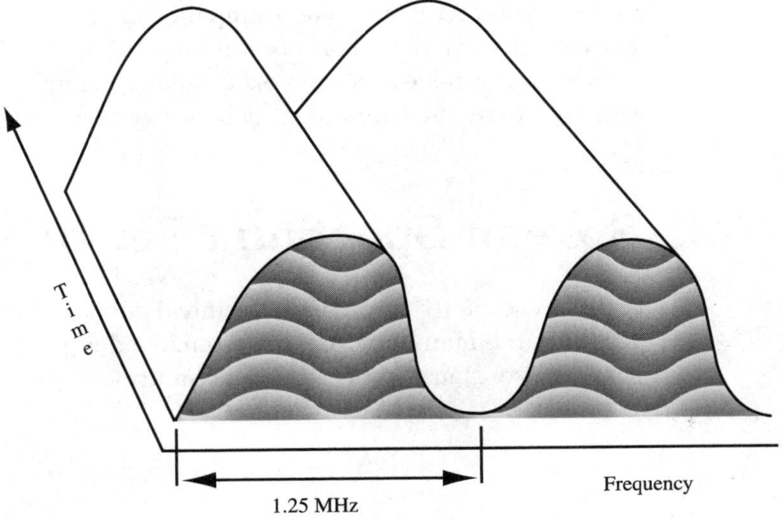

Figure 22-6 CDMA uses a channel that is 1.25 MHz wide for all simultaneous callers.

CDMA changes the nature of the subscriber equipment from a predominantly analog device to a digital device. CDMA receivers do not eliminate analog processing in its entirety, but they separate the communications channels by a pseudorandom modulation technique that is applied to the digital domain, not based on frequencies. In fact, multiple users will occupy the same frequency band simultaneously.

Spread Spectrum Goals

Much of the action in the wireless arena includes the use of the frequency spectrum to its fullest while preserving its efficiency. The primary goal of spread spectrum systems is the substantial increase in bandwidth of an information-bearing signal, greater than required for basic communications.

This increased bandwidth, though not used for carrying the signal, can mitigate possible problems in the airwaves, such as interference or inadvertent sharing of the same channels. The cooperative use of the spectrum is an innovation that was not commercially available in the past.

Regulators around the world have set aside limited amounts of bandwidth to satisfy these services, so that the efficiency is kept high. The limited frequency spectrum allocated preserves upon the goal of using spectral efficiency, which is usually measured with one of the traffic engineering calculations (Erlang or Poisson) per unit in operation in a specified geography and in terms of per MHz. For example, cellular operators use a 25 MHz split between the two directions of communications: 12.5 MHz of transmit and 12.5 MHz of receive spectrum. As technology enhancements occur, practical ways of expanding the amount of coverage become a reality.

Spread Spectrum Services

As the use of RF spectrum continued to put pressure on this limited resource, the manufacturers of systems and regulators were searching for some way to share spectrum among multiple users. Furthermore, sharing is compounded by the need to secure information while on airwaves. These pressures have led to the use of spread spectrum radio. The spreading portion of these systems using a chip set coded for your specific transmitter-to-receiver system uses multiple frequencies (called *hopping*) as one method, or another technique of creating a coded chip set is used.

Both of these services are designed to spread as much energy over a broader range of frequencies to enable less airtime on a specific bandwidth and to ensure the integrity of the information being sent. The technique for the spread spectrum service is called CDMA. Many of the PCS carriers have chosen CDMA as their coding of choice. Because spread spectrum has been introduced at the commercial level, the FCC allocated spectrum in the 1.9 GHz range. This group of frequencies is heavily used by microwave users operating in the 2 GHz range.

Synchronization

In the final stages of the encoding of the radio link from the BS to the mobile, CDMA adds a special pseudorandom code to the signal that repeats itself after a finite amount of time. BSs in the system distinguish themselves from each other by transmitting different portions of the code at a given time. In other words, the BSs transmit time-offset versions of the same pseudorandom code. In order to assure that the time offsets used remain unique from each other, CDMA stations must remain synchronized to a common time reference.

The primary source of the very precise synchronization signals required by CDMA systems is the *Global Positioning System* (GPS) as shown in Figure 22-7. GPS is a radio navigation system based on a constellation of orbiting satellites. Because the GPS system covers the entire surface of the earth, it provides a readily available method for determining position and time to as many receivers as are required. The synchronization is maintained between the transmitter and receiver so that each user device can be isolated in time. A representation of this is shown in Figure 22-8.

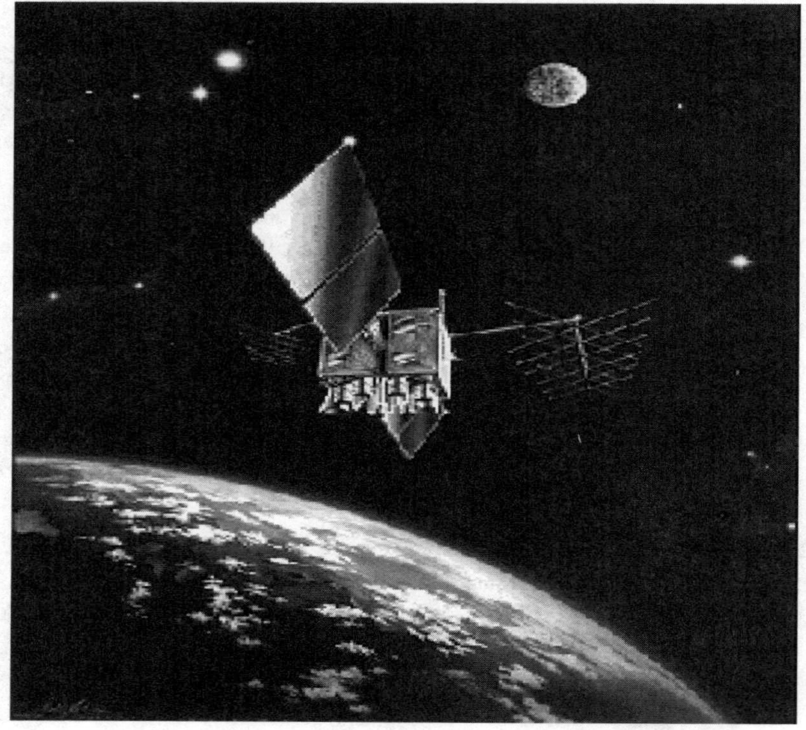

Figure 22-7
CDMA uses GPS to maintain synchronization of the systems.
Source: NASA

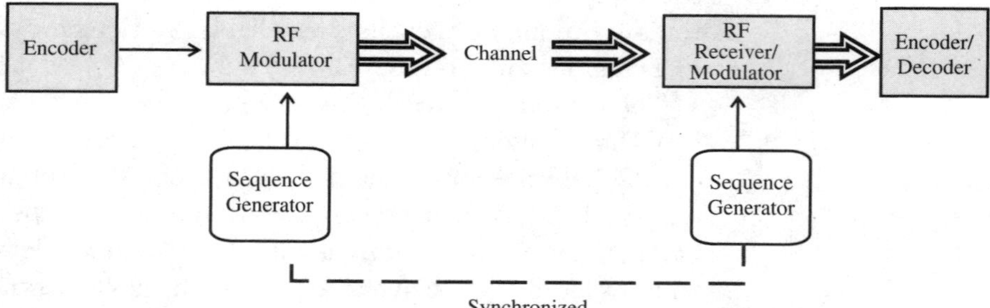

Figure 22-8
The systems maintain synchronization to isolate each user.

Balancing the Systems

CDMA cell coverage is dependent upon the way the system is designed. In fact, three primary system characteristics—coverage, quality, and capacity—must be balanced off of each other to arrive at the desired level of system performance. In a CDMA system, these three characteristics are tightly interrelated. Even higher capacity might be achieved through some degree of degradation in coverage and/or quality. Because these parameters are all intertwined, operators cannot have the best of all worlds: three times wider coverage, 40 times capacity, and CD-quality sound. For example, the 13 Kbps vocoder provides better sound quality but reduces system capacity as compared to an 8 Kbps vocoder.

Operators will have the opportunity to balance these parameters to best serve a particular area. The best balance point may change from cell site to cell site. Sites in dense downtown areas may trade off coverage for increased capacity. Conversely, at the outer edges of a system, capacity could be sacrificed for coverage area.

Common Air Interfaces

Two primary air interface standards are in use today: Cellular (824 to 894 MHz) uses the TIA/EIA/IS-95A. PCS (1,850 to 1,990 MHz) uses the ANSI J-STD-008. These two standards are similar in the features that they offer, with the exception of the frequency plan, mobile identities, and message fields. The standards provide some stability in the operation of the systems but may change over time. However, looking at the forward and reverse CDMA channel can shed some added light on what we can expect from the use of CDMA.

The Forward Channel

The forward CDMA channel is the cell site to mobile direction for communications. It carries traffic, a pilot signal, and any overhead information required by the system. The pilot is a spread but otherwise unmodulated DSSS signal. The pilot and overhead channels establish and maintain the system timing and the station identity. The pilot is also used in the *mobile-assisted handoff* (MAHO) process as a signal strength indicator.

Transmission Speeds and Rates IS-95A uses a forward link that supports a speed of 9,600 bps in the data-bearing channels. The forward error correction code rate is 1/2, and the pseudonoise rate is 1.2288 MHz (which is 128 × 9,600 bps). Recent events have seen that Verizon and Sprint PCS are now offering data rates of up to 144 Kbps on their 2.5G wireless networks using CDMA.

Overhead Channels Three different types of overhead channels exist in the forward link. These include the pilot, sync, and paging channels. The pilot is a requirement for every station.

- **Pilot channel** This is always code channel 0. It operates a demodulation reference for the MSs and a handoff level measurement reference.
- **Sync channel** This carries a repeating message that identifies the individual station and the absolute phase of the pilot sequence. The data rate of the sync channel is 1,200 bps. This mobile finds the framing boundary of the sync channel and times to it simply. It carries a single repeating message that conveys timing and system configuration to the MS.
- **Paging channel** This is used to communicate when the mobile is not assigned to a traffic channel. It notifies the mobile of incoming calls and carries responses to the mobile access. Paging channels operate at 4,800 or 9,600 bps.

Traffic channels These are dynamically assigned channels in response to a mobile access. The traffic channel carries its data in a 20 ms frame.

The Reverse Channel

The *reverse channel* is the mobile-to-cell site communication channel. It carries traffic and signaling information. A reverse channel is only active during calls associated with a specific MS or when access channel signaling takes place to the BS.

Transmission Speeds and Rates IS-95A uses a reverse link that supports a speed of 9,600 bps in the access and traffic channels. The forward error correction code rate is 1/3, the code symbol rate is 28,800 symbols per second after six code symbols per modulation symbol are present, and the pseudonoise rate is 1.2288 MHz.

Channelization The reverse CDMA channel consists of $2^{42} - 1$ logical channels. One of these channels is permanently and uniquely associated with each MS. The mobile uses the logical channel whenever it passes traffic. The channel does not change upon a handoff. Other logical channels are used for access with the BS.

Walsh Codes

These are the primary transmission patterns used in CDMA mobile phone systems. A few dozen CDMA handsets can be operating within (sharing) the same 1.25 MHz of radio bandwidth, and it is important that the interference effects are minimized. Walsh codes serve to identify each transmitter to the base (or vice versa), and they spread the code (the chip rate)—which is the spread spectrum effect.

However to reduce interference, each, on average, must counteract the effect of the others—if 50 percent are transmitting a positively phased pulse at any moment, the other 50 percent should be transmitting a negatively phased pulse. This is possible only because all transmissions are synchronized, and all use orthogonal Walsh codes.

If you toss a coin 50 times, you will most likely have an orthogonal Walsh code, which is 50 bits in length. Do this a dozen times, and you'll have a dozen different orthogonal 50-bit codes, each quite distinctive in its pattern. This is the basis of direct sequence code division multiplexing systems.

Traffic Channel

Traffic channels are the reverse of CDMA channels and are mobile unique. The traffic channel always carries data in a 20 ms frame. Frames at the higher rates of rate set 1 and all those in rate set 2 use CRC codes to help assess the frame quality in the receiver.

Direct Sequence Spread Spectrum

CDMA is a direct sequence spread spectrum system. The CDMA system works directly on 64 Kbps digital signals. These signals can be digitized voice, ISDN channels, modem data, and so on.

Signal transmission consists of the following steps:

1. A pseudorandom code is generated, different for each channel and each successive connection.
2. The information data modulates the pseudorandom code (the information data is spread).

3. The resulting signal modulates a carrier. (Steps 1 to 3 are shown in Figure 22-9.)
4. The modulated carrier is amplified and broadcast (see Figure 22-10).

Signal reception consists of the following steps:

1. The carrier is received and amplified.
2. The received signal is mixed with a local carrier to recover the spread digital signal.
3. A pseudorandom code is generated, matching the anticipated signal. (Steps 1 to 3 of the reception are shown in Figure 22-11.)
4. The receiver acquires the received code and phase locks its own code to it.
5. The received signal is correlated with the generated code, extracting the information data. (Steps 4 and 5 are shown in Figure 22-12.)

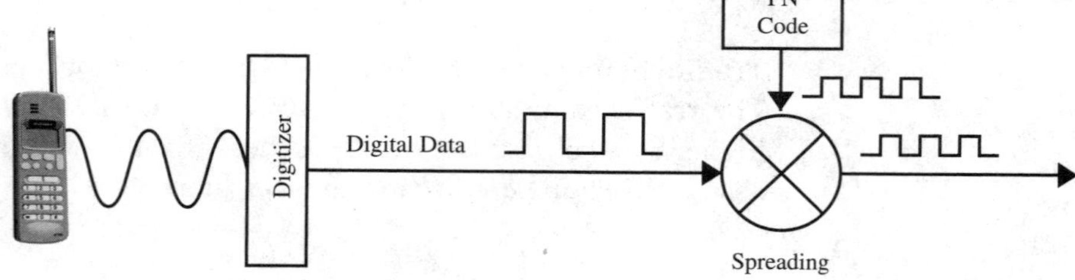

Figure 22-9 The signal is prepared for transmission.

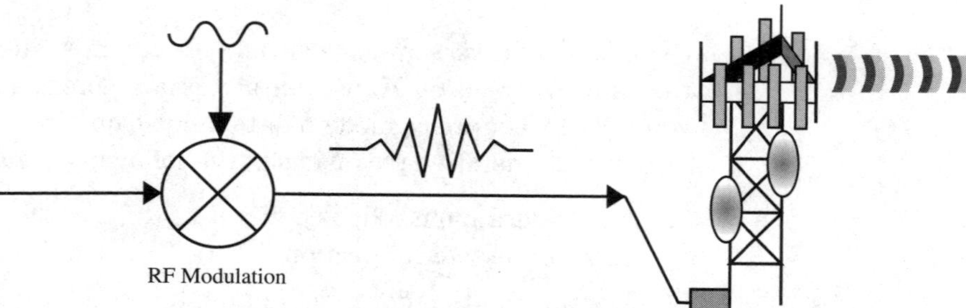

Figure 22-10 The modulated carrier is then amplified and broadcast across the air interface.

Figure 22-11 The reception from the air interface occurs.

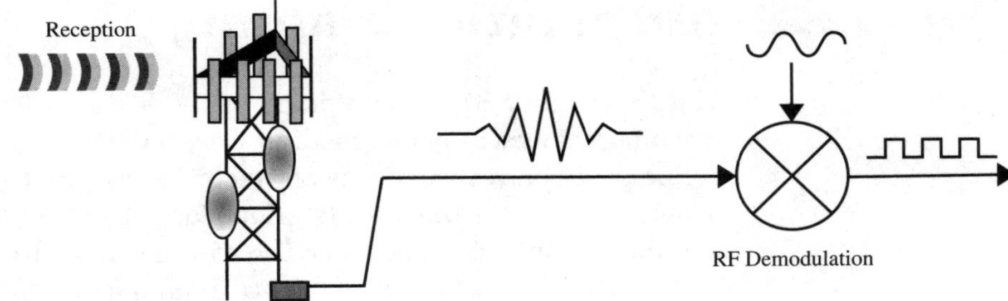

Figure 22-12 The codes are stripped off and the actual data is extracted.

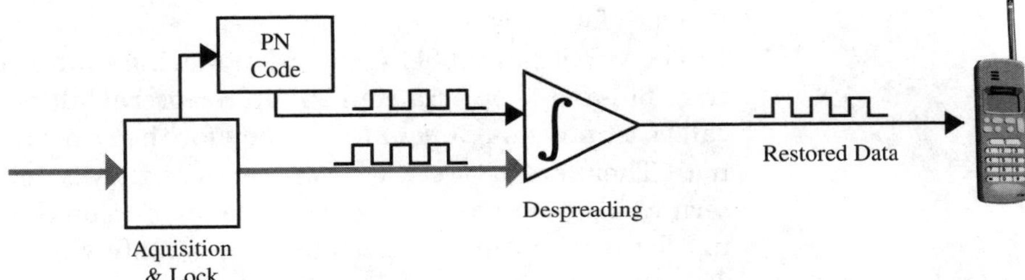

Seamless Networking with IS-41 and SS7

The whole intent of the IS-41 and SS7 interfaces is to enable the wireless carriers to communicate transparently and seamlessly between and among each other. Moreover, with SS7 interfaces to the wireline networks, calls can enter or exit the wireless networks flawlessly. This has been ongoing since 1994 and seems to be moving quite well. If the wireless carriers do not have physical interfaces to the telephone companies, they can use the *Independent Telecommunications Network* (ITN) to provide these services as a service bureau, for a fee. The interconnection between the networks provides industry standards-based internetworking.

Automatic Roaming

The IS-41 and SS7 provide many of the features for automatic roaming among providers, although modifications will be necessary. First, it helps to define financial responsibility between the carriers for their users. Second, it helps to provide the seamless interfaces between the wireless systems and the SS7 interfaces between the wireline and wireless carriers.

As the system operates, automatic roaming allows for the discovery of a roamer. The providing system then learns the identity of the current serving CO or visited system and sends that information to the home system. A profile is established for the roamer in the visited system, enabling the network to find the end user. This, in fact, is how the wireless providers handle location portability. Lastly, the automatic roaming enables the setup for delivery for the calls.

The wireless suppliers have been moving toward transparency through their bill-and-keep arrangement. In reciprocal billing, they assume that the calls cost an equal amount of money for the carriers to originate or terminate. Therefore, between wireless carriers, they do not charge each other for terminating minutes. The wireline companies on the other hand are charging for terminating minutes at approximately $.02 to $.025 per minute. Thus, the carriers are looking for a way to get away from the reciprocal billing and to get to a bill-and-keep arrangement with the wireline providers.

Cellular and PCS Suppliers

The cellular industry has been a success story to behold. Over the past 17 years, they have grown from nothing to over 100 million customers. With very few exceptions, the cellular carriers see themselves as a complementary carrier to the ILEC business. Some of this comes from the fact that two parts of the cellular network started when the licenses were issued: Wireline carriers meant the ILEC operated the regulated portion of the cellular network, and nonwireline meant the competition. In most cases, the ILEC who operated the wireline side of the business then spread out across the country as part of the nonwireline provider (competing for wireless services with their cousins). Therefore, the cellular industry is influenced heavily by the ILEC side of the business. The cellular providers see themselves as complementing the wireline business instead of competing with it.

PCS and SMR providers seem to be the carriers seeking a niche in the market. As they offer services to their customers they view themselves as an alternative to the ILEC wireline services and at the same time as complementary. Depending on how the customer reacts, the PCS providers will move toward one position. In checking with several PCS suppliers, the argument to make the PCS telephone number the sole number came up. The providers offer the same features and functions as the wireline service providers such as the following:

- Call forwarding (busy or don't answer)
- Voice messaging
- Three-way calling
- Caller ID
- Call transfer

The PCS providers are now saying that the customer can use the same number for their home number, their business number, and their traveling number. Why pay for two different lines and service offerings when you can do it all on one telephone (and number)? Their speech is convincing somewhat, but they fail to mention the cost of the airtime for receiving calls and the added cost for making local calls. However, they are now packaging these services in such a way that they are invisible.

Home Location Registers (HLRs) keep track of all network suppliers' users, based on their own network id. The database (HLR) has all the appropriate information to recognize the caller-by-user ESN and mobile telephone number. The database controls the features and functions the user has subscribed to. When a call comes into the network today, the called number is sent to the HLR responsible for the number dialed. These are SS7 messages. The HLR then looks into its database and determines where the caller is located or if the user is on the air.

If the user is located somewhere else, a *Visitor Location Register* (VLR) entry exists at a remote MSC. This entry has been updated by the remote end when the user activated the telephone (powered it on and registered) or when an already powered device rolls into range of a cell site from the new location. Then an SS7 message is sent out to the HLR that the called party is now in someone else's *Automatic Line Information* (ALI) database.

If a call is coming in from the network, the SS7 inquiry comes into the HLR who then sends a redirect message to the remote MSC (the VLR) serving the end user.

Using this roaming capability, the database dips are occurring on a frequent basis. Many of the networks (wireless) are already taxing their databases. They are trying to move more information out to the STP or the local switches rather than constantly hammering into the SCP.

Final Thoughts

PCS mimics the same basic services of GSM discussed in the preceding chapter. The primary difference is that the PCS operators are primarily North American based whereas the GSM standard has been implemented by the rest of the world. Differences exist in the overall operation, but these systems operate in similar fashion. The ability of the PCS system to learn the identity of the roaming device is similar to the smart card used in Europe. The services enabling the CO equipment and databases to communicate with each other and track the roamer simplifies the billing and the service functions. Moreover, the feature transparencies available with PCS creates a seamless network atmosphere.

Although the two technologies are different, they are similar at the same time. The real value is achieved when the user can operate the wireless device (the phone) transparently between supplier networks, with the same look and feel as in the home system.

CHAPTER 23

Wireless Data Communications (Mobile IP)

As the convergence of voice and data continues, a more discreet change is also coming into play. Although data is considered fixed to a location, the end user is now more mobile. This opens a new set of challenges for the industry and manufacturers alike because of the need for mobility. What once was a simple procedure of connecting the user's modem to a land line now poses the need to connect that same user to a device while mobile. Protocols need to be more flexible, accommodating the mobile user as the device is moved from location to location. Moreover, the physical devices (for example, the modems) must be moved often. In a dial-up, circuit-switched communications arrangement, this is not a major problem. The user can unplug a modem, reconnect it to a landline elsewhere, and dial from anywhere.

However, when we use *Internet Protocol* (IP) as our network protocol, data is routed based on a network/subnetwork address. Routing tables keep track of where the user is located and route the datagrams (packets) to that subnetwork (see Figure 23-1). When a mobile user logs on and attempts to dial in to the network, the IP address is checked against a routing table and routed accordingly. Updating the tables can be extremely overhead intensive, and it can produce significant amounts of latency in the Internet or intranet. Using an *ICMP Route Discovery Protocol* (IRDP), which is part of the *Transmission Control Protocol/IP* (TCP/IP) protocol suite, helps. However, when the IRDP process updates its tables, we use a lot of bandwidth. Figure 23-2 is an example of the IRDP process where a message is generated by a host to learn all routes available to get to and through the network.

Something needs to be done to accommodate the use of mobile IP by an escalating number of users wanting to log on anywhere. Of course, one solution is to use the *Dynamic Host Configuration Protocol* (DHCP), which uses

Figure 23-1
Routing table

Destination Address	Next Hop
34.1.0.0	54.34.23.12
79.2.0.0	54.34.23.12
147.9.5.0	54.32.12.10
17.12.0.0	54.32.12.10

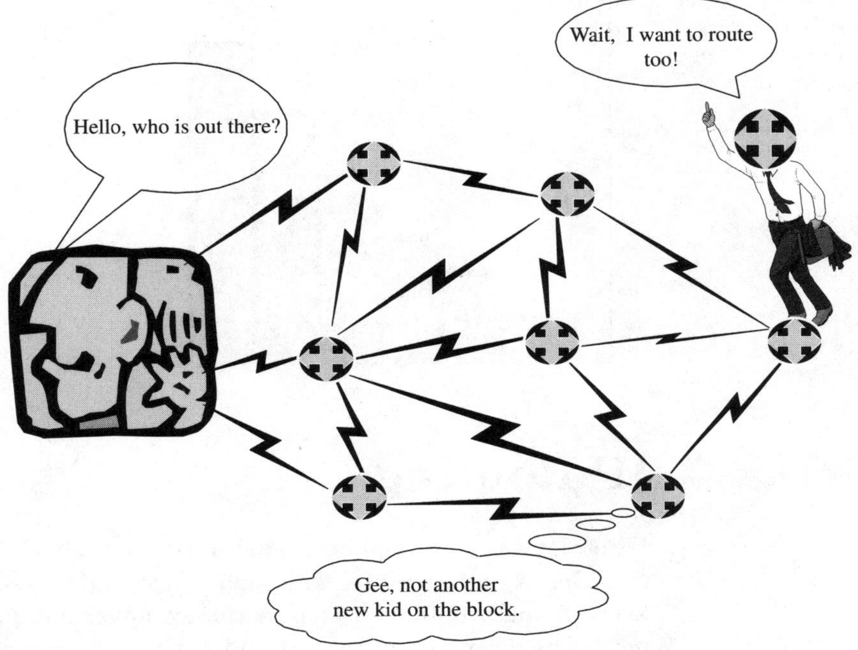

Figure 23-2 IRDP learns the new addresses and routers in the network.

a server-based mechanism to allocate a new IP every time a user logs onto the network or assigns a static IP address to a user who may be using a diskless PC. The purpose of the DHCP is to facilitate the mobile or not-permanently-attached user in an ISP network where addresses are limited and casual users are the norm. So the industry has had to arrive at a solution allowing casual and nomadic users the same access while they travel (roam) as when they are in their fixed office location.

Figure 23-3 shows the growth curve of wireless data users attempting to use mobile IP and wireless data communications over the past couple of years. In this graph we see that the numbers justify the concern and the effort being afforded to the problem. The number of wireless users in the world is escalating and the number of wireless data users shown in this graph will grow from 150 million to 1 billion by 2006 if the carriers can roll out their products reliably. We are living in a mobile society where users want their data, when they want it, where they want it, and how they want it! What percentage of these wireless users will want data over their connection remains to be seen. However, early estimates are that over one-fifth will want their data in a mobile environment.

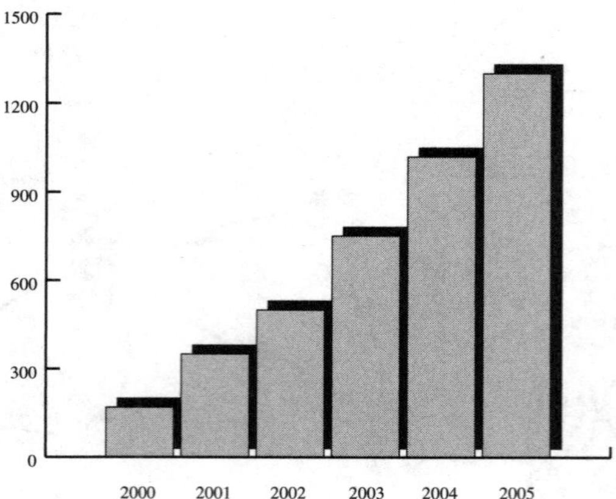

Figure 23-3 Wireless data subscribers (in millions)

IP Routing

When IP was first implemented, the routing scheme was based on tables that were kept in a router. The routers updated these tables periodically to keep the information current. A service advertising protocol was used to notify the downstream routers and hosts of the router's presence. As data was sent through the Internet (intranet or private network), the IP datagrams included both the source and destination address, as shown in Figure 23-4. Routers forwarded the data on the basis of the destination address, the 32-bit address in IP version 4. Routers ignored the source address, looking only at the destination. Using the destination address, the router then concerned itself with the net hop. If a router received a datagram that was destined for a local address, it then sent the data to the physical port for the *Local Area Network* (LAN) subnetwork it resided on. However, if the router received a datagram that was for a destination that was not local, it forwarded the datagram to its next downstream neighbor. The routing process was done on the basis of hop-by-hop forwarding.

After receiving a datagram with an address that was local, the router would forward the data to its local physical port (the LAN). However, if a networked device was moved because the user was mobile, the router continued to send the data to the destination network in its table. A mobile user was therefore not going to receive any datagrams because IP did not guarantee delivery. This broke down the whole process because the devices were mobile, especially in global networks with mobile workers. Conceivably, a user could travel to a remote office and attempt to plug in a laptop computer with a static IP address from corporate headquarters. Because the IP was static, all the IP datagrams were sent along their way to the headquarters' router where the IP subnetwork address was registered.

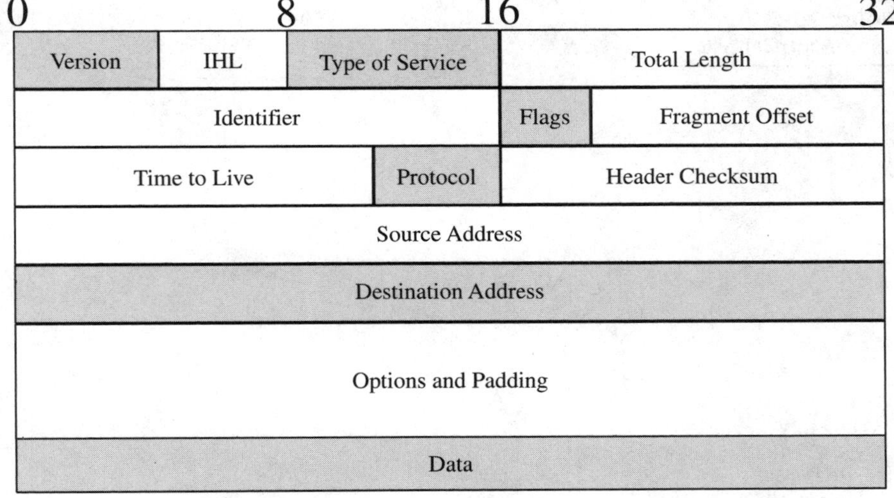

Figure 23-4 Routers look at the IP header for the destination network address.

Part of the Solution

One part of the solution is a change in IP version 4 that allows for mobile users. This mobile IP process is addressed in RFC 2002, which allows for the registration of home agents and foreign agents. A home agent is a registration process whereby a locally attached computer registers with a server or router and gets entered into the routing table. Based on RFC 2002, a mobile device (either a host or a router) can change locations from place to place without devastating effects. If mobile IP is in use by the device, the mobile device will not change its address when it moves from location to location. Instead, it will use an IP forwarding approach by using an "in-care-of" address obtained from some serving device that knows of its presence. Using the home agent concept, the home agent is a server or router on the home network. The home network maintains the mapping of the mobile device's address to its current address.

Think of this in the context of the postal service. If you go on vacation for a month to Phoenix, Arizona, you still need to get your mail (for example, paychecks, bills, and so on). So, before going away, you go to the local post office and tell them to forward all your mail for the next month to Phoenix. The forwarding idea is shown in Figure 23-5. The postmaster will then make a note on the mailbox where he or she sorts your mail. This note will say to forward all mail to P.O. Box 51108, Phoenix, AZ 85076-1108. So even though people are still sending mail to your address, the post office

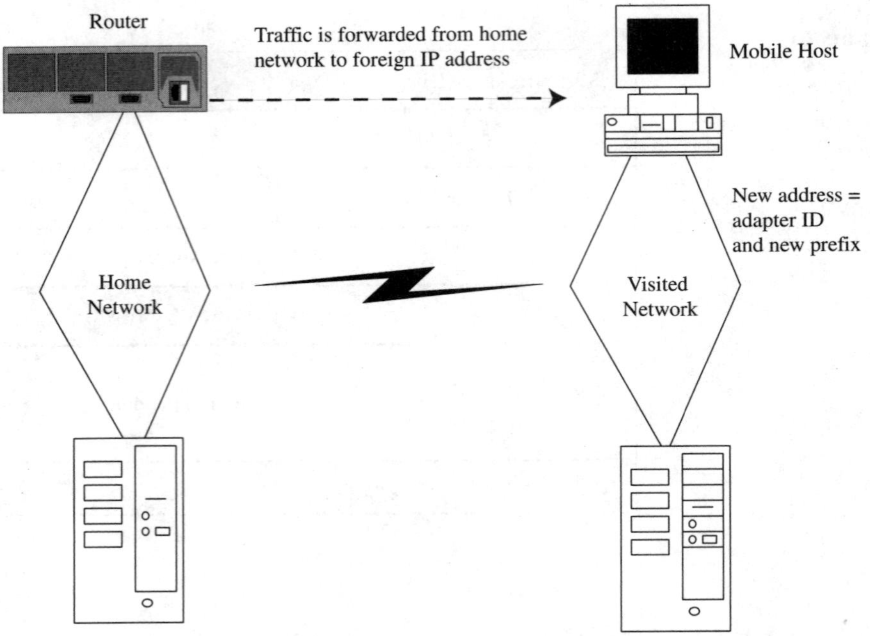

Figure 23-5
Forwarding IP traffic

Router
Traffic is forwarded from home network to foreign IP address
Mobile Host
New address = adapter ID and new prefix
Home Network
Visited Network

employees (your home office agents) are handling the forwarding to you, based on the instructions you left behind.

When the vacation is over and you return to your normal domicile, you notify the local postmaster that you are back. The local folks remove the forwarding address from your normal mailbox and then resume delivery to the home address. The Phoenix postmaster, in the meantime, was responsible for knowing your temporary address while you were in his or her area. Therefore, the local postal people were handling the routing and delivery of the mail while you were in Phoenix. They were acting as your agents in the foreign location. The foreign agent received mail that was marked for your normal home address, but translated this into the current address. This combination of local (home agent) and remote (foreign agent) agents makes the process work.

Now let's assume that after vacation, you leave Phoenix, but do not go home. Instead, you go on an extended business trip to San Francisco. The home agents then label the boxes to reflect the changes that indicate mail is now to be forwarded to the new location.

Applications That Demand Mobile IP

What are the applications that will take advantage of the mobility of the user? Well, the first one is going to follow our example for mail. But e-mail is only one of many applications that users will want to use. Other areas will include transaction processing (orders online), fax capabilities, *File Transfer Protocol* (FTP), and Telnet. Telemetry applications and multimedia are all prime candidates. This is shown in Figure 23-6 where the applications will be paramount to the ability to move around.

Applications are what the data transmission world is all about. So we need to have mobility in the IP world in order to satisfy the applications for a mobile user.

In Table 23-1 the need for these application services becomes more evident. In this table, the driving forces for the implementation of wireless data and mobility within the TCP/IP architectures are more evident.

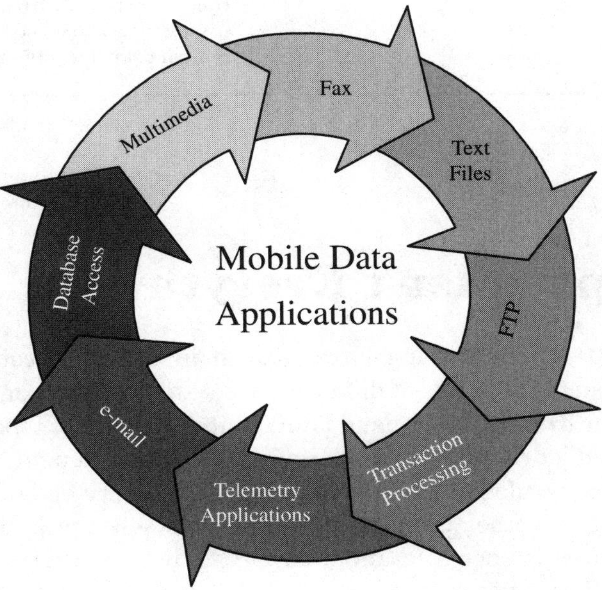

Figure 23-6 Application of mobility in a data world

Table 23-1

Driving forces for mobile data

Market Forces Driving Mobility	Value
Consumer use of computers	Increasing. In North America, the average number of households has climbed to 67 percent and users wired for interconnectivity.
Remote access needs	With more telecommuters, the need for remote access is increasing at a 10 percent or better ratio each year. The number of telecommuters is running at 18 million in 2001.
Users demands for the access they perceive	Users want access to their data when they want it, where they want it, and how they want it. The driving force is that 250 million subscribers will be using wireless data by the end of 2003.
Wireless innovations	Wireless data is still only available at 9.6 Kbps, even though it is advertised at speeds of 33.6 Kbps. In the future, mobile data and IP users will expect 500 Kbps to 2 Mbps speeds from their wireless data devices.
Internet	The growth on the Internet has been so dramatic that more users expect the capability to log onto the Net and get their e-mail, FTP, and transmit other forms of data anywhere and anytime. Internet growth is slotted at 450 million users in the 2003–2004 period, growing to 700 million users by 2005–2006.

Speed Isn't Everything

Some of the driving forces shown in the table couple tightly with the throughput issues of data communications. However, speed is not the only issue driving the demand for mobility. The need for speed and reliability are rapidly driving the users into a mobile environment. In the older days, the speed would hold back the use and acceptance of data transmission. Popularity in other connection devices also sets the pace. An example of the devices is shown in Table 23-2. This lists the various devices people are looking to use.

Table 23-2

Speeds for wireless devices

Devices	Expected Speeds
Personal data assistants	The current wireless connection is 28.8 Kbps. Expect this to grow to 500 Kbps by the end of 2004.
Laptop computers	Current modem technology supports up to 56 Kbps downloadable and 33.6 Kbps uploadable. In the future, users will expect 2–10 Mbps.
Palmtop computers	Same technology as laptops.
Mobile telephones	Current low-speed telephony with speeds of less than 19.2 Kbps circuit-switched technology. The future demand will be speeds of up to 128–170 Kbps and then 384 Kbps.
Paging systems	Low-speed data from *short messaging systems* (SMS).

Variations in Data Communications (Wireless)

Several different flavors of data communications exist for wireless communications. Most people think of the services as being one type only. Earlier the discussion considered the SMS, circuit-switched data, and packet-switched data. The variations in the different technologies depend on the innovations and the air interfaces.

Regardless of the different methods, there are still many variations in using the data transmission characteristics with wireless applications and mobile users. In Table 23-3, a summary of the various methods is shown for the wireless communications techniques.

Table 23-3

Summary of methods for data

Method	Air Interface
Circuit switched	TDMA CDMA GSM
Packet switched	RIM Motient CDPD IP
SMS	Packet-switched separate

Possible Drawbacks with Wireless

Not everything with wireless data is perfect (it's the same for wireline). Occasionally, there will be problems that must be overcome. The use of wired facilities has improved over the years with the use of fiber optics in the backbone. Newer technologies have allowed the industry to improve the performance of data to a 10^{-15} to 10^{-16} bit error rate. This was unheard of before the use of fiber and *Synchronous Optical Network* (SONET) in the backbone. The local loop (last mile) is the weakest link in the equation, producing bit error rates of 10^{-6}. Still, today this is not as bad as it may sound because the distances we run on copper (the local loop) are being shortened daily. The shorter the copper, the better the performance we achieve (because the weakest link is minimized, and the cables are correctable). Therefore, we see improved data performances on the local copper-data transmission systems. However, wireless (air interfaces) have been traditionally error prone, and the amount of frequencies available have always been limited. This interface is limited to the point that many people have not wanted to use air interfaces in the past. With the culture shift and the use of digital techniques to compound the data, air has become much more acceptable.

The wireless medium is also prone to more delay and latency than the wired world. In the wired arena, the average delay for transmitting information across the nation networks is 50 msec or less. In the wired world, this delay can jump to more than 250 msec. At 250 msec, we find that echo begins to get out of control, requiring more equipment to handle this problem. However, while handling the echo control function, we introduce more latency and buffering of the real data. This almost becomes a Catch-22 problem. The more the data is buffered and manipulated across the medium, the greater the risks of introducing errors.

Pros and Cons to Wireless

Some of the inherent problems bring other solutions and benefits to the table. For example, when using a wireless interface, the user is mobile and can go almost anywhere at whim. This allows more flexibility. However, the downside is that the use of cellular and *Personal Communications Services* (PCS) systems causes data and call handover to occur, which can be detrimental to the overall performance. Security issues cannot be ignored with data in the air. On a copper cable, data is slightly more secure because the medium is a little more difficult. The use of an interceptor in the airwaves is much simpler. Digital cellular and PCS make the data more secure than analog, but one still has to be concerned with putting anything in the airwaves.

From the perspective of financial information, the wireless data can be very expensive. In the circuit-switched arena, just as in voice, we pay the carrier a flat rate for a guaranteed amount of usage. The price is fixed whether you use the minutes or not. If you exceed the minutes of usage, you pay a premium added cost for the overage usage. Some plans do allow you to pay for only what you use; however, these plans tend to be at a higher cost per minute for all the minutes of usage. Therefore, it becomes difficult to assess the best deal, depending on variable usage. Packet-switched data, such as *Cellular Digital Packet Data* (CDPD) and RAM Mobile Data services, can be less expensive when small amounts of data are transferred. Yet if the user transmits large and lengthy files, the cost of packet switched data can be as much as 50 to 100 percent higher than circuit-switched data. The variables are still less attractive to use the wireless data.

Other areas where differences exist are in the devices themselves. When dealing with a radio interface (like a cell phone), we have to be aware of the battery life of the device. Power consumption with lengthy data transfers can be critical. The use of the overall battery life is contingent upon the technology used, but current industry standards allow for 2 to 4 hours of talk time on a portable device (transmission time for data). Anything over that is prone to cut off and produce errors. The digital sets are better equipped to handle the data transfers, but dual-mode phones can be problematic when they are in analog mode.

CHAPTER 24

General Packet Radio Service (GPRS)

With the simultaneous gate-opening effects of technological innovation and industry deregulation, the demand for communications and available solutions is exploding. This demand is being fueled by the needs of people and businesses. The most visible evidence of the boom is within Internet traffic and e-commerce or m-commerce. However, it is less appreciated that an unprecedented demand exists from worldwide telephone subscribers. It took a century to get 700 million phone lines installed. Another 700 million will be deployed in the next 15 to 20 years—and that could prove to be a conservative estimate. Although the majority of the new deployments will be wireless phones—700 million of them over the next 10 years—demand for wireline communications is also exploding, driven in part by the need to access the Internet.

This explosion in demand is reflective of the dependence that people have on rapid, reliable communications to keep up with the fast pace of business. The success of the *Global System for Mobile Communications* (GSM), the ubiquitous presence it has garnered, the emerging Internet, and the overall growth of data traffic in general all point to a significant business opportunity for GSM operators. The number of subscribers to the Internet worldwide is growing exponentially, as seen in Figure 24-1, and the growth has been dramatic. The following statistics from the middle of 2001 add some credibility to the overall concept of a data-centric community that is also mobile:

- Number of Internet users—400 million
- Number of wireless users—700 million

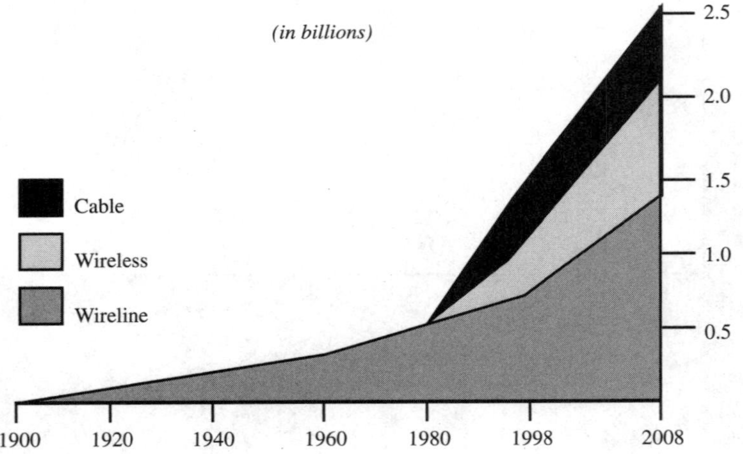

Figure 24-1 The number of telephone and Internet users

By the year 2005, we can expect that more than 1 billion people will be mobile wireless users on a worldwide basis, and by 2006, more than 1 billion people will be Internet users. We are adding over 100 million new Internet users per year. This growth is unheralded in the past century of the telecommunications industry. The phenomenal growth in wireline Internet subscribers points to the possibility of wireless operators capturing some of this market. This assumes that they can offer comparable price-performance capabilities. The factors that we can consider in the process are

- Wireless users are ideal target subscribers for Internet providers.
- Internet users are ideal target subscribers for GSM operators.

Several movements in the Internet community will also force changes. The demographics of the user will change dramatically as the world expands its wireless and Internet presence. The average user will be looking for developments in the new Internet that will provide a broader band communications speed, the capability of using a network-enabled appliance, and a full network capable of sourcing all the needs and applications through high-end portals that can offer the goods and services needed. Instant information for a mobile workforce is paramount.

In addition to customer demand, this revolution is greatly accelerated by technology disruptions. Three technologies are at the fundamental science level. One is summarized in the well-known Moore's Law, which states that the capacity of an individual chip will double once every 18 to 24 months. Therefore, silicon is covering the globe. That phenomenon has been going on for three decades now and will be adding as much capacity in the next two years as has been created in the history of the semiconductor business. Nevertheless, the capacity is finally just getting to the point where it's interesting. Two other technologies, although less well known, are changing just as rapidly, if not more so:

- In optical, in the core of the network, *dense wavelength division multiplexing* (DWDM), using multiple colors of light to send multiple data streams down the same optical fiber, is disrupting the rapid growth of *Wave Division Multiplexing* (WDM), further pushing the envelope.
- Wireless capacity is also exploding, enabling higher bandwidth for voice/data without fiber (45 Mbps—up to 2.5 Gbps with certain wireless tools and 10 Gbps are being discussed as a reality over the next decade).

Combined, these advances are making converged networks possible and inevitable, as well as important to plan for in business. Companies that understand and take advantage of this convergence will have a strategic advantage.

The New Wave of Internet User

During the next few years, the *third-generation* (3G) Internet will drive even further innovation and performance. Figure 24-2 provides a summary of the steps. It began with the *first-generation* (1G) Internet, which was PC-driven. During this period, standards were established, and narrowband services were offered. This led to business model experimentation, new companies, and new brands. In both wireline and wireless communications, users were satisfied with the PC-centric services because the networks did not offer anything else. This led to a somewhat frustrated PC and Internet user on wireline networks, but an even greater level of frustration was evident in the wireless arena.

Today, the 2G Internet is upon us. Trends underlying today's Internet include substantial personalization and the emergence of the business-to-business market. Regardless of the downturn in 2001, the networks will reemerge.

Trends underlying the 3G Internet will drive the growth of the new economy in upcoming years:

- First, as more people go online, in fact almost doubling, the online population will begin to normalize, or resemble the overall U.S.

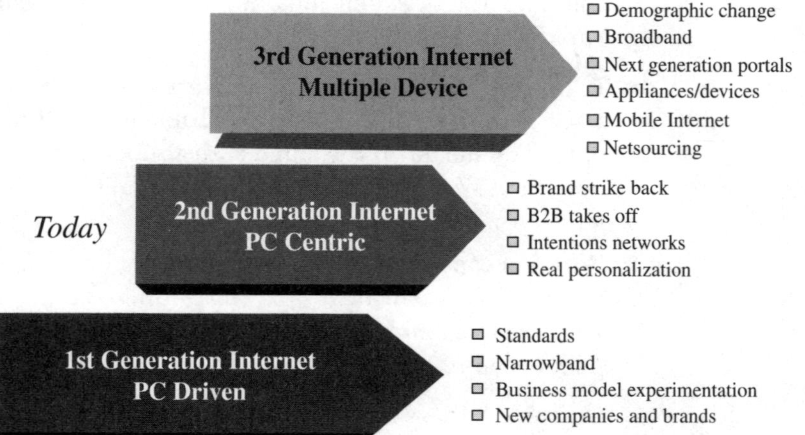

Figure 24-2 The steps in developing the 3G Internet

population. The average age of the online user will go up and his or her income will fall. This means that strong brand recognition increases in importance, greater service levels to support less savvy online users will be required, and convenience and ease of use will become vital. The good news is that these changes will drive a greater comfort level with online shopping for the average online user, more than doubling total spending online.

- Second, broadband will create a personalized, interactive experience. Think of the Web on steroids. Today's interactive experience will be radically enhanced.
  - Instead of instant text messaging, we will have easy access to instant audio and video messaging.
  - Instead of today's chat rooms and discussion lists, we will have far more sophisticated real-time collaboration tools.
  - Today's grainy-streamed audio and video will have broadcast quality tomorrow. The two-dimensional will be three-dimensional.

Although the growth of PCs has slowed to roughly 3 percent per year, new information appliances and communication devices are fast becoming the new power brokers with double-digit growth. Wireless telephones, *personal digital assistants* (PDAs), Blackberry devices, and GPRS terminals are the new devices to behold. The power of the Web will be accessed through mobile phones, PDAs, cable set-top boxes, and even game controllers. Every office and household device in the future will be *Internet Protocol* (IP) addressable, enabling the user and supplier to better service the individualized and customized needs.

This new economic model (although it appears to have gradually slowed) requires that we address the converged world, borrowing from each perspective. Both the world of voice networks and the world of data networks have advantages. Both have a unique profile of strengths. Convergence applications will be practical when you are able to take the best of both worlds and deliver real-world business value. In a nutshell, a converged world requires that our networks and access methods of the future be

- Highly reliable
- Broadband serviced
- Scaleable
- Multiservice oriented
- Flexible and open
- Exceptionally easy to use

These thrusts are driving the data communications market into an explosive situation. The average growth of our voice networks is 4 percent; however, in the data communications arena, the growth is still approximating 30 percent growth per year. This unparalleled growth consists of both goods and services to meet the demands of customers, internal users, and the industry in general.

GPRS

GPRS is a key milestone for GSM data. It offers end users new data services and enables operators to offer radically new pricing options. Using the existing GSM radio infrastructure, up-front investments for operators are relatively low. GPRS solutions began appearing initially in 1999 through 2000 using the infrastructures that are already in place. Pricing for use of the voice side of the network has become commoditized, whereas pricing models for the new data access will cause a revolution. One such threshold looks at an all-you-can-eat model whereby users of wireless phones add a data subscription at $29.95 per month for unlimited use. Another such model is the one used in Japan by DoCoMo, charging a rate of the U.S. equivalent to $.0025 per packet. Others will emerge that will shake the industry mode and create new dynamics in the use of data anywhere.

GPRS services were targeted at the business user. However, the services will soon be available networkwide, targeting both the business and the residential consumer. The widespread adoption and acceptance of GPRS will create a critical mass of users, driving down costs while offering better services. These components will form the basis of a healthy mobile data market with growth figures comparable to GSM voice-only services today. Research by Infonetics indicates that the movement of the user community will also be to a more mobile community. In fact, the study indicates that by 2005, more wireless devices will be used for the Internet than PCs on the Net, as shown in Figure 24-3. This form of growth is again a driver that will force the rapid deployment by carriers and manufacturers alike.

The GPRS Story

The GPRS is a new service that provides actual packet radio access for GSM and *Time Division Multiple Access* (TDMA) users alike. The main benefits of GPRS are that it reserves radio resources only when data is avail-

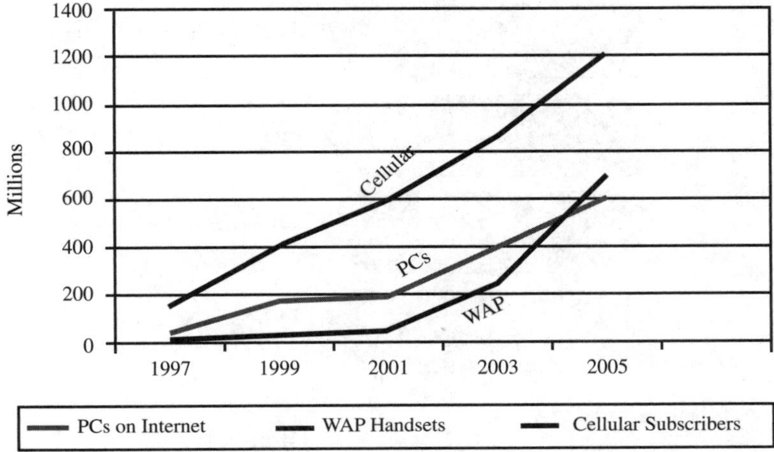

Figure 24-3 Within five years, more wireless devices will be used on the Internet than PCs.

able to send, and it reduces the reliance on traditional circuit-switched networks. Figure 24-4 is a basic stepping-stone description of the GPRS story. The increased functionality expected from GPRS will decrease the incremental costs to provide the data services, which will increase the penetration of the data services, among the consumers and business users alike.

Additionally, GPRS will improve the quality of data services measured in terms of reliability, response time, and features available. Unique applications will be developed in the future that will attract a broad base of mobile users and enable the individual providers to offer differentiated services. One way that GPRS improves upon the capacity capabilities of the network suppliers is to share the same radio resources among all *mobile stations* (MSs) in a cell, thus providing effective use of scarce resources. New core network elements will continue to emerge that will expand services, features, and operations for our bursty data applications.

GPRS also provides an added step toward 3G networks. GPRS will enable the network operators to implement IP-based core architecture for data applications. This will continue to proliferate new services and mark the steps to 3G services for integrated voice and data applications.

What Is GPRS?

As stated previously, GPRS stands for General (or generic) Packet Radio Service. GPRS extends the packet data capabilities of the GSM networks

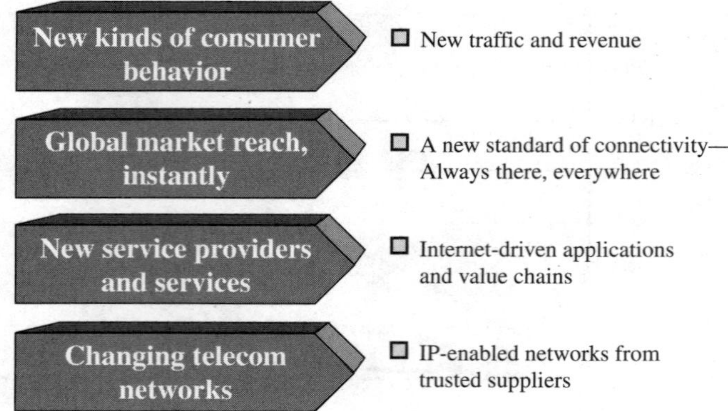

Figure 24-4 The GPRS story

from *Packet Data on Signaling-channel Service* (PDSS) to higher data rates and longer messages. For now, the use of GPRS shall be in the context of GPRS-GSM to distinguish it from the GPRS-136, the North American adoption of GPRS by the IS-136-based systems. GPRS is designed to coexist with the current GSM *Public Land Mobile Network* (PLMN). It may be deployed as an overlay onto the existing GSM radio network. GPRS may also be implemented incrementally in specific geographic areas. An example of this GPRS radio access may be deployed in some cells of a GSM network, but perhaps not all. As the demand grows, coverage can be expanded. A network view of GPRS is shown generically in Figure 24-5.

The GPRS network fits in with the existing GSM PLMN as well as the existing packet data networks. GPRS PLMN provides the wireless access to the wired packet data networks. GPRS shares resources between packet data services and other services on the GSM PLMN. GPRS PLMN also interworks with the *Short Message Service* (SMS) components to provide SMS over GPRS. The intent is to provide a seamless network infrastructure for operations and maintenance of the network.

GPRS is a packet-based data bearer service for GSM and TDMA (IS-136) networks, which provides both standards with a way to handle higher-data speeds and the transition to 3G. It will make mobile data faster, cheaper, and more user friendly than ever before. By introducing packet switching and IP to mobile networks, GPRS gives mobile users faster data speeds and particularly suits bursty Internet and intranet traffic. For the subscriber, GPRS enables voice and data calls to be handled simultaneously. Connection setup is almost instantaneous, and users can have always-on connectivity to the mobile Internet, enjoying high-speed delivery of e-mails with

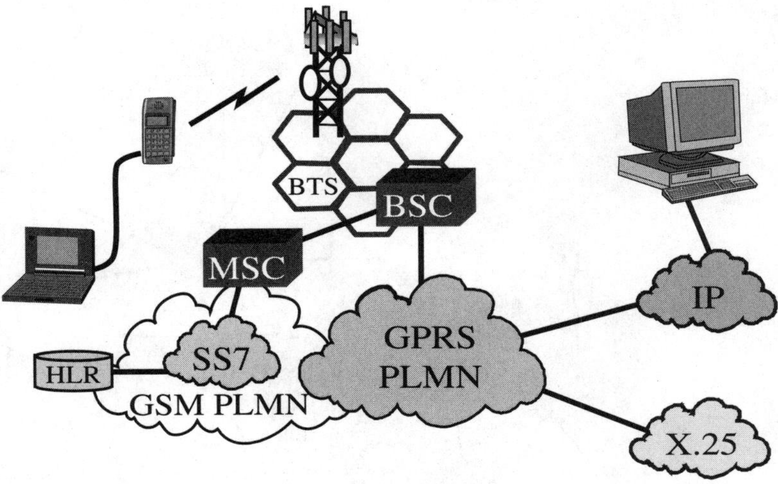

Figure 24-5
A GPRS network view

large file attachments, web surfing, and access to corporate *Local Area Networks* (LANs).

GPRS was defined by the *European Telecommunications Standards Institute* (ETSI) as a means of providing a true packet radio service on GSM networks. GSM equipment vendors are actively developing systems that adhere to the GPRS specifications. At the same time, carriers whose networks are based on *North American TDMA* (NA-TDMA) (IS-136) have decided to deploy GPRS technologies in their networks. Internetworking and interoperability specifications have been developed between ANSI/IS-136 and GSM; therefore, this is a logical extension of the overall scheme. Figure 24-6 is an example of the internetworking arrangements that are planned for use within GPRS.

This creates a coup for the ETSI because up to now, IS-136 networks have been completely based on *Telecommunications Industry Association* (TIA) standards and specifications. Today, GPRS is seen as one of the preliminary steps down a path that will someday lead to the convergence of GSM and IS-136 networks.

Motivation for GPRS

GPRS was developed to enable GSM operators to meet the growing demands for wireless packet data service that is a result of the explosive

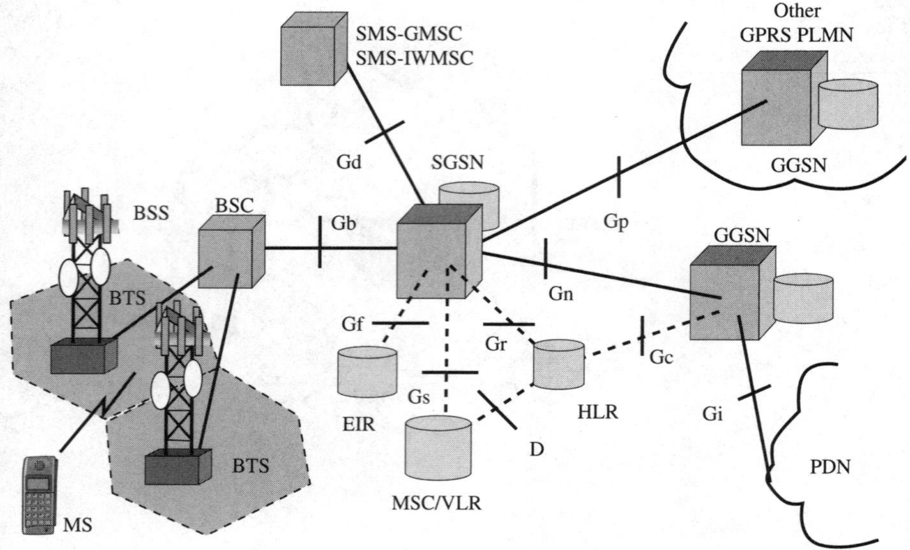

Figure 24-6 Internetworking strategies in GPRS

growth of the Internet and corporate intranets. Applications using these networks require relatively high throughput and are characterized by bursty traffic patterns and asymmetrical throughput needs. Applications, such as web browsing, typically result in bursts of network traffic while information is being transmitted or received, followed by long idle periods while the data is being viewed. In addition, much more information is usually flowing to the client device than is being sent from the client device to the server. GPRS systems are better suited to meet the demand of this bursty data need than the traditional circuit-switched wireless data systems.

GPRS allocates the bandwidth independently in the uplink and downlink. Another goal for GPRS is to enable GSM operators to enter the wireless packet data market in a cost-efficient manner. First, they must be able to provide data services without changing their entire infrastructure. The initial GPRS standards make use of standard GSM radio systems. This also includes GSM standard modulation schemes and TDMA framing structures. By doing this, the cost implications are minimized in the cell equipment. Second, GSM operators must have flexibility to deploy GPRS without having to commit their entire network to it. GPRS provides the dynamic allocation and assignment of radio channels to packet services according to the demand.

Evolution of Wireless Data

Data support over 1G wireless networks started with *Advanced Mobile Phone System* (AMPS) circuit-switched data communications, as shown in the graph in Figure 24-7. This worked by attaching a cellular modem (a standard modem that supports the AMPS wireless interface) with a laptop computer. This began the evolution to the first wireless packet data networks—*Cellular Digital Packet Data* (CDPD), with data rates up to 19.2 Kbps, as shown in the graph in Figure 24-8. CDPD works with AMPS networks and was initially designed for short intermittent transactions, such as credit card verification, e-mail, and fleet dispatch services. According to the Wireless Data Forum, CDPD covered 55 percent of the U.S. population as early as the *third quarter of 1998* (3Q98). It has since grown to cover nearly 87 percent based on the proliferation of more wireless users. In addition, limited support for SMS was introduced to offer paging-like and text-messaging services.

In 2G wireless networks, SMS services became the deployed architecture of choice. They use the existing infrastructure of 2G wireless networks, with the addition of the Message Center component. 2G also introduced asynchronous data and facsimile services over the air interfaces, with initial data rates of up to 14.4 Kbps. This enables users to fax and have dial-up access to an ISP account, corporate account, and the like. Packet data technology gained momentum in 2G, and then in 2.5G networks. This includes

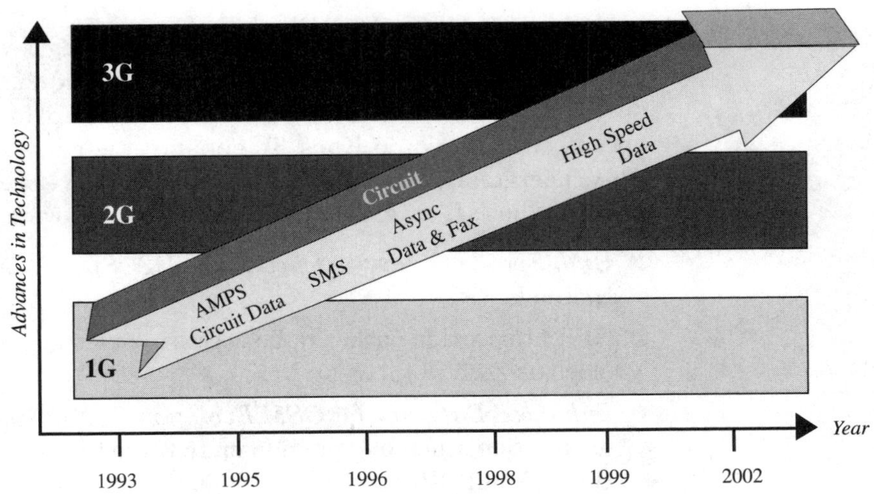

Figure 24-7 The timeline for circuit-switched data

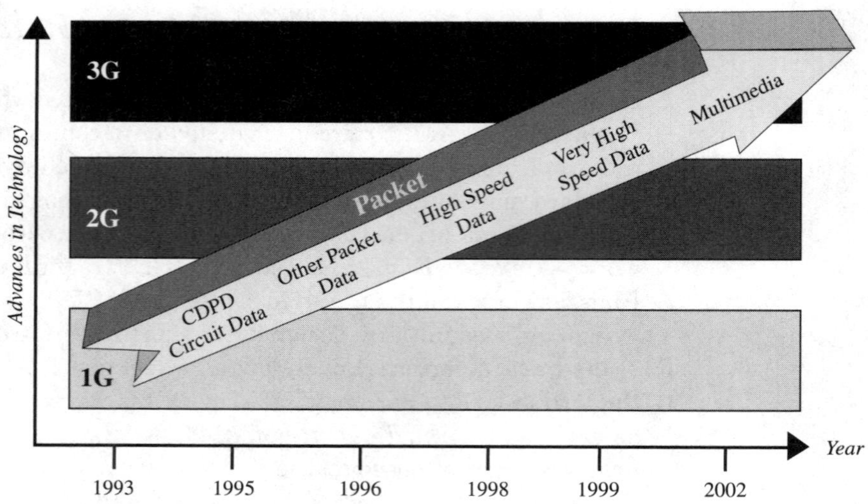

Figure 24-8 The timeline for packet-switched data

GPRS and packet data support in *Code Division Multiple Access* (CDMA). Data rates for the packet switching currently range from 9.6 to 19.2 Kbps. In the future of 2G, we can expect to see data rates at up to 115 Kbps. 3G, when it happens, will support data rates of 384 Kbps to 2 Mbps. Multimedia and high-speed Internet access will be the expected, normalized data access applications.

Wireless Data Technology Options

Today, GSM has the capability to handle messages via the SMS and 14.4 Kbps circuit-switched data services for data and fax calls. The maximum speed of 14.4 Kbps is relatively slow compared to the wireline modem speeds of 33.6 and 56 Kbps. To enhance the current data capabilities of GSM, operators and infrastructure providers have specified new extensions to GSM Phase II, as shown in Figure 24-9, to provide

- *High-Speed Circuit-Switched Data* (HSCSD) by using several circuit channels
- GPRS to provide packet radio access to external packet data networks (such as X.25 or Internet)
- *Enhanced Data rate for GSM Evolution* (EDGE) using a new modulation scheme to provide up to three times higher throughput (for HSCSD and GPRS)

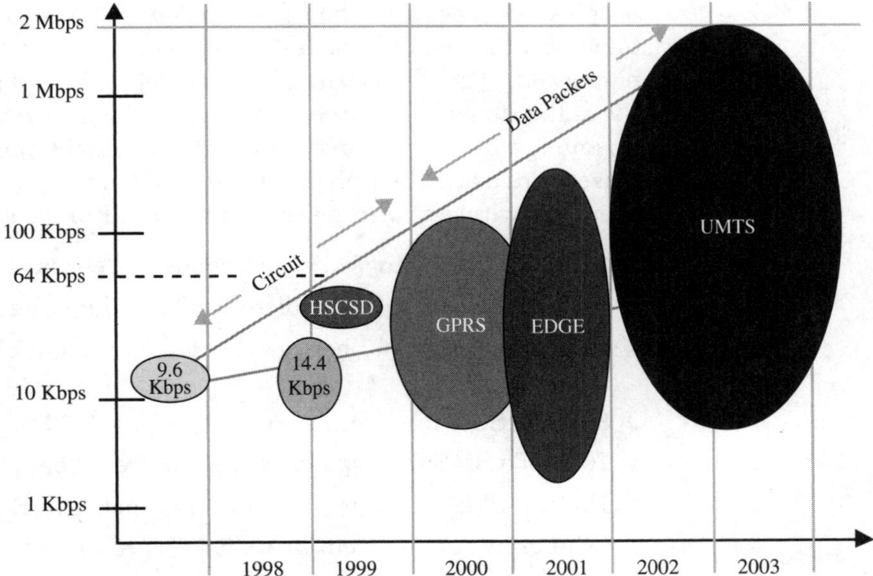

Figure 24-9 Steps of implementation

- *Universal Mobile Telecommunication System* (UMTS), a new wireless technology using new infrastructure deployment

These extensions enable

- Higher data throughput
- Better spectral efficiency
- Lower call set-up times

The way to implement GPRS is to add new packet data nodes in GSM/TDMA networks and upgrade existing nodes to provide a routing path for packet data between the mobile terminal and a gateway node. The gateway node will provide interworking with external packet data networks for access to the Internet and intranets, for example. Few or no hardware upgrades are needed in existing GSM/TDMA nodes, and the same transmission links will be used between Base Transceiver Stations and Base Station Controllers for both GSM/TDMA and GPRS.

GPRS Roaming At the end of June 2001, 551 million GSM customers were on record, and 447 operators in 170 countries have now adopted GSM. The expectation is that GSM growth will continue, and that it will have 700 million GSM customers by June 2002. As the growth continues and more people disconnect from their wired phones, GSM will have 800 million customers by the end of 2003 and 1 billion customers by 2005. Obviously, this places a lot of growth on the providers' networks and puts more demand on the ability to use the service wherever and whenever we want. GSM Voice Roaming was a 15 billion Euro business in 1999, thus indicating that the masses are using their phones today in a roaming manner. Some statistics about the wireless roaming environment in Europe are as follows:

- 540 million roaming calls were made in February 2000.
- 750 million calls were predicted in June, July, and August 2000.
- Data will account for between 20 and 50 percent of all global wireless traffic by 2004.
- 8 billion SMS messages were sent in May 2000.
- 10 billion SMS messages were sent in December 2000.
- 1 billion SMS messages were sent per month in Europe alone in 2001.
- GSM grew at 80 percent in 1999; PCs grew at 22 percent.
- All terminals will be Internet enabled by the end of 2003.
- More GSM terminals will be connected to the Internet than PCs by 2005.
- Wireless devices will be responsible for 30 percent of all Internet traffic by 2005.

The GSM Phase II Overlay Network

The typical GPRS PLMN enables a mobile user to roam within a geographic coverage area and receive continuous wireless packet data services. The user may move while actively sending and receiving data or may move during periods of inactivity. Either way, the network tracks the location of the MS so that incoming packets can be routed to the MS when they arrive. The GPRS PLMN interfaces with the MSs via the air interface. GPRSs will initially be provided using an enhanced version of the standard GSM interface. The operators will evolve their networks to incorporate more advanced radio interfaces in the future so that they can deliver higher data rates to the end user.

The GPRS PLMN interfaces as an overlay to traditional public packet data networks using standard *Packet Data Protocols* (PDPs), as shown in Figure 24-10. The network layer protocols supported for interfacing with packet data networks include X.25 and the IP. Through these networks, the

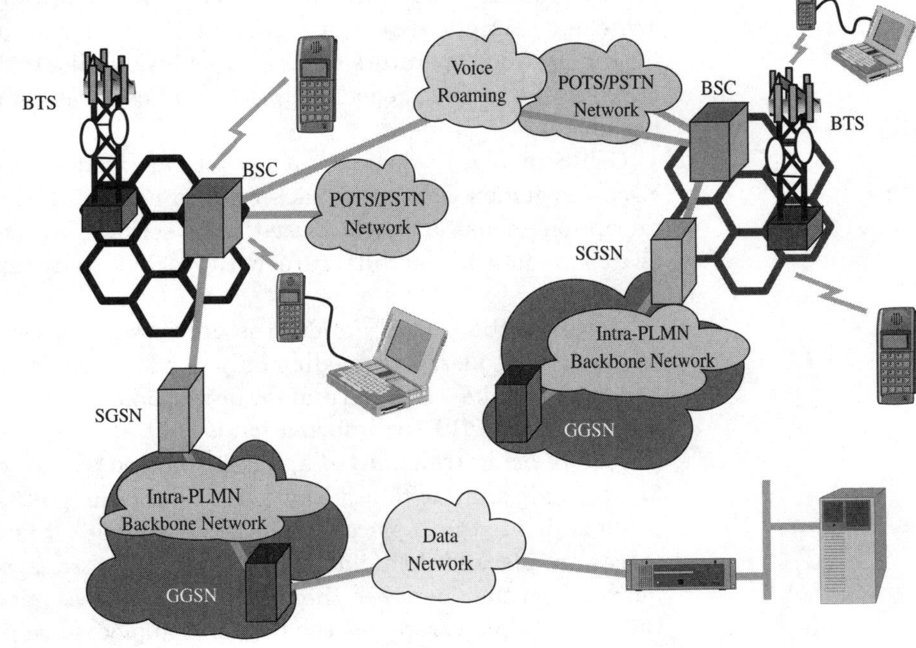

Figure 24-10 The GPRS overlay on GSM

end user is able to access public servers such as web sites and private corporate intranet servers. GPRS can also receive voice services via the GSM PLMN. Voice services and GPRSs may be accessed alternately or simultaneously, depending on the MS's capabilities. Several classes of MSs are possible, which vary in degree of complexity and capability. The actual end-user data terminal used can be a smart phone, a dedicated wireless data terminal, or a standard data terminal connected to a GPRS-capable phone.

Circuit-Switched or Packet-Switched Traffic

An On/Off model characterizes the typical Internet data. The user spends a certain amount of time downloading web pages in quick succession, followed by an indeterminate time of inactivity during which he or she may be reading the information, thinking, or maybe even have left the work space. In fact, the traffic is quite bursty (sporadic) and can be characterized as data packets averaging about 16 Kbps in size with average interarrival times of about seven seconds. If a circuit-switched connection is used to access the Internet, then the bandwidth that is dedicated for the entire duration of the session is underutilized. This inefficient use of the circuit-switched example, shown in Figure 24-11, creates an undesirable scenario for the network operators. Instead, they would like to fill the channels (circuits) to the highest reasonable level and carry as much billable traffic as possible.

GPRS involves overlaying a packet-based air interface on the existing circuit-switched GSM network shown in Figure 24-12. This gives the user an option to use a packet-based data service. To supplement a circuit-switched network architecture with packet switching is quite a major upgrade.

However, the GPRS standard is delivered in a very elegant manner—with network operators needing only to add a couple of new infrastructure nodes and make a software upgrade to some of the existing network elements. With GPRS, the information is split into separate but related packets before being transmitted and reassembled at the receiving end. Packet switching is similar to a jigsaw puzzle—the image that the puzzle represents is divided into pieces at the factory where it is made, and then the pieces are placed into a plastic bag. During the transport of the new, boxed puzzle from the factory to the end user, the pieces get all mixed up. When the final recipient receives the bag, all the pieces are reassembled into the original image. All the pieces are related and fit together, but the way they

Figure 24-11 The circuit-switched traffic example

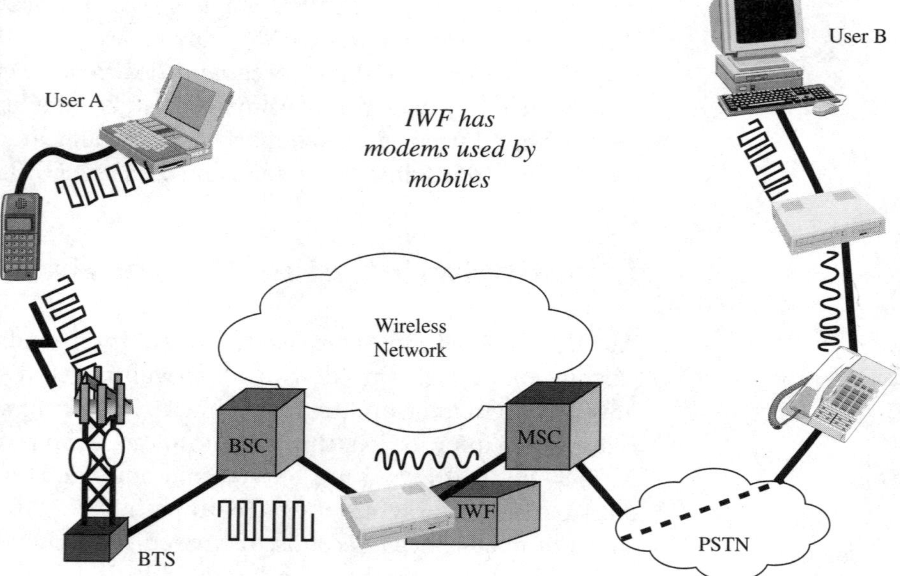

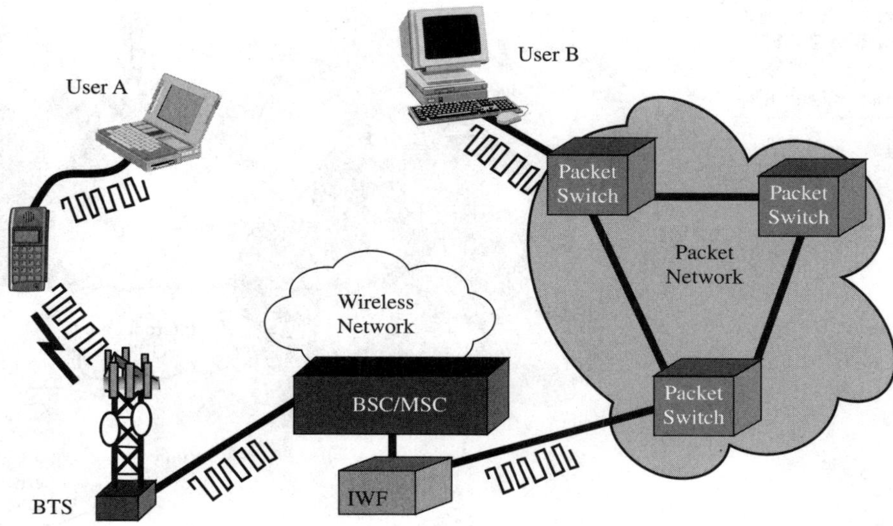

Figure 24-12 The packet-switched example

are transported and reassembled varies by system, as seen in Figure 24-13. The Internet is another example of this type of a packet data network.

GPRS Radio Technologies

Packet switching means that the GPRS radio resources are used only when users are actually sending or receiving data. Rather than dedicating a radio channel to one mobile user for a fixed period of time, the available radio resources can be concurrently shared by several users. This efficient use of the scarce radio resources means that a larger number of GPRS users can share the same bandwidth and be served from a single cell. The actual number of users supported depends on the application being used and how much data each user has to send or receive. Because the spectrum efficiency is improved in GPRS, it is not as necessary to build idle capacity that is only used during peak transmit hours. GPRS, therefore, lets the operator maximize system usage and efficiency in a dynamic and flexible way.

In fact, all eight time slots of a TDMA frame can be made available to each user. However, as the number of simultaneous users increases, collisions will occur between the randomly arriving data packets. This will cause queuing delays on the downlink. Therefore, the effective throughput perceived by each user decreases but more gracefully. The idea of

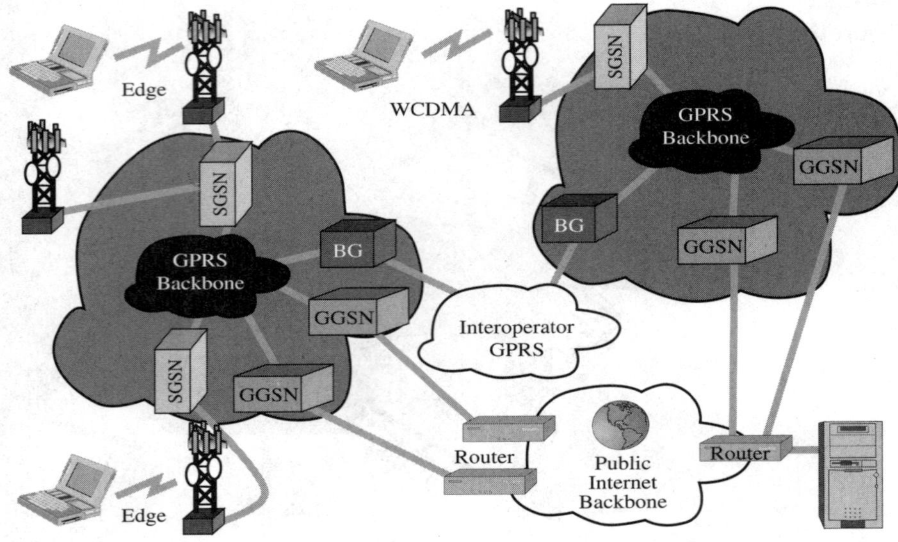

Figure 24-13 The pieces of GPRS and GSM fit together.

concatenation or aggregation of the time slots to be available to one user makes this far more palatable for the end user to understand how he or she can bundle services together and run the data faster.

Cells and Routing Areas

The geographic coverage area of a GPRS network is divided into smaller areas known as *cells* and *routing areas*, as shown in Figure 24-14. A cell is the area that is served by a set of radio *base stations* (BSs). When a GPRS MS wants to send data or prepare to receive data, it searches for the strongest radio signal that it can find. Once the mobile scans for the strongest signal and locates the strongest BS, it then notifies the network of the cell it is receiving the strongest and selects it. At this point, the mobile listens to the BS for news of incoming data packets.

Periodically, the MS uses its idle time to listen to transmissions from neighboring BSs and evaluates the signal quality of their transmissions. If the mobile determines that a different BS signal is received stronger (better) than the current base, then the mobile may begin to listen to the new BS instead. This means that the mobile will listen to a different signal. The process of moving from one BS to another is called *cell reselection*. In some

Figure 24-14 Cells and routing areas

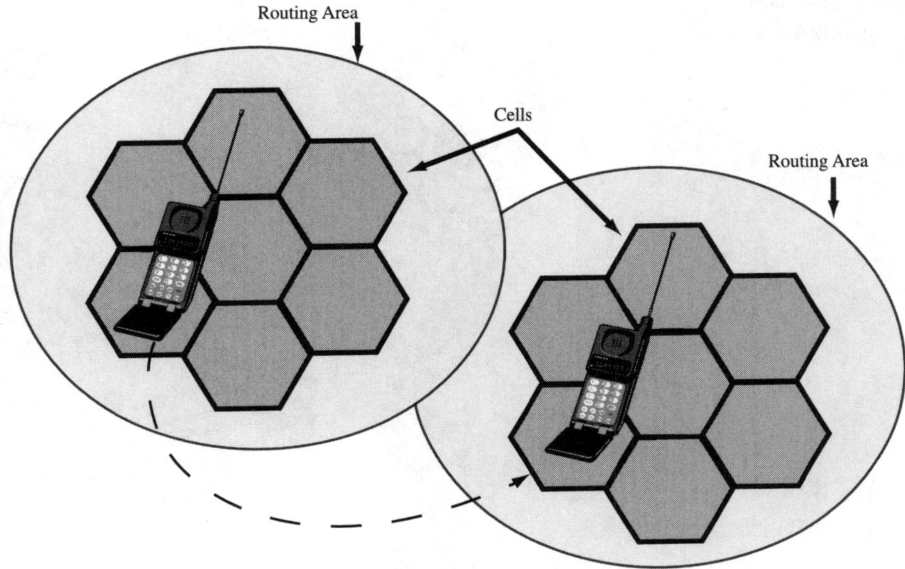

cases, the MS informs the network that it has changed cells by performing a location update procedure.

When data arrives for an idle MS, the network broadcasts a notice that it wants to establish communications with that mobile. This is called *paging* and is very similar to the paging process in wireless voice networks. A group of neighboring cells can be grouped together to form a routing area. Network engineers use routing areas to strike a balance between location-updating traffic and paging traffic. MSs that have been actively sending or receiving data are tracked at the cell level. (The network keeps track of the cell that they are currently using.) MSs that have been inactive (idle) are tracked at the routing area level (the network keeps track of the routing area).

Attaching to the Serving GPRS Support Node

When a GPRS MS wants to use the wireless packet data network services, it must first attach to a *Serving GPRS Support Node* (SGSN), as shown in Figure 24-15. When the SGSN receives a request from a MS, it makes sure

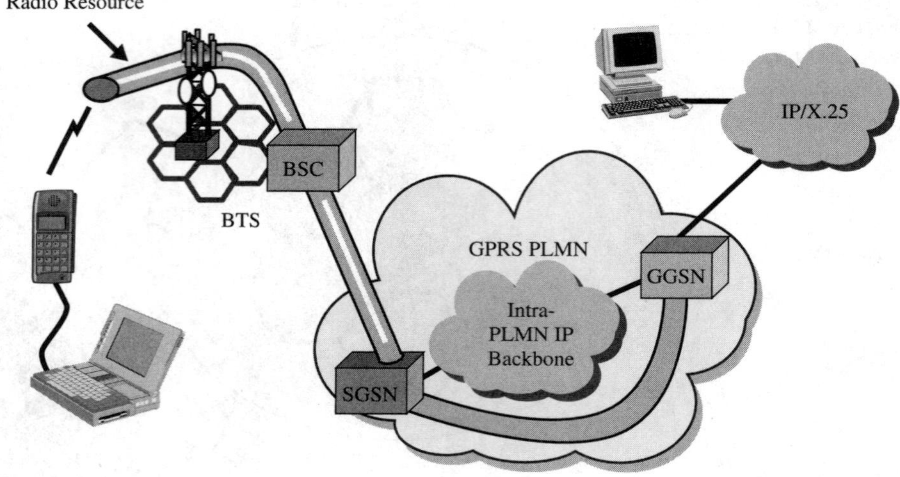

Figure 24-15 Attaching to the SGSN

that it wants to honor the request for service. Several factors must first be considered:

- Is the mobile a subscriber to GPRS? The act of verifying the MS's subscription information is called *authorization.*
- Is the mobile who it says it is? The process of verifying the identity of the MS is called *authentication.*
- What level of *quality of service* (QoS) is the station requesting? Has the mobile subscribed to the level of QoS being requested (is the owner willing to pay for it), and can the network provide this level of service while still providing the levels of service already promised to the other attached users?

Once the SGSN decides to accept an attachment, it keeps track of the MS as the mobile moves around in the coverage area. The SGSN needs to know where the mobile is in case data packets arrive and need to be routed to the mobile. Attaching to the SGSN is similar to creating a logical connection (or pipe) between the mobile and the SGSN. The logical connection is maintained as the mobile moves within the coverage area controlled by the SGSN. The attachment to an SGSN is not sufficient to enable the mobile to begin transferring packet data. To do that, the mobile needs to activate (and possibly acquire) a PDP address (such as an IP address).

PDP Contexts

The PDP addresses are network layer addresses (*Open Standards Interconnect* [OSI] model Layer 3). GPRS systems support both X.25 and IP network layer protocols. Therefore, PDP addresses can be X.25, IP, or both. Each PDP address is anchored at a *Gateway GPRS Support Node (*GGSN), as shown in Figure 24-16. All packet data traffic sent from the public packet data network for the PDP address goes through the gateway (GGSN). The public packet data network is only concerned that the address belongs to a specific GGSN. The GGSN hides the mobility of the station from the rest of the packet data network and from computers connected to the public packet data network.

Statically assigned PDP addresses are usually anchored at a GGSN in the subscriber's home network. Conversely, dynamically assigned PDP addresses can be anchored either in the subscriber's home network or the network that the user is visiting. When a MS is already attached to a SGSN and wants to begin transferring data, it must activate a PDP address. Activating a PDP address establishes an association between the mobile's current SGSN and the GGSN that anchors the PDP address. The record kept by the SGSN and the GGSN regarding this association is called the *PDP context*. It is important to understand the difference between a MS attaching to a SGSN and a MS activating a PDP address. A single MS attaches to only one SGSN; however, it may have multiple PDP addresses that are all

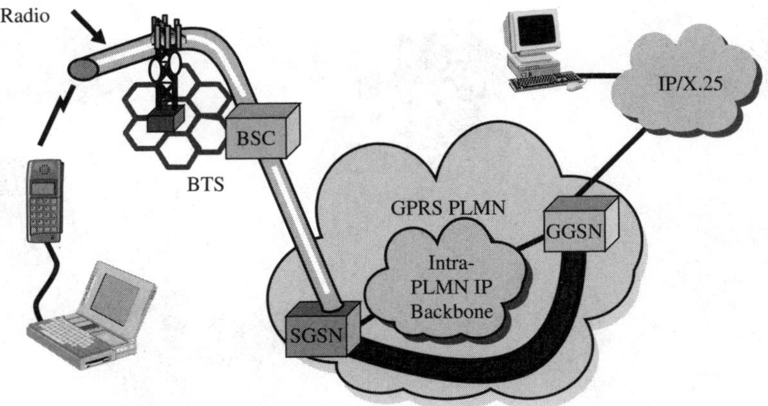

Figure 24-16 Obtaining a PDP context from the GGSN

active at the same time. Each of the addresses may be anchored to a different GGSN. If packets arrive from the public packet data network at a GGSN for a specific PDP address and the GGSN does not have an active PDP context corresponding to that address, it may simply discard the packets. Conversely, the GGSN may attempt to activate a PDP context with a MS if the address is statically assigned to a particular mobile.

Data Transfer

Once the MS has attached to a SGSN and activated a PDP address, it is now ready to begin communicating with other devices. For example, a GPRS mobile is communicating with a computer system connected to an X.25 or IP network. The other computer may be unaware that the MS is, in fact, mobile. It may only know the MS's PDP address. The packets, as shown in Figure 24-17, need to be routed as follows:

Assume that the MS has attached to an SGSN and activated its PDP address. Packets sent from the other computer to the MS first travel across the public packet data network to reach the GGSN that anchors the PDP address. From here, the GGSN must forward the packets to the SGSN to which the MS is currently attached. Obviously, packets flowing in the reverse direction must be first routed through the SGSN and GGSN before being passed to the public packet data network. Communication between the *GPRS Serving Nodes* (GSNs) makes use of a technique known as *tun-*

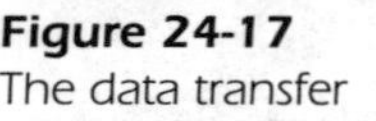

Figure 24-17
The data transfer

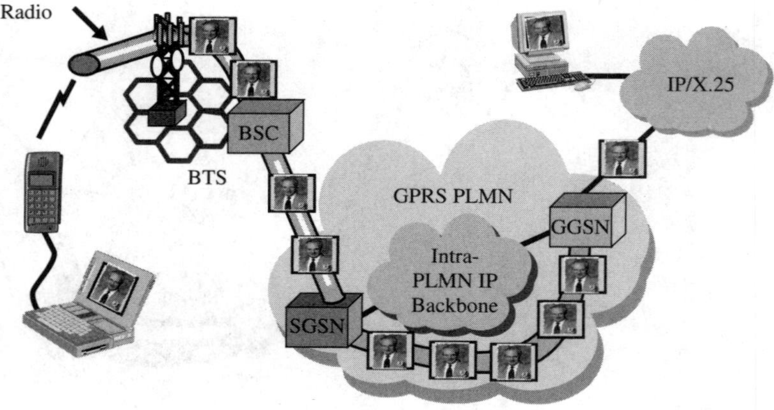

neling. Tunneling is the process of wrapping the network layer packets into another header so that they can be routed through the GPRS PLMN IP backbone network. Inside the network, packets are routed based on the new header alone, and the original packet is carried as the payload. Once they reach the far side of the GPRS network, they are unwrapped and continue along their way through the external network.

From this point onward, they are routed based on their original (internal) header. Using tunneling within GPRS solves the mobility problem for the packet networks and helps to eliminate the complex task of protocol interworking. Mobile IP also makes use of tunneling to route packets to mobile nodes. In mobile IP, packets are only tunneled from the fixed network to the MS. Packets flowing from the mobile to fixed nodes use normal routing. GPRS, by contrast, uses tunneling in both directions.

GSM and NA-TDMA Evolution

Both GSM and NA-TDMA are evolving from 2G to 3G, and GPRS plays an important role in this evolution. Some of the key points to note are as follows:

- As evident, attempts are made to seek synergy between the two TDMA base systems now (as opposed to what happened 10 years ago).
- GSM-GPRS standards and concepts are being adopted in North American TDMA as GPRS-136. The radio interface is being adapted to 30 kHz channels and an IS-136 DCCH channel structure. In fact, many of the North American carriers (AT&T Wireless and Voice Stream, among others) are planning to offer GPRS on GSM as an evolution from the NA-TDMA architecture).
- EDGE has adopted the eight-PSK-modulation scheme that is used for 136+.
- 136HS and EDGE are being developed with synergy in mind. UWC-136 has embraced the GPRS/EDGE architecture for 200 kHz-wide 136HS Outdoors.
- The North American and European proposals differ for the 2.0 Mbps systems. UWC-136 continues to use a purely TDMA scheme, whereas CDMA-based UTRA is the *Radio Transmission Technology* (RTT) of choice for ETSI.

Applications for GPRS

Many applications fit into the mode of GPRS and IPs. These applications are merely a means to an end. In other scenarios, the features and applications can be met with other technologies. The issue at hand is that the use of GPRS facilitates these applications and drives the acceptance ratio. It is easy to say that we can do anything with GPRS, but it is more practical to say at a minimum that we can do the following:

Chat

Chat can be distinguished from general information services because the source of the information is a person with the chat protocol, whereas it tends to be from an Internet site for information services. The information intensity, the amount of information transferred per message, tends to be lower with chat, where people are more likely to state opinions than factual data. In the same way as Internet chat groups have proven to be a very popular application of the Internet, groups of like-minded people, so-called communities of interest, have begun to use nonvoice mobile services as a means to chat and discuss.

Because of its synergy with the Internet, GPRS would enable mobile users to participate fully in existing Internet chat groups rather than needing to set up their own groups that are dedicated to mobile users. Because the number of participants is an important factor determining the value of participation in the news group, the use of GPRS here would be advantageous.

GPRS will not, however, support point-to-multipoint services in its first phase, hindering the distribution of a single message to a group of people. As such, given the installed base of SMS-capable devices, we would expect SMS to remain the primary bearer for chat applications in the foreseeable future, although experimentation with using GPRS is likely to commence sooner rather than later.

Textual and Visual Information

A wide range of content can be delivered to mobile phone users, ranging from share prices, sports scores, weather, flight information, news headlines, prayer reminders, lottery results, jokes, horoscopes, traffic, location-sensitive services, and so on. This information does not necessarily need to be textual—it may be maps or graphs or other types of visual information.

The length of a short message of 160 characters suffices for delivering information when it is quantitative, such as a share price or a sports score or temperature. When the information is of a qualitative nature, however, such as a horoscope or news story, 160 characters is too short other than to tantalize or annoy the information recipient because they receive the headline or forecast but little else of substance. As such, GPRS will likely be used for qualitative information services when end users have GPRS-capable devices, but SMS will continue to be used for delivering most quantitative information services. Interestingly, chat applications are a form of qualitative information that may be delivered using SMS, in order to limit people to brevity and reduce the incidence of spurious and irrelevant posts to the mailing list that are a common occurrence on Internet chat groups.

Still Images

Still images such as photographs, pictures, postcards, greeting cards, presentations, and static web pages can be sent and received over the mobile network as they are across fixed telephone networks. It will be possible with GPRS to post images from a digital camera connected to a GPRS radio device directly to an Internet site, enabling near real-time desktop publishing.

Moving Images

Over time, the nature and form of mobile communication is getting less textual and more visual. The wireless industry is moving from text messages to icons, picture messages to photographs, blueprints to video messages, movie previews being downloaded, and on to full-blown movie watching via data streaming on a mobile device.

Sending moving images in a mobile environment has several vertical market applications, including monitoring parking lots or building sites for intruders or thieves and sending images of patients from an ambulance to a hospital. Videoconferencing applications, in which teams of distributed salespeople can have a regular sales meeting without having to go to a particular physical location, are another application for moving images.

Web Browsing

Using circuit-switched data for web browsing has never been an enduring application for mobile users. Because of the slow speed of circuit-switched data, it takes a long time for data to arrive from the Internet server to the browser. Alternatively, users switch off the images, just access the text on the Web, and end up with text layouts on screens that are difficult to read. As such, mobile Internet browsing is better suited to GPRS.

Document Sharing/Collaborative Working

Mobile data facilitates document sharing and remote collaborative working. This lets different people in different places work on the same document at the same time. Multimedia applications combining voice, text, pictures, and images can even be envisaged. These kinds of applications could be useful in any problem-solving exercise, such as fire fighting, combat (to plan the route of attack), medical treatment, advertising copy setting, architecture, journalism, and so on. This collaborative working environment can be useful anytime a user can benefit from having the ability to comment on a visual depiction of a situation or matter. By providing sufficient bandwidth, GPRS facilitates multimedia applications such as document sharing.

Audio

Despite many improvements in the quality of voice calls on mobile networks, such as *Enhanced Full Rate* (EFR), they are still not broadcast quality. In some scenarios, journalists or undercover police officers with portable professional broadcast-quality microphones and amplifiers capture interviews with people or radio reports that they have dictated and need to send this information back to their radio or police station. Leaving a mobile phone on or dictating to a mobile phone would not give sufficient voice quality to enable that transmission to be broadcast or analyzed for the purposes of background noise analysis or voice printing, where the speech autograph is taken and matched against those in police storage. Because even short voice clips occupy large file sizes, GPRS or other high-speed mobile data services are needed.

Job Dispatch

Nonvoice mobile services can be used to assign and communicate new jobs from office-based staff to mobile field staff. Customers typically telephone a call center whose staff takes the call and categorizes it. Those calls requiring a visit by a field sales or service representative can then be escalated to those mobile workers. Job dispatch applications can optionally be combined with vehicle-positioning applications, so that the nearest available suitable personnel can be deployed to serve a customer. GSM nonvoice services can be used not only to send the job out, but also as a means for the service engineer or salesperson to keep the office informed of progress towards meeting the customer's requirement. The remote worker can send in a status message such as "Job 1234 complete, on my way to 1235."

The 160 characters of a short message are sufficient for communicating most delivery addresses, such as those needed for a sale or service, or some other job dispatch application, such as mobile pizza delivery and courier package delivery. However, the 160 characters require manipulation of the customer data, such as the use of abbreviations like St instead of Street. The 160 characters do not leave much space for giving the field representative any information about the problem that has been reported or the customer profile. The field representative is able to arrive at the customer's premises, but is not very well briefed beyond that. This is where GPRS will be beneficial to enable more information to be sent and received more easily. With GPRS, a photograph of the customer and his or her premises could, for example, be sent to the field representative to assist in finding and identifying the customer. As such, we expect job dispatch applications will be an early adopter of GPRS-based communications.

Corporate E-mail

With up to half of employees typically away from their desks at any one time, it is important for them to keep in touch with the office by extending the use of corporate e-mail systems beyond an employee's office PC. Corporate e-mail systems run on LANs and include Microsoft Mail, Outlook, Outlook Express, Microsoft Exchange, Lotus Notes, and Lotus cc:Mail. Because GPRS-capable devices will be more widespread in corporations than among the general mobile phone user community, more corporate e-mail applications are likely to use GPRS than Internet e-mail applications whose target market is more general.

Internet E-mail

Internet e-mail services come in the form of a gateway service where the messages are not stored or mailbox services in which messages are stored. In the case of gateway services, the wireless e-mail platform translates the message from SMTP, the Internet e-mail protocol, into SMS and sends it to the SMS Center. In the case of mailbox e-mail services, the e-mails are actually stored, and the user receives a notification on his or her mobile phone and can then retrieve the full e-mail by dialing in to collect it, forward it, and so on.

Upon receiving a new e-mail, most Internet e-mail users are not currently notified of this fact on their mobile phone. When they are out of the office, they have to dial in speculatively and periodically to check their mailbox contents. However, by linking Internet e-mail with an alert mechanism, such as SMS or GPRS, users can be notified when a new e-mail is received.

Vehicle Positioning

This application integrates satellite-positioning systems that tell people where they are with nonvoice mobile services that enable people to tell others where they are. The *Global Positioning System* (GPS) is a free-to-use global network of 24 satellites run by the U.S. Department of Defense. Anyone with a GPS receiver can receive his or her satellite position and thereby find out where he or she is. Vehicle-positioning applications can be used to deliver several services, including remote vehicle diagnostics, ad hoc stolen vehicle tracking, and new rental car fleet tariffs.

The SMS is ideal for sending GPS position information such as longitude, latitude, bearing, and altitude. GPS coordinates are typically about 60 characters in length. GPRS could alternatively be used.

Remote LAN Access

When mobile workers are away from their desks, they clearly need to connect to the LAN in their office. Remote LAN applications encompass access to any applications that an employee would use when sitting at his or her desk, such as access to the intranet, his or her corporate e-mail services, such as Microsoft Exchange or Lotus Notes, and to database applications running on Oracle or Sybase. The mobile terminal, such as a hand-held or laptop computer, has the same software programs as the desktop on it or cut-down client versions of the applications accessible through the corporate LAN. This application area is therefore likely to be a conglomeration of remote access to several different information types —e-mail, intranet, databases. This information may all be accessible through web browsing tools or require proprietary software applications on the mobile device. The ideal bearer for remote LAN access depends on the amount of data being transmitted, but the speed and latency of GPRS make it ideal.

File Transfer

As this generic term suggests, file transfer applications encompass any form of downloading sizeable data across the mobile network. This data could be a presentation document for a traveling salesperson, an appliance manual for a service engineer, or a software application, such as Adobe Acrobat Reader, to read documents. The source of this information could be one of the Internet communication methods such as *File Transfer Protocol* (FTP), telnet, http, or Java, or from a proprietary database or legacy platform. Irrespective of the source and type of file being transferred, this kind of application tends to be bandwidth-intensive. Therefore, it requires a high-speed mobile data service, such as GPRS, EDGE, or UMTS, to run satisfactorily across a mobile network.

Home Automation

Home automation applications combine remote security with remote control. Basically, you can monitor your home from anywhere—on the road, on vacation, or at the office. If your burglar alarm goes off, not only are you alerted, but also you can go live and see live footage of the perpetrators. You can program your video or switch on your oven so that the preheating is complete by the time you arrive home (traffic jams permitting). Your GPRS capable mobile phone really becomes the remote control device for your television, video, and stereo. Because the IP will soon be everywhere, these devices can be addressed and fed instructions. A key enabler for home automation applications will be Bluetooth, which enables disparate devices to interwork.

These features and the driving motivators will propel the operators into the implementation of GPRS. Moreover, the applications will offer many new opportunities to users that were previously unavailable. It is no wonder that the hype of GPRS is strong now. The next approach we will look to will be the architecture of the GPRS infrastructure. This will help the reader to understand the overall architectural model used for GPRS.

CHAPTER 25

Third-Generation (3G) Wireless Systems

Wireless continues to develop around the world. Several different standards committees are working on integrating wireless architecture into the overall fold of the network.

Get ready! As the convergence of wireless technology and the Internet continue at an escalating pace, the new possibilities created by 3G and 4G technologies appear endless. Preparing for the revolution, existing *Time Division Multiple Access* (TDMA) operators must evolve their networks to take advantage of Mobile Multimedia applications and the eventual shift to an all-IP architecture. One way to do that is through the evolution of *General Packet Radio Services* (GPRS). However, soon after we see the installation of GPRS, some operators will begin the next step in the evolution process to *Enhanced Data for Global Environment* (EDGE). With EDGE, existing TDMA networks can host a variety of new applications, including

- Online e-mail
- Access to the World Wide Web
- Enhanced short message services
- Wireless imaging with instant photos or graphics
- Video services
- Document/information sharing
- Surveillance
- Voice messaging via Internet
- Broadcasting

At the same time, some operators will skip the step to EDGE and go directly to *Universal Mobile Telecommunications Services* (UMTSs), or what we consider to be a 3G technology. The steps are shown in Figure 25-1 as the carriers choose which way to proceed.

Using a timing window, the evolution of wireless to 3G systems is shown in Figure 25-2, using the evolution of the various techniques that emerged over the years.

GPRS

Probably the most important aspects of GPRS are that it enables data transmission speeds up to 170 Kbps, it is packet based, and it supports the leading data communications protocols (IP and X. 25).

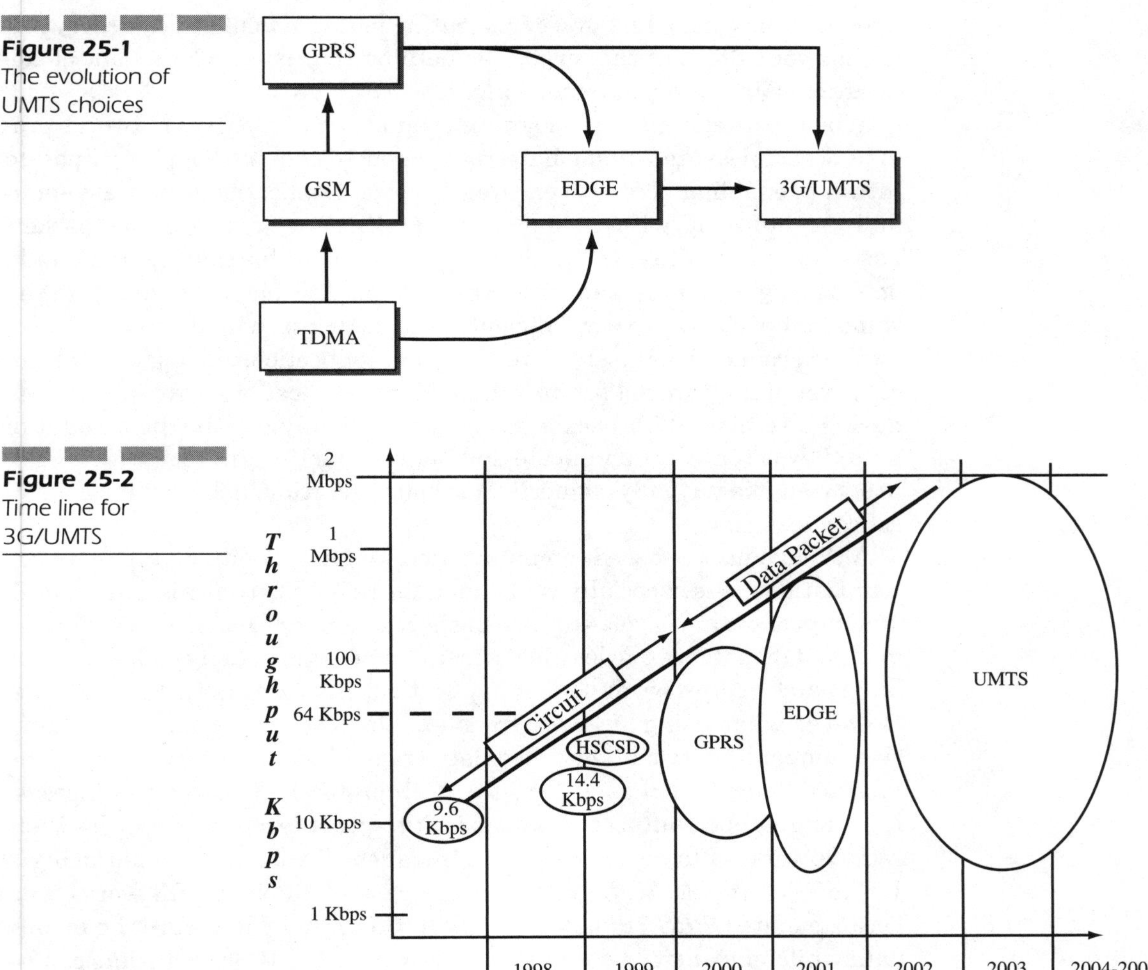

Figure 25-1 The evolution of UMTS choices

Figure 25-2 Time line for 3G/UMTS

GPRS operates at much higher speeds than current networks, providing advantages from a software perspective. Wireless middleware currently is required to enable slow speed mobile clients to work with fast networks for applications such as e-mail, databases, groupware, or Internet access. With GPRS, wireless middleware will probably be unnecessary, making it easier to deploy wireless solutions.

Although current wireless applications are text oriented, GPRS' high throughput finally makes multimedia content, including graphics, voice, and video, practical. Imagine participating in a videoconference while waiting for your flight at the airport, something that is completely out of the question with today's data networks.

Why is packet data technology important? Because packet networks provide a seamless and immediate connection to the Internet or corporate intranet, enabling access to existing Internet applications, such as e-mail and web browsing, without dialing into an ISP. The advantage of a packet-based approach is that GPRS uses the medium, in this case the radio link, only as long as data is being sent or received. Multiple users can share the same radio channel very efficiently. In contrast, with current circuit-switched connections, users have dedicated connections during their entire call, even if they are not sending data. Many applications have idle periods during a session. With packet data, users will only pay for the amount of data they actually communicate and not the idle time. In fact, with GPRS, users could be virtually connected for hours at a time and only incur modest connect charges.

Although packet-based communication works well with all types of communications, it is especially well suited for frequent transmission of small amounts of data. We refer to this as short and bursty, such as real-time e-mail and dispatch (vehicles, field service). Packet is equally well suited for large batch operations and other applications involving large file transfers. However, when using large file transfers, the cost can become very expensive compared to circuit-switched data transmissions. GPRS supports the *Internet Protocol* (IP) as well as the X.25 protocol. IP support is increasingly more important as companies look to the Internet as a way for their remote workers to access corporate intranets. This is true when using a *Virtual Private Network* (VPN). In the case of VPNs, GPRS works well because of its *GPRS Tunneling Protocol* (GTP) that can secure the mobile data while in transit on the wireless networks, and IPSec transfers can be used when transiting the wireline networks. The GTP is shown in Figure 25-3.

The IP protocol is ubiquitous and familiar, but what is X.25, and why is support for it important? X.25 defines a set of communications protocols that, prior to the Internet, constituted the basis of the world's largest packet data networks. These X.25 networks are still widely used, especially in Europe and the Far East. Wireless access to these networks will benefit many organizations. Any existing IP or X.25 application will now be able to operate over a *Global System for Mobile Communications* (GSM) cellular connection. Think of cellular networks with GPRS service as wireless extensions of the Internet and existing X.25 networks as similar to a *Local Area*

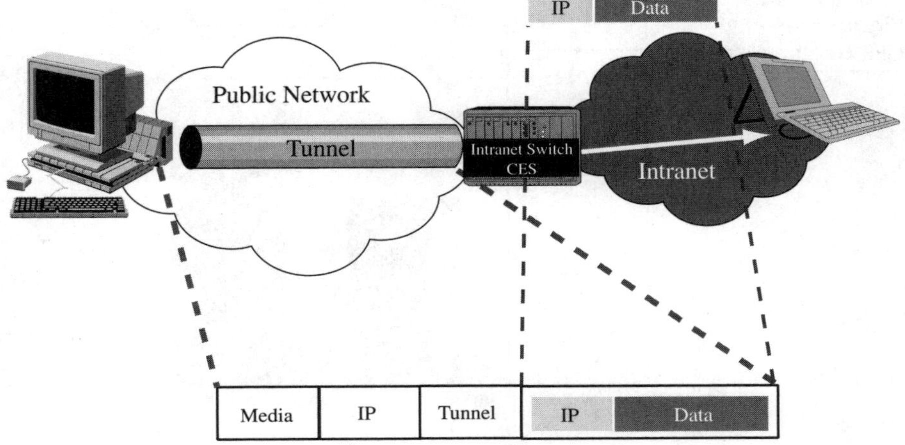

Figure 25-3
GTP with VPNs

Network (LAN) connection. As a LAN connection, once a GPRS *mobile station* (MS) registers with the network, it is ready to send and receive packets. A user with a laptop computer could be working on a document without even thinking about being connected, and then automatically receive new e-mail—although this is not 100 percent available today, it is coming. The user may decide to continue working on the document, then half an hour later read the e-mail message and reply to it. Throughout this time, the user has had a network connection, yet never dialed in. Furthermore, some versions of GPRS terminals allow for simultaneous voice and data communication. The user can receive incoming calls or make outgoing calls while in a data session.

Because there is minimal delay before sending data, GPRS is ideally suited for applications such as

- Extended communications sessions
- E-mail communications
- Chat
- Database queries
- Dispatch
- Stock updates

In addition, GPRS will remove many of the obstacles from the use of multimedia, graphical web-based applications because of the higher

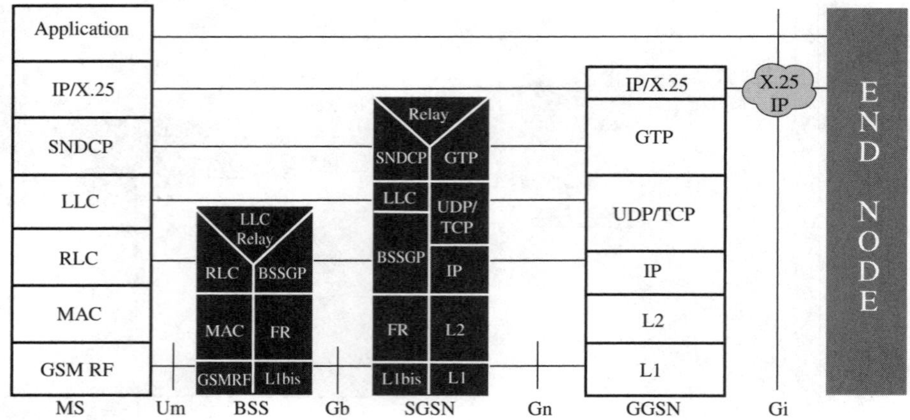

Figure 25-4 GPRS protocol stack

throughput that is possible. Mobile users will easily use graphically intensive web-based applications (Map Quest) to obtain directions. The protocol stack supports a variety of interfaces and links multiple networks together as shown in Figure 25-4.

Because GPRS supports standard networking protocols, configuring computers to work with GPRS will be very straightforward. In the case of IP communications, one can use existing TCP/IP protocol stacks. TCP/IP stacks are readily available for most other platforms as well. With all the developments in the handheld computer area, expect a multitude of hardware platforms to take advantage of GPRS:

- Laptops or handheld computers connected to GPRS-capable cell phones or external modems
- Laptops or handheld computers with GPRS-capable PC Card modems
- Smart phones that have full screen capability
- Cell phones employing microbrowsers using the *Wireless Application Protocol* (WAP)
- Dedicated equipment with integrated GPRS capability, such as mobile credit-card swipes

GPRS coincides with another important technology development: the replacement of a cable connection to a cell phone by a short radio link called *Bluetooth*.

EDGE

Beyond GPRS, EDGE takes the cellular community one step closer to UMTS. It provides higher data rates than GPRS and introduces a new modulation scheme called *8-Phase Shift Keying* (PSK). The TDMA community also adopted EDGE for their migration to UMTS. The data rates allocated for EDGE are started at 384 Kbps and above as a second stage to GPRS. EDGE uses the same modulation techniques as many of our existing TDMA infrastructures using *Gaussian Minimum Shift Keying* (GMSK) 8-PSK. Moreover EDGE uses a combination of FDMA and TDMA as the multiple access control methods. If we look at this from an OSI stack model, EDGE uses FDMA and TDMA at the MAC layer (bottom half of layer 2 OSI). The protocol stack for EDGE is shown in Figure 25-5.

The channel separations are 45 MHz, and the carrier spacing is a 200 kHz channel capacity, the same as GSM and GPRS. The number of TDMA slots on each carrier is the same (eight) as the GSM and GPRS architecture. When a MS wants to transmit its data, it can request and use from one to eight time slots per TDMA frame. Connectivity is handled via a packet-switched data network such as IP and X.25. These can be public data networks or private data networks.

Figure 25-5
EDGE protocol stack

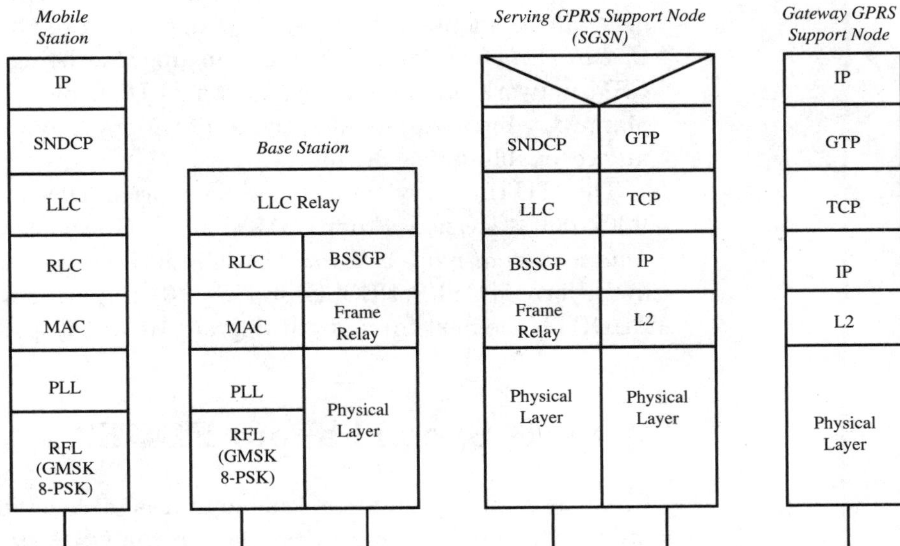

Although most carriers and service providers have plans to deploy enhanced mobile wireless services at higher speeds, the rollout of high-bandwidth wireless transport technology still faces many challenges. On a positive note, widespread demand will be sufficient enough to support cellular enhancements like high-speed data services and expanded voice capacity. Competitive pressures will also compel service providers to upgrade. The *Radio Communications Sector of the International Telecommunications Union* (ITU-R) has actually established five different standards that fall into the category of 3G/UMTS. Moreover, the telecommunication industry is growing increasingly impatient to test the world markets for high-bandwidth wireless communication services. The ITU's IMT-2000 initiative may one day converge, but today many 3G proposals are still under consideration, including

- cdma2000 (an upgrade to cdmaOne)
- UMTS
- *Wideband-Code Division Multiple Access* (WCDMA)
- Universal wireless communications (UWC-136)

UWC-136 is based on TDMA as are Europe's GSM, Japan's *personal digital cellular* (PDC), and the *digital advanced mobile phone system* (D-AMPS) used in the United States.

Existing 2G service providers have already applied for licenses to operate 3G networks around the globe. Although it's unclear what 3G technologies will be adopted, the most 2.5G upgrades are GPRS and *High-Speed, Circuit-Switched Data* (HSCSD), an upgrade being considered by some GSM network operators. Beyond that, EDGE modulation extensions are planned, which will enable service providers to offer even higher performance, enabling true 3G-like services.

The ITU currently embraces various proposed schemes to attaining the IMT-2000 3G vision. From TDMA-based 2G providers of GSM and *North American dual-mode cellular* (NADC) services, interim upgrades will come in the form of GPRS, HSCSD, and IS-136+, and will eventually converge at EDGE for the next throughput upgrade (to 384 Kbps) before 3G.

What Is Special about EDGE?

EDGE is a new modulation scheme that is more bandwidth efficient than the GMSK modulation scheme used in the GSM standard. It provides a promising migration strategy for HSCSD and GPRS. The technology defines a new physical layer: 8-PSK modulation, instead of GMSK. 8-PSK enables each pulse to carry 3 bits of information versus the GMSK 1-bit-per-pulse rate. Therefore, EDGE has the potential to increase the data rate of existing GSM systems by a factor of three.

EDGE retains other existing GSM parameters, including a frame length, eight time slots per frame, and a 270.833 kHz symbol rate. The GSM 200 kHz channel spacing is also maintained in EDGE, enabling the use of existing spectrum bands. This fact is likely to encourage deployment of EDGE technology on a global scale.

UMTS

UMTS is a part of the ITU's IMT-2000 vision of a global family of 3G mobile communications systems. UMTS will play a key role in creating the future mass market for high-quality wireless multimedia communications that will approach 2 billion users worldwide by the year 2010.

UMTS is a modular concept that takes full advantage of the trend of converging existing and future information networks, devices, and services, and the potential synergies that can be derived from such convergence. UMTS will move mobile communications forward from where we are today into the 3G services and will deliver speech, data, pictures, graphics, video communication, and other wideband information direct to people on the move. UMTS is one of the major new 3G mobile communications systems being developed within the framework, which has been defined by the ITU and is known as IMT-2000.

Over the past decade, UMTS has gained the support of many major telecommunications operators and manufacturers because it represents a unique opportunity to create a mass market for highly personalized and user-friendly mobile access to tomorrow's untethered society.

UMTS will build on and extend the capability of today's mobile technologies (like digital cellular) by providing increased capacity, data capability, and a much greater range of services. The launch of UMTS services will see the evolution of a new, open communications universe, with players from many sectors coming together to deliver new communications services, characterized by mobility and advanced multimedia capabilities. The successful deployment of UMTS will require new technologies, new partnerships, and the addressing of many commercial and regulatory issues.

UMTS will enable tomorrow's wireless knowledge worker, delivering high-value broadband information, commerce, and entertainment services to users via fixed, wireless, and satellite networks. UMTS will speed the convergence between voice, data, and multimedia to deliver new services and create fresh revenue-generating opportunities. UMTS will deliver low-cost, high-capacity mobile communications, offering data rates up to 2 Mbps with global roaming and other advanced capabilities.

The next decade will see the emergence of 3G networks to fully realize mobile multimedia services. Enabling anytime, anyplace connectivity to the Internet is just one of the opportunities for 3G networks. The major market opportunity builds on mobile networking to provide

- Group messaging
- Location-based services (GPS)
- Personalized information
- Infotainment

Many new 3G services will not be Internet-based (they will be truly unique mobility services). Data will increasingly dominate the traffic flows. Pent-up latent demand for mobile data services will jump start 3G networks. By 2005, more data than voice will flow over mobile networks. This is an amazing statistic considering that mobile cellular networks today are almost exclusively voice.

WCDMA

WCDMA is an ITU standard derived from CDMA and is officially known as IMT-2000 direct spread. WCDMA is a 3G mobile wireless technology offering much higher data speeds to mobile and portable wireless devices than commonly offered in today's market.

WCDMA can support mobile/portable voice, images, data, and video communications at up to 2 Mbps (local area access) or 384 Kbps (wide area access). The input signals are digitized and transmitted in coded, spread-spectrum mode over a broad range of frequencies. A 5 MHz wide carrier is used, compared with a 200 kHz wide carrier for narrowband CDMA.

WCDMA is the radio access technology selected by the *European Telecommunications Standards Institute* (ETSI) in January 1998 for wideband radio access to support 3G. WCDMA has since become the most popular global 3G air interface mode and is being implemented across the GSM world, but also in the TDMA world and in Japan by J-Phone and NTT DoCoMo. The competing air interface mode to WCDMA is cdma2000, which is being implemented mainly in North America and in Japan by KDDI.

WCDMA can be added to the existing GSM core network. This will be particularly beneficial when large portions of new spectrum are made available, for example, in the new-paired 2 GHz bands in Europe and Asia. It will also minimize the investment required for WCDMA rollout—it will, for example, be possible for existing GSM sites and equipment to be reused to a large extent.

WCDMA Features

WCDMA offers very high capacity with 50 to 80 voice channels per 5 MHz carrier (compared with 8 channels per carrier with 200 kHz for GSM). To achieve the very high data rates, WCDMA requires a wide frequency band of between 5 MHz and 10 MHz (compared with a 200 kHz carrier for regular GSM).

With WCDMA, two frequency bands have been allocated—one for sending data from the terminal and one for receiving data on the terminal. This technique is called *symmetric*, that is, needing the same amount of radio resources in the uplink and downlink.

- WCDMA can be overlaid onto existing GSM, TDMA (IS-136), and CDMA (IS-95) networks.
- WCDMA provides higher capacity and increased coverage: up to eight times more traffic per carrier compared to a narrowband CDMA carrier such as cdmaOne. This is achieved by up to 100 percent better usage of the frequency spectrum.
- WCDMA supports *Hierarchical Cell Structures* (HCSs) by employing *Mobile Assisted Inter Frequency Handover* (MAIFHO).
- WCDMA enables the optional use of adaptive antenna arrays, a concept that enables antenna pattern optimization, and hence extended range and reduced interference on a per-link basis.
- WCDMA supports multiuser detection—a mechanism to reduce multiple access interference and enhance capacity.

WCDMA consists of a set of *Radio Network Subsystems* (RNSs), connected to the Core Network through the l_u interface. Each RNS is composed of one *Radio Network Controller* (RNC), controlling a group of logical elements, called Node B (or BTS), through the l_{ub} interface. Each Node B can serve one or several cells. Each RNS is responsible for the resources of its set of cells and performs functions related to the overall system access, security and privacy, handover detection, decision and execution, and radio resource management and control, including connection set-up and release, as well as transfer of data packets.

WCDMA mobile terminal/phone users can be connected via the U_u (air) interface to one or more Node B entities (this is called *macro diversity*). In the later case, those cells may be allocated to different RNSs. In such a case, one RNC is acting as the serving RNC and is supported by a second RNC through the l_{ur} interface.

The protocols over the l_u interface are divided into user plane protocols and control plane protocols. User plane protocols are required for the implementation of the radio access bearer service (that is, to carry the user data). The control plane protocols are required for controlling the connection between the *User Equipment* (UE), the network, and the radio access bearers.

Mobile Internet—A Way of Life

The mobile Internet is about to enter our daily lives in a big way. It will change the way we keep in touch with our friends and family, the way we do business, the way we shop, the way we access entertainment, and the way we conduct our personal finances.

The Internet is already a part of daily life for most of us, giving us access to a vast range of information and online services from our desktop computers. As a way of conducting business, it is also of growing importance to the global economy. Unlike today's fixed Internet, the mobile Internet will give us access to these services and applications wherever we are, whenever it suits us, from personal mobile devices.

By 2004, there will be as many as 600 million users of mobile Internet services. This means that more people will use mobile Internet than fixed Internet. The market is already taking off. The chart shown in Figure 25-6 represents the 3G subscribers projected for the future.

Twenty billion SMS messages are sent worldwide every month. In Japan, there are more than 10 million users of the iMode service—which is comparable with basic WAP service—and each week another 150,000 new iMode users are added. In a few years, many of us will wonder how we managed without the mobile Internet: It will become an invaluable part of our everyday lives. Giving us more opportunities to keep in touch with

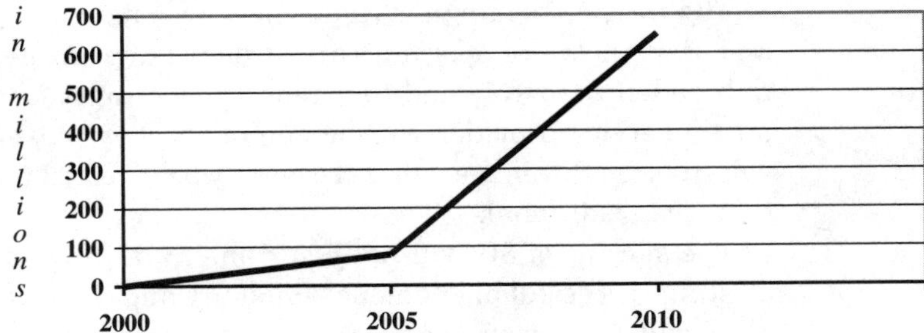

Figure 25-6 Worldwide 3G subscribers in millions

friends, family, and colleagues, empowering us to make fast, yet well-informed business decisions, giving us instant access to information and services and enabling us to purchase the things we need or desire—all in a handy, pocket-sized device. We can expect to see the following types of services from 3G products:

- Customized infotainment
- Multimedia messaging service
- Mobile intranet/extranet Access
- Mobile Internet access
- Location-based services
- Rich Voice

Rich Voice

Rich Voice is a 3G service that is real time and two way. It provides advanced voice capabilities (such as Voice over IP [VoIP], voice-activated net access, and web-initiated voice calls), while still offering traditional mobile voice features (such as operator services, directory assistance, and roaming). As the service matures, it will include mobile videophone and multimedia communications.

At present, the mobile network value chain is centered on the network operator who captures more than 90 percent of market revenues, dominated by income from voice-based services. It is widely recognized, however, that advancing technology, growth of Internet services, and new end-user demands are challenging this traditional value chain.

The new, fast-changing value chain will have new players and entities, and many network operators are already adopting new business strategies to broaden their role and to defend their competitive position. The multimedia service provider will be one of the key players in the multimedia value chain. Revenues will increasingly be diverted to other market players than the traditional.

The success of 3G will not just come from the mere combination of two existing successful phenomena—mobility and the Internet. The real success of 3G will result from the creation of new service capabilities that genuinely fulfill a market need. Meeting market demand is not just a question of technological capability and service functionality. Creating and meeting market demand requires services and devices to be priced at acceptable levels. This requires economies of scale to be present. The ability to benefit from economies of scale is one of the strongest market drivers for 3G services. *Universal Terrestrial Radio Access* (UTRA) now includes both the Direct Sequence and Time Code components of IMT-2000, and so it embeds both the *Frequency Division Duplexing* (FDD) access mode previously known as WCDMA as well as the *Time Division Duplexing* (TDD) modes previously known as TD-CDMA and TD-SCDMA. UTRA is now applicable to the major markets of Europe, China, South Korea, and Japan. UMTS promises significant economies of scale.

The need to protect existing investments in different 2G technologies has shifted the drive toward a single global standard. When you add the significant event of the emergence of the Internet, the additional capabilities of 3G become more focused on the provision of high data rates to deliver multimedia services. The emergence of the Internet as a mass-market content resource had justified the need for such high data rate capabilities and has since shifted the emphasis to packet-switched, IP-based core networks. There is general acceptance within the industry that 3G core networks will eventually be all-IP based.

The solution was the introduction of the IMT-2000 family of systems concept for 3G. One consequence of that solution is that a single global standard does not exist yet. However, the UMTS Forum believes that progress of technology, operational deployments, and market requirements will continue toward convergence. Another consequence—important when considering market perspectives—is that 3G now means different things in different parts of the world.

In Europe, 3G refers to the UMTS technology members of the IMT-2000 family, derived from GSM and deployed on new spectrum. There is a strong focus within the UMTS community on international roaming capabilities and the potential benefits of the economies of scale that result from a common standard deployed across many nations. The same UMTS technology members will be used in South Korea, China, Japan, and most of the Asian region. In the United States, 3G refers to derivatives of existing 2G technologies, deployed largely on occupied spectrum. 3G in the United States focus more on high data rates; international roaming capabilities are not a significant concern. The United States has lagged behind other world regions in the deployment of 2G digital cellular. Industry opinion is that it will continue to lag behind in the deployment of 3G. In Japan and South Korea, 3G means an opportunity to join the worldwide market.

In 2G technologies, GSM currently has 65 percent of the world market, shown in Figure 25-7. Japan has decided that its PDC 2G technology will not be evolved to 3G, but will be replaced by the UMTS/IMT-2000 technologies. The TDMA and GSM communities are working on harmonization procedures for the approach to 3G. The 15 percent of the world market currently using cdmaOne technology, mainly located in the United States and South Korea, has a transition path to the IMT-MC member of the IMT-2000 family, but is limited to existing spectrum.

With UMTS services, providers worldwide will be using multiband rather than multimode handsets—a much more attractive proposition for terminal manufacturers.

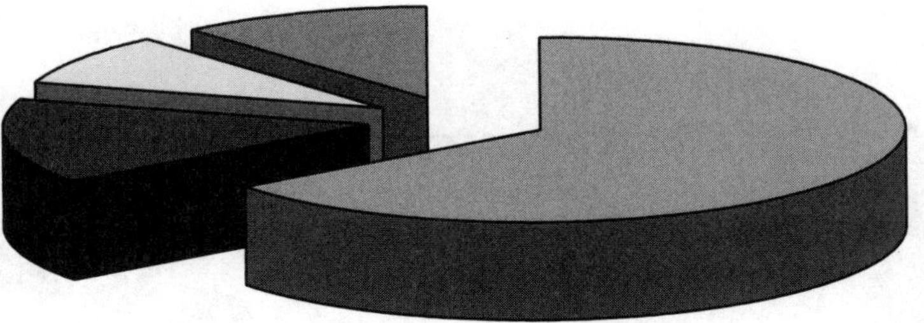

Figure 25-7 GSM user population worldwide

Applications of the Wireless Internet

Using the mobile (wireless) Internet as the model for the growth curves seen, the following are some of the applications for moving to a wireless and a 3G environment:

- **Cutting the umbilical cord** The first wave is making our familiar online services mobile—"cutting the cord" of the Internet. An example is using a laptop computer together with a mobile phone to send and receive e-mail or surf the Net. Users can access Internet-based services at the airport, on a train, or in the park.
- **Pocket WWW** The second wave, which has already begun, brings Internet services to pocket mobile devices. Applications are specially adapted to work on mobile devices with small screens, for example, using WAP. Although this wave brings full, convenient mobility, it is still largely based around traditional Internet services, such as online banking, e-mail, and web access.
- **Real mobility in the Internet** In the third wave, the full potential of the mobile Internet is realized. Services, applications, and content are centered on the mobility, location, and situation of the user—they become situation-centric.

This intelligence can be used to create highly valuable, personalized services. Mobile devices will become indispensable tools that enhance our daily lives. Services will be relevant, useful, and timely. Figure 25-8 is a forecast

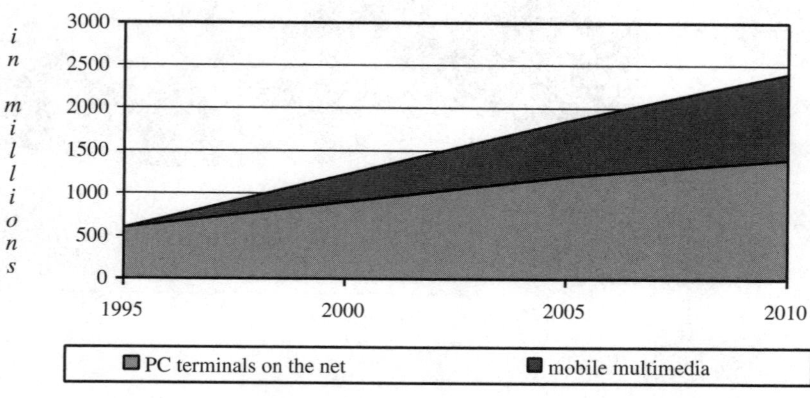

Figure 25-8 Growth figures of mobile multimedia devices on the Internet

of the growth and changes that will occur on the wireless Internet devices over the next decade.

Visions of Wireless

The convergence of high-speed wireless data communications, always-on mobile computing platforms, and instant access to the Internet is driving the biggest shift in mobile computing since the advent of the computer itself. The day is coming when the mobile professional will be linked wirelessly to the power of the Internet, no matter where on the globe he may find himself. He/she will need instant access to critical corporate, personal and public data applications, such as e-mail, e-commerce, stock trading, weather, airline reservations, hotels, car rental agencies, and sports information. The GSM/GPRS community becomes the gatekeeper to the needs of the true mobile professional.

There will be a consolidation in devices as manufacturers seek to meet customer needs while controlling costs. Smart phones and *Personal Digital Assistants* (PDAs) will become talkative PDAs. 3G Laptops and 3G Web Tablets, while keeping their separate identities, will both be handheld Internet devices, hitting the high and low end of 3G users. Like all of the 3G devices, the 3G Web Tablet will have a product lifecycle of its own through 2010. As innovative technology and customer demand cause the accumulation of all new device capabilities, the 3G multimedia device or 3G personal companion will become the sought after all-in-one mobile Internet tool for the middle of the decade. Others have predicted that by 2010, there will be a single wireless gadget that will meet all needs:

- **Smart phones/WAP phones** These early devices provide content and web browsing. They use standard and new operating systems and protocols (like Pocket PC and WAP) and will soon synchronize with other devices (like desktops and mobile phones). As WAP becomes popular and takes advantage of the high data rates and always on capability that GPRS will provide, these devices will naturally evolve into some of the first 3G devices at the even higher 3G data rates. The Smart phone will evolve to a talkative PDA.
- **Talkative PDA** Although there is room for improved coverage and quality, today you can purchase a PDA that also has mobile voice communications (such as, radio modems for GSM, OmniSky). Besides their calendars, address books, and other organizing features, PDAs are

thin and lightweight; many have color screens, and are quickly gaining computer strength due to low power chip designs, screen miniaturization, and evolving operating systems. As they grow in computing capability while maintaining their hand-held form factor, they will continue to distinguish themselves from 3G laptops as less expensive, less powerful solutions. Examples are numerous with Palm, Casio, HP, and others leading the pack. The Kyocera phone and PDA combination are a new twist in the movement toward the PDA.

- **3G laptop (handheld Internet)** Laptops today have modems and *Personal Computer Memory Card International Association* (PCMCIA) cards that enable wireless communications. They continue to get smaller, lighter, and with more powerful computing. With the bandwidth offered with 3G, these powerful, portable computers will thrive with the custom graphics, two-way video, and large file transfers for the future.
- **3G Web Tablet (handheld Internet)** Appearing in 2000, Wireless Web Tablets, these devices offer portable Internet access by plugging into power and access at home and gaining limited mobility via a short wireless connection. As low-cost, lightweight, thin Internet appliances the size of magazines, these devices offer e-mail, robust Internet access, and web browsing. Eventually, they will gain both full mobile access and synchronization with other devices via more powerful 3G spectrum.
- **3G multimedia device (personal companion)** Today's slow connections based on low bandwidth cause jerky video images. Compression techniques cannot overcome the need for speed and capacity. 3G will answer this problem in the mobile world. There are many visions of the ultimate 3G devices, with some saying it will evolve from phones, others from computers. Because there will be different 3G services addressing specific user needs, all of the previous devices will develop from both worlds. However, there will be a need for an all-powerful device that does quality VoIP, full Internet access, and two-way video.

In order to understand the role of next generation wireless services in the broader technology landscape, it is important to understand the current state of the Internet industry and other enabling technologies that shape its development. The Internet is transitioning from an inexpensive medium for advertising, marketing, and customer support to a common platform for transactions and business applications. At the same time, technological and commercial developments are melding together information, communications, commerce, and entertainment into one large, consolidated industry. Part of the reason for this evolution is because more consumers are accessing the Internet using multiple devices and over multiple communications networks. They are also changing their behavior and consumption patterns. In addition, tools and facilities are available that improve the consumer Internet access experience.

Wireless access to the Internet is going to drive the overall development of the Internet for several reasons:

- Wireless enables service providers and Internet businesses to increase their mobile culture and total service consumption.
- The mobility and immediacy offered by wireless enables Internet content delivery and commerce to be location independent.
- The person-specific nature of wireless enables companies to develop customer profiles that enable them to narrowcast and distribute better value-added information to customers.
- Location-based facilities and services provide another tier of customer knowledge that enables Internet businesses to deliver context-specific services that also improve customer value.

In short, wireless is an opportunity for Internet businesses to learn more about their customers, understand their customers' consumption patterns, strengthen their customer relationships, and provide more personalized services. This is a critical component of Internet business strategies and what wireless operators/service providers bring to the table in a full Internet solution.

The most important lifestyle users who draw on wireless access are as follows:

- **Business professionals** This includes the high-value mobile users (such as busy decision makers). The services used will include intranet access, messaging, and scheduling systems.
- **Product managers** These are users with specific occupational requirements for high volumes of information while in a mobile mode. Requirements include remote and mobile access to corporate and external information.
- **Youth** Often, early adopters of technology will be inclined to use instant messaging, games, and entertainment-oriented services.
- **Parents** In many countries, both parents work and share the household responsibilities.
- **Senior citizens** 3G will enable more reliable support electronically and reduce the requirement for labor-intensive support services such

as medical monitoring, location-based medical service, family, caretakers and caregivers, and social workers.

Positioning the Mobile Industry

In the light of the new business chain, the issue is to consider whether to simply provide a wireless IP pipe to a service offering hosted elsewhere on the Internet, or to go for an interoperable end-to-end solution. The wireless IP pipe business using tunneling protocols will become a commodity operation, where cost, coverage, and data rate are the only competitive dimensions. By carefully developing and preselecting useful Internet-based mobility services with competitive tariffs, the user will be encouraged to buy into UMTS services.

At issue is the location of the subscriber profile records, which reflect the personalized service choices of the end user: message filtering options, choice of mobile information, and type of mobile device—correlated with name, billing address, mobile phone number, and e-mail address. This store of data will permit additional returns through selectively targeted mobile e-commerce and advertising. UMTS operators have three separate or combined possibilities:

- They charge the subscriber on a metered call basis. This makes it possible to get a revenue stream via a small additional charge for mobile Internet services. It produces a return on investment well before mobile commerce and advertising become feasible. Added to the traffic revenue, the operator can capture or selectively share this revenue with value chain partners on its own terms.
- The operator provides IP-packet transport (such as GPRS based), which is necessary to integrate Internet services with Intelligent Networking, voice, data, and fax services. Volume discounts will become a possibility beyond billing for just time.
- The operator will know the subscriber's location using emerging cellular positioning technologies. Positioning adds end-user value through information customization, such as details of the nearest restaurant or automatic conversion of e-mail to speech for a driver of a moving car. In the future, location information will enormously increase the revenues.

Wireless Internet will become one of the media channels for content providers, and wireless network operators will join in on the service values.

The WAP-portal offerings of GSM operators and iMode offerings of NTT DoCoMo are examples of a new strategy. The mobile portals are unique because they are a solution in which operators and service providers can manage content and integrate with communication and transactions.

The arrival of WAP and iMode are generally seen as the first steps in the convergence between mobile telecommunications, the Internet, and content industries.

Key Technologies

Some of the critical technologies essential for the successful introduction of UMTS include the following.

UTRA

The ETSI decision in January 1998 on the radio access technique for UMTS combined two technologies—WCDMA for paired spectrum bands and TD-CDMA for unpaired bands—into one common standard. The network architecture is shown in Figure 25-9. This powerful approach promises an optimum solution for all the different operating environments and service needs. The transmission rate capability of UTRA will provide at least 144

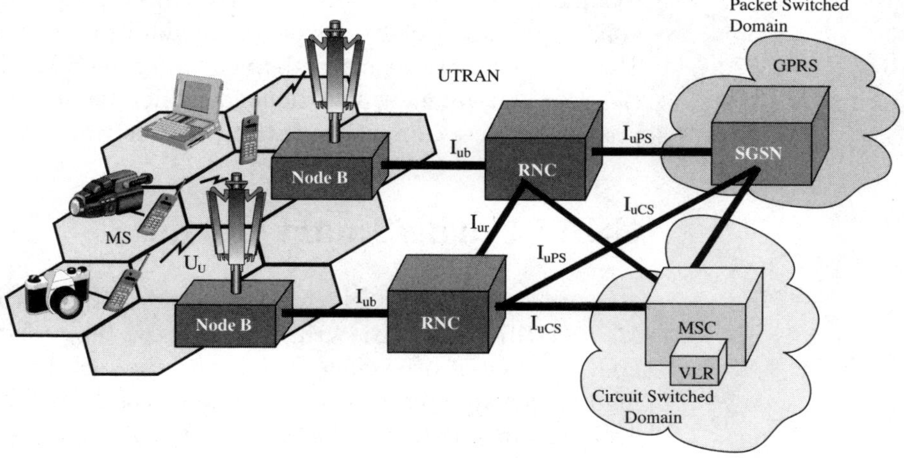

Figure 25-9 The architecture for 3G

Kbps for full mobility applications in all environments, 384 Kbps for limited mobility applications in the macro and micro cellular environments, and 2.048 Mbps for low mobility applications particularly in the micro- and pico-cell environments. The 2.048 Mbps rate may also be available for short range or packet applications in the macro-cellular environment, depending on deployment strategies, radio network planning, and spectrum availability.

Multimode Second Generation/UMTS Terminals

UMTS terminals will exist in a world of multiple standards, and this will enable operators to offer maximum capacity and coverage to their user base by combining UTRA with *second generation* (2G) and other 3G standards. Therefore, operators will need terminals that are able to interwork with legacy infrastructures, such as GSM/DCS1800 and DECT, as well as other 2G worldwide standards, such as those based on the U.S. AMPS standard, because they will initially have more complete coverage than UMTS. Many UMTS terminals will therefore be multiband and multimode so that they can work with different standards, old and new. Achieving such terminals at a cost that is comparable with contemporary single mode 2G terminals will become possible because of technological advances in semiconductor integration, radio architectures, and software radio.

Satellite Systems

At initial service launch in 2002, the satellite component of UMTS will be able to provide a global coverage capability to a range of user terminals. These satellite systems are planned using the S-band *Mobile Satellite Service* (MSS) frequency allocations identified for satellite IMT-2000 and will provide services compatible with the terrestrial UMTS systems.

USIM Cards/Smart Cards

A major step forward, which GSM introduced, was the *Subscriber Identity Module* (SIM) or smart card. It introduced the possibility of high security and a degree of user customization to the mobile terminal. SIM requirements, security algorithms, card, and silicon IC technology will continue to evolve up to and during the period of UMTS deployment.

The smart card industry will offer cards with greater memory capacity, faster CPU performance, contactless operation, and greater capability for encryption. These advances will enable the *UMTS Subscriber Identity Module* (USIM) to add to the UMTS service package by providing portable high-security data storage and transmission for users. As well as configuration software for the operation of any UMTS terminal, images, signatures, personal files, and fingerprints or other biometric data could be stored, and then down- or up-loaded to or from the card.

Contactless cards will permit much easier use than with today's cards, for example, enabling the smart card to be used for financial transactions and management, such as electronic commerce or electronic ticketing, without having to be removed from a wallet or phone. It is expected that all fixed and mobile networks will adopt the same or compatible lower layer standards for their subscriber identity cards to enable USIM roaming on all networks and universal user access to all services. Electronic commerce and banking using smart cards will soon become widespread, and users will expect and be able to use the same cards on any terminal over any network.

New memory technologies can be expected to increase card memory sizes, making larger programs and more data storage feasible. Several applications and service providers could be accommodated on one card. In theory, the user could decide which applications/services he wants on the card, much as he does for his computer's hard disk. This is the challenge and opportunity for service industries that evolving smart card technology presents.

IP Compatibility

UMTS is a modular concept that makes use of the trend towards the convergence of fixed and mobile networks and services, enabling a huge number of applications to be developed. As an example, a laptop with an integrated UMTS communications module becomes a general-purpose communications and computing device for broadband Internet access, voice, video telephony, and conferencing for either mobile or residential use.

The number of IP networks and applications is growing fast. Most obvious is the Internet, but private IP networks (intranets) show similar or even higher rates of growth and usage. UMTS will become the most flexible broadband access technology because it allows for mobile, office, and residential use in a wide range of public and nonpublic networks. UMTS can support both IP and non-IP traffic in a variety of modes, including packet, circuit switched, and virtual circuit.

UMTS will be able to benefit from parallel work by the *Internet Engineering Task Force* (IETF) while they further extend their basic set of IP standards for mobile communication. New developments like IP version 6 enables parameters, such as *quality of service* (QoS), bit rate, and *bit error rates* (BERs), vital for mobile operation, to be set by the operator or service provider. Developments on new domain name structures are also taking place. These new structures will increase the usability and flexibility of the system, providing unique addressing for each user, independent of terminal, application, or location.

UMTS has the support of many major telecommunications operators and manufacturers because it represents a unique opportunity to create a mass market for highly personalized and user-friendly mobile access to the information society. UMTS seeks to build on and extend the capability of today's mobile, cordless, and satellite technologies by providing increased capacity, data capability, and a far greater range of services using an innovative radio access scheme and an enhanced, evolving core network.

Spectrum for UMTS

WRC 2000 identified the frequency bands 1885 to 2025 MHz and 2110 to 2200 MHz for future IMT-2000 systems, with the bands 1980 to 2010 MHz and 2170 to 2200 MHz intended for the satellite part of these future systems. CDMA is characterized by high capacity and small cell radius, employing spread-spectrum technology, and a special coding scheme.

The capabilities of cdmaOne evolution have already been defined in standards. IS-95B provides ISDN rates up to 64 Kbps. The next phase of cdmaOne is a standard knows as 1XRTT and enables 144 Kbps packet data in a mobile environment. Other features available include a two-fold increase in both standby and talk time on the handset. All of these capabilities will be available in an existing cdmaOne 1.25 MHz channel. The next phase of cdmaOne evolution will incorporate the capabilities of 1XRTT, support all channel sizes (5 MHz, 10 MHz, and so on), provide circuit and packet data rates up to 2 Mbps, incorporate advanced multimedia capabilities, and include a framework for advanced 3G voice services and vocoders, including voice over packet and circuit data. Many of the steps have already been taken.

There are now a number of flavors of CDMA as shown in Table 25-1.

Table 25-1

Variations of CDMA

CDMA Type	Description
Composite CDMA/TDMA	Wireless technology that uses both CDMA and TDMA. For large-cell licensed band and small-cell unlicensed band applications. Uses CDMA between cells and TDMA within cells.
CDMA	In addition to the original Qualcomm-invented N-CDMA (originally just CDMA) also known in the United States as IS-95. Latest variations are B-CDMA, WCDMA, and composite CDMA/TDMA. CDMA is characterized by high capacity and small cell radius, employing spread-spectrum technology and a special coding scheme. B-CDMA is the basis for 3G UMTS.
CdmaOne	*First-Generation* (1G) Narrowband CDMA (IS-95).
cdma2000	The new 2G CDMA *Memorandum of Understanding* (MoU) specification for inclusion in UMTS.

The cdma2000 Family of Standards

The cdma2000 family of standards includes core air interface, minimum performance, and service standards. The cdma2000 air interface standards specify a spread spectrum radio interface that uses CDMA technology to meet the requirements for 3G wireless communication systems. In addition, the family includes a standard that specifies analog operation to support dual-mode MSs and *base stations* (BSs).

Purpose

The technical requirements contained in cdma2000 form a compatibility standard for CDMA systems. They ensure that a MS can obtain service in a system manufactured in accordance with the cdma2000 standards. The requirements do not address the quality or reliability of that service, nor do they cover equipment performance or measurement procedures. Compatibility, as used in connection with cdma2000, is understood to mean any cdma2000 MS is able to place and receive calls in cdma2000 or IS-95 systems. Conversely, any cdma2000 system is able to place and receive calls for cdma2000 and IS-95 MSs. In a subscriber's home system, all call placements are automatic. Similarly, call placement is automatic when a MS is roaming. To ensure compatibility, radio system parameters and call processing procedures are specified. The sequence of call processing steps that the MSs and BSs execute to establish calls is specified, along with the digital control messages and, for dual-mode systems, the analog signals that are exchanged between the two stations.

The BS is subject to different compatibility requirements than the MS. Radiated power levels, both desired and undesired, are fully specified for MSs in order to control the RF interference that one MS can cause another. BSs are fixed in location, and their interference is controlled by proper layout and operation of the system in which the station operates. Detailed call processing procedures are specified for MSs to ensure a uniform response to all BSs. BS procedures that do not affect the MSs' operation are left to the designers of the overall land system. This approach to writing the compatibility specification is intended to provide the land system designer with sufficient flexibility to respond to local service needs and to account for local topography and propagation conditions. cdma2000 includes provisions for future service additions and expansion of system capabilities. The release of the cdma2000 family of standards supports Spreading Rate 1 and Spreading Rate 3 operation.

CHAPTER 26

Satellite Communications Networking

Satellite communications have developed through the years. The first use of satellite technology was in the military for voice communications in the early 1960s. Despite the advancements in the technology, commercial providers are prohibited from constructing and launching satellites at will. Despite these limitations, commercial satellites provide valuable information on meteorology, agriculture, forestry, geology, environmental science, and other areas.

Uses of Satellites in Agriculture

Many countries use satellite technology to improve agriculture. Satellites are used to determine productive from nonproductive yields. In many cases, they use other services such as infrared vision. Satellite-based navigation, such as the *Global Positioning System* (GPS), a military invention, also assists in the process of reviewing agricultural production. Farmers can keep track of what they plant, fertilize, and spray. A farmer can now account for different kinds of soil, varying in acidity, organic content, and nitrogen levels. In addition, differences in drainage and treatment can be determined. All this improves yield and prevents the waste of valuable chemicals. This practice is well established in America. More than a dozen companies sell equipment used in agriculture.

Uses of Satellites in Oceanography

Satellites have found a niche in oceanography. In regional and local seas, satellites provide remote sensing, which gives a unique approach to monitoring the environment for the management of living resources, fishing, navigation, and coastal erosion. All maritime nations now have access to valuable information needed to manage the ocean boundaries that comprise their economic zone.

Countries use satellites in many adaptations. From military applications, agriculture, and oceanography, satellites have improved the overall performance of individual countries. Each of the applications listed, however, uses significant amounts of data to perform and monitor these functions.

Commercial Providers

NASA has been working with the U.S. commercial satellite industry to achieve seamless interoperability of satellite and terrestrial telecommunications networks. The high latency and noise characteristics of satellite links cause inefficiencies when using *Transmission Control Protocol* (TCP) in an environment for which it was not designed. These problems are not limited to airborne communications and may appear in other high latency and noisy networks. Considerations, such as higher data rates and more users on data networks, can affect future terrestrial networks similar to those posed by satellites today.

In the future, telecommunications networks with space stations, lunar stations, and planetary stations with their inherent latency, noise, and bandwidth characteristics need to be addressed. For now, the emphasis is on earth-orbiting communications satellites.

History of Satellites

The regular use of satellite communications flourished in the 1960s. In April 1960, Tiros I was sent out to space. Tiros was the first weather satellite that sent pictures of clouds to earth. In August 1960, the United States launched Echo I, a satellite that made satellite communication possible by reflecting radio signals to earth. By 1965, more than 100 satellites were being launched into orbit each year. The 1970s produced innovation in the satellite world. Newer and more efficient instruments were developed. These innovations used computers and miniature electronics in designing and constructing satellites. During the 1980s, satellites were used to save people and other satellites. The uses of satellites in the 1990s rapidly grew for common, everyday tasks. TRW Inc. announced a plan to create a satellite system that would dominate satellite communications. This system called Odyssey would be used for the telephone business. TRW's satellites would focus on populated areas and would cover the earth uniformly. Based on current development, the satellite industry will be productive for decades to come.

How Do Satellites Work?

The basic elements of a satellite communications system are shown in Figure 26-1. The process begins at an earth station—an installation designed to transmit and receive signals from a satellite in orbit around the earth. Earth stations send information via high-powered, high-frequency (GHz range) signals to satellites, which receive and retransmit the signals back to earth where they are received by other earth stations in the coverage area of the satellite. The area that receives a signal of useful strength from the satellite is known as the satellite's footprint. The transmission system from the earth station to the satellite is called the *uplink*, and the system from the satellite to the earth station is called the *downlink*.

Satellite Frequency Bands

The three most commonly used satellite frequency bands are the C-band, Ku-band, and Ka-band. C-band and Ku-band are the two most common fre-

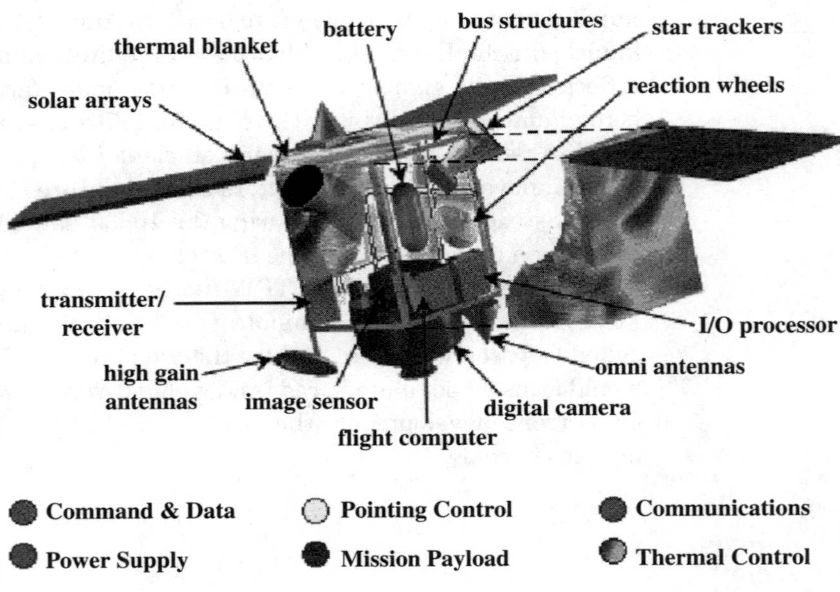

Figure 26-1 Satellite communication basics

quency spectrums used by today's satellites. To help understand the relationship between antenna diameter and transmission frequency, it is important to note that there is an inverse relationship between frequency and wavelength—when frequency increases, wavelength decreases. As wavelength increases, larger antennas are required to receive the signal.

C-band satellite transmissions occupy the 4- to 8-GHz frequency ranges. These relatively low frequencies translate to larger wavelengths than Ku-band or Ka-band. These larger wavelengths of the C-band mean that a larger satellite antenna is required to gather the minimum signal strength. The minimum size of an average C-band antenna (shown in Figure 26-2) is approximately 2 to 3 meters in diameter.

Ku-band satellite transmissions occupy the 11- to 17-GHz frequency ranges. These relatively high frequency transmissions correspond to shorter wavelengths, and therefore a smaller antenna can be used to receive the minimum signal strength. Ku-band antennas can be as small as 18 inches in diameter. Figure 26-3 shows the Ku-band antenna.

Ka-band satellite transmissions occupy the 20- to 30-GHz frequency range. These very high frequency transmissions mean very small wavelengths and very small diameter receiving antennas.

Figure 26-2 C-band satellite antenna

Figure 26-3 Ku-band satellite antenna

Geosynchronous-Earth-Orbit (GEO) Satellites

Today, the overwhelming majority of satellites in orbit around the earth are positioned at a point 22,300 miles above the earth's equator in *Geosynchronous Earth Orbit* (GEO). At a distance of 22,300 miles, a satellite can maintain an orbit with a period of rotation around the earth exactly equal to 24 hours (see Figure 26-4). Because the satellites revolve at the same rotational speed of the earth, they appear stationary from the earth's surface. That's why most earth station antennas (satellite dishes) don't need to move after they have been properly aimed at a target satellite in the sky.

Medium-Earth-Orbit (MEO) Satellites

During the last few years, technological innovations in space communications have given rise to new orbits and totally new system designs. *Medium-Earth-Orbit* (MEO) satellite networks have been proposed that will orbit at distances of about 8,000 miles. Signals transmitted from a MEO satellite travel a shorter distance, which translates to improved signal strength at the receiving end. This means that smaller, lighter receiving terminals can be used. Also, because the signal is traveling a shorter dis-

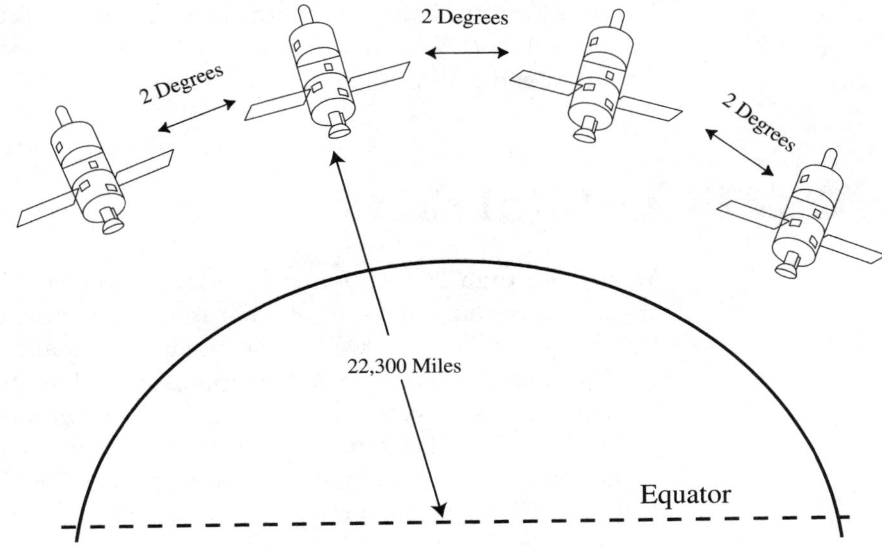

Figure 26-4 Geosynchronous Orbit

tance to and from the satellite, there is less transmission delay. Transmission delay is the time it takes for a signal to travel up to a satellite and back down to a receiving station. For real-time communications, the shorter the transmission delays, the better. For example, a GEO satellite requires .25 seconds for a round trip. A MEO satellite requires less than .1 seconds to complete the job. MEOs operate in the 2 GHz and above frequency range.

Low-Earth-Orbit (LEO) Satellites

Proposed *Low-Earth-Orbit* (LEO) satellites are divided into three categories: little LEOs, big LEOs, and mega-LEOs. LEOs will orbit at a distance of only 500 to 1,000 miles above the earth. This relatively short distance reduces transmission delay to only .05 seconds and further reduces the need for sensitive and bulky receiving equipment. Little LEOs will operate in the 800 MHz (.8 GHz) range, big LEOs will operate in the 2 GHz or above range, and mega-LEOs will operate in the 20 to 30 GHz range. The higher frequencies associated with mega-LEOs translate into more information carrying capacity and the capability of real-time, low-delay video transmission. Microsoft Corporation and McCaw Cellular (now known as AT&T

Wireless Services) have partnered to deploy 288 satellites to form Teledesic, a proposed mega-LEO satellite network. Speeds are expected at 64 Mbps downlink and 2 Mbps uplink.

Orbital Slots

With more than 200 satellites in geosynchronous orbit, how do we keep them from running into each other or from attempting to use the same location in space? To tackle that problem, international regulatory bodies like the ITU and national government organizations like the FCC designate the locations on the geosynchronous orbit, where communications satellites can be located. These locations are specified in degrees of longitude and are known as orbital slots. In response to the huge demand for orbital slots, the FCC and ITU have progressively reduced the required spacing down to only 2 degrees for C-band and Ku-band satellites.

Communications

The communications subsystem of a satellite is essential to the function of all satellites. There are many different components used in spaceborne communications' subsystems including the following:

- Special antennas
- Receivers
- Transmitters

All components must be highly reliable and low weight. Most satellites are fitted with beacons or transponders, which help with easy ground tracking. Global receive and transmit horns receive and transmit signals over wide areas on earth. Many satellites and ground stations have radio dishes that transmit and receive signals to communicate. The curved dishes reflect outgoing signals from the central horn and reflect incoming signals. A satellite's ability to receive signals is also necessary in order to trigger the return of data or to correct a malfunction if possible.

Satellite Installations

In Table 26-1, a list of the satellite installations by country is shown. The list shows the main providers and users of satellite transmission systems.

The technological and regulatory hurdles to create true high-speed satellite networks are fast becoming past tense. Low- and mid-bandwidth systems such as Motorola's Iridium and Hughes' DirecPC can handle some of the needs immediately. These systems are nothing compared to the promise of 2 Mbps, 20 Mbps, and even 155 Mbps streaming down from the sky. All that is necessary is a small antenna, a black box, and a service provider. This will approximate the way we buy service from an ISP today.

Is it ready for prime time yet? Not yet! Iridium's universal telephone didn't kill the cellular telephone. So broadband satellite systems won't kill terrestrial lines. Broadband satellite creators agree that broadband satellite systems will complement terrestrial networks. These satellites will provide high-speed service where terrestrial infrastructure does not exist. However, high-speed, low-cost landlines are here to stay.

Is there an application for the high-speed satellite networks? What makes them different from each other? Each of the main systems is very different. Some of the most visible ones may prove the most difficult to implement. Some of the most staid-looking systems may beat every other system to the punch.

Satellite communications are nothing new. For years, you could hook up a *Very Small Aperture Terminal* (VSAT) system and buy time on a satellite. The VSAT can deliver up to 24 Mbps in a point-to-multipoint link (for example, a multicast) and up to 1.5 Mbps in a point-to-point link. These systems are fine for predictable and quantifiable data transmission, but if you need to conduct business on the fly, they can be a problem. New techniques are required to handle the on-demand usage. Primary among them are more tightly focused beams and digital signal technology, which together can increase frequency reuse (and thereby increase bandwidth) and reduce dish size from meters to centimeters. Accordingly, you also need a large unused chunk of the electromagnetic spectrum.

In 1993, NASA launched its *Advanced Communication Technology Satellite,* (ACTS). ACTS is an all-digital, Ka-band (20 to 30 GHz), spot-beam, GEO satellite system capable of delivering hundreds of megabits per second of bandwidth. NASA showed that such a system could work. The FCC has

Table 26-1

Number of satellites in orbit by country around the world

Country	In Orbit
Argentina	1
Australia	6
Brazil	6
Bulgaria	1
Canada	16
China	15
CIS *(Former Soviet Union)*	1,322
Czechoslovakia	1
France	24
Germany	15
Hong Kong	1
India	11
Indonesia	6
Israel	1
Italy	4
Japan	55
Korea (South)	2
Luxembourg	4
Mexico	4
Portugal	1
Spain	3
Sweden	4
Thailand	2
Turkey	1
United Kingdom	18
United States	658
Arab States	3
Europe	27
International	51
NATO	8
TOTAL	2,271

since granted orbital locations and Ka-band licenses to 13 companies including the following:

- EchoStar
- Hughes
- Loral
- Motorola
- Ka-Star
- NetSat 28
- PanAmSat
- Teledesic

All these companies aim to bring us bandwidth up to 155 Mbps to our home and office. These broadband systems are not going online before 2000.

What will we do with this speed and capacity? Anything you want! Whatever landlines can do, the satellite systems of the new millennium will also be able to do. In Table 26-2, the list of applications covered by broadband satellite communications is shown. These are representative applications; others will be developed.

Table 26-2

Applications for high-speed satellite communications

Applications
Desk-to-desk communications
Videoconferencing
High-speed Internet access
E-mail
Digital and color fax
Telemedicine
Direct TV and video
Transaction processing
Interactive distance learning
News and financial information
Teleradiology

Most of the market that needs data services seems to be well served by landlines. So, why is the emphasis on the use of airborne technology? An obvious market is in places that have underdeveloped communications infrastructures. In some countries, stringing copper or fiber is out of the question because of the initial cost and the terrain where it is needed. Still, a wireless telephone has some merit. You don't need a broadband satellite to make telephone calls; Iridium and other LEO systems will likely serve that market.

So who *does* need this new class of broadband satellite communications? The answer is multinational corporations. The main problem that satellite systems solve is getting high-bandwidth access to places without a high-bandwidth infrastructure. It's unlikely that a satellite system could compete with *Digital Subscriber Line* (xDSL) to the home or fiber to the office.

LEO Versus GEO

However, bandwidth is only half the story. The other half is latency. It's easy to talk about high-bandwidth satellite systems, but that technology has existed in VSATs for years. GEO satellites located at 22,300 miles above the equator induce 250 milliseconds (ms) of round trip delay. With that kind of latency built into the system, (not counting latency added by the various gateways and other translations), a telephone conversation is annoying. Any interactive data intense application has to be nonlatency-sensitive. Online transaction processing will have a problem using a GEO satellite system.

Moving the satellites closer to earth will help significantly. That's just what systems such as Teledesic[1], Skybridge[2], and Celestri[1] will do. With LEOs under 1,000 miles, these systems reduce latency to .1 second. While GEOs are a well-known technology, LEOs are new. The biggest problem is that you need a lot of them to get global coverage. At one point, Teledesic planned a constellation of more than 842 satellites.[3] Until recently, the concept of launching dozens or hundreds of multimillion-dollar satellites was a pipe dream. Each of Teledesic's 288 satellites is estimated to cost $20 million.

---

[1] Teledesic is a joint venture between Microsoft and McCaw Cellular.

[2] Skybridge is a venture of Alcatel of Belgium.

[3] Teledesic dropped the number of required LEO satellites to 288 since the first announcement.

Price is only one issue. Finding a company to launch all these satellites poses another obstacle. Teledesic set an 18-month to 2-year launch window to get its 288 satellites airborne. LEO system planners are talking about putting more satellites into orbit in the next 5 years than the world has put into orbit over the past 40 years. Once the LEO satellites are in orbit, there's an entirely new set of problems. There's the matter of space junk.

Niches in the GEO Sphere

LEOs will be good for high-speed networking, teleconferencing, Telemedicine, and interactive applications. GEOs will be better for information downloading and video distribution, such as broadcasting and multicasting. We're able to use GEO satellites to transport at least 24 Mbps of broadcast IP data and over 2 Mbps of point-to-point TCP/IP data. The latter uses technologies such as TCP spoofing. Several vendors have been using this technique for years to deliver Internet and intranet content at high speed. Ground terminals can use similar TCP spoofing technologies. Nevertheless, there's still the 250 ms delay that you just can't get around. Any lossless protocol is going to have problems with this latency. Even if TCP spoofing works, TCP's 64 Kb buffer makes this somewhat suspect, there's the matter of other protocols such as IBM's SNA and other real-time protocols that are designed around landline performance.

LEO Meets GEO

Motorola's Celestri plans an initial LEO constellation of 63 satellites coupled with one GEO satellite over the United States. The LEO constellation and the GEO satellites will be able to communicate directly through a satellite-to-satellite network.

The hybrid configuration will enable Celestri to take advantage of LEO's shorter delays for interactive uses and GEO's power in the broadcast arena.

Space Security Unit

Once you get beyond the latency and bandwidth issues, there is another challenge: security. If data is being packaged up and broadcast into space, anybody with a scanner can just tune in. The air interface technologies that these systems use will make it more difficult for anyone to eavesdrop. Combined systems will use *Code Division Multiple Access* (CDMA), *Time Division Multiple Access* (TDMA), *Frequency Division Multiple Access* (FDMA), and a bunch of other xDMA protocols. On top of that, many of the networks will offer some internal security encryption system.

The Market for the Network

The total global telecommunications market is about $750 billion, and that's going to double in 10 years, chiefly due to expanded use of data communications. We'll use whatever service we have available to meet that demand: fiber, ATM, *Synchronous Optical Network* (SONET), xDSL, Gigabit Ethernet, cable modems, satellites, and probably a few that haven't even been thought of yet.

All telecommunications systems will compete on availability, price, and speed. That means there are going to be two big winners: whoever gets its broadband service to consumers first, and whoever can offer the most bandwidth and at least reasonable latency. A sample of the techniques and the services being offered are shown in Table 26-3.

With the discussion of the various bands that these services will operate in, the bands are continually referred to. As a means of showing the bands and the frequency they fall into, Table 26-4 is a summary of some of the bands used in the satellite arena.

Table 26-5 shows a sample of the satellite technologies that use the various bands and the size of the terminal and dish in satellite technologies.

Satellite Characteristics

As already mentioned, satellite communications have three general characteristics that lead to interoperability problems with systems that have not been designed to accommodate them:

Table 26-3

Sampling of services planned or offered

Service Offering	Astrolink	Teledesic	Celestri
Companies (parties) involved	Lockheed	Bill Gates, Craig McCaw, Boeing	Motorola
Service offerings	Data, video, rural telephony	Voice, data, video conferencing	Voice, data, video conferencing
Satellite orbit	22,300 miles	435 miles	875 and 22,300 miles
Satellite band	Ka	Ka	Ka and also 40–50 GHz
Dish size	33–47 inches	10 inches	24 inches
Bandwidth	Up to 9.6 Mbps	16 Kbps−64 Kbps or up to 2.048 Mbps	Up to 155 Mbps transmit and receive
Available (approx.)	Late 2000	2002	2002
Number of satellites	9 GEOs	288 LEOs	63 LEOs and up to 9 GEOs
Access technologies	FDMA, TDMA	MF-TDMA, ATDMA	FDMA, TDMA
Intersatellite communications	Yes	Yes	Yes

1. Delay (or latency)
2. Noise
3. Limited bandwidth

It is important to understand each of these in general terms and in more detail. Since the invention and installation of fiber optics, communications systems have been treated as ideal with very low latency, no noise, and nearly infinite bandwidth (the capacities are rapidly moving to terabit bandwidth from current fiber systems). These characteristics still have room for doubt and development, but for satellite communications, the characteristics of fiber can make it very difficult to provide cost-effective interoperability with land-based systems. The latency problem is foremost among the three. Noise can be handled by the application of error control coding. Bandwidth efficiency is an important goal for satellite systems today, but will become increasingly important as the number of users and data requirements increase.

Table 26-4
Sample of the frequency bands

Band	Frequency Ranges Used
HF-band	1.8–30 MHz
VHF-band	50–146 MHz
P-band	0.230–1.000 GHz
UHF-band	0.430–1.300 GHz
L-band	1.530–2.700 GHz
FCC Digital Radio and PCS	2.310–2.360 GHz
S-band	2.700–3.500 GHz
C-band	Downlink: 3.700–4.200 GHz Uplink: 5.925–6.425 GHz
X-band	Downlink: 7.250–7.745 GHz Uplink: 7.900–8.395 GHz
Ku-band (Europe)	Downlink: FSS: 10.700–11.700 GHz DBS: 11.700–12.500 GHz Telecom: 12.500–12.750 GHz Uplink: FSS and Telecom: 14.000–14.800 GHz DBS: 17.300–18.100 GHz
Ku-band (America)	Downlink: FSS: 11.700–12.200 GHz DBS: 12.200–12.700 GHz Uplink: FSS: 14.000–14.500 GHz DBS: 17.300–17.800 GHz
Ka-band	Roughly 18–31 GHz

Latency

We have seen that there is an inherent delay in the delivery of a message over a satellite link, due to the finite speed of light and the altitude of communications satellites. As we stated, there is approximately a 250 ms propagation delay in a GEO. These delays are for one ground station-to-satellite-to-ground station route (or *hop*). The round-trip propagation delay for a message and its reply would be a maximum of 500 ms. The delay will be proportionately longer if the link includes multiple hops or if intersatellite links are used. As satellites become more complex and include on-board processing of signals, additional delay may be added.

Table 26-5 Summary of bands and types of terminals used

System Type	Frequency Bands	Applications	Terminal Type/Size	Examples
Fixed satellite service	C and Ku	Video delivery, VSAT, news gathering, telephony	1 meter and larger fixed earth station	Hughes Galaxy, Intelsat
Direct broadcast satellite	Ku	Direct-to-home video/audio	0.3–0.6 meter fixed earth station	DirecTV, Echostar, USSB
Mobile satellite (GEO)	L and S	Voice and low-speed data to mobile terminals	Laptop computer/ antenna-mounted, but mobile	Inmarsat, AMSC
Big LEO	L and S	Cellular telephony, data, paging	Cellular phone and pagers; fixed phone booth	Iridium, GlobalStar
Little LEO	P and below	Position location, tracking, messaging	6 inches, omnidirectional	OrbComm
Broadband GEO	Ka and Ku	Internet access, voice, video, data	20 cm, fixed	Spaceway, Cyberstar, Astrolink
Broadband LEO	Ka and Ku	Internet access, voice, video, data, video conferencing	Dual 20 cm tracking antennas, fixed	Teledesic, Skybridge, Celestri, Cyberstar

Other orbits are possible including LEO and MEO. The advantage of GEO is that the satellite remains stationary over one point of the earth, allowing simple pointing of large antennas. The lower orbits require the use of constellations of satellites for constant coverage, and they are more likely to use intersatellite links, which result in variable path delay depending on routing through the network.

Noise

The strength of a radio signal falls in proportion to the square of the distance traveled. For a satellite link, this distance is very large and so the signal becomes very weak. This results in a low, signal-to-noise ratio. Typical bit error rates for a satellite link might be on the order of 10^{-7}. Noise becomes less of a problem as error control coding is used. Error performance equal to fiber is possible with proper error control coding.

Bandwidth

The radio spectrum is a finite resource, and there is only so much bandwidth available. Typical carrier frequencies for current point-to-point (fixed) satellite services are 6/4 GHz (C band), 14/12 GHz (Ku band), and 30/20 GHz (Ka band). Traditional C- and Ku-band transponder bandwidth is typically 36 MHz to accommodate one color television channel (or 1,200 voice channels). One satellite may carry two dozen or more transponders.

New applications, such as personal communications services, may change this picture. Bandwidth is limited by nature, but the allocations for commercial communications are limited by international agreements so that the scarce resource can be used fairly. Applications that are not bandwidth efficient waste a valuable resource, especially in the case of satellites. Scaleable technologies will be more important in the future.

Advantages

Although satellites have certain disadvantages, they also have certain advantages over terrestrial communications systems. Satellites have a natural broadcast capability. Satellites are wireless, so they can reach geographically remote areas.

There are major changes occurring in the design of communications satellite systems. New bandwidth-on-demand services are being proposed for GEO. The Federal Communications Commission constantly receives filings for new Ka-band satellite systems each year.

TCP/IP over Satellite

There has been a fair amount of press about TCP/IP, the networking protocol that the Internet and the *World Wide Web* (WWW) rely on, not working properly over satellite. The claims are that the TCP/IP reference implementation's 4K buffer size limits channel capacity and data throughput to only 64 Kbps. The argument is then used that the maximum buffer of 64K limits maximum throughput to 1 Gbps. Therefore, the argument continues that GEO Ka-band satellite services are unsuitable for high-bandwidth applications, as the increased latency of a GEO connection decreases the available bandwidth.

TCP buffers are dimensioned as:

$$\text{bandwidth} \times \text{delay} = \text{buffer size}$$

With a limited buffer size, a longer end-to-end delay decreases the space available to hold spare copies of unacknowledged data for retransmission. This limits the throughput on a lossless TCP connection.

However, this argument ignores the work done on larger buffer sizes for TCP in RFC1323, the *large windows* effort. Work to expand TCP beyond its original, 16-bit buffer space has been going on for several years. Moreover, the effort and result of large windows is supported by several versions of UNIX. The TCP buffer limit isn't as bad as it is made out to be. TCP copes with GEO delays quite nicely today. Individual, high-bandwidth GEO TCP links are possible with the right equipment and software. Military applications have been using TCP/IP in many GEO-based networks for years.

GEO links are suitable for seamless intermediate connections in a TCP circuit. There is nothing denying the use of many small TCP connections over a broadband link; GEO, fiber, and most broadband Internet connections contain a vast number of separate small pipes. The real issue with GEO versus LEO is the acceptability of the physical delay for two-way, real-time applications such as telephony or video conferencing. Even then, the physical delay of GEO is balanced somewhat by the increased switching times through a packet-based LEO network. Having multiple, narrow TCP pipes across satellites works well. Wide pipes with large buffer sizes can suffer from the higher *bit error rate* (BER) of satellites.

For TCP, implementing Selective Acknowledgements (RFC2018) and fast recovery (RFC2001) also improves performance in the face of errors. There is much work in progress with the IP-over-Satellite and the TCP-over-Satellite working groups.

IP-over-satellite's primary advantage is that it can deliver large amounts of bandwidth to underdeveloped areas where fiber does not exist. IP-over-satellite is also subject to a number of adverse conditions that can significantly impact the effective throughput rates and network efficiency. BERs, congestion, queue management, window size, and buffer status can all have a serious impact on the overall IP-over-satellite performance curves.

Another problem is the asymmetry encountered in back channel bandwidth, which in many IP-over-satellite systems is only a fraction of the forward channel bandwidth. Because TCP emerged in advance of IP-over-satellite, TCP often gets mistakenly identified as the source of the problem. It is the operating environment and not TCP alone that is the problem. The design of TCP is not optimized for conditions encountered during satellite transmissions.

The standard GEO round trip hop of 44,600 miles creates problems with the inherent transmission-related delay, breaking down the flow of packets between the sender and receiver, especially at very high transmission speeds. TCP works well as a general-purpose protocol for a congested environment, such as the Internet. However, TCP perceives a satellite delay and bit error as congestion and inherently slows down.

Satellite and ATM

Work has also been done on ATM over satellite. Standardization of ATM over satellite is underway and several organizations are working in parallel. The ATM Forum's *Wireless-ATM* (WATM) group has done some work on ATM over satellite. The *Telecommunications Industries of America* (TIA) Communications and Interoperability group (TIA-TR34.1) works on interoperability specifications that facilitate ATM access and ATM network interconnect in both fixed and mobile satellite networks. TIA is collaborating with the ATM-Forum WATM group. The standard development for satellite ATM networks is provided in ATM Forum contribution 98–0828. Other groups working on ATM and satellite transmission include the following:

- **NASA** NASA is working on the interoperability of satellite networks with terrestrial networks. ATM over satellite QoS results for Motion-JPEG is now publicly available.
- **European Space Research and Technology Center** This group is also interested in ATM-over-satellite research. They have carried out a broadband communications over satellite study.
- **COMSAT** COMSAT is one of the major players in providing ATM-over-satellite services.
- **Communications Research Center** Communications Research Center Broadband Applications and Demonstration Laboratory is

testing and demonstrating information highway applications by using ATM via satellite.

- **RACE Research and Development in Advanced Communications Technologies** In Europe, this organization has also funded various projects for ATM over satellite. The results of the RACE research projects are published as *Common Functional Specifications* (CFS).

Charting the Rules for the Internet

Everyone knows about the Internet. Fewer people, however, understand how it works. Very few people know where the standards come from that shape how the Internet works. Those rules come from five groups, four of them loosely organized and consisting primarily of volunteers. For more details see Chapter 32 in this book, "The Internet".

The groups don't issue standards in the traditional sense, although their conclusions become the regulations by which the Internet operates worldwide. Instead, they agree on how things should be done through discussion. The result is a surprisingly effective, if sometimes convoluted, system for dealing with the fast-growing and often unpredictable nature of the Internet. Like the Internet itself, the groups that oversee it have made recent innovations that continue to evolve and restructure.

The basic rules governing the Internet come out of two task forces:

1. The *Internet Research Task Force* looks at long-term research problems.
2. The *Internet Engineering Task Force* (IETF) concentrates on more immediate issues such as data transfer protocols and near-term architecture.

IETF is the oldest group dedicated to Internet issues, having held its first meeting in 1986. At the time, the nascent Internet was still primarily a federal research activity evolving out of ARPANET, a modest network developed by the U.S. *Defense Advanced Research Projects Agency* (DARPA). The IETF's work is divided among eight functional areas related to such issues as network management, routing, security, applications, and so on.

The activities of IETF and the Internet Research Task Force are coordinated by the Internet Architecture Board. The board has no official role in operational matters and contributes in only a minor way to policy issues.

The Internet Architecture Board and two Internet task forces fall loosely under the umbrella of a fourth organization that is more formal, called the Internet Society.

The Internet links thousands of networks and millions of users who communicate by using hundreds of different types of software. They are able to do so because of *Transmission Control Protocol / Internet Protocol* (TCP/IP), the general procedure for accurately exchanging packets of data adopted by ARPANET in the early 1980s. Other networks soon adopted TCP/IP as well, thus paving the way for the global Internet of today. TCP/IP has been updated several times. It continues to undergo modifications, including those designed to speed throughput, especially for long terrestrial and satellite links.

Tailoring IP Can Accelerate Throughput

Everyone who uses the Internet wants faster connections. The quest for speed has become a big marketing issue for terrestrial and satellite systems alike. However, no one system is inherently best for every application. Finding the most efficient way to connect is a matter of matching communication needs to the unique characteristics of each option. The hardware involved is an obvious factor in system speed. Nevertheless, the protocols used by computers to talk to each other ultimately control throughput. Internet protocols are constantly being revised, and faster versions are coming that promise quicker data delivery over links of all kinds. Primarily the IETF coordinates the work. The IETF seeks continual improvement in the software that makes the Internet work, and it prefers solutions that benefit terrestrial as well as satellite network links.

The Internet works because all computers using it abide by the same rules, or protocols. The most fundamental of these rules is the *Internet Protocol* (IP). Its primary function is to provide the *datagram* that carries the address to which a computer is sending a message.

Guaranteeing that messages are received accurately over the Internet is the primary job of the TCP. TCP sends data in segments and waits for a confirmation message from the receiving computer before sending more. It reacts to traffic loads on the network, dynamically regulating the allowable rate of data transfer between the two computers in an attempt to maximize data flow without overwhelming relay points or the capacity of the receiving computer.

TCP facilitates sending large data files accurately over transmission routes that may be plagued with bit errors or other forms of poor link quality. It also reduces throughput when it doesn't receive acknowledgment as quickly as it expects. TCP assumes that any delays are due to traffic congestion within the network. It responds by cutting back on the transmission rate; it then slowly speeds up again as long as no further delays are detected. The idea behind its *slow-start* mechanism is to match the output of the sending computer to the maximum throughput allowed by the network at any moment. For the most part, TCP works just fine. However, pushing the ultimate throughput capabilities of TCP has revealed shortcomings, especially on high-capacity links with high latency. The problem is that high latency slows TCP's slow-start mechanism dramatically, causing the protocol to restrict the amount of data that can be "in flight" between the sending and receiving computers and preventing the link from being fully utilized.

The problem affects any high-delay path, whether terrestrial or by satellite. For high-bandwidth links, the result is a sharp drop in efficiency. For example, a message controlled by the most commonly used form of TCP can move through a cross-continent, OC-3 fiber link at only about 1.1 Mbps, even though the line can handle 155 Mbps of throughput.

A packet of data can propagate across the United States and back through a dedicated copper wire or fiber in about 60 ms. It takes eight to nine times as long—500 ms for the same packet to make the round trip via satellite.

Individual LEO spacecraft promise very short propagation delays—as little as 12 ms from ground to satellite and back—because they fly closer to the ground. However, a single LEO spacecraft can't "see" from coast to coast, so several such "birds" are required to relay a packet across the United States. LEO birds also are in motion relative to the ground, so their packet latency varies continually, a characteristic that forces current forms of TCP to reassess the allowable throughput rate constantly. The net result is cross-country transmission delays for LEO systems akin to those of long terrestrial links.

Signal latency isn't the only factor in selecting an appropriate transmission medium, however. Each alternative has its own attractions and challenges. More importantly, continuing adjustments to TCP promise to improve performance for any high-latency network link.

CHAPTER 27

Low-Earth-Orbit Satellites (LEOs)

Quite a bit of discussion regarding wireless communications has already been covered in earlier chapters. However, a combination of two separate services and technologies are merging as new services for broadband communications. These two services include long haul communications and the use of *personal communications services* (PCSs). The technologies include the use of satellite and the cellular concepts combined. Worldwide communications services can be achieved by these two combined services, therefore some diligent effort should be made to understand just what is happening in this arena. This truly brings home the concept of communications from anywhere to anywhere. The thought of being out in the middle of a lake and receiving a call, or rafting down a river and making a call, boggles the mind. This is especially true when thinking about some of the more rural areas in the world where no telephone service infrastructure exists today. Yet, in a matter of a few years these remote locations, on mountaintops, in forests, in valleys, or on the sea will all be reachable within a moment's notice. The infrastructure of a wired world will not easily lend itself to this need, due to timing and cost issues. Therefore, the use of a wireless transmission system is the obvious answer.

However, the use of cellular and personal communications devices still leaves a lot to be desired. First, the deployment of these services is always going to be in the major metropolitan areas, where the use and financial payback will be achieved. Thus, in the remote areas it will be decades, if not longer, until the deployment ever works its way close to the remote areas. Enter the ability to see the world from above the skyline! The industry decided to attempt servicing remote areas from a satellite capacity. This is not a new concept; the use of satellite transmission systems has been around for over thirty years. However, the application for an on-demand, dial-up satellite service is new. This will have to be a lucrative business venture because the costs are still quite high. Look at the Iridium project that took over five years to build and launch, only to meet with lax reception from the marketplace. This drove Iridium Inc., into bankruptcy in its first year of operation.

There are still approximately one dozen suppliers competing for space segment and frequency allocation to offer voice, broadband data, paging, and determination services. In each case, the organizations have selected various approaches on how to launch their service offerings and the use of orbital capabilities to provide the service. In general, this discussion will cover the most widely discussed satellite technology today called *Low-Earth Orbit* (LEO). However, some added discussion on the use of *Mid-Earth Orbit* (MEO) and *Geosynchronous Orbit* (GEO) satellite-based communica-

Table 27-1

A summary of the number of competitors and the various orbits being sought

Orbit	Number Of Competitors	Status
Low-Earth	8	Pending licenses granted Orbit (LEO) based on very specific areas of coverage
Mid-Earth	4	Experimental licenses Orbit (MEO) granted for specific areas of coverage
Geosynchronous	4	Licenses have already been Orbit (GEO) issued for some, are experimental with others.

tions, offerings, and applications must be addressed. As mentioned, there are over a dozen organizations applying for the rights to use various orbital slots and frequencies to provide service, in the various orbits as shown in Table 27-1. The licenses requested are being discussed at great length around the world as the future interoperable service.

The table reflects the fast pace that is being created since the obvious train of thought is that the first company in the business, regardless of the orbit used, will gain the market share. Unfortunately, this did not quite happen with the Iridium network, which opened for service in 1998–1999. This is an expensive situation if some other carrier gets there first and takes the market by force. Of course, there are still those who feel there is no need and that these systems will go bankrupt within the first few years.

Low-Earth Orbit

In December 1990, Motorola filed an application with the FCC for the purposes of constructing, launching, and operating a LEO global mobile satellite system known as Iridium. This was the hot button that sparked the world into a frenzy. Iridium was a concept of launching a series of 66 satellites[1] around the world to provide global coverage for a mobile communications service operating in the 1.610 to 1.6265 GHz frequency bands. The

[1]Originally the Iridium proposal was for 77 satellites, but Motorola amended this number after the World Administrative Radio Council meeting in the spring of 1992.

concept was to use a portable or mobile transceiver with low profile antennas to reach a constellation of 66 satellites. Each of the satellites would be interconnected to one another through a radio communications system as they traversed the globe at 413 nautical miles above the earth in multiple polar orbits[2]. This would provide a continuous line-of-sight coverage area from any point on the globe to virtually any other point on the globe, using a spot beam from the radio communications services on-board each of the satellites. The use of this spot beam concept, which had been discussed for years in the satellite industry, allowed for high frequency reuse capacities that had not been achieved before. Iridium wanted to provide the services outlined in Table 27-2. Motorola also suggested that an interconnection arrangement would be set up with all providers around the world through an arrangement with the local *Post Telephone and Telegraph* (PTT) organizations. The concept was sound, and the approach would have provided for the coverage that was lacking in the past to remote areas. In the table, the two columns are used as exclusive of each other. The services can be provided in any of the coverage areas regardless of which service is selected.

Each of the satellites was relatively small in terms of others that had been used. The electronics inside each were very sophisticated, and the use of gateway controllers on earth provided the command and control service for the administration of the overall network. With the Iridium network, providing coverage in areas where cell sites would not have been practical expanded basic cellular communications. This is summarized in Table 27-3, which shows a feature comparison of the cellular concept versus the Iridium concept.

Table 27-2

Services and coverage through the Iridium network

Type Of Service	Coverage Areas
Voice Communications	Air, land, water
Data Communications	Air, land, water
Paging	Air, land, water
Radio Determination Services	Air, land, water

[2]The original concept was to use 7 polar orbits with 11 satellites in each. This would provide worldwide coverage, much similar to an orange slice concept

Table 27-3

Summary of cell sites versus Iridium cells

Cellular Networks	Iridium Network
Sites are fixed	Sites are the moving targets
Users move from site to site	User stays put, sites move the user from satellite to satellite
Areas of coverage are 3–5 miles across	Areas of coverage are 185–1100 miles across
Coverage sporadic, not totally ubiquitous	Worldwide coverage

The areas of coverage in the cellular and PCS networks primarily served populated geographic areas. There were areas that were not financially or operationally prudent to install services. Consequently, very rural areas, mountain terrain, and the like did not receive coverage at all. Under the LEO concept, all areas not previously served could be easily accommodated. Iridium even provided coverage in the extreme rural areas, and during the Kosovo, Yugoslavia crisis. The services addressed by these satellites included the obvious, such as two-way voice but also other services that were not as evident. Table 27-4 is a summary of the actual services that were provided. This includes a minor description of the service and a potential target market. The list is straightforward in a single line description.

In Figure 27-1, the concept of the LEO arrangement is shown. In this particular case, the satellites are traversing the earth's surface at a height of 400+ nautical miles above the earth, in a polar orbit. In the polar orbit, the satellite moves around the earth's poles and passes over any specific point along its path very quickly. The satellites move at approximately 7,400 meters per second in different orbits. Therefore, as one target site moves out of view, a new one comes into view at approximately the same time. A handoff will take place between the individual satellites (using the Ka band).

In Figure 27-2, the ground telemetry and control services are represented, called gateway feeder links. These also use spectrum in the Ka band. Iridium used approximately 16.5 MHz of bandwidth in the L band. The L band is also used from the handset to the satellite, whereas the Ka band is used from satellite to satellite communications, as shown in Figure 27-3. The use of this L band allows low-powered handsets to communicate within the 413 nautical mile distances with the satellites.

Table 27-4
A summary of the initial features available on LEO networks

Feature	Description
Radio Determination Services (RDSS)	Will allow for the location of vehicle fleets, aircraft, marine vehicles, etc. RDSS will also be an integral locator service for all voice communications devices.
Voice communications (VC)	Will allow on demand, dial-up digital voice communications from anywhere in the world.
Paging (P)	A one-way paging service. The paging service includes an alphanumeric display for up to two lines, but will expand to *short messaging services* (SMS) at 160 characters.
Facsimile (Fax)	A two-way facsimile service.
Data communications (DC)	An add-on device will allow the transmission of two-way data. This capability will also allow for two-way messaging (e-mail) service across the network.

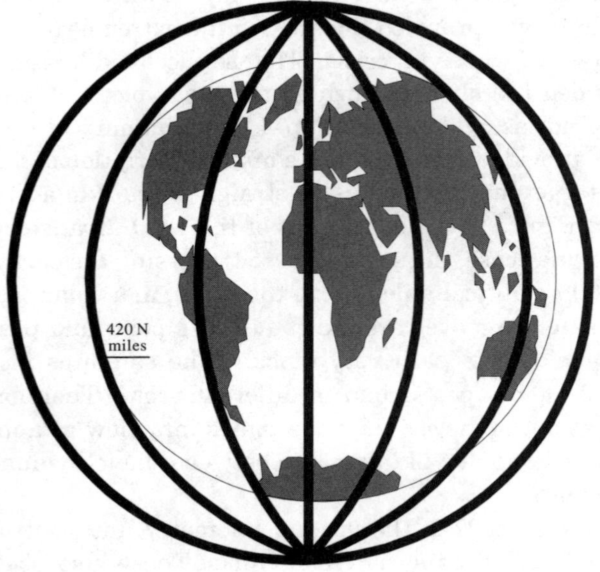

Figure 27-1
The LEO concept

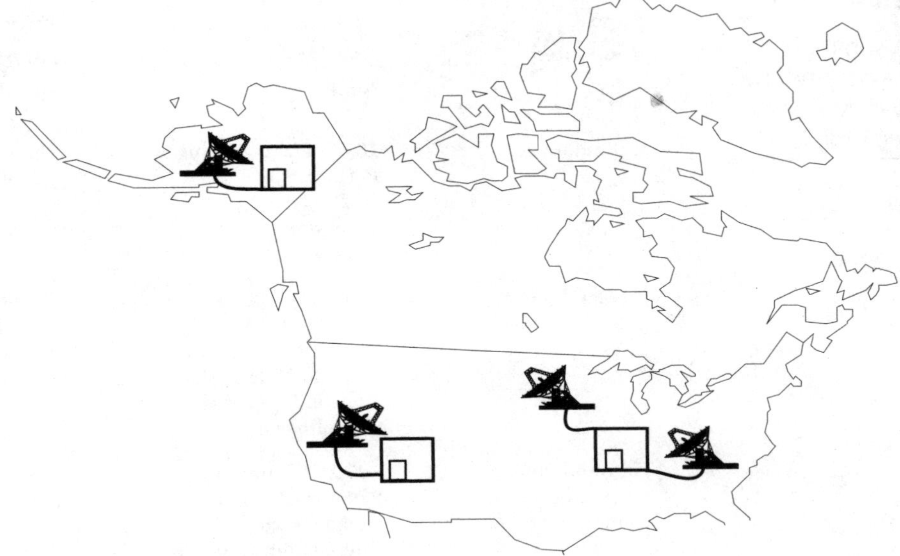

Figure 27-2
Ground station telemetry and control

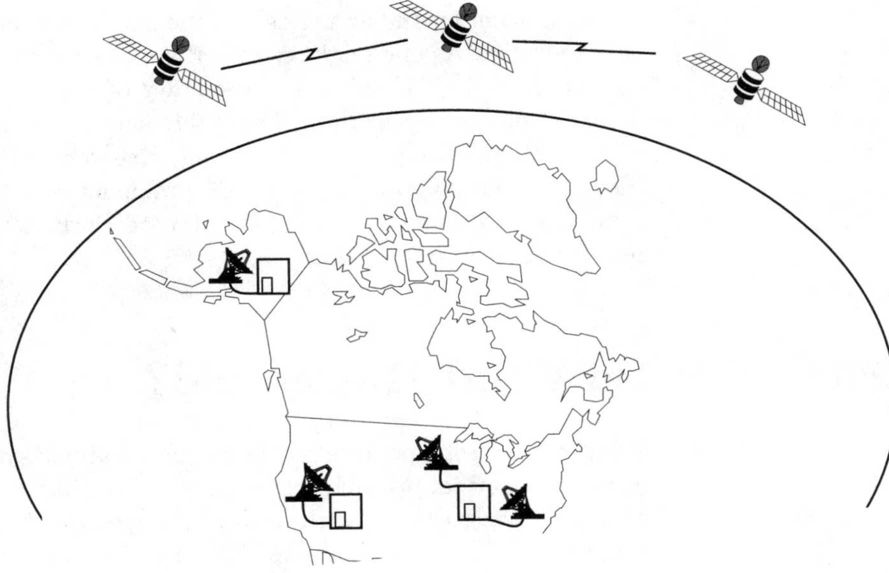

Figure 27-3
Satellite-to-satellite communications is handled on a Ka band.

Table 27-5

A summary of the bands and the bandwidth for Iridium

Band Requested	Immediate Bandwidth Needed	Future Additional Bandwidth
L Band for set	16.5 MHz (operating in the 1.610–1.6265 GHz to cell communications range)	Up to 100 MHz in the L band for the future
Ka Band for gateway feeder links	200 MHz (100 MHz in the 18.8–20.2 GHz range for downlink; and 100 MHz in the 27.5–30.0 GHz range for uplink)	Possibly remain at the 200 MHz, or additional 100–200 MHz for the future
Ka Band for intersatellite communications	200 MHz (all in the 22.55–23.55 GHz range for the intercommunications connection)	Possibly remain at the 200 MHz or additional 100–200 MHz for the future

The market for the handsets and the add-on devices was supposed to be very lucrative. As one might expect, the handsets operated in a dual-mode capability. Table 27-5 shows a summary of the spectrum and frequency bands requested by Iridium. This table shows the immediate request for bandwidth and the projected additions that were projected over time. This may well change with the financial problems now faced at Iridium. The issue here is the ability to create such a network, not to precisely pinpoint how it will look in five years.

So What Happened?

With the market ripe for a global communications system, and the lead in getting the service up and running, how did Iridium fail? One can only speculate, but the cost of launching Iridium was over $5 billion (US) and the project was nearly a year late. Moreover, during launch, several factors plagued Iridium (lost satellites, explosions, and failed launches), putting significant financial strain on the operating budget. When Iridium became available, the initial costs were touted as being $5 to $7 (US) per minute for an international call. In 1992 when the model was created, this price may have been realistic. However, in the time it took to get off the ground, the costs for dial-up voice communications plummeted. No one was willing to pay the prices asked by Iridium. A better price may have been one I predicted in 1995 stating that they will have to charge from $1 to $3 (US) per minute. That may have been one of the total downfalls for the Iridium network. Moreover, the handset price was too high, ranging upwards of $3,000 each. This does not mean that they cannot come back and reestablish themselves as the leader in the LEO industry, but there are many providers right on their heels (such as GlobalStar and Teledesic). These companies and their networks are aggressively chasing behind Iridium and plan to offer the broadband communications services initially when they open their doors. Iridium felt that they could do this later, possibly because of their lead in establishing the market niche.

The Benefits of These Service Offerings

Motorola established a list of benefits from the deployment of the Iridium network services, which at first glance may look biased toward their services. Upon further evaluation, these benefits can be derived from any network of this type. Therefore, generically these are addressed and kept in the context of any LEO network. The benefits lean toward the end user as shown. These are summarized as follows:

- **Ubiquitous services** With continuous and global coverage, any-to-any connections can occur. As users travel either domestically or abroad, the service travels with them. It will eliminate the need for special access arrangements and special numbers that must be dialed. Users should never be out of range from their network. Remote areas with limited demand and finances now have the capability to connect anywhere in the world.
- **Spectral efficiency** As already mentioned, the frequency reuse patterns for the bandwidth allocation will be significant. No other satellite system has achieved these reuse ratios. Iridium was first to claim this capability of efficiency. RDSS portion of the Iridium network is contained in the same spectral arrangement, freeing up 16.5 MHz of spectrum. This is a quantum leap in the efficient use of the spectrum.
- **Public benefits due to flexible design** The digital technology deployed allows the total connection for all voice and data services on a

seven day by twenty-four hour basis. This allows the flexibility of service provisioning. The LEO overcomes some of the limitations of the higher transport systems, such as the delay in the round trip transmission. Because the satellites are low, the user set needs a lower power output device. This orbit has been selected to be the most flexible.

- **The potential to save lives** It's common for the news media to publish stories of people stranded in remote areas with no life support systems who died because of their inability to communicate. The press today is filled with stories of cellular and PCS users notifying authorities of casualties. If only people in remote areas had a means of notifying authorities and/or rescue parties, their lives could be spared in the event they get into a life-threatening situation.
- **Capabilities of the vendor** Motorola states that they are uniquely qualified to provide these types of services due to their background in the development and sales of other ancillary equipment that works in the wireless world. Specifically, they have been one of the major developers in the production, research, and development of private mobile-radio services.
- **LEO deployment promotes international communications** The LEO networks deliver modern digital-transmission services to remote areas of the world. The FCC and the U.S. government are attempting to use telecommunications as a strategic and economic tool to foster development in these areas. Their goals are to
  - Promote the free flow of information worldwide
  - Promote the development of innovative, efficient, and cost-effective international communications services that meet the needs of users in support of commerce and trade development
  - Continuous development and evolution of a communications services and networks that can meet the needs of all nations, and specifically those of developing nations.

The above goals can be met with a mobile communications network such as that proposed in the LEO networks. An alternative is the *Global Services Mobile* (GSM) standard that is emerging throughout the international arena.

Deployment and Spacing of Satellites

Motorola's concept of seven different orbits on a polar orbital path would lend itself to the potential of collisions in mid-air if the spacing were not correct. Therefore, the spacing design had to take into account that the satellites were all traversing the same end-points where the paths would all cross. One can imagine that the two poles of the earth are the midpoints in the orbit. The Iridium network originally consisted of 77 satellites, however, a modification to the plans dropped the number of craft to 66. To assign or determine the best approach, Motorola took into account six critical criteria in determining the best approach:

1. The need to provide a single global coverage over the entire earth's surface for availability at all times. This means that each subscriber must have at least one satellite in view at all times to provide coverage.
2. Some portion of each orbit must be available to allow for low-power outputs, which accommodates the recharging of the communications power subsystems. This minimizes the size requirements of the craft.
3. The relative spacing of the satellites and the line-of-sight relationships had to allow the on-board systems to control cross system linkage.
4. Costs for the entire constellation was a concern in the selection of the orbit and spacing requirements. Minimizing costs was a portion of the decision-making process.
5. The angle of incidence from the end user to the spacecraft as measured from the horizon to the line-of-sight communications process allowed link margins to accommodate the low-powered user device. The slant angle was selected at 10 degrees to meet these criteria.
6. The final criteria was the operational latitude of the spacecraft. Systems operating over 600 nautical miles were affected more by radiation, which drives the cost up for the systems. Altitudes lower than 200 nautical miles required much more fuel and control over the positioning of the craft, thereby driving up the cost. At 413 nautical miles, the fuel consumption and command and control over the craft position coupled with the better radiation performance; the cost and coverage ratios provided a better solution.

The satellites were spaced at 32.7 degrees apart, traveling in the same basic direction and moving at approximately 16,700 miles per hour from north to south and 900 miles per hour westward over the equator. Given this path, each satellite was designed to circle the earth approximately every 100 minutes. At the equator, a single device would provide coverage. However, as the craft moved toward the poles, overlap occurred, increasing the levels of coverage above and below the equator. The expected life cycle of the satellites was around five years.

The basic building blocks of the Iridium system provided the model for other networks ready to launch (such as Globalstar systems). In each case, the systems used proven technology for radio transmission in well-established frequency bands. The basic system was composed of the following:

- The space segment comprised of the constantly moving constellation of satellites in a LEO.
- A gateway segment comprised of earth station facilities around the world.
- A centralized system control facility.
- The launch segment to place the craft in the appropriate orbit.
- A subscriber unit to provide the services to the end use.

The Space Segment

The space segment consisted of small satellites, operating in the LEO, which were all networked together as a switched digital communications system. In Figure 27-4, a sample outline of the beam coverage with Iridium is shown. Each satellite used up to 37 separate spot beams to form the cells on the earth's surface. Multiple relatively small beams allowed the use of higher satellite antenna gain and reduced RF power output. The system was designed to operate with up to 250 independent gateways, although in its initial deployment only between 5 and 20 gateways. Figure 27-5 is an overall view of the network communications through the gateways and the handsets.

The Cell Patterns

The satellites had the capability of projecting 37 spot beams on the earth. The spot beams formed a series of overlapping, hexagonal patterns that would be continuous. The center spot beam was surrounded by three outer rings of equally sized beams. The rings worked outward from the center

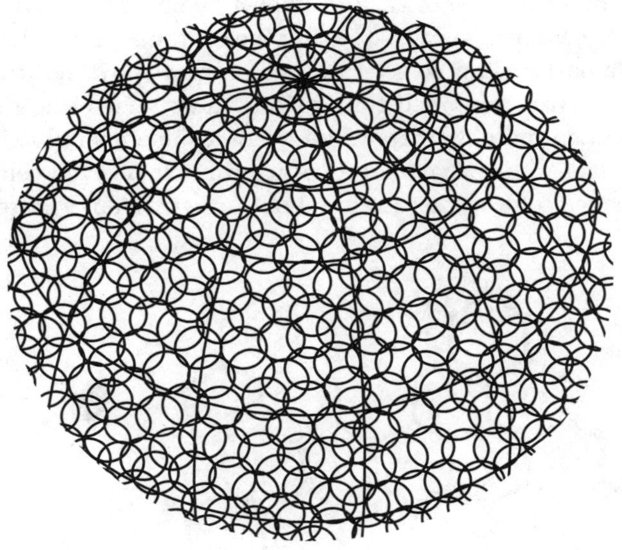

Figure 27-4
The spot beam pattern from Iridium

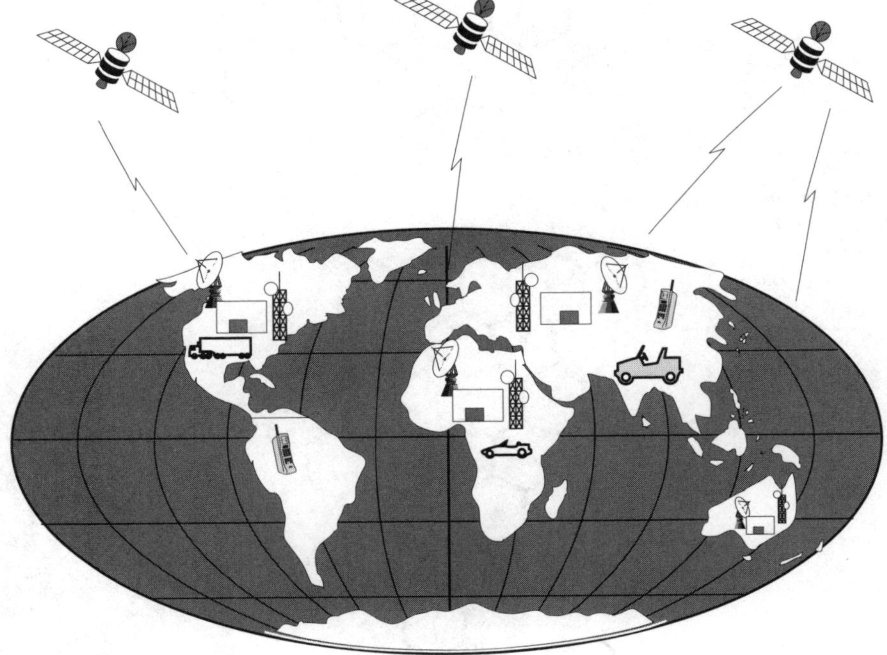

Figure 27-5
The overall system through the gateways

beam in rings of 6, 12, and 18 spot beams. This pattern of a 37-cell pattern is shown in Figure 27-6, using the center beam as the starting point where the pattern can be seen. A spot beam was 372 nautical miles in diameter, and when combined they covered a circular area of approximately 2,200 nautical mile diameter. The average time a satellite was visible to a subscriber was approximately nine minutes. A seven-cell frequency reuse was formed to produce the actual pattern shown in Figure 27-7.

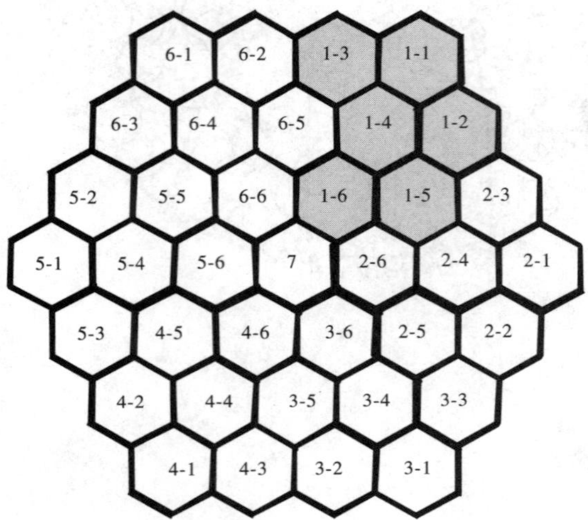

Figure 27-6
The 37-cell patterns

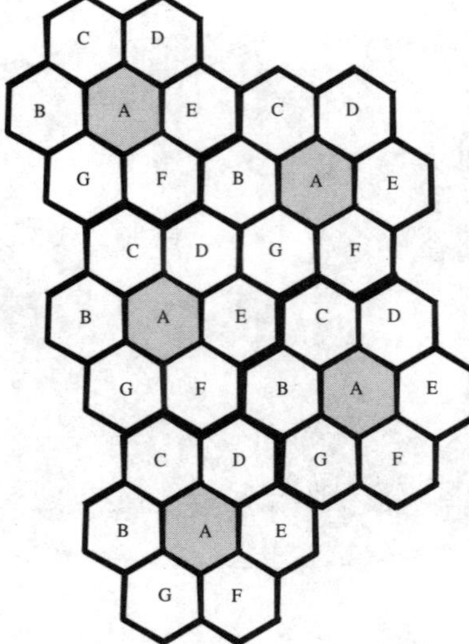

Figure 27-7
A seven-cell frequency reuse pattern

Traffic Carrying Capacity

A 14-slot TDMA format is used, allowing each Iridium cell to be assigned an average of two time slots. The average traffic capacity with a utilization of 16.5 MHz is 174 full-duplex voice channels per cell.

Modulation Techniques

The modulation process and multiple access capabilities of Iridium were modeled after the conventional terrestrial cellular networks, and particularly after the international GSM standard. Combining frequency and time division multiple access, the system also used a data or vocoder voice and digital modulation technique (that is, QPSK, MSK, and so on). Each subscriber unit operated in a burst mode transmission by using a single carrier. The bursts were controlled to appear at the precise time necessary to be properly integrated into the TDMA frame.

The Gateway Segment

The gateway segment controlled user access to the network and provided interconnection to the *Public Switched Telephone Networks* (PSTN). Two initial gateways in the United States divided the eastern and western parts of the country. Two added sites will be located in eastern and western Canada to provide coverage throughout Canada and Alaska. Each of the gateways had the necessary terminal and switching equipment to support the operations.

The Earth Terminal

The key to the earth terminal function was the use of up-and-down link capabilities with three separate RF front-ends supporting continuous service and high up-time availability. One of the RF front-ends was used as an active up/downlink with one satellite. A second RF front-end was used to establish communications with the next active satellite. The third provided a backup capability in case of an equipment failure and provided for diversity against any of the atmospheric conditions that can plague a typical satellite communications system (that is, sunspots, equinoxes, and so on), causing a degradation of service. As the satellites were the moving targets and the gateways were fixed, the antennas followed and tracked the nearest two satellites to them. The communications active channels were handed off from the current satellite to the next active one coming into view, while the first disappeared from view. This handoff was designed to be transparent from the user's perspective.

The Switching Equipment

Each gateway housed the necessary switching equipment to interface between the communications payload in the Ka band and the voice/data channels from the PSTN. The switching systems performed the following functions:

- Transferred common channel signaling information from the PSTN to the RF portion of the Iridium network
- Transferred line and address signaling information from the PSTN to establish circuit-switched calls
- Supplied in-band tones and announcements to PSTN users calling onto the Iridium network with necessary progress tones and conditions
- Digitally-switched PCM signals between channels derived from the terminal channels to the PSTN and provided channels to support the necessary in-band signaling capability for call control and progress.

Interconnecting to the PSTN

Voice connections were designed to be fully compatible with applicable ANSI T1 standards for the United States and the CCITT G and Q Recommendations (International T1/E1 standards) for digital transmission systems using either SS7 or R1 signaling. The data channel specifications were compatible with the OSI standards and with CCITT V and X series recommendations.

The System Control Portion

The *system control segment* (SCS) of the network provided the necessary control over the entire satellite constellation. The SCS managed and controlled all the system elements, providing service on both the short- and long-term picture. The functionality of the SCS was divided into either active control of the satellites and control over the communications assets within the satellites. The SCS monitored and managed the overall network control sequencing, traffic load balancing, and peak conditions, as necessary.

Other Competitors to Iridium

As already mentioned earlier, several other suppliers have all the rights and licenses to launch similar communications services, using constellations of LEOs, either MEOs or Geostationay Orbits. This is the PCS of the future. The companies (at least seven others in the United States) are all vying for the licenses to use the L band communications bandwidth to provide global communications. These include some of the other big names in the business such as Qualcomm, Loral, TRW, Microsoft, and McCaw Communications. Each of the companies has a different approach to use, but they all share a common goal. Get there soon and offer high-speed digital communications with Internet access, e-mail, voice over, IP, and demand video services.

Loral-Qualcomm

Loral-Qualcomm's (located in Palo Alto, California) approach is to use approximately 48 satellites in a big LEO arrangement similar to Iridium, but with less craft. Their network handle is Globalstar systems. Qualcomm expects to use its expertise in the specialized mobile radio business and a CDMA arrangement that they pioneered. This service is currently two years behind schedule, but in the process of launch with expected availability in 2000. Loral-Qualcomm will draw from its experience in the business by deploying their CDMA technology and attempt to reuse the frequency spectrum as much as possible. Many of the initial service offerings from this company will include fleet tracking, locator services, and ultimately the extension of voice and data communications. The uniqueness of the Loral-Qualcomm services is that the system is designed to fully complement the cellular and PCS industries by providing single service coverage from a single telephone to a single number, through an intelligent network. Loral-Qualcomm did not plan to use a dual mode operating set, but a single set that can interface to any network service without having the user make a decision. Globalstar systems constellation is shown in Figure 27-8. These folks stand to be a big contender in the LEO business.

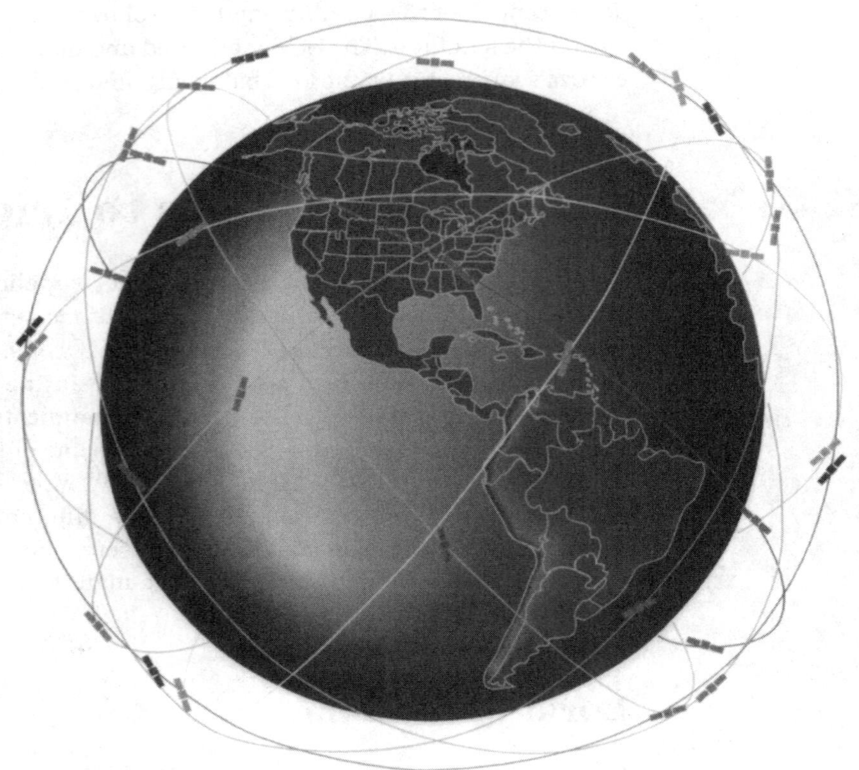

Figure 27-8 Globalstar systems constellation

The Globalstar™ 48-satellite low-Earth orbit (LEO) constellation will provide worldwide wireless communications via voice and data, paging and messaging, and radio-determination satellite services (RDSS) to over 98% of the world's population.

Regardless of which suppliers succeed or fail, the gist of the topic is that satellites orbiting close to the earth have a lot of merit. The orbit will be short lived (5 to 7 years) and the rotation in a polar orbit requires more handoff techniques, but the end result is that we achieve any-to-any connections regardless of where we are in the world. Moreover, the initial implementations usually suffer the pains of growth and sporadic coverage, whereas the newer implementations fix the ills of the initial systems. This means that changes will likely take place in whom the providers are, but the overall concept of using the LEO orbits will remain.

CHAPTER 28

The T Carrier Systems (T-1/T-2 and T-3)

Meeting the high demands for voice, data, and multimedia communications, the larger corporations in the world are moving to the high-speed digital communications of the T Carrier system. This includes many of the copper and optical fiber-based systems. The hottest installation method today is still the T-1. Although T-1 has been around since 1958, when it was first created, and then rolled out into the carrier communities in 1960, one would think this technique became old. However, T-1 and T-3 are still the hottest method of connecting some of the services discussed in other chapters, such as

- *Frame Relay* (FR)
- *Asynchronous Transfer Mode* (ATM)
- Point-to-point data lines
- Internet access
- Signaling systems
- Computer-to-telephony integration
- ISDN
- xDSL (HDSL, CDSL)

Why has this become such a popular method? The implementation of digital high-speed communications can be directly linked to the cost and convenience factors. T-1 became a very price competitive access method for nearly all the communications services needed.

Yet, there has been other activity in the form of T-3 or DS3 services to support the higher demands for the various multimedia and video applications. In both cases, T-1 and T-3 are now very popular in the commercial industry as a means of providing the connectivity desired to meet the continuing communications development within organizations. The pricing models now make these services more readily affordable. Moreover, the enhancements to the digital circuits make this a tried and true technology. For example, today's tariffs make both T-1 and T-3 very price efficient and attractive. Even if a customer does not have enough volume to justify the use of a full T-1 or T-3, the prices enable the consumer to use portions of the services while holding excess capacity for new applications, without causing undue budgetary hardships. A T-3, for example, costs the same as between five and eight T-1s terminated at the same location. This is beyond a financial decision and moves us to think of these services as application supportive.

Once a consumer or organization makes the leap into the digital circuit, the availability of bandwidth to host other services that was previously unavailable becomes paramount. Newer graphics and video services can be turned on, as needed, if the bandwidth is no longer the constraint.

This chapter will discuss the way the multiplexing and access of the digital services is facilitated. There are many other excellent publications that go deeper into the technologies. The reader can choose to do further study if so desired.

The Difference Between T-x and DS-x

In many cases, the terms become problematic because we tend to intersperse them. Therefore, when discussing the different services, it becomes imperative to discuss the difference between a T-x and a DS-x.

T-x (such as T-1 and T-3) refers to the services acquired from a carrier or local provider by physical layers one and two. The T carrier is a physical set of wires, repeaters, connectors, plugs and jacks, and so on. When referring to a T-1, we are actually describing the physical layer interface to the provider networks. A T-1, for example, is a four-wire circuit (unless one of the HDSL or SDSL techniques is used with only one pair of wires). The four-wire circuit is installed between the local provider and the customer's premises, as shown in Figure 28-1. In this example, the provider will install the physical wires or use four wires from a 50 pair bundle on the local loop. From the *Central Office* (CO), the four wires will be cleaned up to remove splices, bridge taps, and load coils from the wires. At approximately 3,000

Figure 28-1
A typical T-1 carrier installation

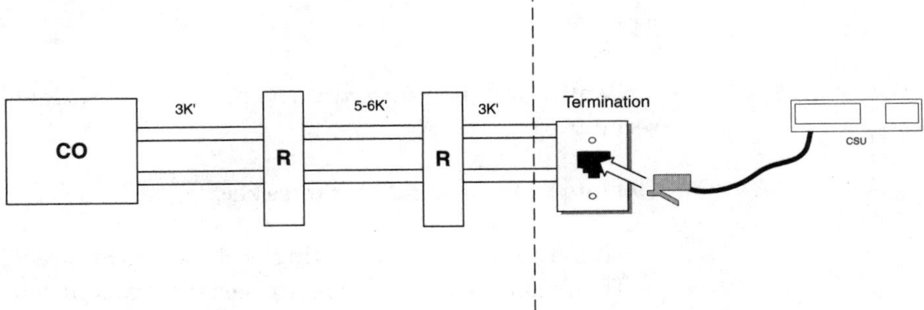

feet from the egress point at the CO, the provider will install a digital repeater (or *regenerator*, as it is also called). Thereafter, every 5,000 to 6,000 feet, another repeater is required until the last leg of the circuit. At approximately 3,000 feet from the customer's entrance point, the last repeater is installed. The provider then terminates the circuit at the demarcation point in a jack. The customer uses a plug to connect the CPE to the circuit. These are all mechanical and electrical devices enabling the installation of the T carrier.

After the circuit is installed and the carrier is terminated, the customer then generates traffic (voice, data, video, and so on) across the circuit. This traffic is carried on the carrier at a digital rate, or what is called the *Digital Signal level 1* (DS-1). DS-1 operates at a digital signaling rate of 1.544 Mbps. The traffic is therefore called a DS-1. (Note in many cases this may be annotated as DS1.)

One can therefore see that there is a difference between the physical carrier and the signaling rate of the traffic carried on the circuit.

DS-1 Framing Review

DS-1 operates in a framed format (some cases exist where a customer may request a nonframed format, but this is rare). The framer establishes the frame in a device called a *Channel Service Unit* (CSU). A frame consists of 193 bits of information created in a 125 μs. In a DS-1, there are 24 time slots associated with the signal. Each time slot carries 8 bits of user information. When the CSU sets up the frame and has 24 slots of 8 bits each, it then adds one overhead bit for framing delimitation. This creates the 193 bits referred to previously.

24 channels	@ 8 bits per channel	+ 1 framing bit	= 1 frame
24	× 8	+ 1	= 193 bits/frame

Continuing the math, the 193 bit frames are generated 8,000 times per second:

193 bits/frame × 8,000 frames/sec. = 1.544 Mbps

This framing and formatting is shown in Figure 28-2.

The frames of information are serially transmitted across a four-wire circuit and operate full duplex. This gives the end user a 1.544 Mbps in each direction simultaneously.

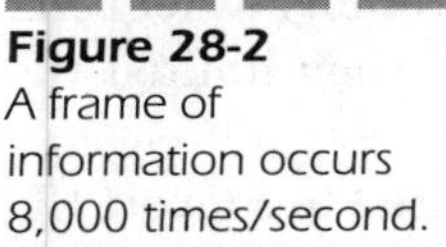

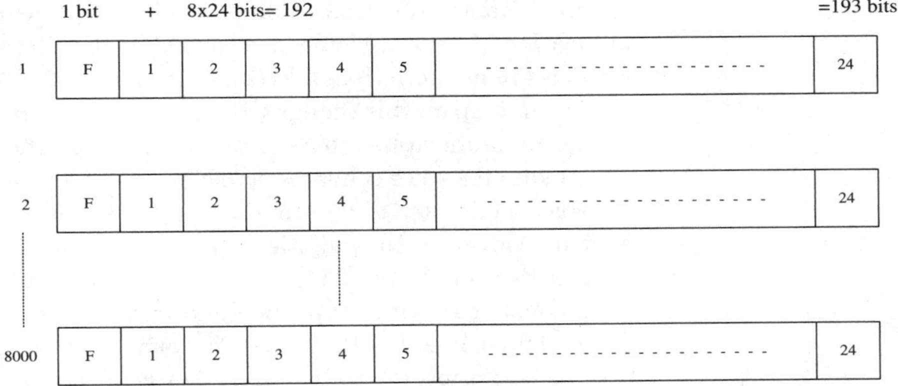

Figure 28-2
A frame of information occurs 8,000 times/second.

Pulse Coded Modulation (PCM)

Before going any further into the DS-1, it may be appropriate to review the modulation technique used to create the digital signal. When DS-1 was first created, it was designed around converting analog voice communications into digital voice communications. To do that, voice characteristics were analyzed. What the developers learned was that voice operates in a telephony world in a band-limited channel operation. The normal voice will produce both amplitude and a frequency change ranging from 100 to approximately 5,000 times a second. These amplitude and frequency shifts address normal voices. However, the telephone companies decided long ago that carrying true voice would be too expensive and would not provide any real added value to the conversation. They then determined that the normal conversation from a human actually carries the bulk of the information when the frequency and amplitude shifts vary between 100 and 3,300 times per second. Armed with this information, the developer determined that reasonable and understandable voice could be handled when carried across a band-limited channel operating at 3,000 cycles of information per second (what was termed as a 3 kHz channel). Taking all the electromagnetic spectrum available to them, the developers then channelized the frequency spectrum (in *radio frequencies* [RF] and in electrical spectrum available on the cabling systems they had) to smaller capacities. The norm was set at 4 kHz channels. This is the foundation of the voice telephone network.

From there, the providers of the infrastructure (the telephone companies) placed bandpass filters on the facilities to limit the amount of electrical information that could pass across their wires (or any other communications facility). Using a standard 4 kHz channel, they limited the bandpass to no more than 3 kHz (see Figure 28-3).

Following up on this thought, the developers then wanted to convert the analog communications to a digital format. Here they studied the way the user modulates voice conversation. Now they had something to work with. The voice modulated at a normal rate operates in the 3 kHz range but now must be converted to a digital signal consisting of 1s and 0s. Using the Nyquist theorem from 1934, the developers used a formula to convert the continuously changing amplitude and frequency shifts to a discreet set of values represented by the 1s and 0s. Using a three-step process, as shown in Table 28-1, they determined that they could carry digital voice. The table represents the three steps followed to do this conversion.

Once the values were determined, the developers used another process. Setting the values in place, they had to determine where along the wave the sample fell. They created a value system to show the 256 levels by using the table shown in Table 28-2. First, the digital PCM signal is representative of both positive and negative values. To reflect where along the wave the sample fell, the eighth bit in each sample is used as a sign (+/−) value. The other seven bits, therefore, represent the actual sample value. There are 128 points on the positive side of the wave and 128 points on the negative side of the wave. In this table, we look at the major stepping points. Two different formats are shown in the table, the PCM used in North America (and Japan) is Mμ-Law, whereas the rest of the world uses an A-Law method. These are different as shown.

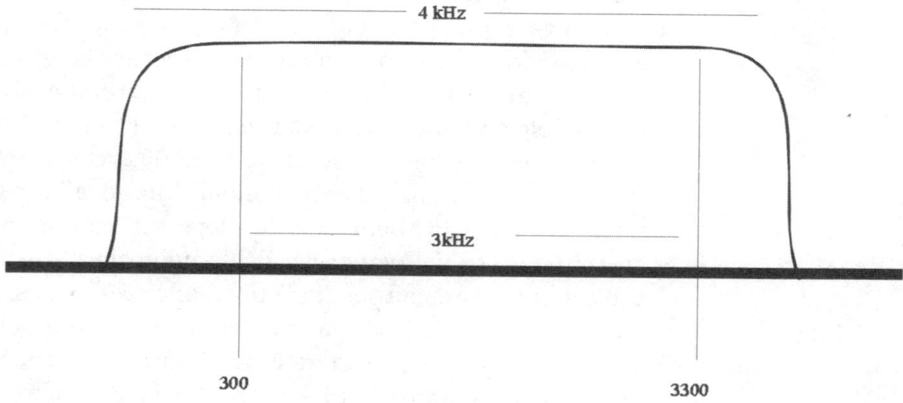

Figure 28-3 Band-limited channel limits the amount of information carried to 3 kHz.

Table 28-1

The three-step process used to create the digital signal

Process	Result
1. Sample the analog wave at twice the highest range of frequencies that can be carried across the line. Using a 4 kHz channel capacity, the highest range of frequencies allowable is 4,000.	A sampling rate of 8,000 times per second: 4,000 × 2 = 8,000.
2. Quantify the values using a logical pattern of 1s and 0s to represent the height of the signal at any point in time. This deals with the amplitude shifts only.	Using an 8-bit sequence, the result is a total of 256 combinations of amplitude that can be represented. Although there may be more amplitude heights (values), the 256 quantities were determined to be sufficient. $2^8 = 256$ possible combinations.
3. After the values are determined from the samples, the final step is to encode the signal into a digital format and transmit information in its digital format onto the wires.	A sample value of 5 on the positive side of the wave will then be represented as an 8-bit data stream. Binary 5 = 00000101.

Table 28-2

Summary of values for PCM in Mu-law and A-Law formats

Coded Numerical Value	Bit Number		Comments
	Mu-Law	A-Law	The left-most bit (bit number 1) is transmitted first. It is the most significant bit. This bit is used as the sign bit; it is a 1 for positive values, and it is a 0 for negative values.
	12345678	12345678	
+127	10000000	11111111	
+ 96	10011111	11100000	
+ 64	10111111	11000000	
+ 32	11011111	10100000	
0	11111111	10000000	
0	01111111	00000000	
−32	01011111	00100000	Note that 0 has two different values. Bits 2 to 8 are inverted between A-Law PCM and Mu-Law PCM. In A-Law, all even bits are inverted prior to transmission.
−64	00111111	01000000	
−96	00011111	01100000	
−126	00000001	01111110	
−127	00000000	01111111	

The E-1 Pattern

Because the A-Law was shown in the table for comparative purposes, it is also wise to introduce the concept of the E-1. In North America, the DS-1 is the way we use our digital services. However, most of the world adopted a different standard called the *E series*. The E-1 is the European equivalent of a DS-1. There are many differences between the two that make them incompatible. The first is the companding method of creating the digital values as shown in the previous table. Above and beyond the companding values (using A-Law), there are differences between the formats used. An E-1 is a digital transmission rate that operates at 2.048 Mbps compared to the DS-1, which operates at 1.544 Mbps. Moreover, the E-1 carries 32 channels of 64 Kbps, whereas the DS-1 only carries 24 channels of 64 Kbps. Other multiplexing forms create higher speed E series just as we do with the DS signals. The following table summarizes the DS-n series compared to the E-n series. See Table 28-3 for the channel speeds of the digital transmission systems.

The Framing Protocols: D4 Framing

Before transmitting the frames across the circuit, the CSU device uses protocols that describe other activities. One protocol is a framing convention to preserve timing, signaling, and check for errors. The older version of this protocol is called *D4 framing*. D4 is comprised of 12 samples from each of the 24 inputs. The samples are held for a short time in a buffer by the CSU.

Table 28-3

Comparison of North American and European digital services

DS-n	Channels	Speed	E-n	Channels	Speed
DS-0	1	64 Kbps	E-0	1	64 Kbps
DS-1	24	1.544 Mbps	E-1	32	2.048 Mbps
DS-2	96	6.312 Mbps	E-2	128	8.448 Mbps
DS-3	672	44.736 Mbps	E-3	512	34.368 Mbps

When all 12 passes are completed, the CSU then places the framing bit on each of the frames and transmits the information across the link.

When preparing the 12 frames, however, the CSU will also do a few other things. In the sixth and twelfth frames, the CSU will steal (rob) one bit from each of the 24 inputs and use this bit for signaling. This means that the CSU steals the least significant bit from the sample and reassigns it. The unfortunate result of this framing convention is that it relegates the end user to only getting 7 data bits and 1 signaling bit every sixth frame. The 7 bits are trusted, whereas the eighth bit is suspect for use as data (because it is signaling). This results in a data carrying capacity of 56 Kbps (7 bits × 8,000 frames/sec). This is shown in Figure 28-4, using the robbed bit signaling approach.

Contrasting the E-1 and DS-1 Frame

Once again differences exist in the format of the framing patterns for the E-1 and the DS-1. The DS-1 is shown with 24 channels operating at 64 Kbps, plus a framing bit (1 bit added to the 24-byte samples) to create a

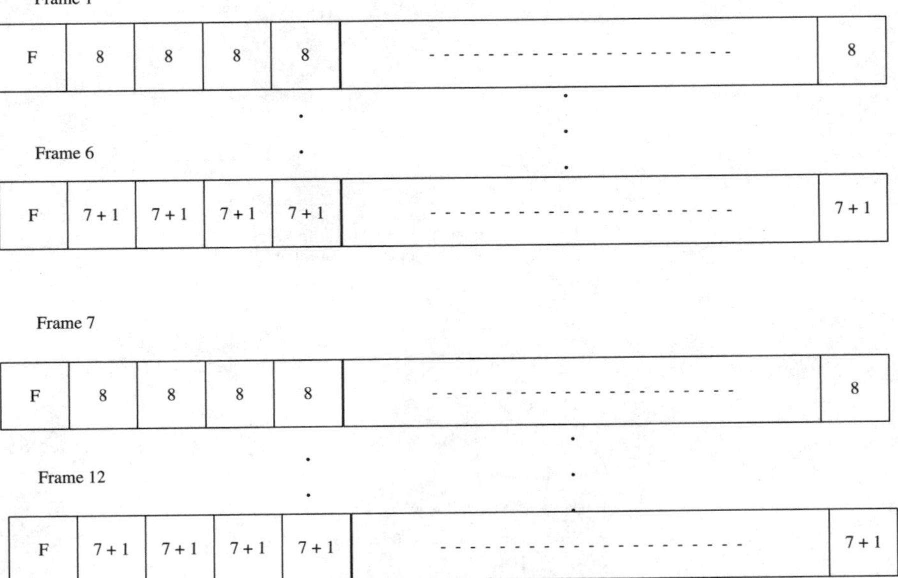

Figure 28-4 Robbed bit signaling in the D4 framing format

frame of 193 bits, whereas the E-1 uses 32 channels of 64 Kbps. However, the first channel of an E-1 frame is used for synchronization, alarms, and international carrier use combined. This means that only 31 channels are available to carry voice or data traffic. This frame format is shown in Figure 28-5. The use of the pattern shown in this frame format uses *Time Slot 0* (TS0) for framing and synchronization. Then TS16 is used for signaling for the various channels in the payload, whereas the DS-1 uses robbed bit signaling as discussed previously or common channel signaling on TS24. These differences are significant enough to prevent an even mapping from a DS-1 to an E-1.

Figure 28-5
An E-1 frame format

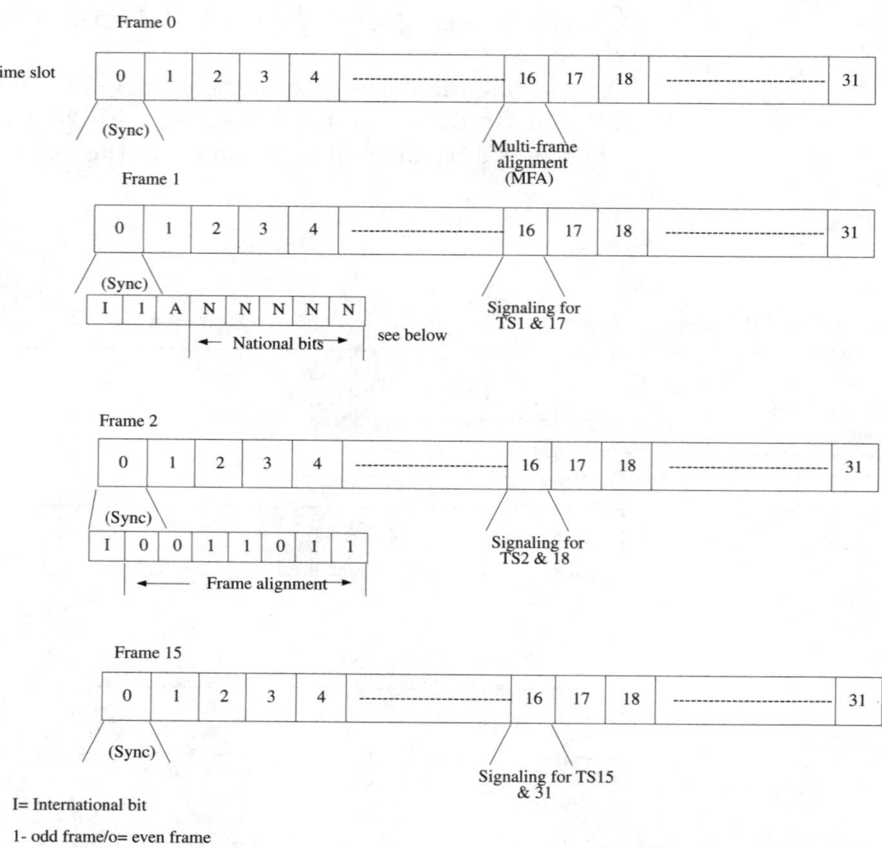

Extended Superframe Format (ESF)

As a means of overcoming the limitations of the D4 framing protocols, the industry developed a newer protocol called *ESF*. This is an extension of the 12-frame format in order to double the size in the buffers at the CSU. When the CSU has 24 frames of information, the CSU will use the framing bit for more than just a locator bit. It will define where the signaling is handled and use this framing bit for other things such as the following remember there are 8,000 framing bits per second:

- 2 K bits are used for framing.
- 2 K bits are used for error checking (CRC−6).
- 4 K bits are used for a maintenance and diagnostic capability to troubleshoot the circuit and improve availability.

The framing format of the ESF is shown in Figure 28-6.

After the framing bits had been addressed, another choice was used to handle the T-1 circuit. As described previously, the robbed bit signaling gets in the way of data transmission. It is not a problem for voice communications but definitely impacts the data side of the business. To solve this problem, a technique called *common channel signaling* was introduced. The use

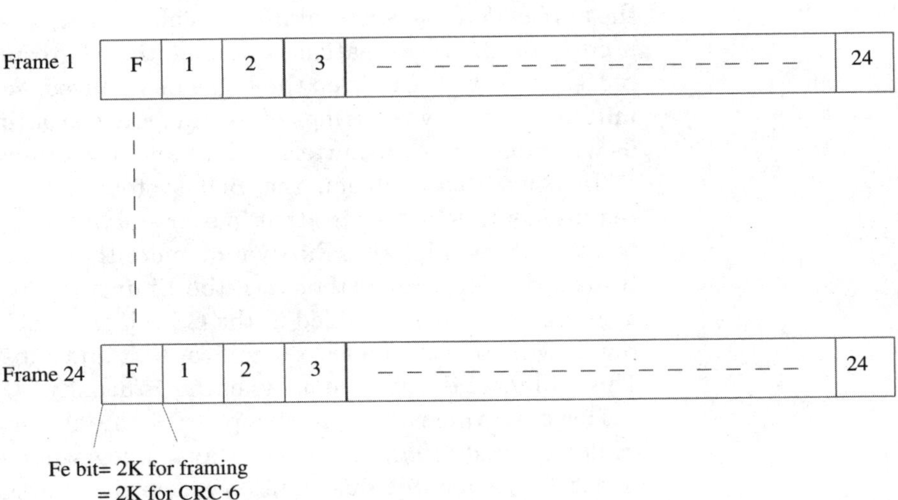

Figure 28-6
ESF improves uptime and error checking.

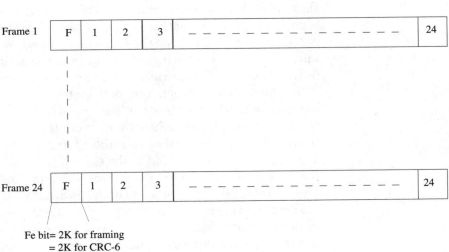

Figure 28-7 Common channel signaling uses channel number 24.

	TS1	2	3	4		23	Comm Channel
F	8	8	8	8	------------------------------	8	SIG

of channel number 24 was assigned for strictly signaling (call setup and teardown) for the other 23 channels. By using a dedicated out-of-band signaling channel, the other 23 channels can carry all 64 Kbps for data, yielding a higher throughput per channel, but with a penalty of losing one channel from the T-1. The format of the common channel signaling arrangement is shown in Figure 28-7. This is a choice a user has to make. If all the user wants is voice communications, the robbed bit signaling will suffice. If, however, the user wants to transmit data, then the common channel signaling will be a potential benefit to the data transmission.

Other Restrictions

Another problem surfaced when T-1 was introduced for data transmission. In order to synchronize the carrier services, T-1 uses a byte synchronization plan. In every eight bits, there must be at least one pulse (represented as a digital one) for the transmitters to derive their timing on the line. If voice is the application, there are minimal problems with this constraint. As long as a party on the conversation is talking, there are sufficient numbers of 1s pulses transmitted to keep the line synchronized. Yet, when data is transmitted, there may be strings of 0s transmitted continuously. This string of 0s will cause the transmitters to drift and lose timing.

To solve this problem, the Bell system introduced a "ones density requirement," which states that in every eight bits transmitted there must be at least one digital 1. Further, no more than 15 consecutive 0s may be transmitted in a row; otherwise, the timing will be lost. The solution to meeting this rule is handled in the CSU. If the CSU receives eight 0s in a row, it will strip off the least significant bit and substitute a 1 in its place. This will meet the requirements of the 1s density.

However, when the 0 is stripped off and a 1 substituted in its place (called *pulse stuffing*), there is no way to know when this is done or when it is not. Consequently, this solution meets the timing and synchronization demands of the network but leaves the user with the risk that the eighth bit

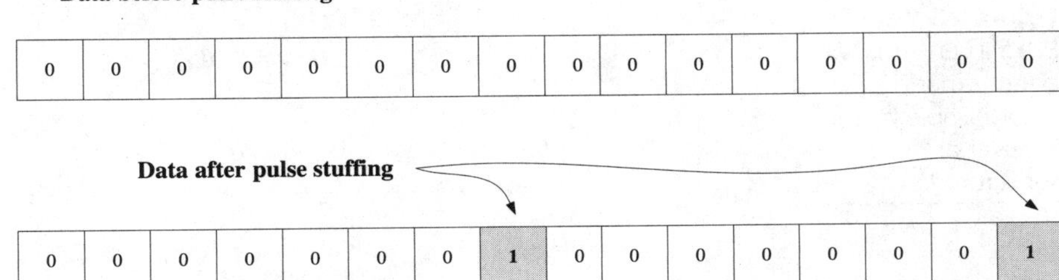

Figure 28-8 Pulse stuffing enables the timing to be preserved.

is wrong. This leads the user to only trusting the first 7 bits, but ignoring the eighth bit. Ultimately, the result is seven usable bits instead of eight or a 56 Kbps throughput on the line instead of 64 Kbps (seven bits × 8,000 samples = 56 Kbps). This results in a lot of wasted bandwidth and a limited throughput for data transmission. The pulse stuffing mechanism is shown in Figure 28-8.

B8ZS

It did not take long before the user began to demand better efficiency of the T-1 line. Therefore, a newer technique is used called *Bipolar* (or binary) *8 Zero Substitution* (B8ZS). With B8ZS, the CSU is responsible for substituting a long string of 0s, eight at a time, with a fictitious 8 bits to meet the demands of the line synchronization, while at the same time enabling the receiving device to recognize that eight 0s were originally sent. The B8ZS inserts a bit pattern of 0001 1011, easily satisfying the demands of the synchronization plan. By inserting these bits, the fourth and the seventh bits will be set as violations to the bipolar rule (*alternate mark inversion* [AMI]). Receiving a bit pattern of 0001 1011 with a bipolar violation in positions 4 and 7 are recognized as a flag that eight 0s were intended. Therefore, the receiving CSU will strip off the fictitious word and reinsert all 0s to the receiving device. This allows for clear-channel 64 Kbps data transmission. The B8ZS is shown in Figure 28-9.

With all these problems and solutions, one can only wonder why everyone uses the T-1. The primary reason is the 1.544 Mbps throughput (or slightly less based on the restrictions) for voice and data transmission. The second reason is the cost/benefits ratio associated with bundling the

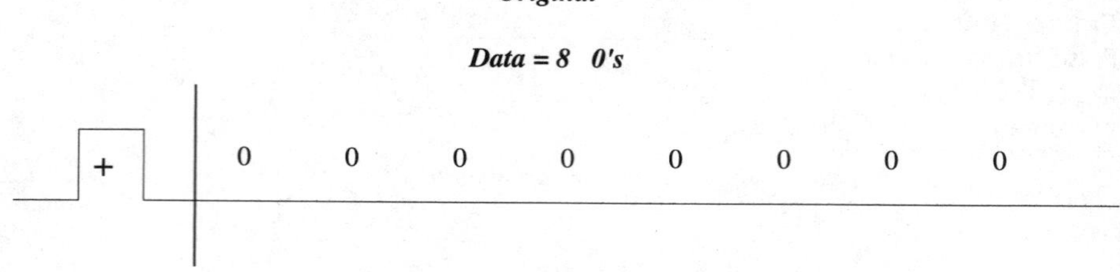

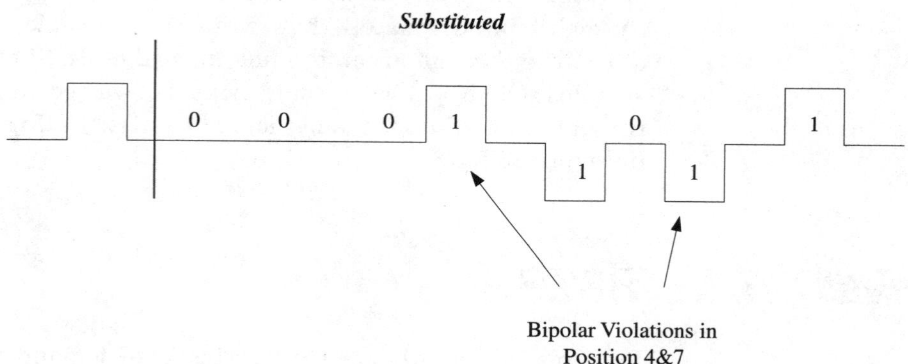

Figure 28-9 B8ZS inserts an 8-bit pattern that is recognizable by the receiver as substituted.

services onto a single, four-wire circuit. Third is the benefit of all digital circuitry to the door. It is because of these three reasons that T-1 is the most widely used digital transmission service in North America.

T-2 Transmission (or DS-2)

For all its work, the T-1 was a beginning step to the North American Digital Hierarchy. From there, other multiplexing steps were instituted to get higher speed communication services. A second means of achieving the high-speed communication is to multiplex four DS-1s into a DS-2. The DS-2 (physically, it is a T-2) is a data channel carrying 96 channels of 64 Kbps each, plus overhead yielding a nominal 6.312 Mbps data transmission speed. To create the DS-2, four DS-1 signals are combined. This creates a DS-2 frame as shown in Figure 28-10. The DS-2 frame is often referred to as the DS-2 M frame and consists of four subframes. Each of the subframes are labeled: M1 though M4. A subframe consists of six blocks of information; each block contains 49 bits.

The first bit in each block is an overhead bit. A DS-2 frame has 24 overhead bits. The remaining 48 bits in each block are DS-1 information bits. Carrying this out then there are

$$48 \text{ DS1 bits/block} \times 6 \text{ blocks/subframe} \times 4 \text{ subframes/DS-2 frame} = 1{,}152 \text{ DS-1 information bits}$$

The four subframes do not represent each of the separate DS-1 signals; instead a bit interleaving of the four DS-1 signals forms the DS-2 frame.

The overhead bits precede the data bits in each of the blocks. The data bits are interleaved, where 0i designates the time slot devoted to DS-1 input i. After 48 information bits, 12 from each of the DS-1s, a new DS-2 overhead bit is inserted. The total number of DS-1 bits transmitted per second in a DS-2 frame is therefore

$$\text{DS-1 rate} \times 4 \text{ DS-1 signals/DS2} = \text{information bits}$$

$$1{,}544 \text{ Mbps} \times 4 \text{ DS-1 signals} = 6.176 \text{ Mbps}$$

Figure 28-10
A DS-2 M frame

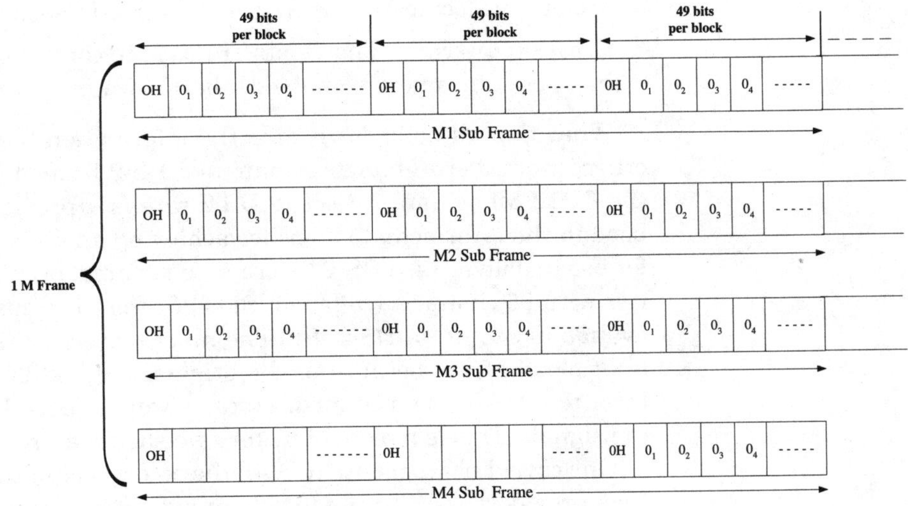

The rate chosen for the DS-2 is 6.312 Mbps. This enables extra overhead for bit stuffing to synchronize the signals.

DS-2 Bit Stuffing

The four DS-1 signals are synchronous to themselves but asynchronous to each other. This means they may be operating at different rates. To synchronize the signal, the multiplexers use a bit stuffing technique. (This was referred to as pulse stuffing in the DS-1 discussion.) Bit stuffing is used to adjust the incoming rates because they all differ.

Framing Bits for the DS-2

Framing bits (F-bits) form the frame alignment much the same as the DS-1 framing bit. There are a total of eight framing bits in a DS-2 frame (two in each subframe). F-bits are located in the first bit position in blocks 3 and 6 of each subframe. The frame alignment pattern is "01," which repeats every subframe. The F-bits are shown in Figure 28-11. Other overhead bits are also shown in the framing pattern. These include the following

- **M-bits (multiframe)** There are four M-bits per DS-2 frame. These are always located in the first bit position in each subframe.
- **C-bits (control)** The C-bits are used to control bit stuffing. Three C-bits per subframe are enabled, also shown in Figure 28-11.

What this all boils down to is the information bits, overhead bits, and other inserted stuff bits all equate to a DS-2 framed format that produces the 6.312 Mbps data rate. The DS-2 rate is a provider service, primarily used in the intermediate steps for achieving a DS-3 (called also a T-3) and for use in delivering a DS-2 to a remote terminal over a subscriber loop carrier with 96 channels (called an SLC-96). Very few customers ever had the availability to use a DS-2. But one must not view this as an unnecessary multiplexing rate because of the original method of multiplexing from a DS-1 to a DS-3; the intermediate rates were needed to control and format the signals. This is a piece of history we should all remember.

One caveat bears mention. With the emergence of the ADSL services, the local providers were trying to deliver a download rate from the network to the consumer for a 6 Mbps video rate and 2 Mbps for data and POTS. The upload rates were lower, but the download rates approximated the DS-2

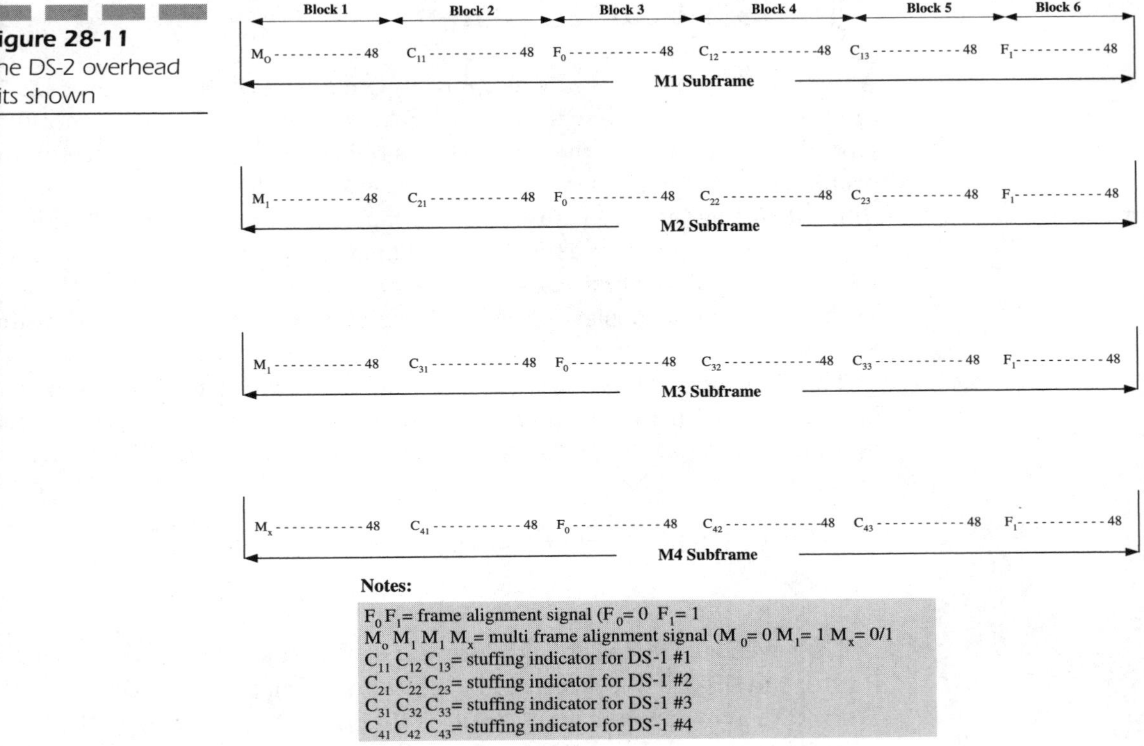

Figure 28-11 The DS-2 overhead bits shown

rate. (It is not classified as a DS-2; the reference here is for analysis to the speeds and aggregates we are now seeing from the suppliers).

DS-3 Service (T-3)

The next step in the evolution of digital transmission services is the T-3 or DS-3. This is a high-speed communication signal used by the larger organizations and the public carriers (local and long distance carriers). Because of its pricing and delivery, this was never practical except for the largest of corporations and providers. The delivery on a copper-based plant used coaxial cable to reach and support the 44.736 Mbps rate of speed. Newer versions deliver the DS-3 on fiber optics or microwave radio, but these are strictly media used to carry the signal. Getting the signal is what is important.

The DS-3 Frame Format

A two-step process is used when multiplexing to a DS-3. First, four DS-1s are multiplexed together to form a DS-2. Second, seven DS-2s are multiplexed together to form the DS-3. This is referred to as the M13 format and protocol. Literally, the same method is used to multiplex the seven DS-2s into a DS-3 as was used to multiplex four DS-1s into a DS-2. In Figure 28-12, the framing format is shown, which uses the M-13 Asynchronous Protocol. The DS-3 frame consists of seven subframes, labeled as 1 through 7. Each subframe consists of eight blocks of data and each block contains 85 bits.

The first bit in each block is an overhead bit, which make 56 overhead bits in the DS-3 frame. The remaining 84 bits in a block are the DS-2 information bits. Doing the math on this, the total DS-2 bits in a DS-3 frame is as follows:

$$84 \text{ DS-2 bits/block} \times 8 \text{ block/subframe} \times 7 \text{ subframes/DS-3}$$

$$84 \times 8 \times 7 = 4704 \text{ bits}$$

The seven subframes do not represent each of the separate DS-2 signals. Rather, the DS-3 frame is formed by bit interleaving the seven DS-2 signals. The interleaving process is the same as that used for the DS-1s to the DS-2. After 84 information bits, 12 from each DS-2, an overhead bit is inserted. To

Figure 28-12
A DS-3 frame

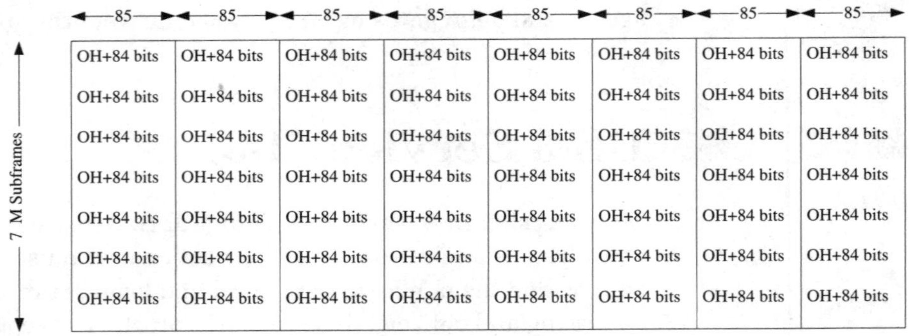

7 M Subframes

85	85	85	85	85	85	85	85
OH+84 bits	OH+84 bits	OH+84 bits	OH+84 bits	OH+84 bits	OH+84 bits	OH+84 bits	OH+84 bits
OH+84 bits	OH+84 bits	OH+84 bits	OH+84 bits	OH+84 bits	OH+84 bits	OH+84 bits	OH+84 bits
OH+84 bits	OH+84 bits	OH+84 bits	OH+84 bits	OH+84 bits	OH+84 bits	OH+84 bits	OH+84 bits
OH+84 bits	OH+84 bits	OH+84 bits	OH+84 bits	OH+84 bits	OH+84 bits	OH+84 bits	OH+84 bits
OH+84 bits	OH+84 bits	OH+84 bits	OH+84 bits	OH+84 bits	OH+84 bits	OH+84 bits	OH+84 bits
OH+84 bits	OH+84 bits	OH+84 bits	OH+84 bits	OH+84 bits	OH+84 bits	OH+84 bits	OH+84 bits
OH+84 bits	OH+84 bits	OH+84 bits	OH+84 bits	OH+84 bits	OH+84 bits	OH+84 bits	OH+84 bits

8 blocks/M Subframe × 85 bits	=	680 bits per M Subframe
NOTES: 7 Subframes × 8 blocks of 85 bits	=	4760 bits per M Frame
1 O/H bit per block × 8 blocks x 7 rows	=	56 O/H bits
Data bits = 5460 - 56	=	4704 per M Frame

get to the nominal number of DS-2 information bits in a DS-3 frame in a one-second period, we see the math as follows:

$$\text{DS-2 rate} \times 7\ \text{DS-2} = \text{Number of bits carried}$$

$$6.312\ \text{Mbps} \times 7 = 44.184\ \text{Mbps}$$

The overall rate of the DS-3 is 44.736 Mbps, enabling extra overhead for bit stuffing and the overhead bits used in the framing format.

DS-3 Bit Stuffing

The seven DS-2 signals are asynchronous to each other (because they may not have been formed in the same multiplexer). They may be operating at different rates. Bit stuffing is used to adjust the different incoming signals. This is the same reasoning behind the bit stuffing in each of the preceding discussions. Stuffing bits are used to justify the signals within the DS-3 frame, and they provide enough consistency for the multiplexers to derive the channels from within the DS-3 signal. In the eighth block of each subframe, an added stuffing slot is reserved to do this justification. The last block therefore may contain the following two choices:

- 1 overhead bit plus 84 information bits
- 1 overhead bit, plus 1 stuffing bit, plus 83 information bits

The stuffing bits reserved in the eighth block will occur in each of the seven subframes randomly. However, a stuff bit will occur in 7 of 18 frames but not always in the same place or the same pattern. This helps to make the transmitters asynchronous because one never knows how many bits will be used for overhead in each of the seven subframes.

The DS-3 Overhead Bits

As already stated, there are overhead bits allocated within the DS-3 frame for the purposes of aligning the frames, providing indication of stuffing and other low-level functions. These are labeled as follows:

- **F-bits** The framing bits form the frame alignment for the DS-3 signal. The F-bits are located in the first bit slot in blocks 2, 4, 6, and 8 of each subframe. The frame alignment pattern is "1001" and repeated in every subframe.

- **M-bits** The multiframing bits form the multiframe alignment signal. Three M-bits are used in each DS-3 frame. The M-bits are located in the first bit slot in block 1 of subframes 5, 6, and 7. The DS-3 equipment uses the M-bit pattern "010" to locate the seven subframes.
- **C-bits** These are used to control the bit stuffing. There are three C-bits per subframe, thus 21 C-bits per DS-3 frame. These are numbered a C_{ij}, with i designating the subframe and j designating the position of the C-bit in the subframe.
- **X-bits** Two X-bits are designated for each DS-3 frame. These are located in the first bit slot of block 1 in subframes 1 and 2. The X-bits must be identical in any DS-3 frame (either 00 or 11). The source may not change the X-bits more than once per second. These bits are used for service messages as a low-level, transmitter-to-receiver form.
- **P-bits** The P-bits (or parity bits) carry parity information. There are two P-bits per DS-3 frame. The P-bits are located in the first bit slot of the first block in subframes 3 and 4. A bit-interleave parity check computes parity over the 4,704 information bits contained in the DS-3 frame. The state of the parity bits is always identical. The two bits are set to a "1" if the previous frame contained an odd number of 1s, or the two bits are set to "0" if the previous frame had an even number of 1s.

CHAPTER 29

Synchronous Optical Network (SONET)

SONET is a standard developed by the *Exchange Carriers Standards Association* (ECSA) for the *American National Standards Institute* (ANSI). This standard defines an optical telecommunications transport for U.S. Telecommunications. SONET standards provide an extensive set of operational parameters for optical transmission systems throughout the industry. The North American industry uses the SONET specifications, whereas the rest of the world uses a close cousin defined as *Synchronous Digital Hierarchy* (SDH), which we will discuss in the next chapter. Between the two sets of standards, the industry attempted to define the roles of transport for the telecommunications providers using optical fibers as the transport medium.

SONET provides more, though. It defines a means to increase throughput and bandwidth through a set of multiplexing parameters. These roles provide certain advantages to the industry, such as the following:

- Reduced equipment requirements in the carriers' network
- Enhanced network reliability and availability
- Conditions to define the overhead necessary to facilitate managing the network better
- Definitions of the multiplexing functions and formats to carry the lower level digital signals (such as DS-1, DS-2, and DS-3)
- Generic standards encouraging interoperability between different vendors' products
- A flexible means of addressing current as well as future applications and bandwidth usage

SONET defines the *Optical Carrier* (OC) levels and the electrical equivalent rates in the *Synchronous Transport Signals* (STS) for the fiber-based transmission hierarchy.

Background Leading to SONET Development

Prior to the development of SONET, the initial fiber-based systems used in the PSTN were all highly proprietary. The proprietary nature of the products included the following:

- Equipment
- Line coding
- Maintenance

- Provisioning
- Multiplexing
- Administration

The carriers were frustrated with the proprietary products because of interoperability problems, sole-source vendor solutions (which held the carriers hostage to one vendor), and cost issues. These carriers approached the standards committees and demanded that a set of operational standards be developed that would allow them to mix and match products from various vendors. In 1984, a task force was established to develop such a standard. The resultant standard became SONET.

Synchronizing the Digital Signals

SONET involves the synchronization of the digital signals arriving at the equipment. Keeping in mind that the signals may be introduced in one of three ways, it is important to attempt to get everything on a common set of clocking mechanisms. In digital transmission, the normal way of synchronizing traffic is to draw a common clocking reference. In the hierarchy of clocking, systems use a stratum clocking architecture. The stratum references in North America come in a four-level architecture. These are shown in Table 29-1.

In a set of synchronous signals, the digital transitions in the signals occur at the same rate. There may be a phase difference between the transitions in the two signals, but this would be in specified ranges and limits.

Table 29-1

Summary of clocking systems

Stratum Reference	Location Used	Accuracy
1	Primary Reference drawn from a GPS or the National Reference Atomic clock	± 1 pulse in 10^{11}
2	Toll offices (Class 1 to 4)	± 1.6 pulses in 10^8
3	End offices (Class 5)	± 4.6 pulses in 10^6
4	Customer equipment (multiplexer, channel bank, and so on)	± 32 pulses in 10^6

The phase differences can be the result of delay in systems, jitter across the link, or other transmission impairments. In a synchronous environment, all the clocks are traceable back to a common reference clock (the Primary Reference Clock).

If two signals are almost the same, they are said to be plesiochronous. Their transitions are close (or almost the same), and variations are contained within strict limits. The clocking between the two different sources, although accurate, may be operating at a different rate.

Finally, if two signals are randomly generated and do not occur at the same rate, they are said to be asynchronous. The difference between two clocks is much greater, possibly running from a free running clock source.

Any one of these signals, synchronous, plesiochronous, or asynchronous, may arrive at a SONET multiplexer to be formulated and transmitted across the network. SONET defines the means of synchronizing the traffic for transmission.

The SONET Signal

SONET defines a technique to carry many signals from different sources and at different capacities through a synchronous, optical hierarchy. The flexibility and robustness of SONET are some of its strongest selling points. Additionally, in the past, many of the high-speed, multiplexing arrangements (DS-2 and DS-3) used bit interleaving to multiplex the data streams through the multiplexers. SONET uses a byte-interleaved multiplexing format. This is a strong point because it keeps the basic DS-0 intact throughout the network, making it easier to perform diagnostics and troubleshooting. Byte interleaving simplifies the process and provides better end-to-end management.

The base level of a SONET signal is called the *Synchronous Transport Signal Level1* (STS-1), operating at 51.84 Mbps. The first step in using the SONET architecture is to create the STS-1. Other levels exist in multiples of the STS-n to create a full family of transmission speeds. The SONET hierarchy is shown in Table 29-2.

Why Bother Synchronizing?

In the past, transmission systems have been primarily asynchronous. Each terminal device in the network was independently timed. In a digital syn-

Table 29-2

Summary of electrical and optical rates for SONET

The SONET Hierarchy			
Electrical Signal	**Optical Value**	**Speed**	**Capacity**
STS-1	OC-1	51.84 Mbps	28 DS-1 or 1 DS-3
STS-3	OC-3	155.520 Mbps	84 DS-1 or 3 DS-3
STS-12	OC-12	622.08 Mbps	336 DS-1 or 12 DS-3
STS-24	OC-24	1.244 Gbps	672 DS-1 or 24 DS-3
STS-48	OC-48	2.488 Gbps	1344 DS-1 or 48 DS-3
STS-192	OC-192	9.95 Gbps	5376 DS-1 or 192 DS-3
STS-768	OC-768	40 Gbps	21504 DS-1 or 768 DS-3*

Other rates exist, but these are the most popularly implemented
*new rates being defined

chronous transmission system, clocking is all important. Clocking uses a series of pulses to keep the bit rate constant and to help recover the ones and zeros from the data stream. Because these past clocks were independently timed, large variations occurred in the clock rate, making it extremely difficult (if not impossible) to extract and recover the data. A DS-1 operates at 1.544 Mbps ± 150 pps, whereas a DS-3 operates at 44.736 Mbps ± 1789 pps. These differences mean that one DS-1 may be transmitting at up to 300 pps different than the other (assuming that DS-1 is at -150 pps, and the second one is at $+150$ pps). The differences can make it difficult to derive the actual data across a common receiver.

Back in the section on the T-carriers (see Chapter 28, "The T-Carrier Systems [T1/T2, and T3)"], we discussed the asynchronous method of multiplexing a DS-3. In that section, we saw that four DS-1s were bit interleaved together to form a DS-2 and that bit stuffing occurred. From there, seven DS-2s were bit interleaved together to form the DS-3, but there were several possible steps where bit stuffing occurred at the multiplexing point. The stuff bits were random occurring in seven of 18 frames, causing confusion in delivering and demultiplexing the signal. Moreover, when a problem occurs on a DS-3 using the M13 asynchronous Multiplexing technique, the entire DS-3 must be demultiplexed to find the problem. This is inefficient.

Therefore, the method of synchronously multiplexing in a SONET architecture provides for better efficiency and problem resolution. Using SONET, the average frequency of all the clocks will be the same. Every clock can be traced back to a common reference, which is highly reliable and stable. Bit stuffing can be eliminated in the preparation of the STS-1 signal; therefore, the lower-speed signals are more readily accessible. The benefits outweigh the possible overhead associated with the SONET multiplexing scheme. In SONET, the hierarchy of clocking follows the master-slave clocking architecture. Higher level (stratum 1) clocks will feed the timing across the network to lower level devices. Any jitter, phase shifts, or drifting by the clocks can be accommodated through the use of pointers in the SONET overhead. The internal clock in a SONET multiplexer may also draw its timing from a *Building Integrated Timing System* (BITS) used by switching systems and other devices. This terminal will then serve as the master for other SONET nodes downstream, providing timing on its outgoing signal. The receiving SONET equipment will act in a slave mode (loop timing) with their internal clocks timed by the incoming signal. The standard specifies that SONET equipment must be able to derive its timing at a stratum level 3 or above.

The SONET Frame

SONET also defines a frame format in which to produce the basic rate of 51.84 Mbps (the STS-1). Each of the additions to the multiplexing rates is a multiple of the STS-1. The basic format consists of a frame that is 80 bytes (columns) wide and 9 bytes high (rows). The basic STS-1 signal is then applied into this 810 byte frame. The frame is shown in Figure 29-1. The frame will occur 8,000 times per second. If we calculate the math on this we have the following:

$$810 \text{ bytes} \times 8 \text{ bits/byte} \times 8000 \text{ frames/sec} = 51.84 \text{ Mbps}$$

Overhead

From the 810 byte frame, overhead is enabled in several ways to handle the *Operations, Administration, Maintenance, and Provisioning* services (OAM&P).

The first part of the overhead is defined as the transport overhead. The transport overhead uses the first three columns and all nine rows. This creates 27 bytes of transport overhead. As shown in Figure 29-2, the transport overhead is divided into two pieces. The first three columns and the first

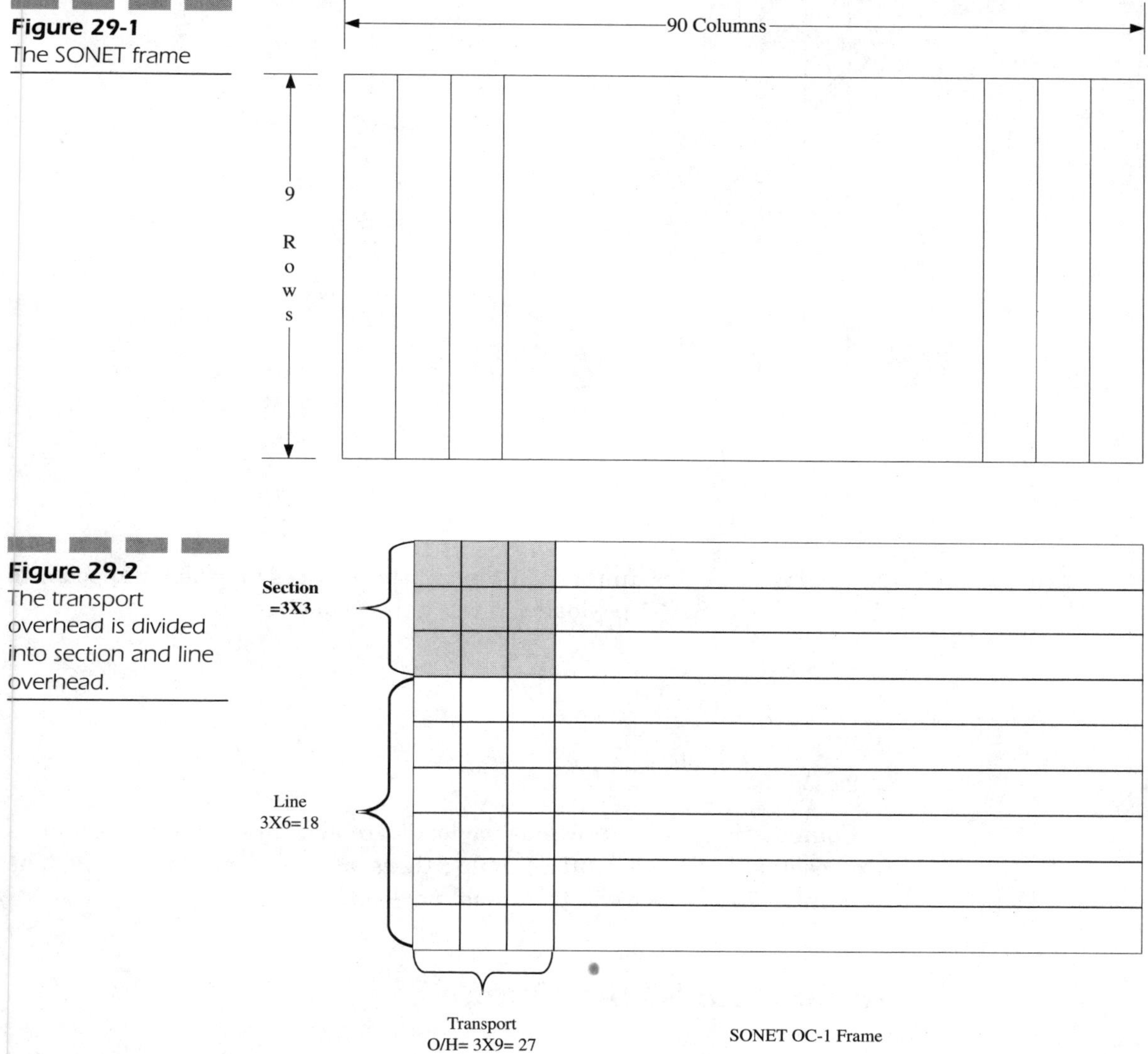

Figure 29-1 The SONET frame

Figure 29-2 The transport overhead is divided into section and line overhead.

three rows (9 bytes) are used for section overhead. The remaining six rows in the first three columns (18 bytes) are used for line overhead.

The remaining 87 columns and nine rows (783 bytes) are designated as the *Synchronous Payload Envelope* (SPE). Inside the SPE, an additional one column, nine rows high (9 bytes) is set aside for *Path Overhead* (POH). This is shown in Figure 29-3. After the POH is set aside, the resulting

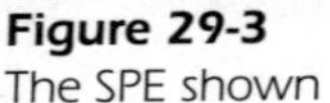

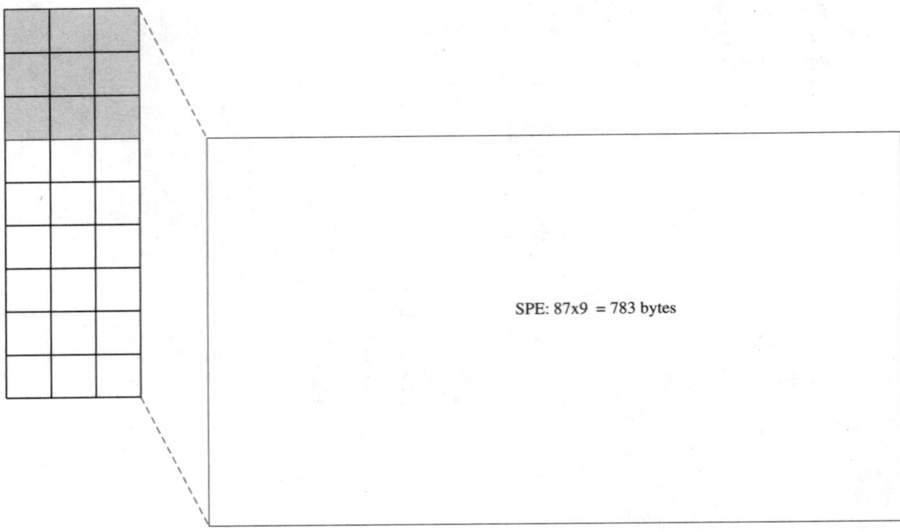

Figure 29-3
The SPE shown

payload is 774 bytes. In these 774 bytes, the services are then mapped into the frame. The STS-1 payload can carry the following:

- 1 DS-3
- 7 DS-2s
- 21 E-1s
- 28 DS-1s

Combinations of the previous payloads are also allowable. Two columns are reserved as fixed stuff columns; these are columns 30 and 59. The remaining 756 bytes carry the actual payload.

Inside the STS-1 Frame

The SPE can begin anywhere inside the STS-1 envelope. Normally, it begins in one STS-1 frame and ends there. However, it may begin in one STS-1 frame and end in another. The STS payload pointer contained inside the transport overhead points to the beginning byte of an SPE. The possibilities then are that the frame can carry a locked payload or a floating payload. Floating, in this regard, refers to the payload floating between two frames. The overhead associated with SONET is designed to let the receiver know where to look for its payload and extract the information at a starting point. This floating frame is shown in Figure 29-4.

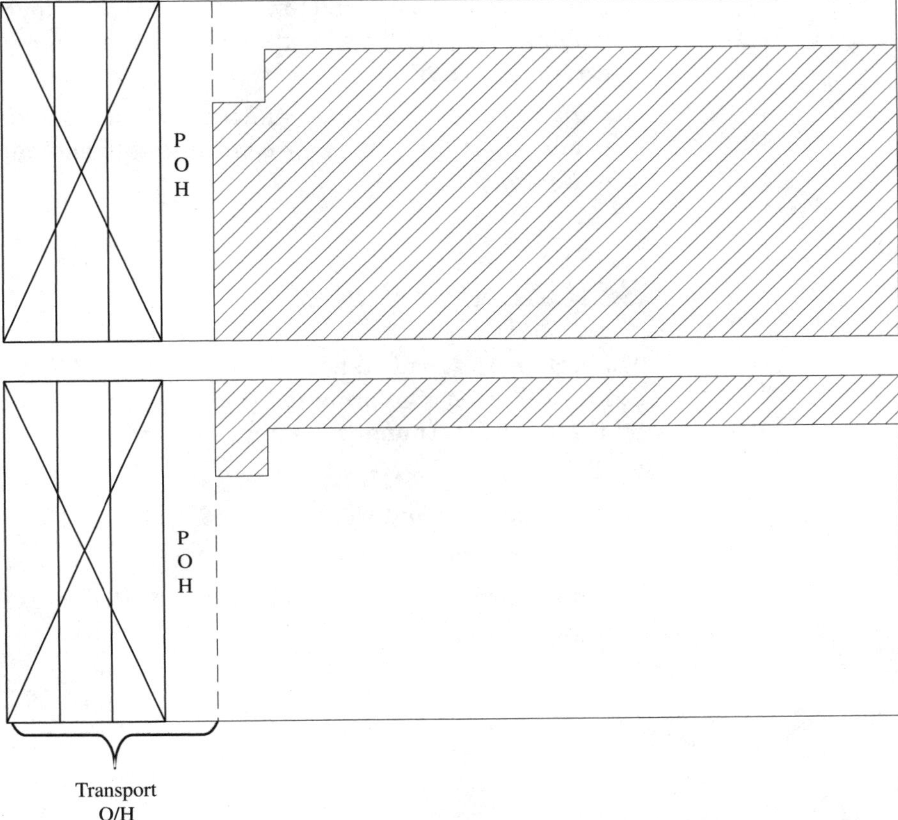

Figure 29-4
A floating payload inside two frames

SONET Overhead

SONET provides significant overhead, enabling the simpler multiplexing techniques and improved OAM&P operations. In the next discussion, the overhead is examined to better understand why so much is dedicated to the overhead functions.

In Figure 29-5, the architecture of the SONET link is shown as defined by ECSA and ANSI. In this reference, the link architecture is broken down into three parts:

- *Section* The section is defined as the portion of the link between two repeater functions or between a repeater and line terminating equipment. Sufficient overhead is enabled to detect and troubleshoot errors on the link between these two points (STE).

- *Line* The line overhead provides sufficient information to detect and troubleshoot problems between two pieces of *Line Terminating Equipment* (LTE).
- *Path* The POH provides sufficient overhead to detect and troubleshoot problems between the end-to-end path terminating *Pieces of Equipment* (PTE).

Section Overhead

The section overhead is shown in Table 29-3 and Figure 29-6. This overhead contains 9 bytes of information that is accessed, generated, and processed by the section terminating equipment. It supports functions such as the following:

- Performance monitoring
- A local orderwire
- Data communications channels for OAM&P messages
- Framing information

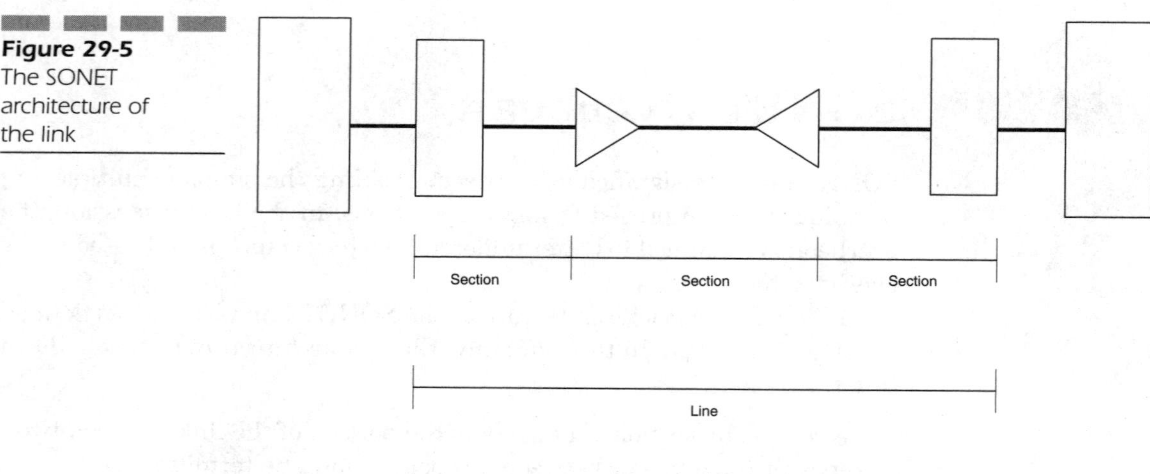

Figure 29-5 The SONET architecture of the link

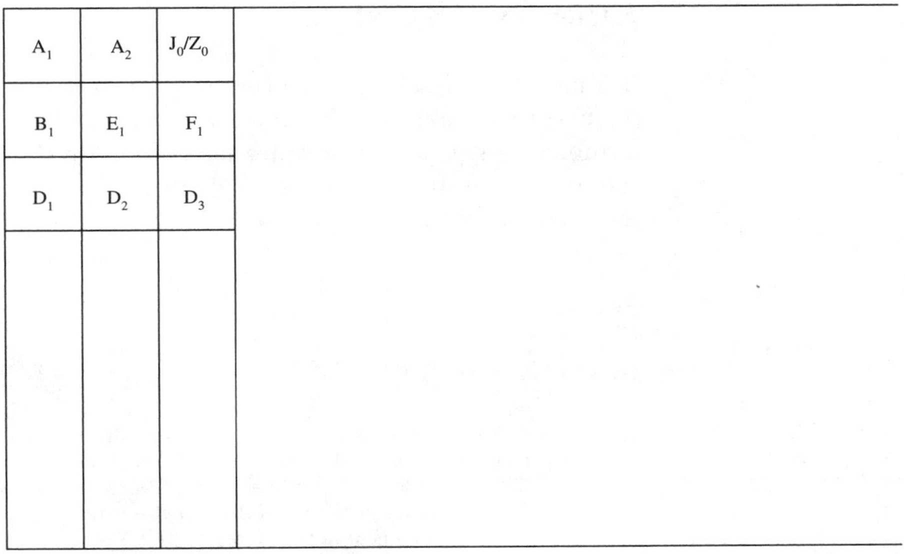

Figure 29-6
The section overhead

A_1	A_2	J_0/Z_0
B_1	E_1	F_1
D_1	D_2	D_3

Table 29-3

The individual bytes are defined in the designation and their use

Byte	Description
A1-A2	Framing bytes: These two bytes provide an indication of the start of an STS-1 frame.
J0/Z0	Section Trace (J0) or Section Growth (Z0): This byte in each of the STS-1s in an STS-n was originally defined as the STS-1 ID (C-1) byte. It has been redefined as either the section trace byte (in the first STS-1 of STS-n) or as a growth byte (used in the second through nth STS-1s.)
B1	Section *Bit Interleaved Parity* code (BIP-8) byte: This is a parity code (using even parity) used to check for transmission errors over a section of the link (a repeater). The value is calculated over all bits of the previous STS-n frame after scrambling and then placed in the B1 byte of the STS-1 before scrambling. This byte only defines for STS-1 number 1 of an STS-n signal.
E1	Orderwire: This is a 64 Kbps channel for a voice communications orderwire between two repeater functions or a repeater and a line terminating equipment. It allows for two technicians to talk to each other while they troubleshoot problems.
F1	Section User byte: This byte is set aside for users' choice. It terminates at all section terminating equipment within a line. It can be read or written at each section terminating equipment on the line. This may be used for proprietary user maintenance and diagnostic systems, or for SNMP functions on a user network.
D1-3	Section *Data Communications Channel* (DCC) bytes: This creates a 192 Kbps data channel for diagnostics and testing for OAM&P. It uses a message-based channel for remote locations to control alarms, monitoring, and other maintenance functions.

Line Overhead

In Table 29-4, we see the line overhead, which occupies rows 4 through 9 in the first three columns. The line overhead used between two pieces of line terminating equipment provides for the OAM&P functions. This is more extensive than the section overhead. Figure 29-7 also shows its position in the overhead of the STS-1 frame.

Table 29-4

The individual bytes of the line overhead

Byte	Description
H1-H2	STS Payload Pointers: These two bytes indicate the offset bytes between the pointer and the first byte of the STS SPE. The pointer bytes are used in all STS-1s within an STS-n to align the transport overhead in the STS-n and perform frequency justification. These bytes can also be used to indicate when concatenation is used or to indicate path *Alarm Indications Status* (AIS).
H3	Pointer Action byte: This byte is allocated for frequency justification. The H3 byte is used in all STS-1s within an STS-n to carry the extra SPE byte to either increment or decrement the pointer.
B2	Line *Bit Interleaves Parity* code (BIP-8) byte: This parity code byte is used to determine if a transmission error has occurred over the line. It uses even parity and is calculated over all bits of the line overhead and STS-1 SPE of the previous STS-1 frame before scrambling. The value is placed in the B2 byte of the line overhead before scrambling.
K1-K2	*Automatic Protection Switching* (APS) channel bytes: These two bytes are used for protection signaling between two line terminating devices for bidirectional automatic protection switching and for detecting *Alarm Indication Signal on the Line* (AIS-L) and *Remote Defect Indication* (RDI) signals.
D4-D12	Line *Data Communications Channel* (DCC) bytes: This creates a 576 Kbps channel capacity for message signaling from a central location for OAM&P. These bytes are available for generic, internal, or external messages, or vendor-specific messages. The use of a protocol analyzer is needed to access the DCC.
S1/Z1	Synchronization Status byte (S1): The S1 byte is used for synchronization status of network elements. It is located in the first STS-1 of an STS-n. Bits 5 to 8 convey the synchronization information. The Growth (Z1) byte is located in the second through nth STS-1s of an STS-n and allowed for growth.

Table 29-4 cont.

The individual bytes of the line overhead

Byte	Description
M0-M1	STS-1 REI-L (M0) byte is defined for STS-1 in an OC-1 or for the STS-1 electrical signal. Bits 5 to 8 report on the *Line Remote Error Indication* function (REI-L, which used to be called the FEBE). STS-n REI-L (M1) byte is located in the third STS-1 (in order of appearance in the byte interleaved STS-n electrical or optical signal) in an STS-n (where n ≥3) and is used for the REI-L function.
Z2	Growth (Z2) byte: This byte is located in the first and second STS-1s of an STS-3, and the first, second and fourth through nth STS-1s of an STS-n (where 12 ≤ n ≤ 48). These bytes are reserved for future growth.
E2	Orderwire byte: This orderwire byte provides a 64 Kbps channel between line equipment for an express orderwire. It is a voice channel used by technicians, and is ignored as it passes through the repeaters.

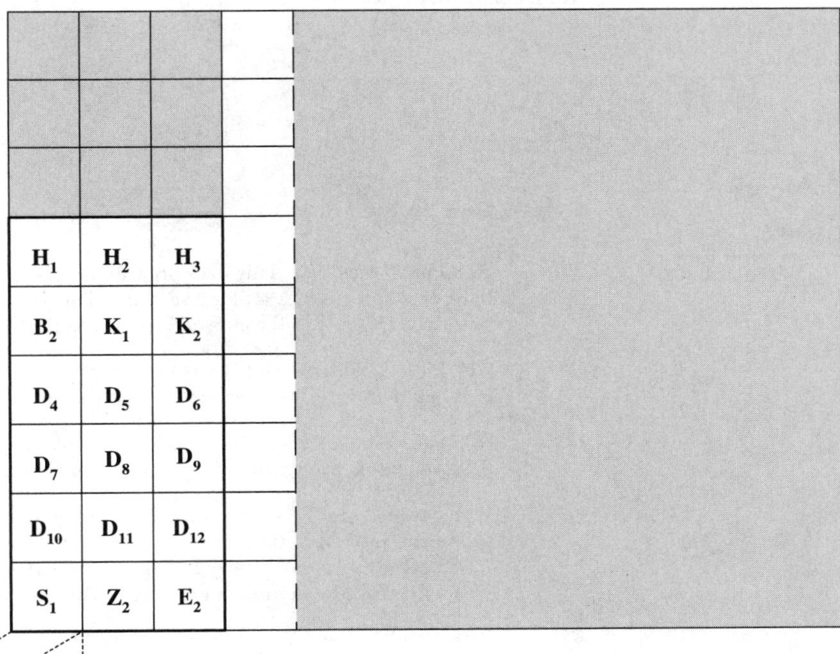

Figure 29-7
The line overhead of a SONET frame

POH

The final piece of overhead is the POH contained inside the SPE. POH contains 9 bytes of information starting at the first byte of the STS SPE. The POH provides for communication between two path-terminating PTE. The PTE is where the actual multiplexing and demultiplexing function takes place as the services are mapped into the SONET frame. The functions supported by the POH are as follows:

- Performance monitoring of the STS SPE
- Signal label the individually mapped payloads
- Path status
- Path trace

The POH is shown in Table 29-5 and in Figure 29-8 to correlate them in the overall function of the STS frame.

Table 29-5

The POH defined

Byte	Description
J1	STS Path Trace byte: This user programmable byte repetitively transmits a 64 byte, or a 16-byte (E.164) format string. This is used so the receiving end of the path to verify it is still connected to the sending end.
B3	STS Path *Bit Interleaved Parity* code (path BIP-8) byte: This byte is even parity to check if an error has occurred across the entire path.
C2	STS Path Signal Label byte: This byte is used to identify the content of the STS SPE, including the status of the mapped payloads.
G1	Path Status byte: This byte is used to convey path terminating status and performance characteristics back to the originating end. Bits 1 to 4 are used for ATA Path REI-P (was referred to as the FEBE). Bits 5, 6, and 7 are for an STS Path RDI (RDI-P) signal, and bit 8 is not defined.
F2	Path User Channel byte: The byte is used for user communications between end-to-end elements. This may be a proprietary network management system or an SNMP system to determine the status and alarms of the far end on the path.
H4	VT Multiframe Indicator byte: This byte provides a general multiframe alignment indicator for payload containers. Currently, it is only used for VT payloads.
Z3-Z5	Growth bytes: These bytes are undefined at this time.

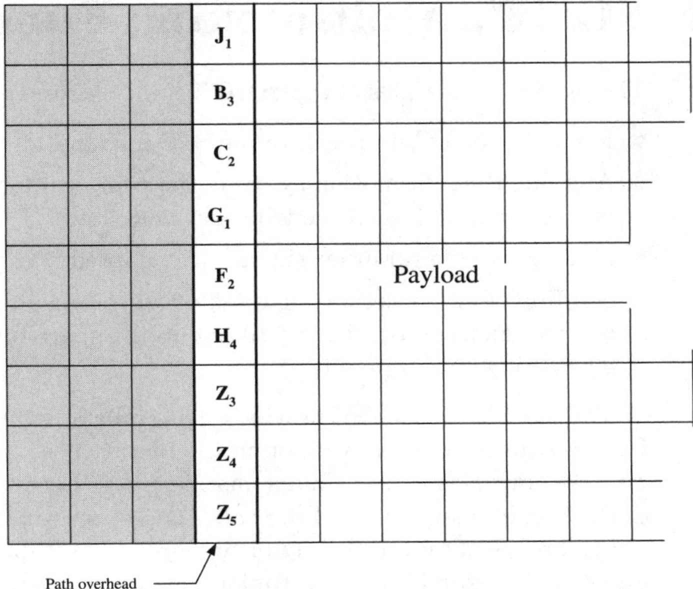

Figure 29-8
POH in the STS frame

Virtual Tributaries

SONET is more than the STS frame; it also defines the sub-STS levels of payload that can be carried. The STS-1 payload can be subdivided into virtual tributaries, synchronous signals to transport lower-speed signals. The normal sizes defined by SONET are shown in Table 29-6.

Table 29-6
The values of the virtual tributaries defined for SONET

VT	Bit Rate	Equivalent DS-n Level	Required Bytes (Rows × Columns)
1.5	1.728 Mbps	DS-1	27 (9 × 3)
2	2.304 Mbps	E-1	36 (9 × 4)
3	3.456 Mbps	DS-1C	54 (9 × 6)
6	6.912 Mbps	DS-2	108 (9 × 12)

SONET Multiplexing Functions

The primary principles of the SONET multiplexers are as follows:

- Mapping the tributaries into the STS-n frame
- Aligning the information by using the pointer information to determine where the first byte of the tributary is located
- Multiplexing lower order signals are adapted to higher order signals
- Stuffing of bits necessary to handle the various lower-speed, asynchronous channels, and filling up the spare bits to keep everything in alignment

The SONET equipment provides these functions. SONET can carry very large payloads, as we have seen in the hierarchy of the data speeds. Up to now, we have seen the primary mapping and layout of an STS-1 (OC-1). SONET equipment can add the value necessary to protect investments by either lower-rate multiplexing or higher-rate multiplexing. If one looks at an OC-3, for example, the multiplexer will produce a larger STS frame. In this case, as shown in Figure 29-9, the frame is three times larger or 270 bytes (columns) wide and nine rows high for a total of 2,430 bytes. Note from this figure the overhead for the STS-n is located in the beginning of the frame, whereas the POH is located at the start of each payload.

Figure 29-9 The STS-3 (OC-3) frame

STS-1			STS-2			STS-3			POH
A_1	A_2	J_0/Z_0	A_1	A_2	J_0/Z_0	A_1	A_2	J_0/Z_0	J_1
B_1	E_1	F_1	B_1	E_1	F_1	B_1	E_1	F_1	B_3
D_1	D_2	D_3	D_1	D_2	D_3	D_1	D_2	D_3	C_2
H_1	H_2	H_3	H_1	H_2	H_3	H_1	H_2	H_3	G_1
B_2	K_1	K_2	B_2	K_1	K_2	B_2	K_1	K_2	F_2
D_4	D_5	D_6	D_4	D_5	D_6	D_4	D_5	D_6	H_4
D_7	D_8	D_9	D_7	D_8	D_9	D_7	D_8	D_9	Z_3
D_{10}	D_{11}	D_{12}	D_{10}	D_{11}	D_{12}	D_{10}	D_{11}	D_{12}	Z_4
S_1	Z_2	E_2	Z_1	Z_2	E_2	Z_1	Z_2	E_2	Z_5

270 Bytes

Add-Drop Multiplexing: A SONET Benefit

Another major benefit of the SONET specification is the ability to perform add-drop multiplexing. Even though network elements are compatible at the OC-N level, they may still differ from vendor to vendor. SONET doesn't attempt to restrict vendors to all providing a single product, nor does it require that they produce one of every type out there. One vendor may offer an *Add-Drop Multiplexer* (ADM) with access to the DS-1 level only, whereas another may offer access to DS-1 and DS-3 rates. The benefit of an ADM on a *Wide Area Network* (WAN) is to drop (demultiplex) only the portions of the optical stream required for a location and let the rest pass through without the demultiplexing process. It would be extremely inefficient to have to demultiplex an entire OC-12, for example, only to drop out one DS-1. The ability to extract only what is necessary helps to prevent errors, loss of data, and other delays inherent with older technologies. The add-drop multiplexer makes this attractive for carriers to use in rural areas where they may bundle many lower-speed communications channels onto a single OC-1 or OC-3 to carry the information back to the central metropolitan areas. Moreover, beyond just dropping a digital signal out of a higher-speed OC-n, the carrier can fill in what has been vacated. (For example, if a DS-1 is dropped off along the optical path, a new DS-1 can be multiplexed back into the OC-3 in its place.) This enables the carriers considerable flexibility. In Figure 29-10, an ADM is shown. Here portions of the bandwidth can be dropped off, and additional new signals can be added in place of the data stream dropped out of the higher-speed signal. A single stage add-drop multiplexing function can multiplex various inputs into an OC-n signal. At an

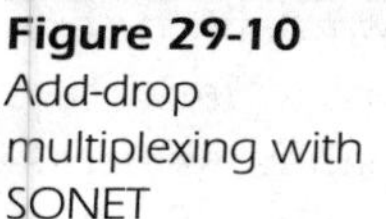

Figure 29-10 Add-drop multiplexing with SONET

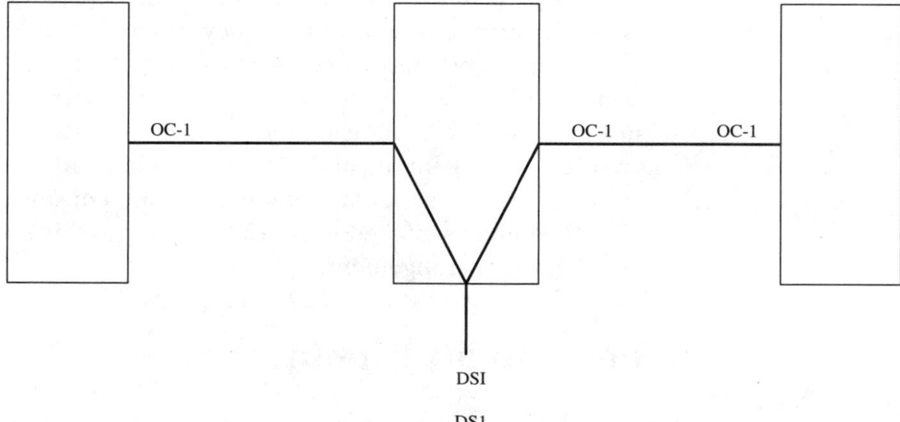

add-drop site, only those signals that need to be accessed are dropped and inserted. The remaining traffic continues through the network switching system without requiring special processing.

SONET Topologies

Several different topologies can be used in a SONET network by using the various multiplexers. These include the normal topologies most networks have been accustomed to over the years. They include the following:

- Point-to-point
- Point-to-multipoint
- Hub and spoke
- Ring

It is the variations that enable the flexibility of SONET in the WANs built by the carriers but are now becoming the method of choice at many large organizations.

In each of the topologies, larger organizations are finding the benefits of installing highly reliable, interoperable equipment at the private network interfaces and access to the public networks.

Point-to-Point

The SONET multiplexer, the entry level PTE for an organization (or the equipment installed by the LEC at the customer's premises to access the network), acts as a concentrator device for the multiple lower-speed communications channels, such as DS-1 and DS-3. In its simplest form, two devices are connected with an optical fiber (with any repeaters as necessary) as a point-to-point circuit. As the entry-level point into a SONET architecture, the inputs and outputs are identical. In this environment, the network can act as a stand-alone environment and not have to interface with the public switched networks. See Figure 29-11 for the point-to-point multiplexing arrangement.

Point-to-Multipoint

The next step is to consider the point-to-multipoint arrangement. This will use a form of add-drop multiplexing to drop circuits off along the way.

Figure 29-11 Point-to-point service with SONET

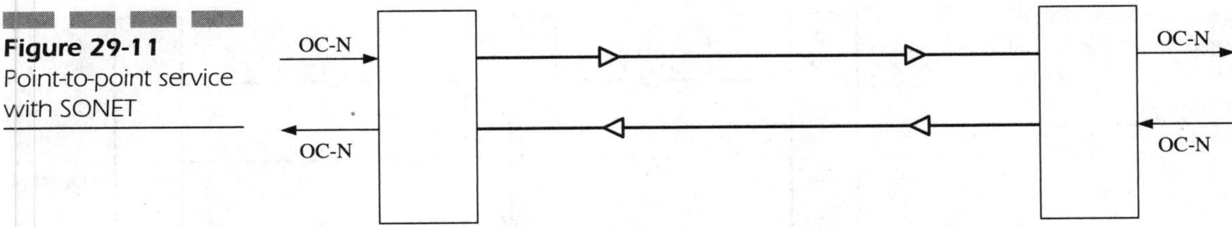

In a large corporate network spanning the country (or any subset), a single, high-speed link may be employed. The SONET ADM is used for the task at hand, dropping circuits out without demultiplexing the entire high-speed signal. In Figure 29-12, the ADM is installed between two far-end locations so that signals can be added or dropped off as necessary. This is a better solution than renting three different circuits between points A-B, A-C, and B-C, which adds to the complexity and cost. By using a circuit from A-B-C with ADMs, the service can usually be more efficiently accommodated.

Hub and Spoke

The hub and spoke method (sometimes referred to as the *star network*) enables some added flexibility in the event of unpredicted growth or constant changes in the architecture of the network. SONET multiplexers can be hubbed into a Digital Cross-Connect where it is concentrated and then forwarded on to the next node (see Figure 29-13). This is used in many larger organizations where regional offices are located, and district or branch offices are tied into the network through the hub. Once again, the flexibility is there if a major change occurs in the network architecture or in the event of major campaigns in the organization. Hubs will act as the cross-connect points to link the various echelons in the network together. These may be developed in a blocking or nonblocking manner. Typically, some blocking may be enabled.

Ring

Next, the glory of SONET! In a ring architecture, where SONET automatic protection switching is employed, the best of all worlds comes to fruition. In this topology, ADMs are used throughout the network, and a series of point-to-point links are installed between adjoining neighbors. The bidirectional

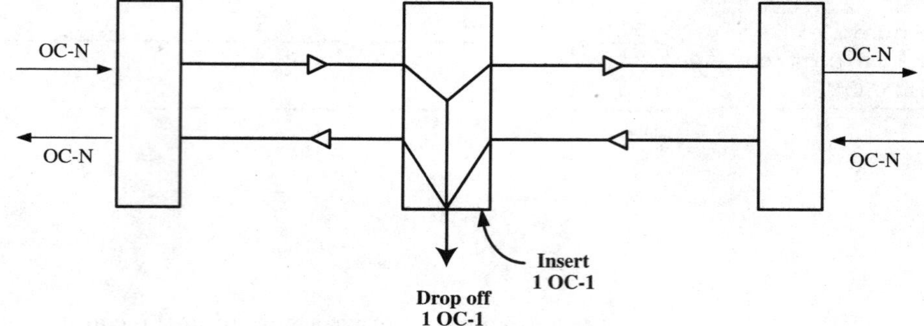

Figure 29-12 ADMs installed along the way

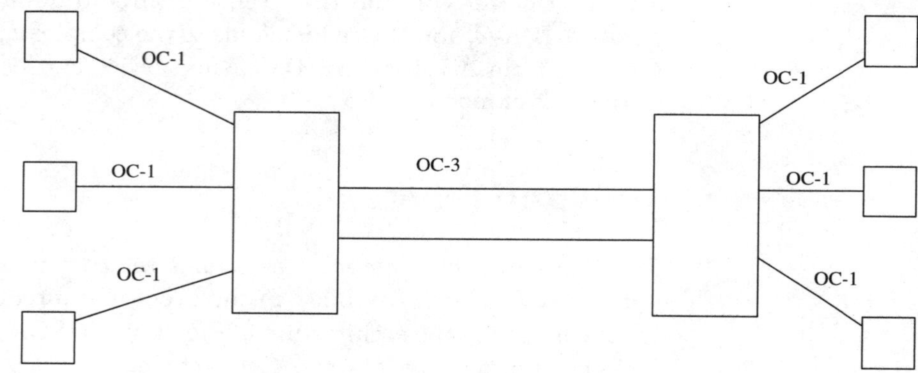

Figure 29-13 Hub and spoke in a SONET multiplexer network

capability places the most robustness into the network; however, unidirectional services can also be installed. The primary advantage of the ring architecture is survivability in the event of a cable cut or a failure in a network node. The multiplexers have sufficient intelligence to reroute or reverse direction in the event of a failure. If more than one fiber link is installed, the systems could use alternate paths, but they must recover in milliseconds (which APS on SONET is designed to do). Figure 29-14 shows the ring topology with dual fibers run (bidirectional service) between the ADMs.

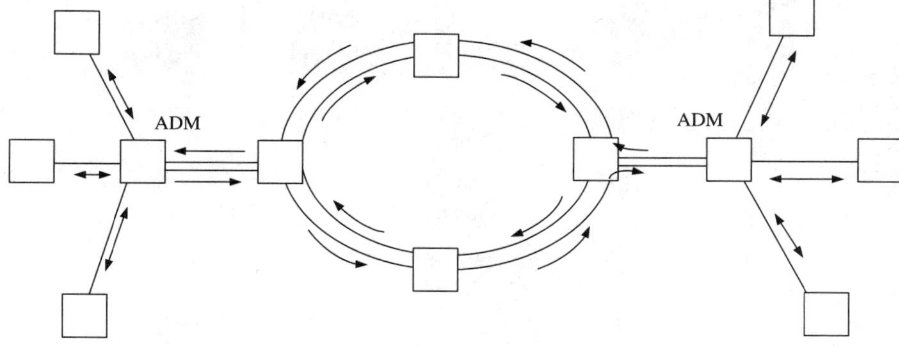

Figure 29-14 Ring architecture of SONET multiplexers

Evolution of SONET in the Rest of the World

Following the work by ANSI in developing the SONET standards, the rest of the world began to move in tandem. The ITU-TSS decided to define a synchronous standard that would address internetworking between the ITU and ANSI transmission hierarchies. In 1989, the ITU published the SDH standards. SDH is a world standard, whereas SONET is a North American standard. However, SONET is considered a subset of the SDH standards. Prior to the publication of the SDH standard, different rates and speeds were available in the North American community (which is the United States, Canada, and Japan) compared to the rest of the world. The ITU was looking for a means of accommodating the differences in a standardized package so that both standards could coexist. SDH marks one of the first steps in making that happen. However, although SDH sets a rate of multiplexing and the derived speeds (and the means of achieving those speeds), some differences between SONET and SDH exist. The packaging of the tributaries is different.

In Table 29-7, a comparison of the SONET and SDH architectures is shown. SDH defines the *Synchronous Transfer Mode* (STM) level N, whereas SONET defines the STS-n and the OC-n.

Table 29-7

Comparison of SONET and SDH rates

SONET Signal	Rate	SDH Signal	SONET Payload	SDH Payload
OC-1	51.84 Mbps	STM-0	28 DS-1 or 1 DS-3	21 E-1
OC-3	155.52 Mbps	STM-1	82 DS-1 or 3 DS-3	63 E-1 or 1 E-4
OC-12	622.08 Mbps	STM-4	336 DS-1 or 12 DS-3	252 E-1 or 4 E-4
OC-48	2.488 Gbps	STM-16	1344 DS-1 or 48 DS-3	1008 E-1 or 16 E-4
OC-192	9.953 Gbps	STM-64	5376 DS-1 or 192 DS-3	4032 E-1 or 64 E-4

The OC-n also has a compatible STS-n rate.

Even though the rates are the same, the packaging of the frames for SDH and SONET is different.

SDH

Ever since the standards bodies approved the recommendations for the SDH (and SONET), the services have been effectively used to improve and revolutionize the industry. Significant cost efficiencies and performance improvements have been shown. SDH provides a means for the rest of the world to use the capabilities of fiber-based transport systems and multiplexing architectures to improve upon the older *Plesiochronous Digital Hierarchy* (PDH), which was inefficient and expensive.

Digital networks continue to expand in complexity and penetration within the carriers' networks, now moving closer to the consumer's door. High-speed communications prior to the formulation of SDH in 1990 operated at speeds of up to 139.364 Mbps. However, the carriers implemented coaxial and radio-based systems operating at 140 Mbps to 565 Mbps. The networks were severely constrained due to the high cost of the transmission medium (coaxial cable, especially). The multiplexing rates used plesiochronous rates, which led to the European PDH.

After the development of fiber and the enhancements of integrated circuitry, the newer transmission speeds and complex networking architectures became realistic. In Europe, the evolution and deployment of ISDN also led to the proliferation of the B-ISDN standards, which enable a simple multiplexing technique. In the United States, the Bell breakup prompted the local carriers to look for interoperability and improvements in network management because of the proliferation of the number of carriers providing long distance services.

The ITU-TSS agreed that something had to be done to improve and standardize the multiplexing and the interoperability, while at the same time taking advantage of the higher capacity of optical fiber. Older bit interleaving of multiplexers should be replaced by byte interleaving to afford better network management. The new standard appeared as SONET in the North Americas, drafted by Bellcore. Later, this same standard developed into the SDH/SONET standard, as approved by the ITU. Although SONET and SDH were initially drafted in support of a fiber, radio-based system, supporting the same multiplexing rates became available.

CHAPTER 30

Synchronous Digital Hierarchy (SDH)[1]

As discussed in Chapter 29, "Synchronous Optical Network (SONET)," SONET is the high-speed optical-based architecture for carriers and users alike. This optical-based networking strategy was developed in North America by the *American National Standards Institute* (ANSI). While the ANSI committees were working on SONET though, another movement was underfoot. In Europe the standards committees were also wrestling with the logical replacement to the *plesiochronous digital hierarchy* (PDH), which is an asynchronous multiplexing plan to create high-speed communications channels. The Europeans came up with a separate multiplexing hierarchy called the *Synchronous Digital Hierarchy* (SDH) in support of the SONET standards.

Ever since the standards bodies approved the recommendations for the SDH (and SONET), the services have been effectively used to improve and revolutionize the industry. Significant cost efficiencies and performance improvements have been shown. SDH provides a means for the rest of the world to use the capabilities of fiber-based transport systems and multiplexing architectures to improve upon the older PDH, which was inefficient and expensive. The PDH evolved in response to the demand for *plain old telephone services* (POTS) and was not ideally suited to deliver the efficient use of bandwidth and high-speed services.

Digital networks continue to expand in complexity and penetration within the carriers' networks, now moving closer to the consumers' door. High-speed communications prior to the formulation of SDH in 1990 operated at speeds of up to 139.364 Mbps. However, the carriers implemented coaxial and radio-based systems operating at 140 Mbps to 565 Mbps. The networks were severely constrained due to the high cost of the transmission medium (coaxial cable especially). The multiplexing rates used plesiochronous rates, which led to the European PDH.

After the development of fiber and the enhancements of integrated circuitry, the newer transmission speeds and complex networking architectures became realistic. In Europe, the evolution and deployment of *Integrated Services Digital Network* (ISDN) also led to the proliferation of the *Broadband ISDN* (B-ISDN) standards, which enables a simple multiplexing technique. In the United States, the Bell breakup prompted the local carriers to look for interoperability and improvements in network management because of the proliferation of the number of carriers providing long-distance services.

---

[1]This chapter comes from the *Optical Switching and Networking Handbook*, R. Bates, McGraw-Hill 2001.

The *International Telecommunication Union-Telecommunication Standardization Sector* (ITU-TSS) agreed that something had to be done to improve and standardize the multiplexing and the interoperability while at the same time take advantage of the higher capacity of optical fiber. The older bit interleaving of multiplexers should be replaced by byte interleaving to afford better network management. The new standard appeared as SONET in North America, drafted by BellCore.[2] Later this same standard developed into the SDH/SONET standard as approved by the ITU. Although SONET and SDH were initially drafted in support of fiber, radio-based systems supporting the same multiplexing rates also became available.

Why SDH/SONET

There are many reasons why SONET and SDH were necessary. The primary reason was that previous technology (PDH) was limited in many ways, such as

- U.S. and European systems had little in common in their mapping and multiplexing systems. Therefore, expensive translators were required for transatlantic traffic on leased lines.
- Standard equipment from different vendors in the same country was incompatible. Everyone produced a proprietary solution that worked with his or her own equipment.
- Systems did not offer self-checking of equipment and network components. Expensive manual checks were required, and extraordinary repair systems were the norm.
- No standard for high bandwidth links existed. Everything maintained the proprietary approach. This created havoc in the industry and needed to be improved.
- Not all of the multiplexing systems were synchronous. In the United States, anything above DS-1 bandwidth was asynchronously multiplexed, timed, and mapped.

[2]The BellCore name has since been changed to TelCordia Technologies.

Synchronous Communications

What does synchronous mean anyway? Why is it so important to the telecommunications industry? The easiest way to describe the need for synchronization is that the bits from one telephone call are always in the same location inside a digital transmission frame such as a DS-1. In the United States, telephone calls using digital transmission systems create a DS-0. The DS-0s are multiplexed, 24 per DS-1 channel. DS-1 lines are synchronously timed and mapped; therefore, it is easy to remove or insert a call. Finding the location creates an easy add-drop multiplexing arrangement.

Plesiochronous

Plesiochronous means "almost synchronous." Variations occur on the timing of the line so bits are stuffed into the frames as padding. The digital bits (1s and 0s) vary slightly in their specific location within the frame creating "jitter." This occurs on a frame-to-frame basis, creating ill timing and requiring some other actions to make everything bear some semblance of timing. An example in the previous chapter with the multiplexing of a DS-3 occurred when

- Four DS-1 lines are bit-interleaved and multiplexed together to create a DS-2.
- Seven DS-2 lines are bit-interleaved and multiplexed to create a DS-3.
- If we need to isolate a particular call from a DS-3, the entire DS-3 must be demultiplexed to the DS-1 level where we can then extract the individual DS-0s.
- Very expensive equipment is needed at every *Central Office* (CO) to demultiplex and multiplex high-speed lines across the backbone networks.

Consequently, the standards committees (both ANSI and the ITU) began working on solutions to the multiplexing problems. The ultimate attempt was to develop a synchronous transmission system that could replace the plesiochronous transmission systems.

In Table 30-1, a summary and comparison of the rates of speed for worldwide speeds and hierarchical arrangements are shown. This table brings the differences to light when you compare the speeds and multiplexed channels combined.

Table 30-1

Comparing the rates of speed for the various levels of the digital hierarchy in Kbps

Hierarchical Level	North American DS-x	European CEPT-x	Japanese Level	International Rates
0	64	64	64	64
1	1,544	2,048	1,544	2,048
2	6,312	8,448	6,312	6,312
3	44,736	34,368	32,064	44,736
4	274,176	139,264	97,728	139,264
5		564,992	397,200	564,992

In 1986, the *International Telegraph and Telephone Consultative Committee* (CCITT) published a standard set of transmission rates for SDH. The SDH standards finally emerged in 1992.[3] These are filed under the following standards:

- G.707
- G.708
- G.709

Using the same fiber, a synchronous network is able to increase the available bandwidth while reducing the amount of equipment in the network. Moreover, the provisioning of SDH for sophisticated network management introduces much more flexibility into the overall networking strategies for the carriers.

SDH

As the synchronous equipment was rolled into the network, the full benefits became more apparent. The carriers experienced significant cost reductions and avoidances, less hardware, and increased efficiencies in the multiplexing of the various rates established by the ITU SDH multiplexing. However, other benefits required fewer spares to be maintained in the network. Additional benefits were had by the use of the SDH standard multiplexing

[3]CCITT is now called the *ITU-TSS*.

formats, which could encapsulate the PDH, multiplexed signals inside the SDH transport. This protects the carriers' investments and prevents the use of forklift technology. In fact, SDH offered the network operators the ability to future-proof their networks, enabling them to offer *Metropolitan Area Network* (MAN), *Wide Area Network* (WAN), and B-ISDN services on a single platform.

The SDH forms a multiplexing rate based on the STM-N frame format. The STM stands for *Synchronous Transfer Mode*. The STM-N general frame format works as follows. Similar to the SONET OC-1 (albeit larger), the basic STM-1 frame consists of

$$270 \text{ columns} \times 9 \text{ rows} = 2{,}430 \text{ octets}$$

$$9 \text{ columns} \times 9 \text{ rows} = 81 \text{ octets section overhead}$$

The remaining 2,349 octets create the payload. Higher rate frames are derived from multiples of STM-1 according to value of N. The standard STM-1 frame is shown in Figure 30-1. This is similar but different from the frame in an OC-3.

Data Transmission Rates

In the standards, a number of transmission rates are defined or possible based on the multiplexing rates. Not all rates are commercially available;

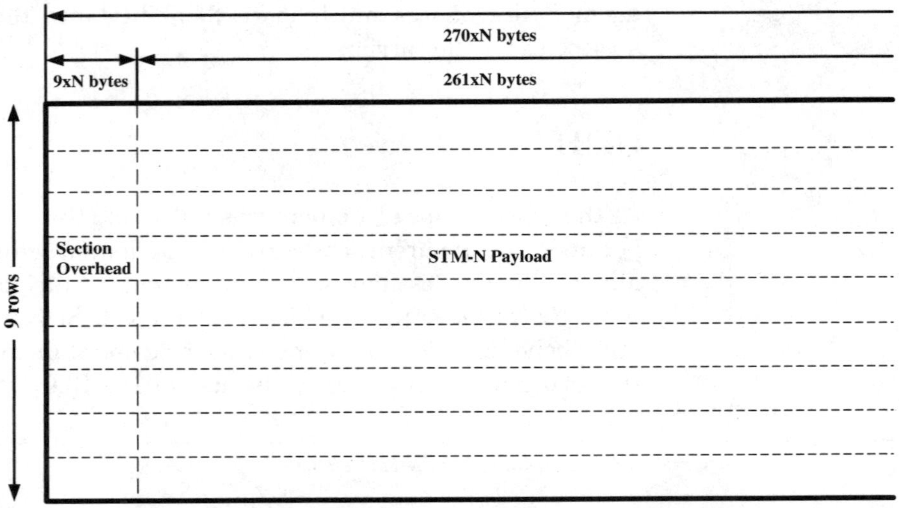

Figure 30-1
The STM-1 frame formats

however, the rates are there in case a new rate is needed. Similar to the SONET standards, SDH defines the *Synchronous Transport Signal level N* (STS-N) and the *SDH level N* (SDH-N). These define the electrical rate of multiplexing and the optical rate of multiplexing, both working at the appropriate rate of speed necessary to map and multiplex the higher rates of speed. The typical rates of speed and the appropriate STS and STM rates are shown in Table 30-2.

Only three of the hierarchical levels are actually defined in the standard and are commercially available. These are the STM-1, STM-4, and STM-16. Other rates will become available as needed. As the multiplexing occurs and the overhead is subtracted from the payload, the actual throughput is shown in Table 30-3. This table reflects the aggregates of throughput in the three commercially available STM payloads.

Some Differences to Note

Many of the actual rates in use are defined under the initial capacities of SONET in the CCITT standards. The original SDH standard defined the transport of 1.5, 2, 6, 34, 45, and 140 Mbps within the transmission rate of 155.52 Mbps (the STM-1). SDH was intended to carry the SONET rates. However, the European manufacturers and carriers carry only the *European Telecommunications Standards Institute* (ETSI) defined PDH rates of 2, 34, and 140 Mbps. (The E-2 specification was deleted in the specification.) SDH really turns out to be more than just a set of multiplexing speeds and transport rates. It is also a *Network Node Interface*

Table 30-2

Comparison of STS and STM rates

Electrical Rate	Optical Rate	Speed
STS-1	STM-0	51.84 Mbps
STS-3	STM-1	155.52 Mbps
STS-9	STM-3	466.56 Mbps
STS-12	STM-4	622.08 Mbps
STS-18	STM-6	933.12 Mbps
STS-24	STM-8	1.244 Gbps
STS-36	STM-12	1.866 Gbps
STS-48	STM-16	2.488 Gbps

Table 30-3

SDH payload compositions for the hierarchical rates

STM Level	Data Rate	Actual Payload
1	155.52 Mbps	150.112 Mbps
4	622.08 Mbps	601.344 Mbps
16	2.4883 Gbps	2.40537 Gbps

(NNI) defined by the ITU-TSS for worldwide use, partially compatible with the ANSI SONET specification. Further, it is one of two options for the *User-to-Network Interface* (UNI) to support the B-ISDN.

The Multiplexing Scheme

When trying to understand the SDH architecture, it is important to remember that the North Americans have always done things one way and the Europeans (actually, the rest of the world) have done things in a different way. The two entities never seem to become harmonized. As a result, the way the mapping and multiplexing was arranged differs from the way the standards and multiplexing techniques work in the North American communities (ANSI and SONET). The language differs from what is used in SONET specifications.

As the framing and formatting begin with the STM-1 frame, using the equivalent of the OC-3 specifications, the language begins to shift in a simple yet confusing manner. The STM-1 frame is shown in Figure 30-2 again but with a few differences. First, the frame does consist of 270 columns and 9 rows, creating a frame of 2430 bytes. Second, there will be 8,000 frames per second (the frame time is 125 μ seconds). This creates the overall throughput of 155.52 Mbps, the same as SONET.

As the data are prepared to place in the frame, several different modes of transport are available. At the input, the data flows into a container with a size designation. For example, the C11 is the equivalent of a T1 transport mechanism at 1.544 Mbps. The C12 will carry the E-1 transport mechanism at 2.048 Mbps. Another transport is the C2, which is the equivalent of a DS2 operating at 6.312 Mbps.

Each of these levels of the container is then input into a *Virtual Container* (VC level N). In Figure 30-3, the inputs are being flowed into the basic container and then mapped into the VC. This may sound very similar to the SONET *virtual tributary* (VT). SDH defines a number of containers,

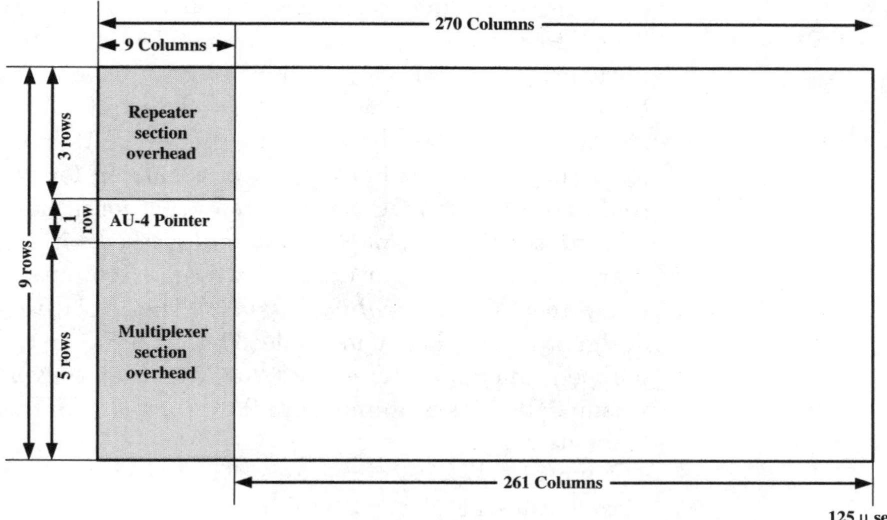

Figure 30-2 Size of the STM-1 frame

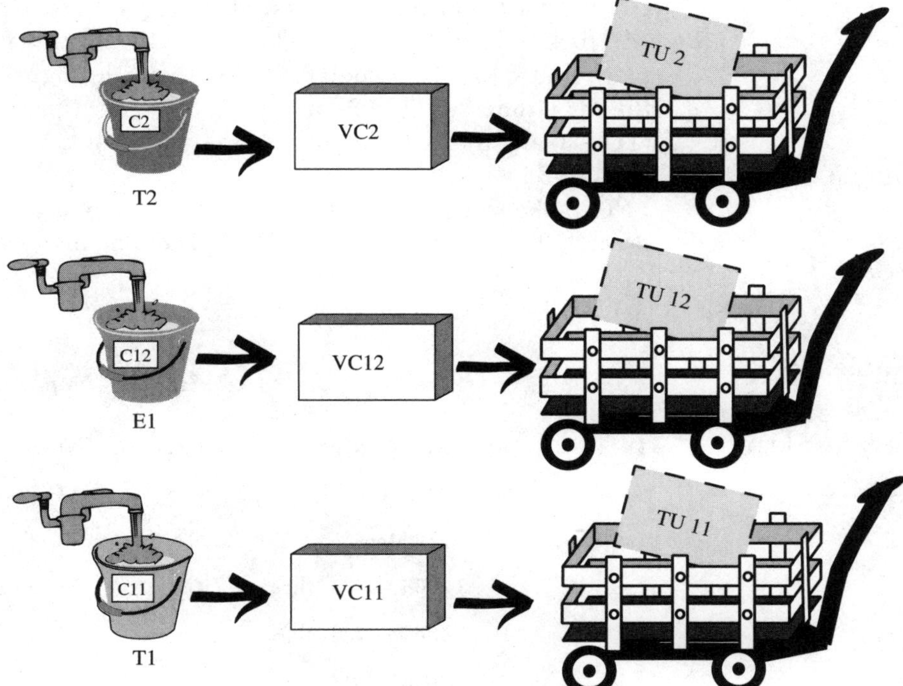

Figure 30-3 The VCs are created.

each corresponding to an existing plesiochronous rate. The information from a PDH signal is mapped into the relevant container. This is done similarly to the bit-stuffing procedure used in conventional PDH multiplexers. Each container has some overhead and control information contained in it, called POH. The bytes associated with the POH enable the carriers to perform end-to-end monitoring and provisioning for performance rates. The container and the POH combined together form the VC.

Next, the VCs are mapped and multiplexed within the frame into a tributary unit level N. Our example will now complete the input from a container to a VC to a *tributary unit* (TU). The levels are pretty much the same designation as shown in Table 30-4. The T1, E-1, and T2 lines can be mapped and multiplexed into the VCs that convert the format needed within SDH. The containers are then aligned with the timing of the system to create a TU.

Four T1s (TU 11), three E-1s (TU 12), or one T2 (TU 2) can be multiplexed into a TU Group 2 as shown in Figure 30-4. The architecture begins this way at the entry levels of the SDH.

The next step in the process is to develop a higher level of multiplexing. Here the TUs or TU Groups are multiplexed into a higher level. Let's think of this as the T1/T2 and T3 architecture in North America. Seven Tributary Group 2s are multiplexed into a TU Group 3. This is a T3 operating at the 45 Mbps (+/-) rate of speed. The alternative to this is to take one T3 and place that into a TU Group 3. This is shown in Figure 30-5 where seven of the TU Group-2s (TUG-2) or one TU 3 are multiplexed into a TU Group 3 (TUG-3) and then carried into the system as an *Administrative Unit Group 3* (AUG-3).

To make this even larger, the next step is to continue the multiplexing process. Therefore, three of the TUG-3s or one mapped E4 into a container

Table 30-4

Levels of input as they map into the TUs

Equivalent	Rate	Input	Mapping	Aligning
DS-1	1.544 Mbps	C11	VC-11	TU-11
E-1	2.048 Mbps	C12	VC-12	TU-12
DS-2	6.312 Mbps	C2	VC-2	TU-2
E-3/T3	34.368/44.736 Mbps	C3	VC-3	TU-3
E-3/T3	34.368/44.736 Mbps	C3	VC-3	AU-3
E4	139.264 Mbps	C4	VC-4	AU-4

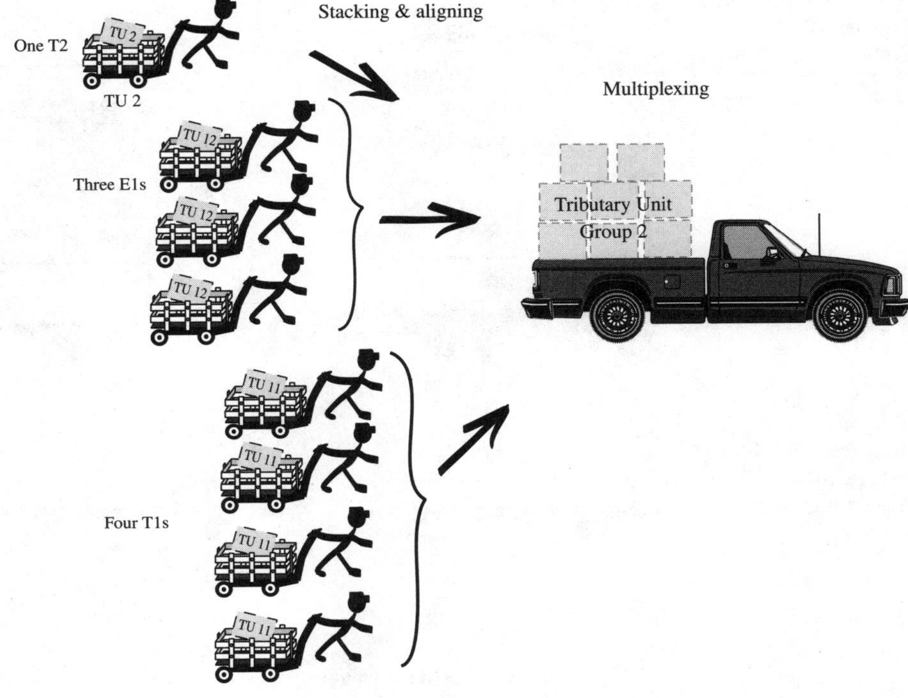

Figure 30-4 Mapping and aligning into the TU Group 2

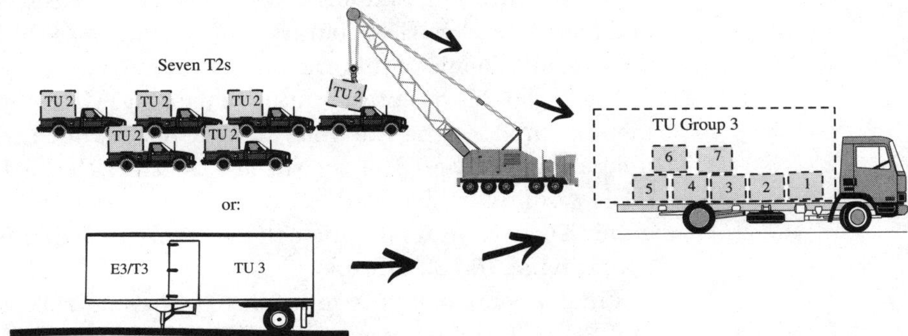

Figure 30-5 The process of creating a TUG-3

4 is then multiplexed into the next higher level in the hierarchy called a *Virtual Container 4* (VC-4). This is shown in Figure 30-6.

When the VC-4 is aligned into a SDH frame, it becomes the Administrative Group 4 carried through the system as *Administrative Unit Group 4* (AUG-4). In Figure 30-7, this occurs and is aligned to the SDH framing format of an E4 rate of 139.364 Mbps inside a 155.52 Mbps framed SDH transport, STM-1.

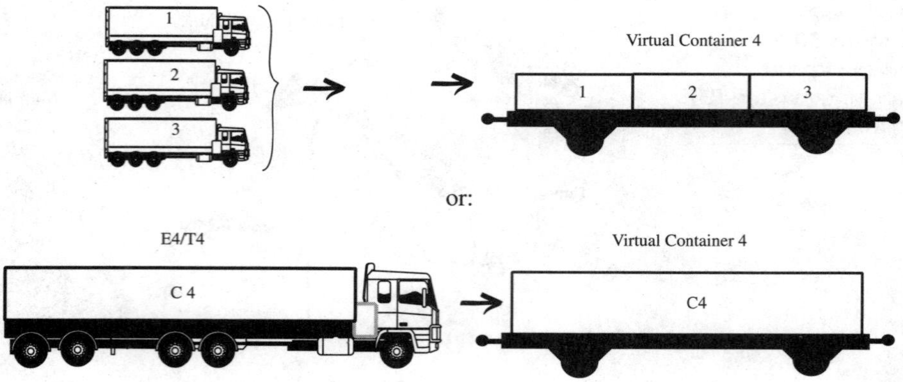

Figure 30-6 Creating the VC-4

Figure 30-7 The E4 is created inside an STM-1.

One can see that the rates of multiplexing are very similar to the SONET specification; however, differences in the bit stuffing and clocking still exist inside this framing and formatting. The comparison now turns to the look and feel of the SONET tributary multiplexing and the SDH multiplexing in the overall scheme of this transport system.

Figure 30-8 is a representation of using a SONET multiplexing arrangement to produce the Administrative Unit 3 (which by all stretches of the imagination is the T3, or the seven T2s). Here the AUG is the equivalent of three Administrative Unit 3s. One can now derive the fact that the three administrative units creating the administrative unit group s is an OC-3 operating at 155.52 Mbps.

Finally, comparing the multiplexed rates into these AUGs in both SDH and SONET, the outcome is the STM-N as shown in Figure 30-9. We can add the services as necessary to get to the STM-4, which is the 622.08 Mbps transport or the STM-16 that is our 2.488 Gbps transport system. One can see that the two systems are very closely aligned to each other; just different multiplexing and formatting arrangements are used in defining the overall platforms.

Table 30-5 is a comparison of the combined services just described, including the two systems, the electrical rates and the payloads. This comparison also includes information regarding the overall frame size. Looking

Figure 30-8
The SONET tributary multiplexing scheme

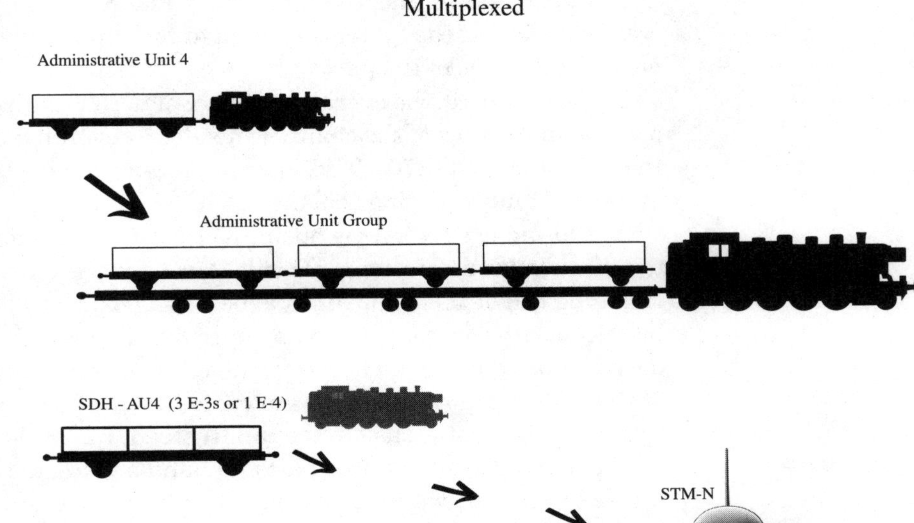

Figure 30-9
The final outcome of the SDH or SONET framing

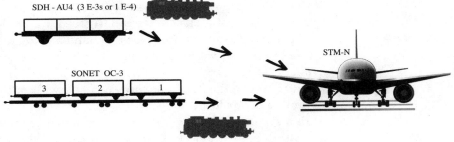

Table 30-5

Comparing SDH and SONET frame sizes and nomenclature

SDH	SONET	Fiber	Mbps	Frame Size	Rows/ Frame	Payload Bytes	Payload[a] Bytes/Row
STM-0	STS-1	OC-1	51.84	810	90	774	86
	STS-3	OC-3	155.52	2,430	270	2,322	258
STM-1	STS-3c	OC-3c	155.52	2,430	270	2,340	260
	STS-12	OC-12	622.08	9,720	1,080	9,288	1,032
STM-4	STS-12c	OC-12c	622.08	9,720	1,080	9,387	1,043
	STS-48	OC-48	2,488.32	38,880	4,320	37,152	4,128
STM-16	STS-48c	OC-48c	2,488.32	38,880	4,320	37,575	4,175
STM-64	STS-192	OC-192	9,953.28	155,520	17,280	148,608	16,152
STM-256	STS-768	OC-768	39.81312	622,080	69,120	594,432	64,608

[a]In some cases, the concatenated payloads will produce more bytes as the payloads are all in one set of *path overhead* (POH). For example, if we eliminate the POH of two OC-1s when using an OC-3c, then the payload increases to 2,340 and the payload bytes per row increase by two.

at the table shows that the industry standardization was somewhat successful in getting the systems to align to each other and prevent some of the age-old problems of the past.

In Figure 30-10, the overall structure of the VC multiplexing is shown as a graph in the flow of the channels inside the overall STM-N as depicted by the ITU standard G.707. This is a simple way to view the overall structure of the SDH multiplexing scheme.

As the end user (vis-à-vis business user) becomes more dependent on the use of the communications infrastructure and the efficiencies that are possible, the use of this bandwidth is becoming explosive. At one time, a gigabit of capacity was considered an enormous amount. Now, the gigabit to the desk is more the norm. Therefore, the use of this bandwidth constitutes one of the fastest segments of the growth in the industry. No longer can we be satisfied with just a basic voice call. Instead, we are looking for the necessary bandwidth, on a global basis, to handle voice, data, video, and multimedia. Moreover, we require a network that is close to being 100 percent available. The PDH networks could not provide this service level or flexibility. However, the SDH networks can and will continue to offer virtually unlimited bandwidth.

In a synchronous network, all the equipment is timed (synchronized) to an overall network clock as we saw in the chapter on SONET. The delay

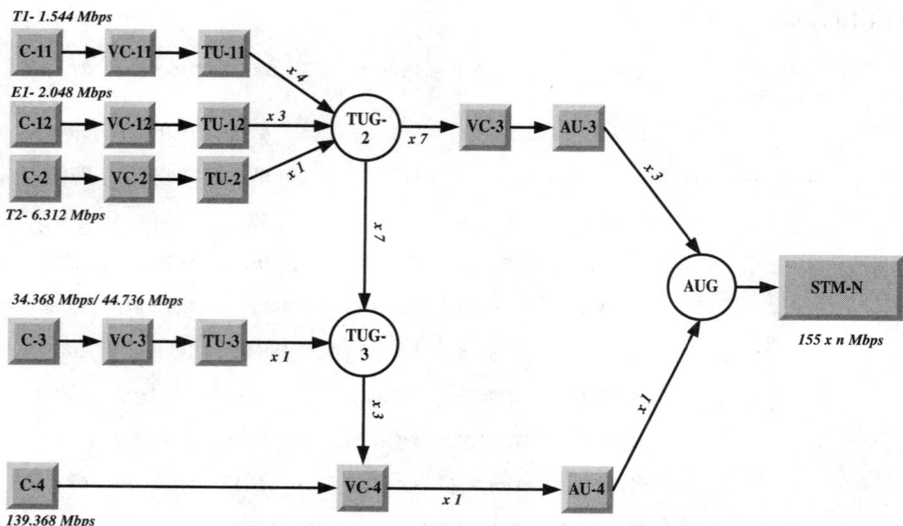

Figure 30-10
The multiplexing structure via G.707

associated with transmission links may vary slightly from link to link. Consequently, the location of the VCs within an STM-1 frame may not be fixed. The variations are handled using pointers associated with each VC. The pointer will point to the beginning position of the VC in relation to the STM-1 frame. This pointer can be increased or decreased as necessary to handle the variations in the position of the VC.

ITU G.709 defines different combinations of the VCs that can fill the payload of the STM-1 frame. The process of data fill in the containers and adding overhead is repeated many times at different levels in the SDH. This results in the nesting of smaller VCs within larger ones. The nesting process is repeated until the largest bundle (VC-4) is filled. After that, the filled VC-4 is loaded into the payload of the STM-1 frame as shown in Figure 30-11. The pointer processing is different here in the TSM-1 frame and a STS-3 frame in SONET.

After the payload envelope of the STM-1 frame, additional control information is added to the frame to form the section overhead. This overhead

Figure 30-11 The data fill based on G.709

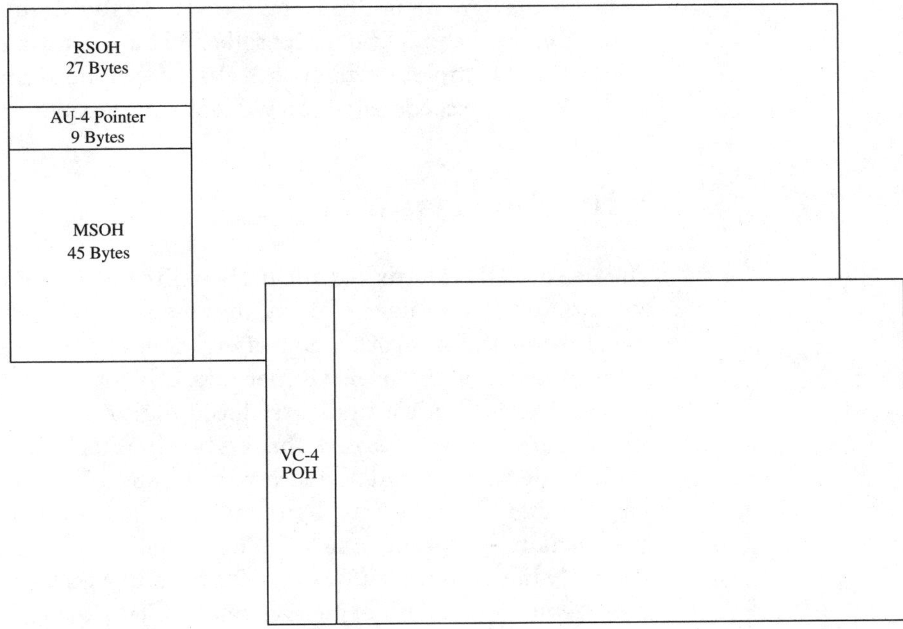

stays with the payload as it propagates along the fiber route between two multiplexers. The reasons this overhead is used include

- It aids in the provisioning for *Operations, Administration, and Maintenance* (OA&M).
- It creates user channels for diagnostics.
- It provides protection switching.
- It aligns on frame boundaries.
- It determines performance at the section level.
- Other associated chores.

Table 30-6 reviews the common rates and interfaces we use in our daily communications lives. These rates include the common interfaces used in an organization for various connectivity solutions.

In the chart shown in Figure 30-12, we see that the use of the SONET and SDH equipment became prevalent in the early 1990s. However, the movement is away from the older PDH technologies and more to the installation of newer SONET and SDH equipment. Within the carrier communities, the movement is dramatic with virtually no older PDH equipment remaining in the pipeline for installation for the future. This market (PDH) is just about completely gone with the synchronous marketplace exploding at a rate that exceeds anyone's wildest expectations.

Why the Hype?

The use of SDH affords significant benefits to the carrier, but there are also benefits to the end user. Clearly, any technology that is used in the infrastructure of the network must offer benefits to the end user and carrier alike. Otherwise, the user will not take advantage of the potential share of the services. When we first introduced digital networking in the network, many users wanted to access the digital circuits. Bearing in mind, the carriers installed the digital services in the 1960s. Yet 1984 was the first time they tariffed the use of the digital architecture to the customer's door. Mind you, certain exceptions existed. The author personally installed a special assembly in the mid-1970s using a T1 from Boston to California. This special arrangement was both expensive and difficult to work with because of the newness to end users. However, the carriers had the services in place and converted the backbone networks (wide-area) to all digital-standards-based infrastructures. They needed to operate in this fashion to meet the demand for higher-speed communications, reliability expectations, and cost ratios.

Table 30-6

Comparisons of common interfaces

Interface	Medium Used	Data Rate Mbps	Capacity in DS-0s	VT/VC Designator
DS-0	Twisted Pair	.064	1	
ISDN BRI	Twisted Pair	.144	2B+D (2)	
DS-1/ISDN PRI	Twisted Pairs (1 or 2)	1.544	24	VT1.5/VC-11
E-1	Twisted Pairs (2 or 3)	2.048	32	VT-2/VC-12
ADSL	Twisted Pair	1.5 up 6.384 down		
Ethernet	Coaxial/twisted pair (4)	10		
E-3	Coaxial/fiber	34.368	512	VC-31
T3	Coaxial/fiber/MW	44.736	672	VC-32
STS-1	Coaxial	51.84	672	
Fast Ethernet	Twisted pairs (4) or fiber	100		
FDDI	Fiber	100		
Gigabit Ethernet	Fiber	1000		
STS-3	Coaxial	155.52	2,016	
E-4	Coaxial/fiber	139.368	2,048	VC-4
OC-1	Fiber	51.84	672	
OC-3/ STM-1	Fiber	155.52	2,016	
OC-12/STM-4	Fiber	622.08	8,064	
OC-48/STM-16	Fiber	2,488.32	32,256	
OC-192/STM-64	Fiber	9,953.28	129,024	

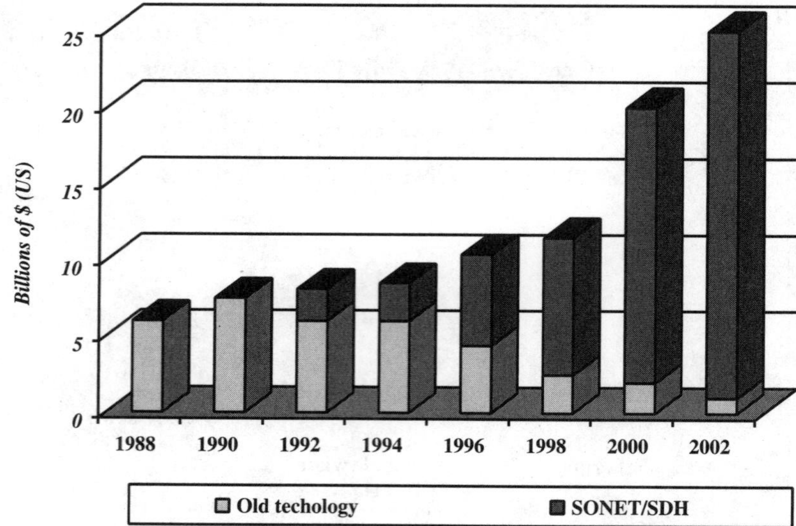

Figure 30-12 Market for SONET/SDH equipment

Unfortunately, with the introduction of their networks and standards, there were still problems. The North American carriers installed standards-based equipment in their networks whereas the European carriers installed their own standards-based equipment. The major problem was that the two did not speak with each other. To solve the problem, gateways were required to interface between the different pieces of equipment. However, when the SDH and SONET standards were completed, a new beginning was possible. The intention of both standards was to enable interoperability and transparency between systems. SDH brings the following advantages:

- High-speed transmission rates of up to 10 Gbps in today's backbone. SDH is suitable for the overall carrier networks, which are the information superhighways of today.
- Simplified process for add-drop multiplexing. When we compare this to the older systems, the PDH networks were extremely complicated. PDH required that we demultiplex an entire DS-3, for example, in order to get at an individual DS-0. It is now much easier to extract and insert low-bit rate services into or from the higher-speed services.
- High reliability means high availability. With the SDH networks, providers can meet the demands of their customers faster, better, and cheaper! The providers can now use a standard set of equipment to

meet the need, eliminating the need for multiple spares and different operations systems. Moreover, with the *automatic protection switching* (APS) services of the SDH, the network can heal itself when a component fails. This means that the customer never realizes that a network error or failure occurred.

- Future networking equipment will be based on multiples of the SDH and SONET equipment. The perfect platform for the carriers is one that satisfies the demand without changes en masse. SDH can satisfy the basic telephony needs, low- and high-speed data communications demands, and the newer demands of the Internet, specifically for streaming audio and video capabilities. This also meets the need for video or multimedia on-demand services, only recently becoming more popular.

Three words are crucial to the network operators and end-users alike, they are

- Bandwidth
- Bandwidth
- Bandwidth

The need for bandwidth cannot be overstated in the network today. There is never too much, only enough to satisfy the current need. As we see the use of the networks increasing, the carriers and manufacturers are continually developing new SDH speeds and multiplexing rates to meet that demand. Recently, the OC-768 in SONET and the SDH-256 were developed to support up to 40 Gbps on a multiplexed circuit. Add to that the capability to perform multiple wavelength multiplexing possibilities on a single piece of fiber, and we have terabits per second possibilities. That will hold us for a short term, but telecommunications services are like water. Water seeks its own level; telecommunications services increase when we have more bandwidth until we consume it all. Then, it is back to the drawing board to develop a new technique to increase the capacities again. This is a vicious circle, but one that is also extremely exciting as the thresholds are pushed to new levels and limits.

The Model as It Pertains to SDH

Everything we do usually ties back to a reference to the international standards. For this we use the *Open Systems Interconnect* (OSI) as a reference model. In general, SDH operates at the bottom layer (Layer-1 Physical) of the OSI. This physical layer is subdivided into three separate components. When we deal with this mapping on the OSI, we usually refer to four sublayer components.

Similar to the SONET architecture, SDH networks are divided into various layers that are directly related to the network topology:

- The lowest layer is the physical transmission section, the medium in the form of the glass (fiber), but can also represent radio systems such as microwave or satellite.
- Actually, the photonic layer is below the physical layer, where we turn the pulses into light (photons) and propagate them on the fibers. In many cases, the photonic and the physical layers are combined.
- The regenerator section is a path between repeaters. This is also at the physical layer because it involves that length of fiber between two repeaters. The overhead is called the *repeater (or regenerator) section overhead* (RSOH). Signaling across the medium also occurs at this level.
- Following the repeater portion, we have the *multiplexer overhead* (MSOH) used by the multiplexers for the necessary overhead to track the operation of the circuits and OAM&P. The multiplex section deals with the part of the SDH link between two multiplexers where we map and multiplex our services (DS-1) into the SDH transport.
- Sitting on top of the MSOH are two VC layers. These two layers are a part of the mapping process whereby the individual tributary signals are placed inside the SDH payload. The VC-4 is the mapping for the 139 Mbps payload, whereas the VC-12 are the individual E-1 signals mapped into the SDH.

The overall picture of this is shown in Figure 30-13, the SDH layer model. In this figure, the photonic layer is shown as a subset to the physical interface. Note that the *Asynchronous Transfer Mode* (ATM), POTS, and *Internet Protocol* (IP) networks are also shown in this figure in as much as they fit into the overall scheme of the model.

In Figure 30-14, the mapping of the payloads occurs at the SDH multiplexer, whereby the next portions of the circuit are added. This is shown in the path description in this figure.

SDH brings the harmony to the overall multiplexing of the various signals and transport rates. It also acts as the gateway between SONET and SDH structure. It has been over a decade since the actual ratification of the standards, and now more than ever, the benefits are visible. Transconti-

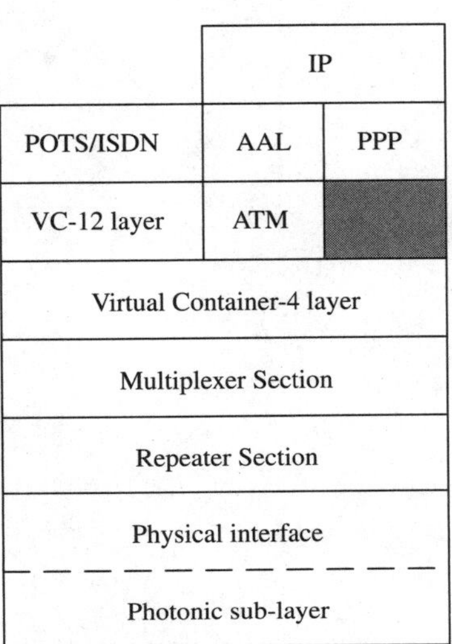

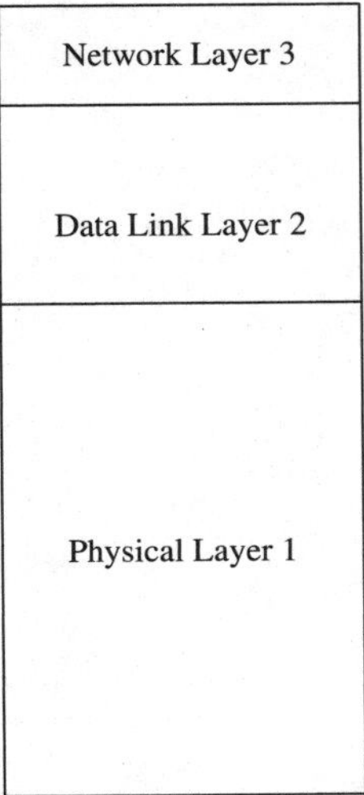

Figure 30-13 SDH layer model contrast to OSI

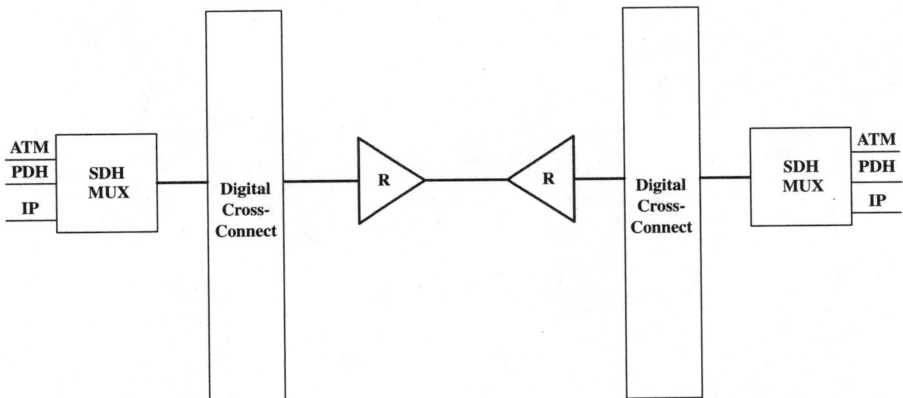

Figure 30-14 Path mapping of SDH

nental links are now in heavy use across the world on fiber-based architectures. The costs of circuits between and among countries are at an all-time low due to the bandwidth. As more equipment and carriers are deployed across the world, the World Wide Web of carriers and Internet access will definitely become the norm.

CHAPTER 31

Wave Division Multiplexing (WDM)

In the past few years, several major improvements have been made in the use of optical fiber communications. In earlier discussions regarding the use of SONET (see Chapter 29, "Synchronous Optical Network [SONET]"), the issue of bandwidth surfaced. Not only does the issue of bandwidth keep coming up but also the problem of just how quickly we consume all the bandwidth that is made available. No matter how much or how fast we improve our spectrum availability, newer applications crop up that literally *eat up* all the capacity available to us.

To solve this problem, the use of *frequency division multiplexing* (FDM) with our light-based systems became a topic of research. What ensued was the ability to introduce various wavelengths (frequencies) of light on the same fiber cable and the resultant increase in possible throughput. We have seen increases that approximate between 16 to 30 times the original capacity of a single fiber. These capacities are now being pushed to the limit with variations being promised of up to 128 times the capacity of the existing fiber technologies. What this means is that the old days of having to replace the in-place fibers with new technology have been replaced with newer technology that uses the in-place fiber, necessitating only the change in electronics on the line. We can expect to see these advances provide virtually unlimited bandwidth without en masse changes in the infrastructure. The future holds the promise of reduced costs, increased bandwidth utilization, and ease of implementation that meet the demands of our higher speed communications.

WDM

Ten years ago, the implementation of the OC-48 SONET specification had the industry believing that limitless bandwidth was available. One can just imagine that a mere decade ago the 2.5 ± Gbps capacities of the optical fiber networks was innovative and exceeded our wildest imaginations about how we would ever fill these communications channels. Yet, the industry began to recognize the changes in consumption patterns. In the demand for multimedia communications, video WAN started to erode even the highest capacities available. To solve this problem, researchers began to experiment with the use of more than one light beam on the same cable. Light operates in the frequency spectrum similar to the older cable TV systems, employing FDM, as shown in Figure 31-1.

By using different radio frequencies on a cable TV system, the carriers were able to expand the number of TV channels available to them on the

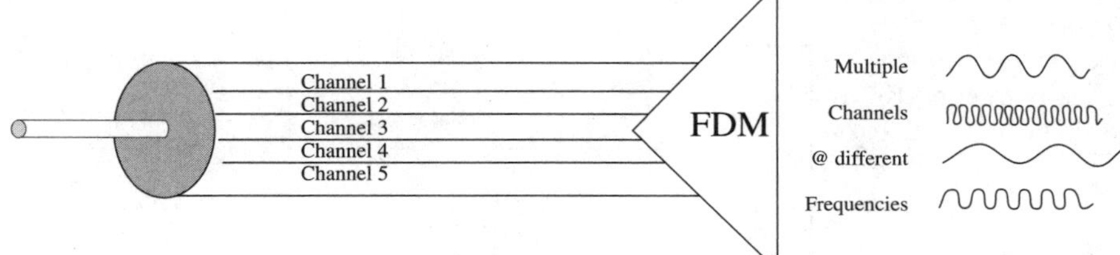

Figure 31-1
FDM has been used by cable TV operators to carry more information on their coaxial cables.

same coaxial systems. Why not do the same things with the various frequencies of light? Thus, a new era was born in the use of wavelength division multiplexing. By adding a tightly separated wavelength of light on the same cable, more capacity could be obtained.

In normal fiber transmission, the use of two standard wavelengths was originally deployed as shown in Figure 31-2. First, a red band of light was used to transmit the signal from the originating end. A blue band was used at the opposite end. Therefore, the same cable could carry send-and-receive traffic by using different color bands. The original OC-48 transmission (operating at 2.5 ± Gbps) was exciting, using a single wavelength of light and driving the signal over 20 to 30 miles of fiber. However, the carriers soon recognized that skyrocketing demand for the bandwidth would outpace the throughput of this single wavelength transmission.

Depending on the usage of various types, the life cycle of the OC-48 was anticipated to be approximately two years. Clearly, the rapid increase in Internet access, cellular communications, high-speed data, and the multimedia improvements were on a collision course with the capacity of this OC-48 architecture. Choosing to move to a higher capacity multiplexing technique in *time-division multiplexing* (TDM) was a definite consideration. This led to the higher rate multiplexing at the OC-192 level (9.953 Gbps±), a four-fold increase in the speed and throughput of the SONET networks. Yet, even though the increases were achieved to the OC-192 level, the initial implementation was with one wavelength of light.

At the same time, OC-48, using two wavelengths, produced a 5 Gbps throughput on the same fibers, proving that the technology could work. Shortly after the introduction of the OC-192, strides were taken to introduce OC-48 running four wavelengths (10 Gbps) or a single OC-192 using one wavelength.

Shortly after 10 Gbps were demonstrated, the designers began to experiment with a 20 Gbps capacity, using eight wavelengths of OC-48 or two wavelengths at the OC-192 rate. Now the stage was set to push the

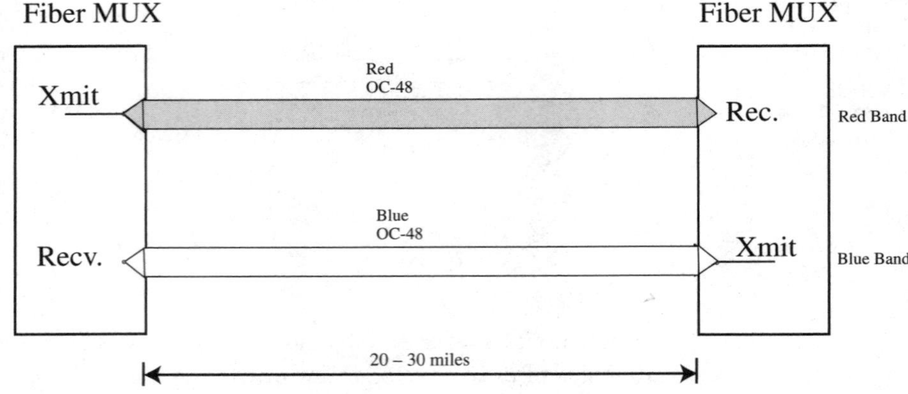

Figure 31-2 The colors of light separate the bands carried on the cable.

envelope as far as possible! It was only a matter of a few years before the developers began introducing quantum leaps in their multiplexing ability. Now with multiples of OC-48 and OC-192, the capabilities of fiber-based transmission exceeded their developers' wildest imagination. Capacities for the *dense wavelength division multiplexing* (DWDM) service were now ranging from 160 Gbps to as much as 400 Gbps. This is the first of many steps we can expect to see in the near future. Some rumors indicate 128 wavelengths of OC-192 being possible shortly after the turn of the century. That is 1.2 Tbps!

Table 31-1 summarizes the capacities of the DWDM today and what the future may hold. This table shows the level of multiplexing, as well as the possible throughput we can expect for the future. This table also shows the number of wavelengths on the single mode fiber.

Table 31-1 A summary of current capacity of the DWDM services

DWDM Level	Number λ	Number OC-48	Number OC-192	Total Throughput
1	1	1	—	2.5 Gbps
	2	2		5.0 Gbps
2	4	4		10 Gbps
	1		1	
3	8	8		20 Gbps
	2		2	

Table 31-1 cont.

A summary of current capacity of the DWDM services

DWDM Level	Number λ	Number OC-48	Number OC-192	Total Throughput
4	4 hybrid	1	2	25 Gbps
5	16 4	16	 4	40 Gbps
6	8 hybrid	4	4	50 Gbps
7	32 hybrid	32	8	80 Gbps
8	16	12	4	100 Gbps
9	16	—	16	160 Gbps
10	32 hybrid	16	16	200 Gbps
11	32	—	32	320 Gbps

Fiber Optics Summarized

Before proceeding too far from the beginning of DWDM, it may be appropriate to look at the fiber optic world and its relationship to the demand of the industry. Moreover, the types of fiber in use determine the capacity being delivered today, so a quick review of the variations of fiber may help to define the overall benefits of the multiplexing schemes on the glass.

Fiber has been used in North America since the 1960s. The carriers recognized the limitation of the copper cable they had been installing for years. First, the copper was very expensive. Moreover, it was also highly susceptible to noise and interference (both *Radio Frequency Interference* [RFI] and *Electromagnetic Interference* [EMI]). The carriers tried to get more capacity on the wires to make the cable plant more efficient and cost effective. However, modulation techniques left the carriers wanting more. Most of the cabling systems were installed by using some form of analog transmission system. In the late 1950s and into the 1960s, the Bell System developed a digital modulation technique, which could take advantage of the limited bandwidth and capacities of the copper cables.

Radio-based systems were also in heavy use on the carriers' networks. The limited frequencies, radio interference, and line-of-sight demands of radio all led to frustration. As a result, the use of fiber optics became very attractive to the carriers. Fiber is made up of glass, which is an inert material that is unaffected by electromagnetic and RFI. The thickness of the glass is used to determine the throughput and the light capacities in a carrier-based network. Two primary types of fiber were developed over the years to support the communications demands of the network. These two types are multimode fiber and single mode fiber. Multimode fiber is thicker and therefore limited in its bandwidth. Single mode is thinner and purer, so it has greater capacities.

Multimode Fiber

When fiber was first being used in the communications area, the technology was still developing. Multimode fiber was used as a means of carrying communications signals on the glass more reliably. Multimode, as the name implies, has multiple paths where the light can reach the end of the fiber. Single mode fiber, on the other hand, has only one path that the light can travel to get to the other end. Using a large piece (chunk) of pure glass, the developers extruded the fiber into much thinner glass strands.

The multimode fiber is the thicker glass by today's standards. The multimode was developed in two different types, step index and graded index, both operating differently. Figure 31-3 is a representation of the multimode fiber, using a step index mode. In this case, the glass is very thick at the center core (approximately 120 to 400 microns). The thickness of the glass is crucial to the passage of the light and the path used to get from one end to another. The step index is the thickest form of fiber; using the core of 120 to 400 microns thick, the light is both refracted and reflected inside the encased fiber. Because of the density of the glass, the center core refracts the light in different angles. At the same time, there is an outer cladding on the glass used to reflect the light into the center of the glass. This combination of refraction and reflection of the light, along with the density of the glass, causes the light to take different paths (or bounces) to the end of the cable. Different light beams inserted into the fiber will take different lengths of path and therefore different timing periods to get to the end of the cable.

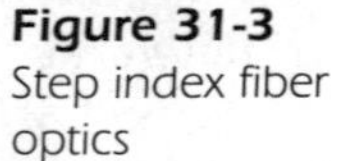

Figure 31-3 Step index fiber optics

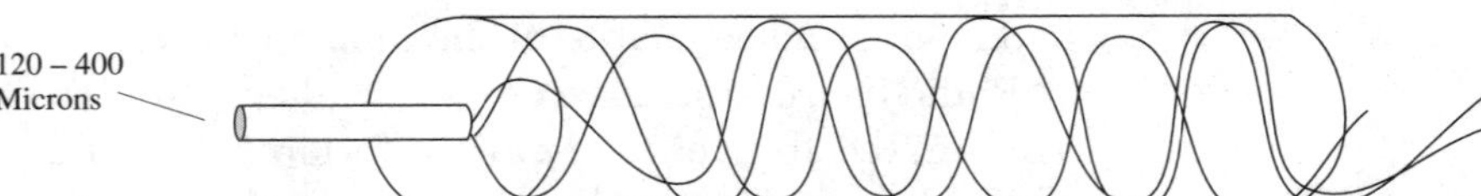

Graded Index Fiber Optics A second form of multimode fiber was the use of a graded index of the glass, as opposed to the step index. Essentially, the grading of the refraction levels of the light was changed to compensate for the amount of refraction and the density of the glass. In the step index, modal dispersion (spreading out of the light) was in part responsible for a reduction of the throughput on the fiber. By using a grading of the density of the glass from the center core out, the capacity of the fiber was increased. Simply stated, the more dense the glass was, the more refractive the surface of the glass was, or the more refraction taking place, the longer the path. By having a step index, the path of the outer part of the glass was longer than the path in the center of the glass. This meant that the light arrived in different times because the length of the path was longer. Grading the center core to be a higher level of refraction and the outer parts of the glass to be thinner (and less refractive), we could use the characteristics of the glass to get approximately the same length of a wave on the cable and therefore increase the speed of throughput. The graded index is shown in Figure 31-4. The better the grading of the index, the more throughput we can expect. Currently, the two forms of graded index fiber use either a 62.5 micron or a 50 micron center conductor with a 125 micron outer cladding on the glass.

Single Mode Fiber

As fiber became more popular and research was stepped up, a newer form of glass was developed. If the glass could be made so thin and so pure in the center, the light would have no choice but to follow the same path every time. A single path (or single mode) between the two ends enabled the developers to speed up the input because there was no concern about varying lengths of the path as shown in Figure 31-5. Using a single mode fiber today, the thickness is approximately 8.3 to 10 microns thick at the center. There is still an outer cladding on the outer edges to reflect the light back into the center of the glass. The outer cladding is still approximately 125 microns. The single mode fiber is where most of the activity is being generated today. Many of the telephone carriers deployed a multimode fiber in

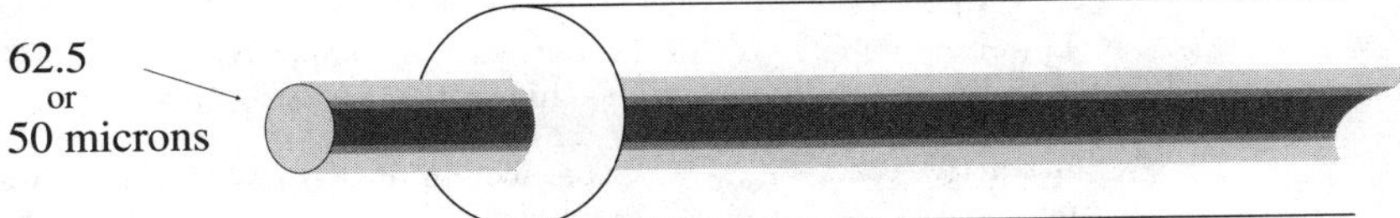

Figure 31-4 A graded index fiber

Figure 31-5 Single mode fiber is so pure and so thin that the light has only one path from end to end.

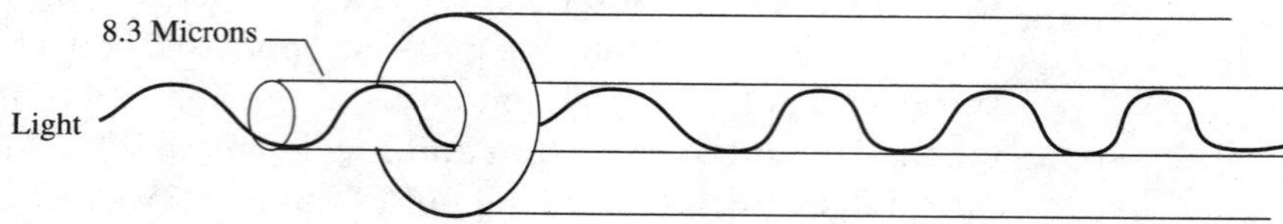

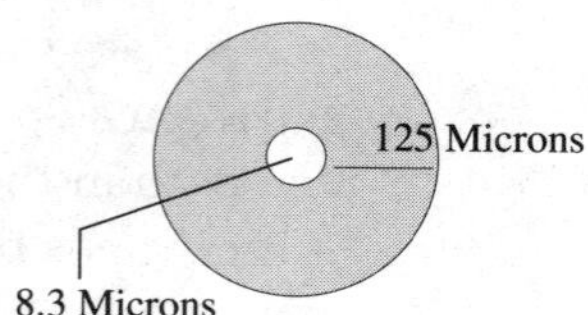

their networks when it was first introduced. However, over the years the multimode has given way to single mode fiber throughout the public carrier networks.

Benefits of Fiber over Other Forms of Media

The benefits of fiber over other forms of media are shown in Table 31-2. However, it is important to note that the fiber has been used extensively in the long distance telecommunications networks and the local telephone company networks. It is more recently that a single mode fiber has worked its way into the end user networks (LANs and CANs). With single mode fiber, the speeds are constantly being upped, while the error performances are continually being improved. This table concentrates on the bit error rates and the speed of the cabling systems.

Table 31-2 Fiber-based advantages over other media

Advantage	Description
Lower errors	*Bit error rate* (BER) approximates 10^{-15} for fiber, whereas copper will be in the 10^{-4} to 10^{-8} range.
Attractive cost per foot	Cost per foot on fiber is now approximately \$.20 compared to \$.13 for copper.

Table 31-2 cont.

Fiber-based advantages over other media

Advantage	Description
Performance	Immune from RFI and EMI without extra cost of shielding on copper.
Ease of installation	Ease of installation due to lower weight and thickness.
Distances	Greater distance with less repeaters. Now can achieve 30 to 200 miles without repeaters. Copper and radio limited to less than 30 miles.
Bandwidth improvements	Fiber nearing 1 Tbps, copper 100 Mbps, and coax 1 Gbps.
Capable of carrying analog and digital	Using TDM and WDM, the fiber is both digital and frequency multiplexed increasing capacity.

Back to WDM

Now, back to the DWDM concept. There is no mistake that the demands and needs for telecommunications services are growing. The combination of voice, data, and multimedia applications that are constantly putting pressure on the infrastructure add to the growing problem. Where normal digital transmission systems use TDM and cable TV analog technologies use FDM, WDM is a combination of the two schemes combined. The use of TDM in a multiplexer breaks the bandwidth down to specific timeslots such as those found in SONET based networks. However, by using the combination of frequency (different wavelengths) and time (timeslots), the fiber can be used to carry orders of magnitude more than traditional fiber-based systems. Figure 31-6 shows the combination of the two multiplexing arrangements in a single format. Furthermore, there are several different wavelengths and colors of light that can be used to produce far more capacity.

No one can mistake the fact that telecommunications capacity needs are rapidly growing. DWDM was developed to produce even better results on a single fiber pair than the original techniques deployed on the backbone networks. Expanding capacity with DWDM can produce some significant improvements for today's technology. An example of this is the fact that some DWDM multiplexers can produce up to 240 Gbps on each pair of

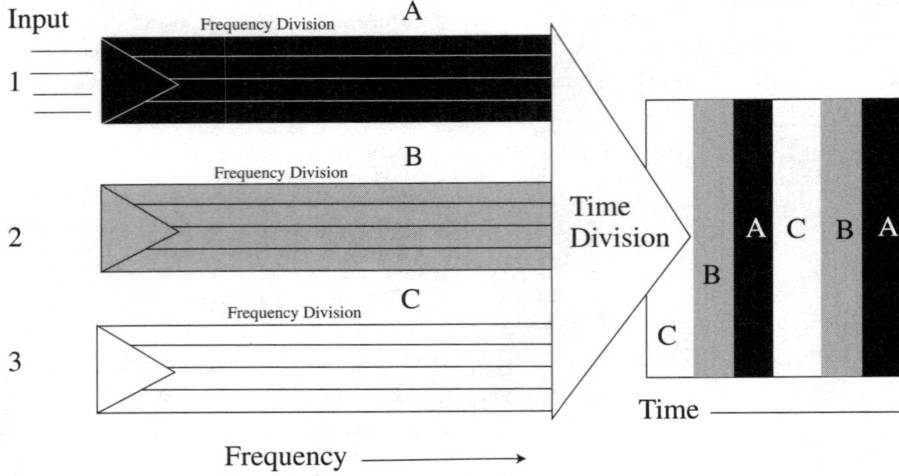

Figure 31-6 Combinations of frequency and time multiplexing produce the results in WDM.

fibers. This equates to a call carrying capacity of up to 3.1 million simultaneous calls on a single pair of fiber. Now many of the carriers have installed at least 12 pairs of fiber in their backbone networks. Performing the math yields a total of 37.2 million calls simultaneously on the fiber, using current technology before more fibers are required. This is obviously very attractive to the carriers who would prefer to get more service on their existing infrastructure rather than installing more glass in the ground. WDM is particularly useful when congestion begins to build up on the existing carrier-based networks. Surely, the equipment is expensive, but the cost of adding multiplexing equipment or changing it out is far less expensive than digging up the ground to lay more fibers. The electronics for the DWDM multiplexers are rapidly becoming very cost efficient.

Using different light wavelengths, DWDM simultaneously transmits densely packed data streams on a single fiber. By combining DWDM with special amplifiers and filtering equipment on the links, the carriers can achieve unprecedented throughput on their existing single mode fiber. Current technology supports approximately 16 different lengths on a single fiber (OC-192 is 10 Gbps, using 16 wavelength produces up to 160 Gbps in each direction). As stated earlier in this chapter, 128 wavelengths are targeted for early in the new millennium. Table 31-3 is a summary of some of the benefits of the DWDM usage.

Table 31-3

Summary of the demand to justify the use of DWDM

Industry Demand	What Causes the Demand
Need for bandwidth	Internet access, *Personal Communications Services* (PCS), and data/voice integration.
Need for reliable communications	In order to guarantee the reliability those customers are demanding, the carriers have been committing alternate routes and spare fiber capacity to back up existing infrastructure.
More capacity	In order to get the *service level agreements* (SLA) and maintain the network in a fashion expected by the consumer, the carriers have to install backup circuits and fibers in their highest density routes.
Higher performance on the network	Network dependency has become the norm. All forms of traffic must run on the existing infrastructure, and the carriers must provision capacity to meet the ever-changing demands of the network.

Why DWDM?

Clearly, the use of the fiber technologies has taken over the industry. However, the more information that can be generated over a single fiber, the better the carriers like it. To be specific, before digging up the ground to lay more fibers, it is far less expensive to use the multiple wavelengths of light on a single fiber. Thus, as the carriers seek to gain more utilization on the existing fiber, the incentive will be for the manufacturers to continue to proliferate this technology.

The major manufacturers have obviously put all their resources into the development of the DWDM techniques to meet the rising demand from the carriers and the end users alike. Using multiple wavelengths enables the carriers to achieve the 320 Gbps capabilities today, with terabit speeds being discussed in the very near future. Using a few different techniques, the major suppliers have developed keen interest in the ability to utilize DWDM. *Multiwavelength Optical Repeaters* (MOR) help in deploying this technology out to the carrier networks. A MOR (or MOR plus) is an amplifier used to support at least 16 wavelengths on the same fiber in each direction. Using 16 wavelengths (16 λ) operating at the OC-192 rate yields a total of 160 Gbps in each direction or 320 Gbps in one direction. In Figure 31-7, a typical multiplexer is used to generate a bidirectional flow of information through the fiber multiplexer. This figure shows the major components of the systems used by Nortel but can be the same as used by others (for example, Lucent Technologies, Siemens, or Alcatel). In either case, the major point here is that the use of a DWDM coupler aggregates the capacity of the fiber by using two different light bands (red and blue) and multiple wavelengths to achieve the results we have grown to expect.

By using the fiber-based example for the increased capacity of the fiber, the DWDM method appeals to many of the providers and carriers alike. The use of fiber-based multiplexing also adds some other enhancements that were not traditionally available in the past. Because the signal never terminates in the optical layer, the interfaces can be bit rate and format independent, enabling the service providers the opportunity to integrate DWDM easily with their existing equipment and infrastructure, while still gaining the access to the untapped resources and capacities in the existing fiber.

DWDM combines multiple optical signals so they can be amplified as a group and transported over a single fiber, increasing the capacity. Each signal carried on the DWDM architecture can operate at a different rate (for

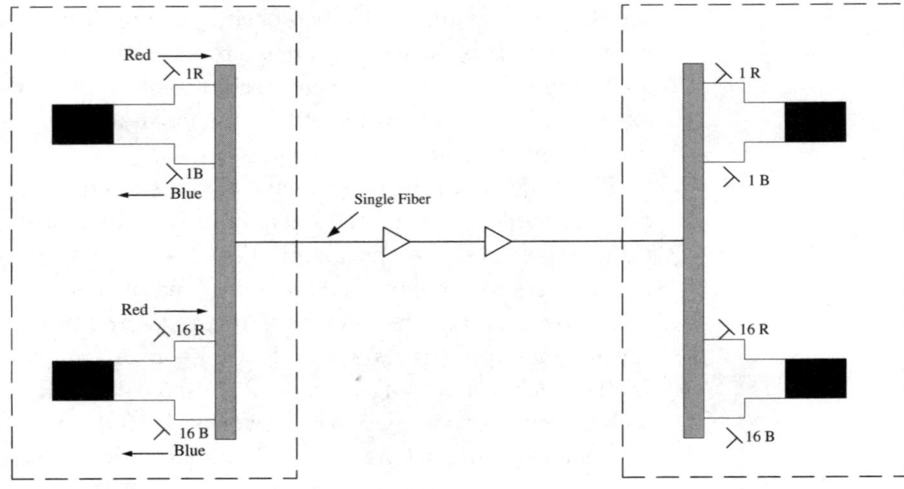

Figure 31-7 A typical DWDM multiplexer

example, OC-3, OC-12, OC-48, and so on) and in a different format (for example, SONET, ATM cells, data, and so on). Using a mix and match approach enables the carriers to achieve different rates of speed and aggregated throughput, depending on the multiplexing equipment used. Future DWDM developments are touted as being able to carry up to 40 Lambda of OC-192 or a total of 400 Gbps. As a comparison, 400 Gbps will be 90,000 volumes of the encyclopedia in one second!

DWDM requires that the transmission lasers have a very tight wavelength tolerance so that the signals do not interfere with each other. After all, the systems carry 16 or 32 different wavelengths on a single fiber. It is imperative that the tolerances be held tightly to prevent mass destruction of the data signals being generated on the fibers. The *International Telecommunications Union* (ITU) has specified 100 GHz standard spacing between the wavelengths, and most of the vendors are now manufacturing lasers at the standards-based wavelengths for the combination into DWDM.

The most advanced today, using OC-192 lasers, has a tolerance of 0.1 nm, which is small enough to preclude any interference between two adjacent wavelengths. Designing OC-192 DWDM systems is ostensibly more difficult to design than the older OC-48 systems. The OC-192 systems being deployed in the networks also use both unidirectional and bidirectional systems in support of the fiber and DWDM multiplexers. However, the future may change things very quickly. Where OC-192 offers a 10 Gbps operation, the future offers an OC-768, which will create a fourfold increase in the capacities in the network. At the OC-768 rate, a multiplexing scheme will produce 40 Gbps on a single fiber rate (SONET rate), which then can use the DWDM capabilities to create a throughput of up to four times the 320 Gbps rate currently available. That equals terabit speeds (1.28 Tbps)! However, the use of spacing on the various lambdas will actually enable up to 400 Gbps and offer future speeds of up to 1.68 Gbps. What OC-768 will also bring to the table is a capability to concatenate the OC-192 levels. Today's current technology can use a concatenated OC-48 but stops there in the bonding of higher speeds. In the future, as ISPs and traditional telephone carriers need to expand their capacities, the need to draw a 10 Gbps concatenated throughput will appear. Today, this is not a big concern. However, in three to five years, this will become a commonplace occurrence. Table 31-4 is a summary of some of the other things that we are expecting to occur over the next five to ten years in the area of fiber transport systems. This table takes advantage of the work being done in many of the labs and research facilities around the world. The ability to push the envelope is what the communications industry is all about.

Table 31-4 A look at what DWDM and fiber rates will bring over the next decade

Technology and Capacities	Current versus Future Technology
DWDM at OC-192 and 40 λ	Current technology capable of carrying 40 different wavelengths (using ITU 100 GHz spacing) at 10 Gbps each or a total of 400 Gbps. Currently, the industry has achieved a total of 320 Gbps with the 400 Gbps rate in the very near future.
DWDM at OC-192 and 80 λ	The current spacing of 100 GHz, as specified by the ITU, is under attack. The near future holds the promise of doubling the number of wavelengths by using a 50 GHz spacing, enabling up to twice as many wavelengths on the same fiber (80) each operating at 10 Gbps.
DWDM at OC-768 and 40 λ	This is a turn-of-the-century technology with up to 40 Gbps per wavelength and 40 wavelengths or a total of approximately 1.6 Tbps.
DWDM at OC-768 and 80 λ	By the year 2002, we can probably expect to see the OC-768 plus the use of 80 wavelengths or a 3.2 Tbps throughput on the fiber.
DWDM at OC-192 and 160 λ	By the year 2005, we can expect to see that the wavelengths' decreased spacing and tighter tolerances on the fiber lasers will yield a total of 160 different wavelengths at 10 Gbps each or back to the 1.6 Tbps rate.
DWDM at OC-768 and 160 λ	This is likely a 2008 to 2009 technology that will again double the capacity and yield a total of 160 different wavelengths at the OC-768 rate (40 Gbps) or netting 6.4 Tbps on a single fiber.

These rates are all projected to be available in the next decade. However, as the technology seems to be doubling every six to twelve months, the rates shown previously could happen as soon as 2005! The issue is not how long it will take but the ability to drive better and faster data and other communications needs on the same fibers. The future holds a lot of excitement with these speeds.

CHAPTER 32

The Internet

We are accustomed to succinct definitions of systems and equipment. The Internet is a little difficult to describe in 25 words or less. The Internet is really a collection of privately owned networks. By private, we mean that it is not owned by the government but by private and public companies. The *Internet Architecture Board* (IAB) and the *Internet Engineering Task Force* (IETF) set standards. The Internet Society elects the IAB members. The IETF membership is voluntary.

A Brief History

The Internet began as *Advanced Research Projects Agency Network* (ARPANET). Because universities were working on research for the military, a means was desired to be able to share research papers. Some say it started before ARPANET by informal agreements between universities to dial each other up and exchange e-mail. At the time, there were no formal protocols such as *Internet Protocol* (IP). These informal dial-up connections were using acoustic-coupled modems at 1200 bps. Actually, there was no real network—only these agreements to forward mail.

ARPANET started with the first packet-switching technology. The first packet software was called an *Intermediate Message Processor* (IMP) and was built under contract by Bolt, Bernak, and Newman. The goal was to remotely log into computers in Berkeley from UCLA. The initial experiment didn't work, but it provided the basis for continued improvement, which lead to ARPANET. The Internet began then as an interconnection of *Interface Message Processors* (IMPs) that could relay traffic between hosts. (A host was a mainframe or a minicomputer. Today, any device can interface to the network and run the networking protocols.)

The military changed the name of the project to *Defense Advanced Research Projects Agency* (DARPA). Under DARPA, the Internet continued to grow and expand. Outgrowths of this work were *Windows Microsoft Internet Commercial System* (WMICS), *Windows Internet Naming Service* (WINS), *Military Network* (MILNET), and eventually *Defense Data Network* (DDN). The military, therefore, had no need any longer for DARPANET and handed it over to the *National Science Foundation* (NSF) and the Internet became NSFNET. The NSF was funding most of the research anyway, and the universities were using some of that funding to support their Internet habit. The goal of the NSF was to provide shared access to supercomputers. Each university was naturally lobbying for its own supercomputer. Providing remote access via the Internet would be a lot less expensive for the NSF. The NSF funded the Internet from the mid-1980s until 1992 when benefits previously reserved for the academic community became apparent to all. What was happening was that students, who had had access to the Internet while on campus, were cut off upon graduation. They were therefore very creative in figuring out ways to hang onto their Internet accounts.

This created a set of haves and have-nots. The pressure built for the commercialization of the Internet. Continuing the funding became the pressing question. Under the NSF, the Internet had grown to include virtually all major schools. Since funding was from the government, access was easy. If your school wanted access to the Internet, you simply called up your nearest associated school who had access to the Internet and asked for a port on their router. You then leased to communications lines of the appropriate bandwidth. You sent some money to the next school to help them pay for the additional router and communications capacity that they would need to have to support your additional traffic going through their system.

Notice that we said that the Internet was a loose collection of many networks, and here we just added another network. (By the early 1990s, the potential of the Internet was well recognized, and everyone was clambering for access.)

The initial purpose of the Internet was to exchange data and make research papers available to the entire academic community. Universities would therefore have open hosts connected to the Internet. This meant that anyone could essentially log onto the host as an anonymous user, search through the directories on the host server, and download whatever files were wanted. It's interesting to compare this free and easy access with today's world of networks protected by firewalls. Actually, it wasn't that free and easy because in the Unix operating system, files and directories have permissions as to who can read and write to them.

The mechanism for doing this is a utility called *File Transfer Protocol* (FTP). In order to FTP, your system (host) had to have the FTP client, and the target system had to have the FTP server ready and willing to process your request. If one is familiar with Unix and the directory structure of the target host, this isn't too difficult. For the average techno-peasant, however, it is beyond grasp. Surfing or browsing, as we know it today, was not easy. Only the patient, persistent, and knowledgable browsers were able to find what they were looking for.

The Internet is not a *Bulletin Board Service* (BBS). A BBS is essentially a host into which anybody can dial and collect and receive information or messages. These services abounded in the late 1980s and early 1990s. The Internet is not a timesharing service either, although some hosts provide that service. Timesharing became popular in the 1970s when computers were too expensive to own for smaller companies so they would do remote computing utilizing dumb computers using services such as Tymnet and *CompuServe Information Service* (CIS). CIS grew into an information resource provider and created agreements with Telenet and Tymnet to provide worldwide access while they developed their own worldwide network. The original CIS access was via dumb ASCII terminals at low data rates. They had a character-oriented information interface that was no friendlier than the Internet. Given the period, it offered a powerful service.

In the early 1990s, *America Online* (AOL) saw the benefits of the Internet and began to offer an information service via a *Graphical User Interface* (GUI) that was very easy to use. Although AOL was a proprietary network, it rapidly added various information services in addition to the ever-present e-mail. It is likely that its phenomenal growth is due in significant part to its ease of setup and use. Several information services came and went. We like to maintain that AOL and CIS are not *Internet service providers* (ISPs) because their primary business is providing interactive information services. All their computers are in one room. Rather than being connected to a network of computers, you are connected (for all practical purposes) to one huge computer that offers many different services. By comparison, the Internet consists of millions of computers connected to millions of networks, each accessible from any other network. It is not obvious to the casual observer that there is much difference today because the browser interface cleverly hides the complexities of connecting to multiple different computers and both systems seemingly have endless amounts of information available. Today, both AOL and CIS permit seamless movement between their own information services and those available on the Internet. Selecting a link to information on the Internet causes the automatic launching of a browser.

AOL was the most successful of the information services, eventually buying CIS. This was primarily due to marketing, low cost, and ease of use. (CIS was a little late introducing a GUI and when they did, it was obscure, non-intuitive, and hard to use.)

AOL's membership snowballed because the more users there were, the more advertisers were willing to pay for advertisements and offer services via AOL. The more services they offered, the more attractive they became to more subscribers. The best part was that AOL could now charge more for its banner ads because that many more folks would view them when they signed on.

Early Internet Services

The Internet was, as indicated, difficult to use in the beginning, requiring knowledge of telnet and the Unix command structure (not to be confused with the Telenet packet switching network). Telnet is a Unix utility that lets you remotely log into another host. Once you logged in, you had free reign of that system with two significant limitations:

- You had to have an account on the target system (in the early days most systems would let you log in as anonymous).
- You could only access the directories and files for which general user permission was granted.

This permission system was so clever that a user could be granted permission to see a directory and even run programs in it, but not be allowed to modify its content.

As indicated, when you found the file or program for which you were looking, you launched the FTP utility and downloaded it to your system. Searching for and finding things was difficult at best.

The next step in improving the utility of the information on the Internet was the development of Archie, which was designed at McGill University in Canada. Archie essentially indexed FTP sites. The target machine had to run the Archie server, while your machine ran the Archie client. The good part was that the Archie server was accessible via an Archie client, e-mail, or telnet. Archie was a great catalog on the Internet. Archie returned a list in Unix language that gave you host and file names to use when you set up the FTP session. Archie was a step in the right direction of making the Internet easier to use, but you still needed to know how to use telnet and FTP.

Gopher

Gopher was also a client-server system. It provided a menu of choices, and it allowed you to set up bookmarks of locations you liked. Like previous Internet access, it was designed around the user having a dumb ASCII terminal. Thus, it provided menus and minimalist keystrokes to make a selection. That was the good part. Unfortunately, the single keystrokes were not always intuitive. The intent was that you would launch the Gopher client from your own host and communicate with the Gopher server on the remote host. It was possible to telnet to a remote host and then to launch Gopher, but this was painfully slow and created excessive network traffic. Remember that the interconnection of hosts was often with 56 Kbps lines.

Veronica

There was some speculation as to the source of this name, but most people think it was named after Archie's girlfriend. Others insist it stood for a very easy, rodent-oriented net-wide index to computer archives, but this is unlikely. Veronica is an extension to Gopher that facilitated searches and returned a list of *hits* (hits in this case were file names that matched the name in the original search request). Veronica was also a client-server arrangement, where the Veronica server keeps the database being searched.

By selecting one of the returned hits, you could get more information about it or transfer it back to your computer. Veronica therefore automated the FTP process, but was still dumb terminal- and menu-oriented.

Wide Area Information Service (WAIS)

Wide Area Information Service (WAIS) was another database search engine that allowed you to enter keywords and search a database for entries. Although the number of databases was large, finding this was still not easy because you were using Gopher to search WAIS.

World Wide Web (WWW)

The *World Wide Web* (WWW) has essentially replaced all of these older search engine capabilities. The early WWW often resorted to Gopher or WAIS to actually do the transfer.

The two developments that made the WWW useful were browsers, hypertext, and hyperlinks. Hypertext was a way of encoding formatting information, including fonts, in a document while using plain ASCII characters. Hyperlinks were essentially addresses imbedded in the hypertext web page. By selecting the hyperlink, you were taken directly to a new page.

Browsers

Browsers basically automated the keyword search function that we had done via menus using Gopher. Browsers today are very large and complex programs. Look on your hard drive, and you will find that the Netscape application is about 1.3 MB. With each revision, it gets larger and more complex as new features are added. Microsoft's browser is buried in the operating system, and it is therefore difficult to tell how big it is.

Marc Andreesen was instrumental in creating one of the first browsers called Mosaic. It became the foundation of Netscape's Navigator. Netscape's success can be attributed to the fact that it followed the AOL model of giving away the browser software. Netscape even gave away their source code to its browser on the theory that thousands of heads are better that a few, and eventually it will result in a better product. This same philosophy made Linux such a strong operating system for PC platforms. AOL later acquired Netscape, so what goes around comes around.

Browsers are also part of the client-server world. The money lies in getting the service providers to buy your suite of server software that provides the data to the browsers. This suite of software may provide many other features and capabilities such as calendar, proxy, *Lightweight Directory Access Protocol* (LDAP), and mail server functions as well.

Hypertext

In the Unix world (where this entire Internet started), there were only two kinds of files:

- ASCII text-based files that were all the documentation, source code, and configuration files
- Binary files that were the executable program files

Unfortunately, no one considered fancy formatting, multiple fonts, graphics, and tables. ASCII text was boring in the modern world of animated color. The question became, how can we add this capability?

Individual vendors' products, such as Microsoft Word™, utilize proprietary code sets in which the font and size (for example) are imbedded. Other products, such as Corel WordPerfect™, chose to imbed special tag characters, indicating the beginning and end of special font and size groups of characters. Unfortunately, with special (non-ASCII) characters imbedded in the text, these were no longer text documents, but binary documents that could only be operated on by proprietary vendor-specific programs that were not universally available or free.

How then could we keep the documents on the Internet open, free, standardized, and comprised strictly of ASCII characters so that anyone could read them? How could we extend the capability without making the previous version obsolete? A major problem with a specific vendor's product was that when the new one came out, the older versions couldn't read the newer versions' formats. This was a major inconvenience, designed to force the users to upgrade to the latest version.

The solution was to go with the tag approach. Rather than using special characters as tags, we simply used a sequence of ASCII characters, which meant something specific to the browser and did not impair the capability of a dumb ASCII terminal to read them. For example, the tag <title>Important Subject</title> will cause the browser to display that line as a title. For the curious, you may view the source of a web page and see all these tags. All you have to do is go to the View menu and select Source. You are then presented with all the original ASCII information. Although a little difficult to read because of all the tags, all the text is there, as are all the references to other web pages (HREF) and all the font and formatting information. While we are on the subject, you might try converting one of your text documents to rich text format. Here the formatting stuff is all in the beginning of the document—just another (standardized) way of sending formatting information in plain ASCII text format.

If you are using an older browser, it simply can't properly display the text within the new tag, but the text is still there, and you can read it. Fortunately, new versions of the browser are readily available and free for the downloading.

Hypertext then allows the standard ASCII text characters to define special formatting that the browser can display.

Hyperlink

A hyperlink is simply a pointer to another web page. It is the complete address to find the specified web page. The link visible on the browser-presented web page might say something innocuous like "more info." If you view the source and search for the HREF or the "more info," following the HREF5 will be the actual path to that page. Selecting the hyperlink caused a lot of background processing. The browser took the *Universal Resource Locator* (URL) and fabricates a query to that location just as though you had filled in that value manually in your browser window. It then sets up a connection, downloads the desired page, and terminates the connection.

Universal Resource Locator (URL)

The URL is simply the Internet address of the web page. URLs are displayed in this form: www.tcic.com. This URL takes you to TCIC's main web page. Selecting a hyperlink from that page takes you to a *Universal Resource Identifier* (URI), which points to the files in the directory tree of the host server. Each slash (/) in the name identifies a directory level on the server. In some cases where a document management system is employed to build and provide web pages, these slashes are actually logical divisions within the resource and have nothing to do with actual directories. Today, the trend is to use dynamically built web pages. They can be better customized (see the discussion on cookies) to the user's needs, you don't have to store thousands of different web pages, and the processors are fast enough to create them quickly.

Directory/Domain Name Service (DNS)

One of the most interesting and important parts of the Internet is the *Directory Name Service* or the *Domain Name Service* (DNS). In short, the DNS system permits human-readable names to be quickly translated into IP addresses that are needed to route the packets across the network.

As is described in the addressing section, the IP address is structured by network and then by host. It is read from left to right by the router in trying to find the proper destination network.

The human-readable addresses are hierarchical from right to left. For example, take the address Bud@TCIC.com. First, we know that it is somewhere in the .com domain. If I am George@biguniversity.edu, I need to find Bud's address. It is very likely that the university has no idea of what the address might be. The local e-mail system therefore makes a query of the unnamed "." domain. This annotation is a signal to the Internet servers maintaining the databases that we are looking or a name of a user. There are several of these servers around the Internet, which are updated daily by the *Internet Network Information Center* (InterNIC) as names are registered.

There is some controversy over the InterNIC's absolute control over domain names. (Someone has to be in charge to prevent chaos, but change is underway with the introduction of competition in this area.) To register a domain name, one (or one's ISP) must contact the InterNIC (or a suitable competitor) and pay an annual fee. (Prior to the commercialization of the Internet, *Stanford Research Institute* [SRI] performed this function). There are proposals to open other domains (for example, .biz, .store, and so on) and let other entities administer the names within that domain.

Each country has a domain and administers these domain names themselves. All public entities such as cities, counties, and states are under the .us domain. Some of the country domains are .uk for United Kingdom, .cz for Czech, .de for Germany, .au for Australia, and so on.

Each domain then has its own DNS server, so when George is trying to send e-mail to Bud, his e-mail server asks the ."" domain for Bud's address. The ."" domain replies that it can only provide the address of the .com DNS. We then ask the .com DNS that replies with the DNS address of TCIC. Since TCIC is really under an ISP, what we really get is the address of the DNS server at that ISP. Finally, we get Bud's real address. Now, this address is put into the e-mail packets, and they are sent on their way. In addition, we, the users, never knew about all the fooling around that went into finding the address in the first place.

Java™

The early Internet was strictly ASCII text based. Then came the inclusion of *Graphical Interchange Format* (GIF) files. As indicated previously, these were simply references in the hypertext document to a location that contained the graphic file that was displayed. Next, came the desire to automate or animate web pages.

Sun Microsystems invented a clever (and, to some, controversial) language called Java, which is a registered trademark of Sun Microsystems.

The basic problem was that to animate a page the local machine had to execute some software. This opened the door to viruses. Sun's clever idea was to have the browser execute the program, rather than the host hardware. The good part was that you were somewhat protected from malicious programs. The bad part was that the browser was interpreting the Java language on the fly, and this interpretation was slow. (Faster machines help a lot here.) The original idea was that the browser could prevent the Java script from doing any damage (like wiping out the hard drive). Unfortunately, the more powerful we needed Java to become, the more capability it needed on our host machine. Microsoft, naturally, has its own approach to page automation called Active-X. The good part is that it runs as machine code. The bad part is that it runs as machine code and can therefore contain viruses. The user is given the opportunity to download or not download the Active-X code. If you trust the source, go ahead and run it. If you are not sure, cancel the download and do without the animations.

Surfing the Web

When you select a hyperlink, the browser creates a packet requesting a web page and sends it to the specified URL. Your browser actually sets up a connection to the server. The server replies with the requested file (web page) and the browser displays the page. Your browser now stores this page in its cache memory so that after you have followed several other links, you can easily get back by selecting the Back button. The button retrieves the page out of the cache rather than having to fetch it from the source (which, as you have experienced, could take a while). You should periodically empty your cache. First, depending on the settings of your browser, it may never use the cache again. Second, if the browser does always check the cache first, regardless of age, when you sign onto the site the next time a month later, the page you see is the old one from the cache. Actually, this is an exaggeration because the browser has a setting for how old a page can get before a fresh copy is fetched. You can just throw away the cache folder (it is safer to discard the contents); your browser will build a new one when you launch it. If you do a lot of surfing, this cache can take up a lot of disk space.

You can always force the browser to get a fresh copy of the page by selecting the Reload (or Refresh) button. This causes the browser to ignore whatever was cached. The fact that each page can contain multiple references to other pages anywhere in the world is the reason it is called the World Wide Web (also referred to the World Wide Wait). Links can take you anywhere, including back where you started. There is no hierarchical structure to the Web.

Tracking Visitors

From a commercial point of view, we would like to know how many visitors or hits we have on our web page. (Here hit is defined as someone accessing our page, such as setting up a connection to us.) We can tally each connection and present that to potential advertisers as an indication of the popularity of our page. First, the number of connections or hits to our page is only approximately related to the number of viewers. Here is why: if you happen to set the home page to www.tcic.com, every time you launch your browser, you go there. It looks like a hit! The fact that you immediately go to some other book-marked page doesn't register. Second, depending on your browser settings, you could revisit a page multiple times, but your browser has that page cached. The web page owner then has no way of knowing that you are frequently referring to his page during your online session. Third, if the client or user is surfing from behind a proxy, his local proxy server will provide your page whenever requested, saving network bandwidth, but the web site owner again doesn't know that he has been hit once again.

Cookies

One of the more controversial aspects of the Web is the existence of cookies. Cookies are nothing but an encoded set of information that the web server asks your browser to keep for it. The cookies simply contain information about you and the sites you visit. They may contain information, such as your credit card number, that you have entered while visiting a site. They may also contain links you selected from that web site. If you visit outfitter.com and you look at hiking boots, the next time you log into that site, the first page may have backpacks and tents prominently displayed. The site reads the cookies it left on your machine the last time you visited. It determined that you are an outdoor type (and decided it would be a high probability of sale for you on related outdoor equipment). However, you still have the ability to view all the other parts of the site because the index page always contains an index to other pages on the main page.

Cookies, then, are a convenient way for the vendor to keep information about site visitors without having to keep a huge database of all visitors whether they are casual or frequent visitors.

You can throw away the cookie file anytime you want. The browser will rebuild it when it needs it. You can also read your cookie file by opening it in a text reader. Depending on the browser, the cookie file normally contains ASCII text. The information is normally encoded as a set of numbers that is not meaningful to anyone but the originator of the cookie. Some web pages won't work correctly unless you have the cookies enabled. You may leave "Accept cookies" turned on and discard the file at the end of your session if you are paranoid about the information that might be stored therein.

Search Engines

Even if we have links to follow, there is no good way to find a specific set of information. We still need a database catalog we can search that lists sites that might contain the information we want. The search engines (such as Yahoo! and Google) provide a database to search very similar to the way Gopher and Veronica tools did in the past.

Enterprising individuals also developed "web crawlers" that would follow hyperlinks based on a key word and fetch all the associated pages. You could start these crawlers, go to bed, and wake up with a full disk. Today's databases are a combination of crawling and advertising. The business plan of the search engine provider is to offer advertising along with the database information. Companies that pay more get a better position on each page. In a few cases, the ordering of the hits on the page is a function of how much the information source paid to gets its listing put first. The goal of the game is to entice the web surfers to your site. Once there, you can either sell them something or present them with advertising for which you have been paid. The more visitors to your site, the more you can charge for the so-called "banner advertisements." For folks looking for information, these banner ads are just background noise. Being enticed to look at one will often lead you on an interesting, but irrelevant, wild goose chase.

Standards

Within the Internet, standards are created and managed on a voluntary basis. The IETF makes recommendations to the IAB for inclusion as standards. Remember that the whole Internet started as a volunteer-based network, each building on what was done before. Anyone who determines a new feature is needed or a problem needs fixing creates a solution. That solution is implemented on one's own network, and when ready, it can be submitted to the IETF as a *Request for Comment* (RFC). It is then published on the relevant newsgroups for others to try it and comment on it. After it has survived this torture test, it is ready for formal adoption. Because the whole Internet is voluntary, it is up to the network administrator to decide whether to use that RFC or not. Failure to implement it, however, may mean compatibility problems with other networks.

This process is very practical and very different from that used by the formal international standards organizations, such as the *International*

Standards Organization (ISO) and the *International Telecommunications Union* (ITU). These have a formal membership and a proposal submission and review procedure that is designed to form a consensus.

Internet Operation

We have seen that the Internet is really a collection of independently owned and operated networks. The concept of networks was previously dominated by the telephone model. It had a central switching exchange and all information ran through this exchange. The telegraph network also used centralized switching. Like the telephone switching system, it, too, had telegraph lines that connected the switching centers. The telegraph-switching center was called a message switch. These message-switching centers were originally manually operated copy and relay centers. Later, paper tape with automatic punches and, finally, computers replaced the manual switching centers. It became clear that some mechanism was needed to distinguish between the message content and the control characters that were used by the computer to recognize the beginning and end of the message.

Message switching with its telegraphy history was message integrity oriented. Each message was carefully acknowledged. Losing messages was not acceptable. This formed the basis for reliable message transfer.

About the time the Internet was taking shape, the X.25 network protocol was being developed. One might say that the X.25 protocol is the successor to message switching. It was essentially designed with the Telco model in mind. It was intended to make money by charging for packets, just as the message-switching system was designed to charge for each message. The goal was to provide end-to-end message integrity.

A further impetus to the development of X.25 was the deregulation of the telephone equipment market brought about by the Carterphone decision of 1968. Heretofore, modems were only available from the Telco, and the data rates were slow. At that time, the Telcos controlled telephone calls between countries and all data (telex) messages went through what were known as international record carriers. Because they were the only carriers permitted to transport such international traffic, the cost was high. Modems unfortunately fell into the hands of the traveling public who would make dial-up telephone calls across international boundaries and transmit data at a fraction of the cost of the "record traffic." The Telcos tried to control this, but it was hopeless. Their next attempt was to offer a slightly lower cost *Public Data Network* (PDN), which was the X.25 network. Although it worked well, it really was still more costly (although more reliable) than using the *Public Switched Telephone Network* (PSTN). The X.25 network replaced the telex network to most parts of the world. When traveling, using the X.25 network to contact, for example, CompuServe is often the most cost-effective way to do it.

One of the large steps forward that X.25 provided was that it made the transport mechanism (packet switch) separate from the message content. The network was then optimized for packet transport and was completely independent of the message or its content. The maintenance of end-to-end connectivity is a natural outgrowth of the Telco and message-switching history.

The operation of X.25 is very different from the IP. It is important at some point to draw the distinction between the two protocols.

Without a full-blown discussion of the OSI Reference Model, let us start with the basic concepts of packet switching and the transport of packets across an X.25 network. Figure 32-1 is a picture of the Reference Model to be used as a reference as we discuss the operation of networking protocols. A few quick notes are in order:

- The physical layer attaches to whatever network is at hand. There could be multiple physical connections in any given system.
- The data-link layer handles communications across the physical link only between adjacent machines by putting data into frames.

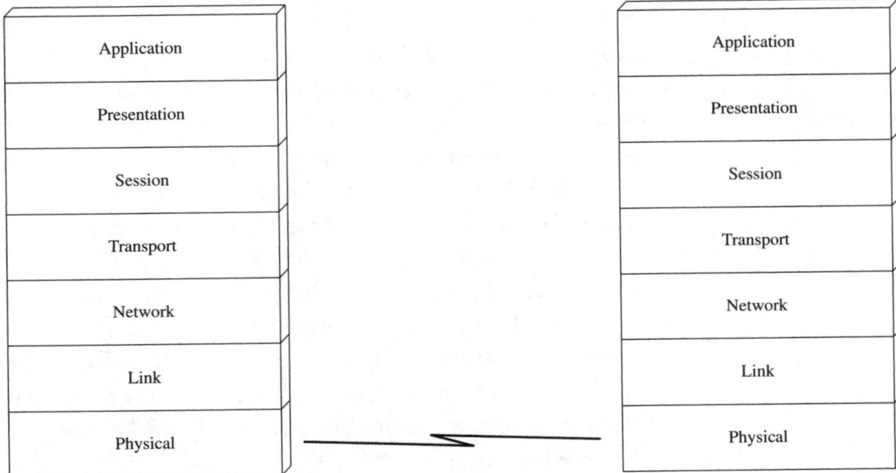

Figure 32-1
The OSI model as the reference

- The network layer must be consistent across the entire network, and it is where packet switching is handled.
- The transport layer handles end-to-end transport and integrity of data. The communicating entities at the transport layer have no idea (very theoretically) that the underlying layers exist or what their characteristics are.

X.25 is a *Connection-Oriented Network Service* (CONS). The IP provides, by contrast, a *Connectionless Network Service* (CLNS). This is the major difference, and as we shall see, it pervades the networking decisions made and the choices made for the functionality of the transport layer. The details of how the packets are handled within the packet switches are very similar in terms of how the queues are handled.

A connection is established by X.25 for the end user across the network. Although we said that transport is responsible for end-to-end connections, the network layer can also choose to provide this capability. If our network layer is going to do this, then the transport layer need not be complex at all. You will note in the following discussion that the reference is to telephone terminology. The reasons for this are now clear.

An X.25 connection is established by the end system across the network. The calling party sends a setup packet containing the address of the source and destination systems. Figure 32-2 shows the basic network configuration. The packets travel on each leg of the journey in a link layer frame that primarily provides error detection and correction by using the *frame check sequence* (FCS) on a link-by-link basis. The packet containing the addresses permits each packet switch along the way to make a routing decision about the best way to send the packet toward its destination.

The frame itself consists of a *Flag* (F) character at the beginning and end as delimiters of the frame. The *address field* (A) is rarely used, but logically contains the physical address of the device on the network. The control field tells what kind of frame it is (data bearing or control) and can be used to contain sequence and acknowledgement numbers. The FCS is a very powerful, 16-bit checksum system that is able to detect 99.998 percent of all errors. Given the quality of the circuits of the day, such a powerful error-detection scheme was necessary. The control field shown as C in Figure 32-3 carries the sequence and acknowledgement numbers.

Figure 32-3 shows the relationship of a packet being carried inside a frame. Note that the frame is terminated (FCS evaluated and acknowledgement sent on each hop). The packet is reencapsulated in a new frame for its hop across the next link. Chapter 11, "Frame Relay," discusses Frame Relay, which is the successor to X.25.

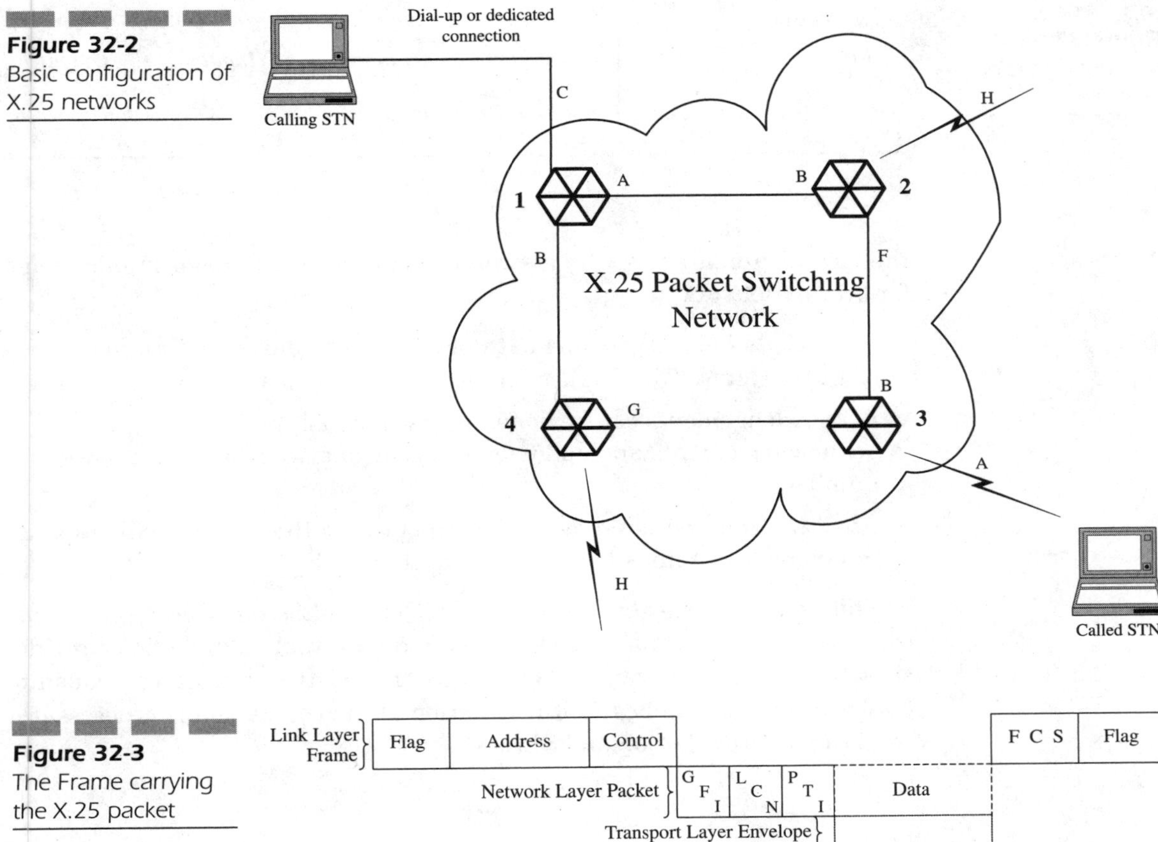

Figure 32-2
Basic configuration of X.25 networks

Figure 32-3
The Frame carrying the X.25 packet

In addition to the routing information, the first packet also contains the very important *Logical Channel Number* (LCN). You may think of it as a random number that is used by each packet switch instead of the address. There are several benefits to doing this. First, you don't need to put the complete address in each packet, and second, you achieve a relationship between packet switches that lasts for the duration of the call. (The importance of this will only become clear later). This relationship between packet switches is called a virtual circuit. You will find further discussions of virtual circuits in both Chapter 11 and Chapter 12, "Asychronous Transfer Mode."

Figure 32-4 shows the routing table (technically a switching table) in each packet switch shown earlier in Figure 32-2 that resulted from the call setup from caller to called system. This association of ports and LCN create

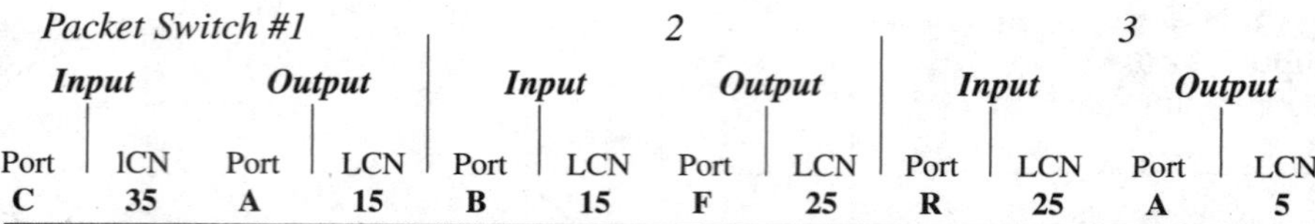

Packet Switch #1				2				3			
Input		Output		Input		Output		Input		Output	
Port	ICN	Port	LCN	Port	LCN	Port	LCN	Port	LCN	Port	LCN
C	35	A	15	B	15	F	25	R	25	A	5

Figure 32-4 The routing table example

the virtual circuit across the network. This has several major implications for overall network design:

- All packets follow the same path between any source destination pair of end systems. (This makes maintaining sequencing of packets easy.)
- Acknowledgements can be done link by link, using frame acknowledgements and end to end by using network-layer sequence numbers.
- At setup time, we could determine the load on the packet switches and interconnecting links before choosing that path.

This would enable *Quality of Service* (QoS) and *Service Level Agreement* (SLA) adherence. It must be clearly stated here that X.25 doesn't provide these services. However, the existence of a virtual circuit relationship between packet switches for the duration of the connection is a necessary condition to provide QoS and SLAs.

Connectionless Network Services (CLNS)

Now that we understand CONS, we can look at CLNS. IP is a CLNS. The apocryphal story, as is oft repeated, is that IP was designed during the height of the Cold War. The military (funded by ARPANET) wanted a bomb-proof network—one in which any of the lines or packet switches could be knocked out at any time and not seriously affect network operation. Therefore, the designers chose a dynamically routed CLNS. The philosophical differences between a CLNS and CONS are large, but each is a valid way to construct a network. The Internet has grown to become a high-volume worldwide network. Even some of the original designers are amazed at its capability to continue to grow yet maintain its robustness. More than a few doubters have been predicting Internet meltdown. Congestion has occurred due to lack of bandwidth connecting routers. Adding more bandwidth has so far alleviated this condition. There was some discussion in the late 1990s that the demand for bandwidth would exceed the supply in the year 2001. The technology breakthrough of *dense wave division multiplexing* (DWDM discussed in Chapter 31) essentially solved the bandwidth problem. The solution isn't inexpensive per se, but it yields tremendous bandwidth for the price.

It is now easy to draw a comparison between X.25 and IP. Each IP packet (called a datagram) is a stand-alone entity containing all the information needed to be routed from one end of the network to the other. The lack of any formal associations between packet switches means that (theoretically) packets can be routed instantly and dynamically without regard for the final destination. Figure 32-5 shows the relationship between frame and packet. Frames in the case of IP are only used for error checking. The *control field* (C) simply says that this frame contains data. Bad frames (for example, those failing the FCS check) are simply discarded. Error recovery will be left to the transport layer.

The contents of the IP packet header are shown in the more conventional representation in Figure 32-6. Note that each row is 32 bits (four bytes) wide. The first thing we notice about IP packets is the large amount of overhead in the packet header. The IP header is a minimum of 20 bytes (octets) long. We will not describe the use of all these fields.

The important fields for our current discussion include the following:

- The version field theoretically enables multiple versions of the protocol to exist in the network, thus permitting seamless upgrading of the protocol. IHL is the *Internet Header Length*, and is the length in bytes of the header.
- TOS is *Type of Service* and was intended to permit priority handling of packets. It is generally not implemented, although some vendors provide proprietary implementations.

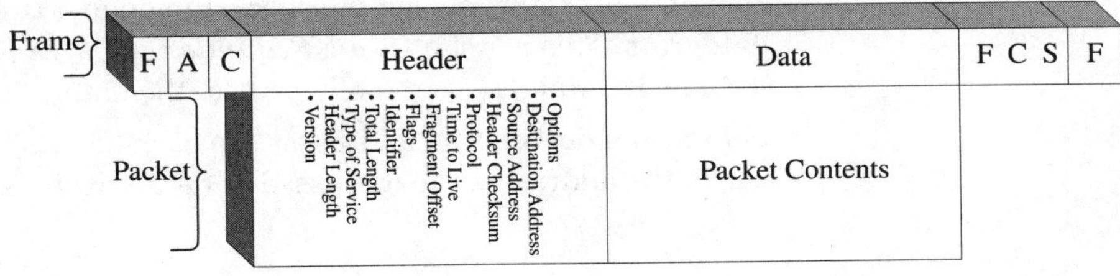

Figure 32-5 The relationship between the frame and packet

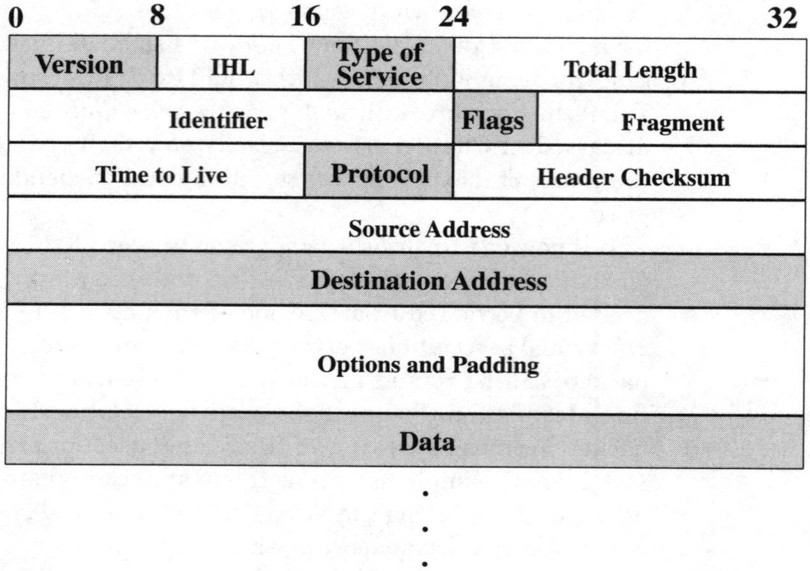

0	8	16	24	32
Version	IHL	Type of Service	Total Length	
Identifier			Flags	Fragment
Time to Live		Protocol	Header Checksum	
Source Address				
Destination Address				
Options and Padding				
Data				

.
.
.

Figure 32-6 The IP header

- Total Length is the total length in bytes of the datagram.
- The next 32 bits, including the identifier flags, and fragment offset, are rarely used.
- TTL is the *Time to Live* indicator and is essentially a hop counter. The originator sets it to a value a little larger than the maximum number of hops that it should take to get the packet to the destination. If the network can't deliver it in that many hops, then it is considered to be undeliverable and discarded. This prevents undeliverable packets from bouncing around the network forever.
- Protocol is a pointer to the layer above the network layer that is using the IP packet. According to OSI purity laws, that should be the transport layer. The designers of IP built in more flexibility so that other layers could use the packets.
- Header Checksum is a simple checksum on just the header of the datagram. This is not a *Cyclic Redundancy Check* (CRC) of the sort used by the link layer to really protect the data.
- Source and destination addresses are each 32 bits long. The format and use of the addresses are discussed in detail in the addressing section.

Options and Padding

IP provides some interesting options, one of which is forced routing. If forced routing is employed, the route to follow is specified in the options field. Padding is used to fill out the last field so that the data field starts on an even 32-bit boundary.

Routers use the source and destination fields to send the packet to the next router. The Internet uses a best-effort, hop-by-hop routing system to keep the handling of the packets simple. Much is made of the fact that the Internet is a CLNS and therefore can route packets almost arbitrarily and those packets arrive out of order or not at all. Although it is theoretically true that a router could choose a different path for the next packet with the same destination address as the previous packet, let us consider the practical realities. First, if the chosen route is good for the current packet, the chances are excellent that it is also good for subsequent packets. Second, the router normally has a limited number of entries in his routing table: the normal route, the secondary route, and the default gateway. Therefore, the number of different paths that the packet could theoretically take is large. In practice, it is relatively small. The practical result is that most packets follow the same path and generally arrive in order, but there are no guarantees.

Transmission Control Protocol (TCP)

For customers needing high-integrity, guaranteed, error-free delivery, a transport layer must be built on top of IP. *Transmission Control Protocol* (TCP) provides the sequenced, acknowledged reliable delivery mechanism. Figure 32-7 shows the format of the TCP packet called a segment. The important parts of the TCP segment for our discussion are the sequence number and acknowledgement numbers.

Although the IP packets carrying these TCP segments may arrive out of order, TCP can easily resequence them. The acknowledgement field provides the reliable delivery part.

Figure 32-7
The TCP header

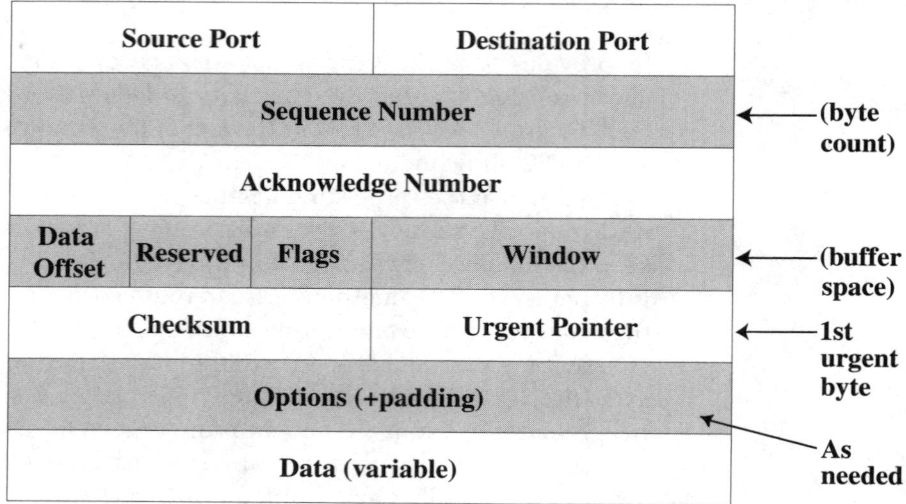

The Fields in the TCP Header

TCP's header is a minimum of 20 bytes (octets). The fields in the TCP header are as follows:

- **Source port and destination port** The source and destination port numbers are used to identify the listening agent that is using TCP to transfer data. IP uses the protocol field; TCP uses the port number. The port address, in conjunction with the IP address of the packet, is referred to as the socket. Therefore, in order for a socket to exist, one needs both the IP address and the port address. Port addresses below 1024 are assigned by (or reserved for assignment) by the IETF. When you surf the Web, your browser always uses TCP port 80. One can provide web servers on other ports (for example, port 8080), but to use that port, the sender has to specify the port number after the host name because the default is port 80. TCP actually has 65,535 ports available, but much less are used.
- **Sequence number** The sequence numbers are not packet numbers, but rather byte numbers such that if there were 100 bytes in a segment, the sequence number will increment by 100 each time. This helps the recipient reserve the proper amount of buffer space if it gets, let's say, the tenth segment first.

- **Flags** The flags section is used to set up and tear down connections, as well as to push the final segment and get acknowledgements for the segments.
- **Window** Window describes the flow control window. Rather than the more conventional mechanism of on and off flow control, TCP uses a credit system where credit (in bytes) is granted to the sender. When the sender has sent that many bytes, it must stop. The recipient keeps this updated to provide for a continuous flow of data.
- **Checksum** As on the IP packet, the checksum is not a CRC, but a simple checksum, and it is run on the entire TCP segment.
- **Urgent pointer** The urgent pointer points to urgent data (if any). Theoretically, this handles higher-priority data within a segment ahead of normal priority data.
- **Options (+ padding)** Here again, options can be added to TCP. These fields need only be added when the options are present. The field is always padded to even 32-bit boundaries.

TCP is an end-to-end protocol that makes sure that the data gets to the destination and in proper order. It accomplishes this by setting up a connection to the other end, much as X.25 sets up a connection, except that it doesn't create a virtual circuit. Therefore, when you visit a web site, TCP sets up a connection so that you reliably get a complete web page. The page may consist of several segments and many packets, since each packet is about 576 bytes by default.

After the page has been delivered and acknowledged, the connection is torn down. When you select the next URL or hyperlink, a new connection must be made. Remember that there is no way to know whether the user will select another link on that same web site or a hyperlink to some other site. Some browsers provide indications of this connection process on the bottom edge of the browser window.

User Datagram Protocol (UDP)

In the interest of thoroughness in a discussion about IPs, we need to mention *User Datagram Protocol* (UDP). It also is Transport Layer Protocol, but does not create connections. It is another "send and pray" protocol like IP. Who then would want such a thing? Consider the problem of a database query. You don't need a connection, but you need an answer quickly. If the answer doesn't come back in a reasonable time, ask again.

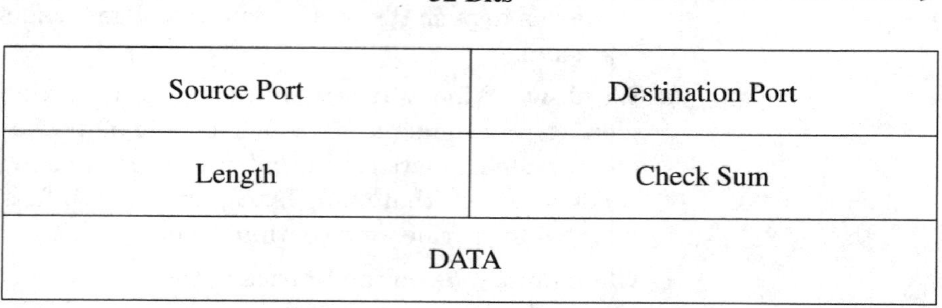

Figure 32-8
The UDP header

Both the DNS and *Simple Network Management Protocol* (SNMP) use UDP. These are both query/response protocols. Figure 32-8 shows the UDP header in all its simplicity. By default, the UDP header is only 8 bytes (octets) long.

The port addresses are the same as described for TCP. The length tells how long the segment is in bytes, and the checksum is, as with TCP, a simple sum over the whole segment.

IP Addressing

Now that we have a handle on the protocols that are going to be used, let's take a look at the structure and significance of addressing and then we will discuss how routing is done.

The addressing scheme is important for obvious reasons. Addressing is used by the routers to find the destination for the IP packets. One could imbue the routers with God-like knowledge of the network, or the router could have a simple table to use to look up the next hop. If there is no entry in the table, then send it to the default router. The whole Internet works because of this default route mechanism. If any device doesn't know how to return a packet, it sends it to the default router, known as the default gateway.

Routers Versus Gateways

A point of clarification is needed about routers and gateways. In the Internet community, routers and gateways get a little confused. In the previous discussion, we use the term router almost exclusively. As pointed out, in the Internet community, it is referred to as the default gateway.

The Internet terminology started before terms were standardized. A gateway was a box that sat between two networks. Therefore, it was natural to refer to routers, especially ones connecting different networks as gateways.

Today, a router is a box that operates at the network layer, reads IP addresses, and chooses the best hop it can find to the destination network. A gateway is a device that handles protocol conversion. A mail gateway is a great example. If you are using IBM's PROFS mail system, you have a mail gateway that reformats the messages and changes the characters in the message from IBM's proprietary *Extended Binary Coded Decimal Interchange Code* (EBCDIC) to standard ASCII characters.

When selecting the addressing scheme, the Internet designers chose their own path, rather than taking a page from the telephone folks.

The designers were thinking in terms of networks for universities, government, and the military. They anticipated that there would be a few very large networks for the military, government agencies, and very large universities; some medium-sized networks; and many little networks for smaller colleges.

Correspondingly, IP has three address formats, known as class A, B, and C addresses, respectively. A class D address format exists for multicasting to groups of destinations. Figure 32-9 shows the IP address formats.

You will note that in the IP packet format that the address is 32 bits long or four (4) bytes, which is represented in a dotted decimal format. Each byte

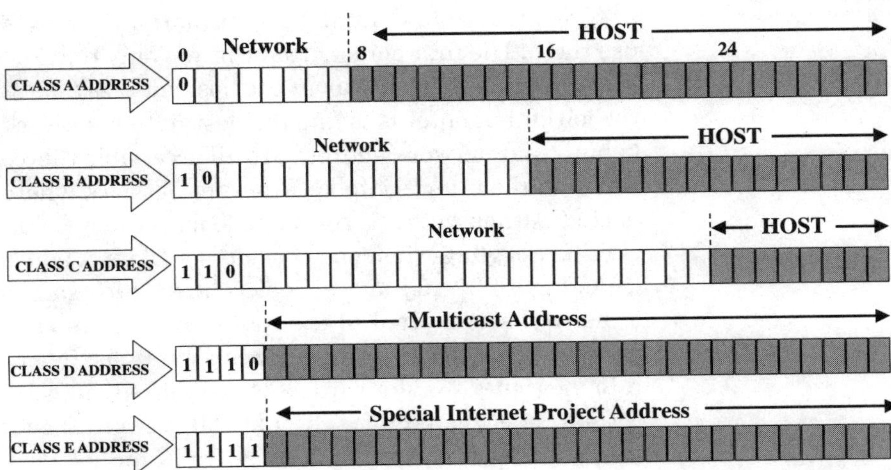

Figure 32-9
The Internet IP addressing scheme

is represented by a decimal number of the value of that byte. This makes it relatively easy for humans to read and use the addresses.

A 4-byte address might look like this: 147.45.193.15. We can tell immediately that this is a Class B address. The first bit of the address is 0 for Class A and 1 for Class B. Class C addresses have the first 2 bits set to 1. Therefore, the class ranges are as follows:

- Class A addresses are from 0 to 128.
- Class B addresses are from 129 to 191.
- Class C addresses are from 192 to 223.
- Class D addresses for multipoint broadcast are from 224 to 239.
- Class E addresses (reserved for special projects) are from 240 to 254.

You will note that certain addresses aren't valid or used. "All zeros" means this network. "All ones" is the broadcast address.

Class A addresses were reserved for the largest organizations. The format of the address has 8 bits reserved to define the network and 24 bits to define the host. Originally, a host was a computer that was smart enough to run IP and TCP. Today, every network-connected device contains a computer, and therefore it has its own IP address. In fact, many devices have multiple IP addresses because they use different addresses for network management functions than the one they use for data transfer. Therefore, it is very easy to run out of IP addresses.

A Class A address provides for 127 networks with over 16 million attached devices. A Class B address provides for 16,000 networks and 65,000 attached devices. A Class C address provides for about 2 million networks and 255 attached hosts on each small network.

Now that we understand the format of addresses, let's see how we can use them. The first point is that routers only route on the network address. The router doesn't care about the host address. The assumption here is that the job of the router is to find the destination network and that all the hosts behind that network address are all accessible thereafter. Imagine that the router is the interface to an Ethernet network where all the hosts are connected. Although this is conceivable for a Class C address with 255 hosts, it is mind boggling when one considers a Class A address.

As Figure 32-10 roughly depicts, there are examples of the three classes of networks connected to the router. If a device on the 145.15.0.0 Class B network wants to send to a device on the 14.0.0.0 network, all it does is send it to the router that forwards it to the Class A network. The router considers the job done when it finds the 14.0.0.0 network. Now, clearly, it is silly not to mention that it is unworkable to have 17 million devices on one Ethernet.

Figure 32-10
Different classes of IP addresses connected to a router

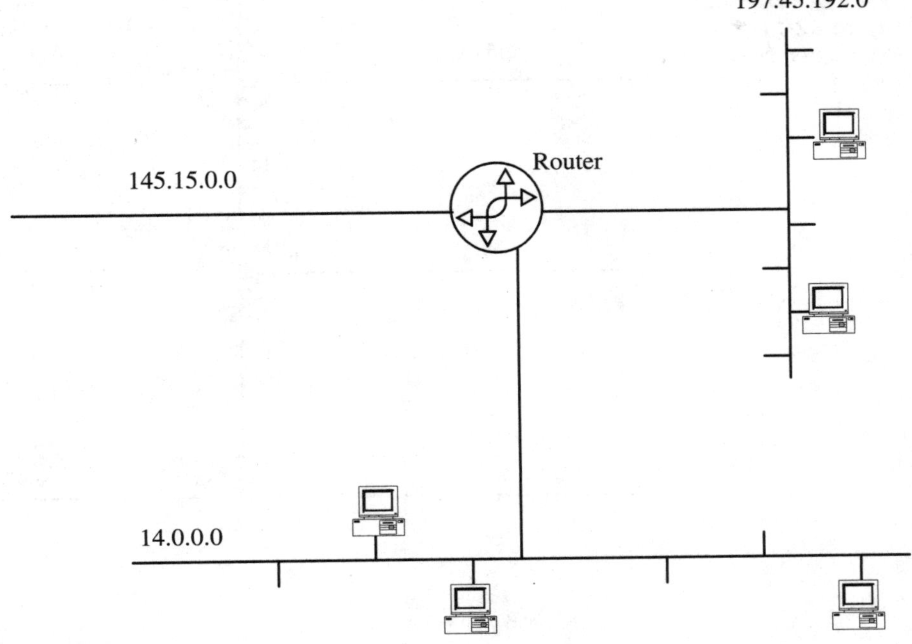

Subnetting

The solution to the problem is to subnet the 14.0.0.0 network into a bunch of smaller networks. We do not have to create subnets on even byte boundaries, but for the sake of discussion, we will break the Class A net into Class-C-like addresses. Figure 32-11 shows (without getting too carried away) what this might look like.

The first thing we note is that we have introduced another router because the only way to get between subnets is by using a router. So now, our first router (R-1) thinks it has found the 14.0.0.0 net when it finds a port on the 14.1.1.0 network just as before. The second router (R-2) has to know that the 14.0.0.0 network has been subnetted into Class C nets, and that it isn't finished routing until it finds the Class C addressed network. The way the router and all hosts know this is by being given a subnet mask that contains all 1's over the portion of the address that specifies the network.

The subnet mask for hosts on the 14.0.0.0 network in Figure 32-10 would be 255.0.0.0. The subnet mask for hosts on any of the networks from routers 2 or 3 would be 255.255.255.0. Thus, these routers and hosts don't care if it is a Class A address; they route until they find the last network.

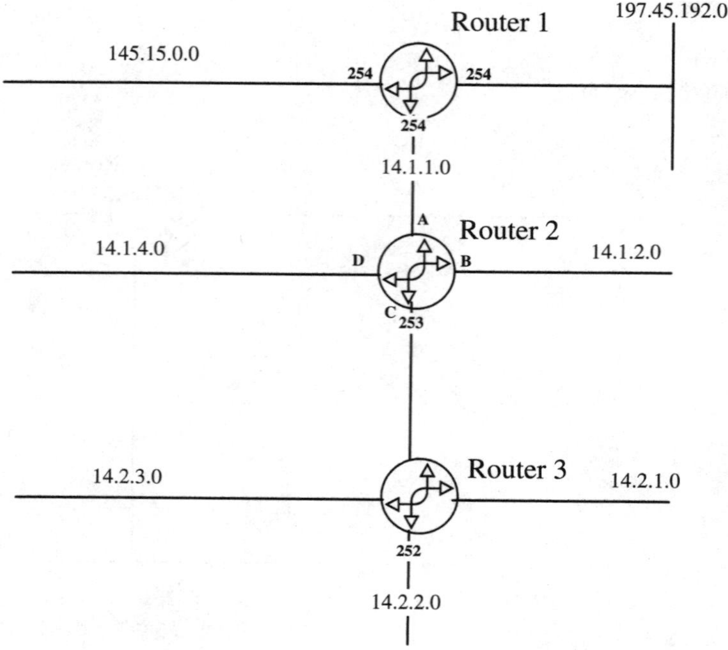

Figure 32-11 A subnetted class A address

The good news is that we can now limit the size of our broadcast domains. Can you imagine what would happen on the 14.0.0.0 network (shown earlier in Figure 32-10) if we sent out a broadcast message looking for a particular station? Everyone everywhere would get it. Since nearly all of the Class A and Class B addresses have been assigned, the only addresses available are Class C, and they are running out.

Network Address Translation (NAT)

The solution to the problem is to use *Network Address Translation* (NAT). NAT permits us to use any network addresses we want internally while looking to the outside world as though we were a Class C network. A few clever tricks make this work. First, we must have only one access point to the Internet.

If we use, for example, IBM's Class A address internally, you can imagine the havoc that would be created if one of our routers advertised that we were now IBM on the Internet.

To use NAT, our interface device to the Internet simply keeps a dynamic table that maps official Class C addresses to the address of our internal user's Class A address. An internal user, 14.1.4.25, for example, wants to surf the Internet. Our real address on the Internet (public address) is 245.123.237.0. The firewall then maps 14.1.4.25 to 245.123.237.1. The firewall will map this in both directions. There is a secondary side benefit to providing NAT. Hackers will never know our real internal network addresses or address structure. They won't be able to tell servers from terminals. It therefore makes it that much harder for them to hack into our network. With a properly configured firewall, it is almost impossible anyway.

There are some obvious limitations, namely that we could theoretically only have 254 folks surfing the Net at one time. If this is a serious limitation, another Class C address is requested from the ISP or the InterNIC.

Back to preventing havoc on the Internet. To prevent any possibility of the fact that we are using the 14.0.0.0 address from ever getting out and creating havoc, we can legally use the 10.0.0.0 address. This address is "illegal" on the Internet.[1] Any Internet router discovering a 10.0.0.0 address in a packet will simply discard that packet. This effectively gives us a free Class A address that we can subnet any way we want. Our firewall still must do NAT because it is not polite to send garbage packets out on the Internet.

CompuServe and AOL use NAT between their internal address scheme and the real Internet.

DHCP, BOOTP, ARP, and RARP

The next dynamic addressing scheme is that used by most ISPs. They don't have enough addresses to be able to give a real Internet address to every subscriber. So, what they do is dynamically assign one to you when you sign on and put it back in the pool of unused numbers when you sign off. This is called *Dynamic Host Configuration Protocol* (DHCP). These all do essentially the same thing, although there are significant technical differences. When you sign on, you contact the terminal server at the ISP, who in your

[1]Certain network addresses are nonroutable across the Internet as they are reserved for private address space (network 10 is an example of this).

stead goes and gets an IP address for you. If you were to sign on from an Ethernet, you would send a broadcast message to all devices on your subnet asking the keeper of unused IP addresses to assign you one. In a large network, there are two ways to handle addressing. One is to permanently assign every device (host) an address. The other is to use DHCP. There are arguments for both approaches.

Address Resolution Protocol (ARP) is a little different. It is used by a router to find the physical address of a destination host when it has found the proper network. The router only has network-layer information in the packet. It needs to have a physical (Ethernet) address in order to actually deliver the packet. It uses the ARP to broadcast to all devices on the destination looking for that device whose IP address matches the one for which it is looking. When the target host responds, the response contains the matching IP address and the physical, Ethernet address as well.

Reverse Address Resolution Protocol (RARP) works just the opposite. The host knows its Ethernet address and sends out a broadcast to learn its IP address.

Routing

As we have indicated, IP routers route the datagrams to the next hop. Routing is accomplished by simply looking in the routing table to find the next hop. This, of course, begs the question of where this table comes from.

Table 32-1 shows the routing table for router R-2 in Figure 32-11 and Figure 32-12. We have taken the liberty of giving all the ports of the router the same final byte address, although it is connected to different networks. Using this convention makes it easier to troubleshoot the network since the router's address is always the same. If you are designing and building your own network, this is handy to remember.

In these diagrams, there is only a single path from any one location to another. This makes the routing table simple and static. We now introduce Figure 32-13. This is Figure 32-11 with two major changes. First, we are using the same diagram so we don't have to learn a whole new set of numbers, but now these are just addresses on the network and have nothing to do with NAT anymore. Second, we introduce router R-4.

Table 32-2, the routing table for router R-2, now has three additional entries marked by * because R-2 now has two ways to get to the 145, 197, and 14.2.3 network. Notice that the hop counts are larger in each case because the connection is not direct. Notice also that we could have put

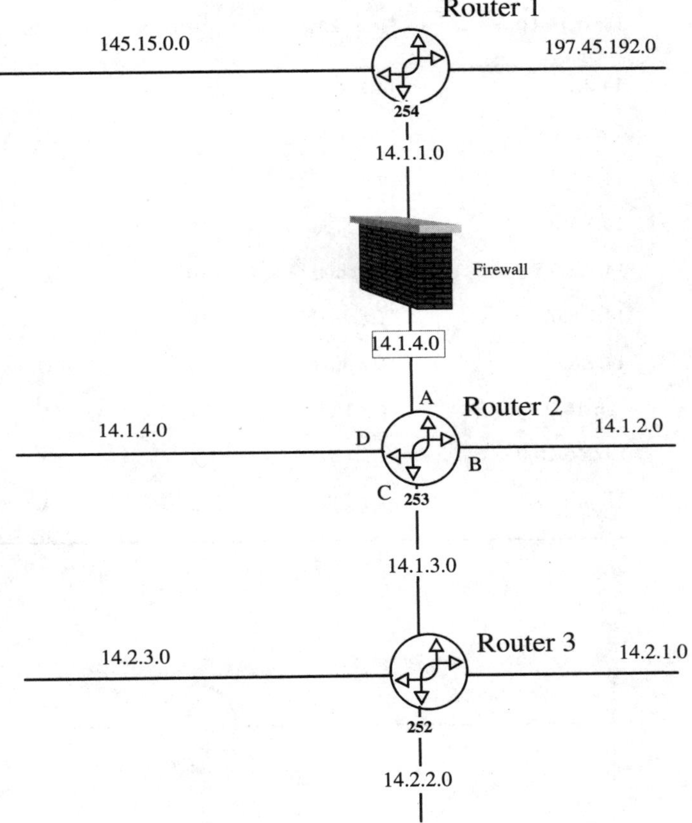

Figure 32-12 Class A subnetted with a firewall to the Internet

entries in for the 14.2.1.0 and 14.2.2.0 networks via R-1 as 3 hops away, but these entries, as well as the one added to the 197, are bogus entries. If R-3 (or R-1 in the case of the 197 network) is down, there is no point in sending packets, although we think that there is a route.

Dynamic Routing Tables

You have seen how much fun it is building routing tables by hand. Imagine doing this for 1,000 routers. What happens if there is a failure in the network? We would have to figure it out and send out new tables. For this reason, networks use dynamic routing, whereby they learn who their

Table 32-1

The routing table for router R-2

Destination	Next Hop	Hops	Port#
14.2.1.0	14.1.3.252	1	C
14.2.2.0	14.1.3.252	1	C
14.2.3.0	14.1.3.252	1	C
14.1.1.0	Attached	0	A
14.1.2.0	Attached	0	B
14.1.3.0	Attached	0	C
14.1.4.0	Attached	0	D
145.15.0.0	14.1.1.254	1	A
197.45.19.0	14.1.1.254	1	A
All others	14.1.1.254	1	A

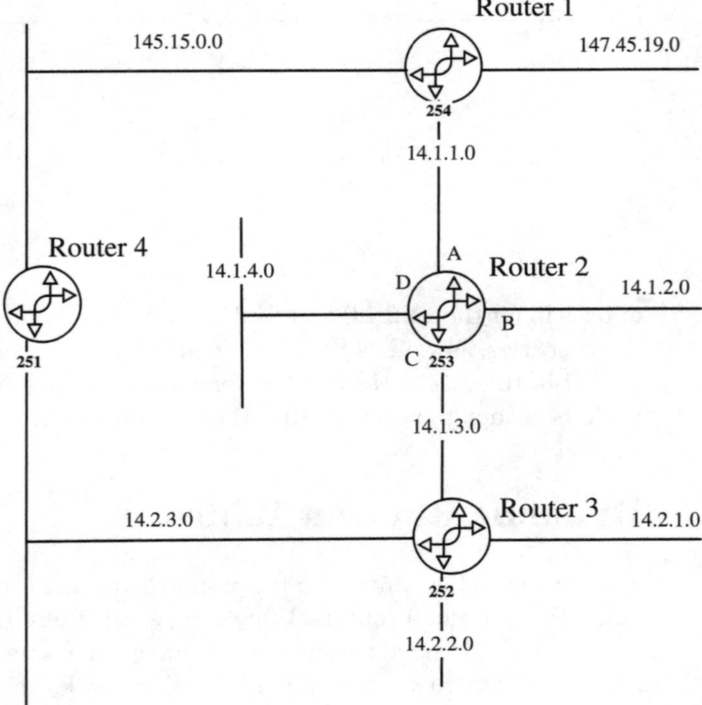

Figure 32-13
A new router is added to the network.

Table 32-2

The routing table for router R-2

Destination	Next Hop	Hops	Port#
14.2.1.0	14.1.3.252	1	C
14.2.2.0	14.1.3.252	1	C
14.2.3.0	14.1.3.252	1	C
14.2.3.0*	14.1.1.254	2	A
14.1.1.0	Attached	0	A
14.1.2.0	Attached	0	B
14.1.3.0	Attached	0	C
14.1.4.0	Attached	0	D
147.15.0.0	14.1.1.254	1	A
147.15.0.0*	14.1.3.254	2	C
197.45.19.0	14.1.1.254	1	A
197.45.19.0*	14.1.3.252	3	C
All others	14.1.1.254	1	A

neighbors are and build their own routing table from that information. Two basic methods for doing this are used on the Internet.

Routing Information Protocol (RIP) was the original mechanism for sending routing table updates. Its operation is very simple. Each router packs up its routing table into packets and sends these packets to all of its neighbors. The neighboring routers then pick through the received table to look for better routes to destination networks than they already have. If they find better routes, they put them in their table. The process repeats itself. Clearly, if we do this once per day, then old or invalid routes will remain in the table for at least a day. The more dynamic we want our network to be, the more often we should do the update. Unfortunately, this consumes lots of network bandwidth when done too often. Every half a minute (30 seconds) to a few minutes seems to be a reasonable range. The irony here is that larger networks have a greater probability of something changing so they have to do it more often than smaller networks.

There are a couple of problems with RIP when the network gets big. The tables grow exponentially, and therefore it takes more and more router processing power to handle the table updates. The second problem is that good news travels fast, but bad news (unavailability of routes) travels slowly. RIP works fine in small networks.

To handle the problem of exponentially growing routing tables, we break the network into routing domains. These are called autonomous systems. The basic idea is that all the routers within a routing domain or autonomous system can exchange routing tables as usual. All routes to locations outside their domain are sent to the boundary router (frequently called a gateway router or simply a gateway). The boundary router is also part of the local domain and is part of the exterior domain. It uses a different protocol to exchange routing tables with its peers called *Boundary Gateway Protocol* (BGP). This keeps the volume of traffic between routing areas or autonomous systems manageable. Figure 32-14 depicts these routing domains. The Internet terminology (using gateway for router) is shown on the top of each figure, while the ISO standard terminology is placed below each figure.

The next approach to the routing problem is to use the *Open Shortest Path First* (OSPF) protocol. This protocol was developed and standardized by the ISO. It differs from RIP in two significant ways:

- Only changes in network status are sent rather than the whole routing table. This significantly cuts down on network overhead.

Figure 32-14
Routing domains

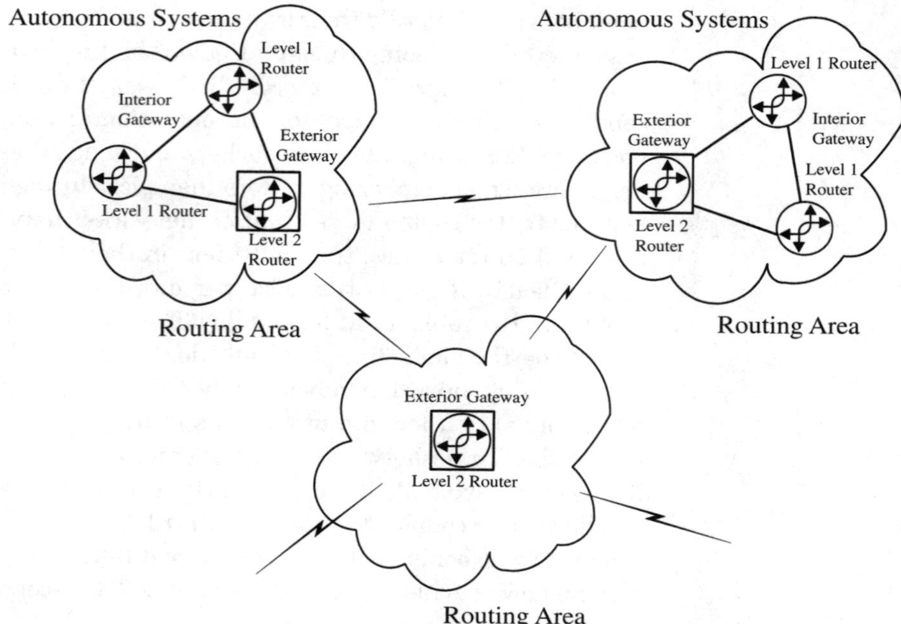

- It creates a hybrid metric to use in lieu of a simple hop count to the route selection process.

In RIP, the router cannot make a distinction between a T1 line and a 9600 bps line if they are both *x* hops to the destination. By adding this hybrid metric that includes a weighted average of hop count, data rate, error rate, delay, and even maximum transmission unit size, a communications cost for that path can be calculated. True, it takes more router processing power to calculate all this, but it allows the router to make better route choices. OSPF isn't the only protocol that does this; each of the router vendors has invented a proprietary routing protocol.

Although OSPF is more efficient in the total volume of update information sent, there still comes a time in network size where the routing update information begins to affect the network bandwidth available. Therefore, ISO has defined a two-stage hierarchy of routing areas called level 1 and level 2 areas. These are the same as autonomous systems with a different name. Level 1 routers are the local routers, and they exchange local data. Level 2 routers are the gateway routers, and they are the interdomain routers.

Routing Versus Switching

There is a good deal of confusion over the issue of routing versus switching. Now that we have discussed and given examples of each, let's see if we can make a clear distinction. This section began with a discussion of X.25 and packet switch. In X.25 we used the first packet to contain the address. That packet had to be routed to find the destination. In the process of doing the routing, a virtual circuit was set up by using the LCN that was now carried in every packet in lieu of the address. This is called a switching system because the table into which the packet switch must look to find out what to do with each packet is small. For example, only those currently active virtual circuits have entries. The packet switch can therefore process each packet very quickly. If the total number of virtual circuits was limited to some number like 1024, you could set aside a block of memory and use the LCN as an index into the table.

A router, on the other hand, must have a complete table of all possible routes including, of course, the default route. Clearly, you could make the table smaller by using the default route more often, but that just moves the problem to another router. As we've discussed, this table grows exponentially with the size of the network. The addresses are neither hierarchical nor structured nicely to facilitate table searches. In addition, we have the problem of keeping the tables updated. The big efficiencies that switching shows over routing is not so much in table searching, where there is an advantage, but in maintaining the routing tables. Most packet-switching systems use static (manually built) routing tables or use OSPF. Some use a separate control network to manage routing tables. Therefore, the table update problem does not have the impact on a packet-switching network as it does on a router-based network.

Real-time Applications

This subject is covered in detail in Chapter 33, "Voice-over IP (VoIP)". However, two points need to be made. All of these packet-switching systems suffer from highly variable interpacket delay. The connectionless systems suffer from greater variability or are rather less predictable.

This arises from two sources: first, the queuing and routing functions in the routers or packet switches, and second from the fact that the output port/path may be busy, requiring that the packet be queued until its turn arrives. The systems we have discussed have no priority schemes for handling one type of packet ahead of any other.

In order to handle real-time applications, such as VoIP, we must be able to control interpacket delay and ideally control absolute end-to-end delay. This is for all practical purposes impossible with the Internet as it stands since it is a CLNS and made up of many different providers.

The virtual, circuit-based networks have a better chance since there is a relationship between packet switches. To these, we need only add the capability to evaluate the internal congestion and that of its external path to the next packet switch. Then, during call setup, the packet switches could determine before setting up the virtual circuit whether it supports the delay or bandwidth requirements of the requestor.

Multi-protocol Label Switching (MPLS)

Multi-protocol Label Switching (MPLS) is an attempt to "Band-Aid" virtual circuits onto a CLNS IP network and is covered in greater detail in Chapter 34, "Multi-protocol Label Switching (MPLS)." The idea is to append a label (called also a tag, in which case it is referred to as tag switching) to each packet. Since we already discussed X.25 and LCNs, this tag is essentially an LCN. Thus, the routers can create a virtual circuit across the network. We now have created an association between routers. The routers can make a determination with their peers as to whether they have the capability to support the QoS requested at virtual circuit setup time.

This additional capability leads to multiple-priority queuing systems inside the routers and packet switches. Unfortunately, it can't address the in-and-out queuing times and the absolute transmission delay between packet handlers. One way to address this problem is to employ an ATM network (refer to Chapter 12).

Summary

One of the hottest vehicles used in the discussion of convergence between voice and data networks is the Internet. Only 10 percent of all the data in the world is running across the Internet. However, 90 percent of the data runs on corporate private and *Virtual Private Networks* (VPNs) in 1999. This will probably be true for the next few years. However, one major difference exists; these corporate networks are modeled after the Internet, creating intranets and extranets. Thus, we must be very aware of what is happening in the world of data communications.

Moreover, the real-time applications such as VoIP, streaming audio, and streaming video applications are all placing added burden and traffic on the Internet. As the convergence continues, many changes will take place. These are the parts of the data network that draw the most attention.

CHAPTER 33

Voice over IP (VoIP)

Internet Protocol (IP) telephony is the emerging communications technology arising from the convergence of the worldwide data infrastructure and the telecommunications network. Until now, the voice world and the data world have been largely separate. Voice traffic moves on circuit-switched networks where a fixed path is established at call setup and maintained for the duration of the call. Network resources are allocated to the call, leading to a logical model of time-based pricing. Over the years, millions of millions of dollars have been invested in circuit-switched technology, and these networks are considered extremely reliable pretty much worldwide. Voice traffic growth is in the single digit category.

Data traffic has moved on packet-switched networks. Messages or files are broken up into small packets, and each packet is sent out onto the network on its own, typically with header information containing the needed destination address and other descriptors. The series of packets moves through the network individually, and different packets may take different paths through the mesh of routers and switches. Packets do not necessarily arrive at the destination in the same order they left the sender, and some packets may be lost along the way. Network resources are shared, however, leading to a very efficient architecture. Often, packet-switched network access is priced at a fixed rate, depending on guaranteed bandwidth, but sometimes the network is priced on a usage basis, with users charged for the number of bytes they transmit.

The latest hype in the industry is the convergence of the voice and data transmission systems. This is nothing new because we have been trying to integrate voice and data for years. However, a subtle difference exists with today's convergence scenarios:

- In the beginning, the convergence operation mandated that data be made to look like voice, then it could be transmitted on the same circuitry. Circuit-switched voice and data used the age-old techniques of a network founded in voice techniques.
- Now, the convergence states that voice will look like data, and the two can reside in packetized form on the same networks and circuitry. Circuit-switching technologies are making way for the packet-switching technologies.

These changes take advantage of the idle space in voice conversations, where it has been determined that during a conversation, only about 10 to 25 percent of the circuit time is actually utilized to carry the voice. The rest of the time, we are in idle condition by either of the following:

- Listening to the other end

- Thinking of a response to a question
- Breathing between our words

In this idle capacity, the compression of voice stream can facilitate less circuit usage and encourage the use of a packetized form of voice. Data networking is more efficient because we have been using data packeting for years through packets, frames, or cells.

The use of a packet-switching transmission system enables us to interleave voice and data packets (video, too) where there is idle space. As long as a mechanism exists to recoup the information and reassemble it on the receiving end, it can be a more efficient use of bandwidth. It is just this bandwidth utilization and effective saving expectations that have driven the world into a frenzy over packetizing voice and interleaving it on a data network, especially the Internet.

A note on the cost of voice and data communications is probably in order here. (It is the opinion of this author.) Currently, the drive is to get free voice on a data network. In the early years of data communications, data always was given a free ride on the voice networks. Telecommunications managers diligently fine tuned their voice networks and allowed the data to run over the voice networks during the off hours (after hours). This use of the circuitry was paid for through the dial-up voice communications. The concept was to use the Internet for voice communications in a PC-to-PC dialogue. If packets are packets, then voice should be as easy to digitize and send via the Internet as any text message. The motive of course was toll avoidance. Because most people can access the Internet via a local phone call, which is typically free in many North American locals, and because many people pay a fixed monthly ISP fee for nearly unlimited usage, the view is that long distance calls would be free. But this was only the beginning. Today, the drivers for IP telephony include:

- Toll avoidance by the end user
- Access charge avoidance by competitive exchange carriers and long distance carriers
- New services, such as the integration of customer records with voice communications
- Preference for and experience with the Internet by the youth population
- Streamlined networks (one network to manage)
- Declining PC prices, which make Internet access more affordable
- Portability of Internet access
- Growing role of electronic commerce

This method of providing data over the voice networks crept into some of the business hours when real-time communications were needed. However, many times the voice tie lines and leased lines (point-to-point circuits) usually had some reserve capacity. Therefore, the data was placed on the point-to-point circuits that were justified by voice. (History!) As the competition in the telecommunications market heated up, we saw the costs for a minute of long distance drop from $.50 to $.65 per minute in North America down to an average cost today of about $.05 to $.10 per minute. This significant drop in cost has been the result of competition and technological enhancements. At the same time, the data convergence took advantage of this falling cost factor. Meanwhile, the data was migrating from a point-to-point technique on a dial-up connection to a more robust packet form of transmission, using X.25 or IP packet-switching techniques. The cost per bit of data transmission was dropping rapidly. Then the introduction of the Internet came to the commercial world. With the commercialization of the Internet in the early 1990s, the culture and the cost factors changed at an escalating pace. The cost of data transmission has been touted as being free on the Internet, the only cost being the access fees. However, the cost of data networking has been rapidly declining on private line (intranets), drawing the attention from both a public and a private networking focus. This convergence has everyone looking for free voice by interleaving voice on the data networks.

I agree with the scenario and would not attempt to detract from that goal. However, the other side of the equation is that the circuit-switched networks are continually driving the cost per minute for voice down. Closely aligned to this paradigm is the fact that the North American long distance companies are now offering specials on their service. For example, in 1999, many of the carriers offered $.05 per minute on nights and weekends, another carrier offered Fridays free, and so on. Looking closely at their advertisements, we see a culture changing slowly, moving its way into the industry. It is this shift that makes one wonder if free voice on the data networks is as good a deal as we think. Alternatively, it may be the movement to draw the crowd in. I believe that the strategy for the future is "give them the voice, they will buy the data!" However, we have to wait and see how this plays out.

Contrasting the motivators, we should also consider a list of restraints, items preventing IP telephony from happening or from taking off. These include

- A lack of technical expertise, certainly compared to circuit-switched technology
- An uncertain regulatory environment

- The capability of the *interexchange carriers* (IECs) to lower their prices, thereby reducing the motivation to move to packet voice
- A lack of user knowledge
- Lower quality of voice signals when carried over packet networks
- A lack of unified standards, leading to noninteroperability between hardware vendors and networks
- Unproven reliability

VoIP

The public telephone network and the circuit-switching systems are usually taken for granted. Over the past few decades, they have grown to be accepted as almost 100 percent reliable. Manufacturers built in all the necessary stopgaps to prevent downtime and increase lifeline availability of telephony. It was even assumed that when the commercial power failed, the telephony business continued to operate. This did not happen accidentally, but through a very concerted development cycle brought about by the carriers and the manufacturers alike.

Access to a low-cost, high-quality worldwide network is considered essential in today's world. Anything that would jeopardize this access is treated with suspicion. A new paradigm is beginning to develop because more of our basic communications occur in digital form, transported via packet networks such as IP, *Asynchronous Transfer Mode* (ATM), and Frame Relay. Packet data networking has matured over the same period of time that the voice technologies were maturing. The old basic voice and basic data networks have been replaced with highly reliable networks that carry voice, data, video, and multimedia transmissions. Proprietary solutions manufactured by various providers have fallen to the side, opening the industry to a more open and standards-based environment.

In 1996, there was as much data traffic running on the networks as there was voice traffic. Admittedly, industry pundits are all still saying that 90 percent of the revenue in this industry is generated by voice applications. This may be an accounting problem because on average 57 percent of all international calls originating in North America and going to Europe and Asia are actually carrying fax, not voice. Yet, they are considered dial-up voice communications transmissions because of the methodology used. Moreover, the voice market is growing at approximately 3 to 4 percent per year, whereas data is growing at approximately 30 percent per year. Because data traffic is growing much faster than telephone traffic, there has been considerable interest in transporting voice over data networks (as opposed to the more traditional data over voice networks). Support for voice communications using IP, which is usually just called VoIP, has become especially attractive given the low-cost, flat-rate pricing of the public Internet. In fact, toll quality telephony over IP has now become one of the important steps leading to the convergence of the voice, video, and data communications industries. The feasibility of carrying voice and call signaling messages over the Internet has already been demonstrated. Delivering high-quality commercial products, establishing public services, and convincing users to buy into the vision are all still in their infancy. The evolution of all networks begins this way, so there is no mystique in it.

In VoIP networks, the packetization of voice happens in real time. VoIP also decreases the bandwidth utilized significantly because multiple packets can be transmitted simultaneously. The *Signaling System 7* (SS7) and *Transmission Control Protocol* (TCP)/IP networks are used together to set up and tear down the calls. *Address Resolution Protocol* (ARP) is also used in this process. The process of creating IP packets is as follows:

1. An analog voice signal is converted to a linear *pulse code modulation* (PCM) digital stream (16 bits every 125 μs).
2. The line echo is removed from the PCM stream. It is further analyzed for silence suppression and tone detection.
3. The resulting PCM samples are converted to voice frames, and a vocoder compresses the frames. G.729A creates a 10 ms long frame with 10 bytes of speech. It compresses the 128 Kbps linear PCM stream to 8 Kbps.
4. The voice frames are integrated into voice packets. First, a *Real-Time Protocol* (RTP) packet with a 12 byte header is created. Then an 8 byte UDP packet with the source and destination address is added. Finally, a 20 byte IP header containing source and destination gateway IP addresses is added.
5. The packet is sent through the Internet where routers and switches examine the destination address, and route and deliver the packet appropriately to the destination. IP routing may require jumping from network to network and may pass through several nodes.
6. When the destination receives the packet, the packet goes through the reverse process for playback. The IP packets are numbered as they are created and sent to the destination address. The receiving end must reassemble the packets in their correct order (when they arrive out of order) to create voice. The IP addresses and telephone numbers must be mapped properly.

Table 33-1

Circuit-switched versus packet-switched network characteristics

	Public-Switched Telephone Network (PSTN)	IP
Designed for	Voice only	Packetized data, voice, and video
Bandwidth assignment	64 Kbps (dedicated)	Full-line bandwidth over a period of time
Delivery	Guaranteed	Not guaranteed
Delay	5 to 40 ms (distance dependent)	Not predictable (usually more than PSTN)
Cost for the service	Per minute for charges for long distance, monthly flat rate for local access	Monthly flat rate for access
Voice quality	Toll quality	Depends on customer equipment
Connection type	Telephone, PBX, switches with Frame Relay and ATM backbone	Modem, ISDN, T1/E1, gateway, switches, routers, bridges, backbone
Quality of Service (QoS)	Real-time delivery	Not real-time delivery
Network management	Homogeneous and interoperable at network and user level	Various styles with interoperability established at the network layer only
Network characteristics (hardware)	Switching systems for assigned bandwidth	Routers and bridges for layer 3 and 2 switching
Network characteristics (software)	Homogeneous	Various interoperable software systems
Access points	Telephones, PBX, PABX, switches, ISDN, high-speed trunks	Modem, ISDN, T1/E1, gateway, high-speed DSL and cable modems

The characteristics of the circuit-switched networks and the IP-packet networks are compared in Table 33-1.

IP telephony will also have to change somewhat. We will expect it to deliver interpersonal communications that end users are already accustomed to using. These added capabilities would include (but not be limited to) the following:

- *Calling Line ID* (CLID)
- Three-way calling

- Call transfer
- Voice mail
- Voice-to-text conversions

Users are very comfortable with the services and capabilities delivered by the telephone companies on the standard dial-up telephone set, using the touch-tone pad. IP telephony will have to match these services and ease-of-use functions in order to be successful.

IP telephony will not replace the circuit-switched telephone networks overnight; this will be a coexistence for the near future. In 2003, analysts expect that the amount of IP telephony will amount to 3 percent of all voice traffic domestically and approximately 10 to 15 percent of international traffic. This amounts to 50 billion minutes of traffic, so it is consequential. One must be prepared for both alternatives to carrying voice in the next decade. Thus, the differences between the two opposing network strategies will be ironed out, and the world may shift into a packet-switched voice network over the next decade.

QoS

An important feature of converged networks is that they support QoS features for the transfer of voice and video streams. QoS refers to a network's capability to deliver a guaranteed level of service to a user. The service level typically includes parameters such as minimum bandwidth, maximum delay, and jitter (delay variation).

One argument against IP-based telephony today is the lack of QoS. The manufacturers and developers will have to overcome the objections by producing transmission systems that will assure a QoS for lifeline voice communications. Mission critical applications in the corporate world will also demand the capability to have a specified grade of service available. Applications that will be critical include some of the technologies we discussed in Chapters 7, "Computer-to-Telephony Integration (CTI)," and 9, "CTI Technologies and Applications." The CTI applications with call centers being web enabled, interactive voice recognition, response, and other speech-activated technologies will demand a QoS to facilitate the use of these systems. Each will demand the grade and QoS expected in the telephone industry.

Another critical application for IP telephony will be the results of quality of voice transmission. Noisy lines, delays in voice delivery, and clicking and chipping all tend to frustrate users on a voice network. Packet data networks carrying voice services today may produce the same results. Therefore, overcoming these pitfalls is essential to the success and acceptance of VoIP telephony applications. Merely installing more capacity (bandwidth) is not a solution to the problem, but it is a temporary fix. Instead, developers must concentrate on delivering several solution sets to the industry such as those shown in Table 33-2.

The QoS requirements for IP telephony can therefore be summarized as shown in Table 33-3, which considers the layered approach that vendors will be aggressively pursuing. IP telephony datagrams entering the network will be treated with a priority to deliver the QoS expected by the end user. The routers and switches in the network will assign a high priority marking on each datagram carrying voice and treat these datagrams specially. Queues throughout the network will be established with variable treatments to handle the voice datagrams first, followed by the data datagrams.

An essential concept with QoS mechanisms is that they must be negotiated up front before the data transfer begins, a process referred to as *signaling*. The purpose of this negotiation process is to give the network

Table 33-2

Different approaches to QoS

Strategy	Description
Integrated Services (IntServ)	IntServ includes the specifications to reserve network resources in support of a specific application. Using the *Resource Reservation Protocol* (RSVP), the application or user can request and allocate sufficient bandwidth to support the short- or long-term connection. This is a partial solution because IntServ does not scale well as each networking device (routers and switches) must maintain and manage the information for each flow established across their path.
Differentiated Services (DiffServ)	Easier to use than IntServ, DiffServ uses a different mechanism to handle the flow across the network. Instead of trying to manage individual flows and per-flow signaling needs, DiffServ uses DS bits in the header to recognize the flow and the need for QoS on a particular datagram-by-datagram basis. This is more scalable than IntServ and does not rely solely on RSVP to control flows.
802.1p prioritization	The IEEE standard specifies a priority scheme for the layer 2 switching in a switched LAN. When a packet leaves a subnetwork or a domain, the 802.1p priority can be mapped to DiffServ to satisfy the layer 2 switching demands across the network.

Table 33-3
QoS requirements for IP telephony

Layer Addressed	Technique	Variable
1	Physical port	Variations of port definitions or the prioritization of port interfaces based on application
2	IEEE 802.1p bits	Dedicated paths or ports for high bandwidth applications, but very expensive to maintain
3	IP addressing	RSVP (IntServ) DS bits in the IP header (DiffServ)

equipment that is responsible for providing QoS an opportunity to determine if the required network resources are available and in most cases reserve the required resources before granting a QoS guarantee to the client. The overall process consists of the following steps:

1. A client requests a resource.
2. The network determines if the request can be fulfilled.
3. The network signals yes or no to the request.
4. If yes, it reserves resources to meet the need.
5. The client begins transferring data.

Another contentious issue in the quest for converged networks is deciding which is the appropriate layer of the protocol stack to merge the traffic. The old way to combine the traffic was at layer 1, using separate *time-division multiplexing* (TDM) circuits for voice and data traffic. The problem with this approach is that it is cumbersome to configure and made inefficient use of bandwidth because there was no statistical multiplexing between separate circuits.

Until recently, the vision for voice data convergence was through the use of ATM at layer 2. The convergence of voice and data was the reason that ATM was developed over a decade ago. ATM has built-in QoS features that were defined just for this application. However, ATM has a fixed cell length and that leads to added overhead. Also, one must manage ATM and IP networks.

The most recent trend is to merge voice and data traffic at layer 3 over IP networks. This approach takes advantage of new IP QoS features such as the RSVP and DiffServ technology. These technologies can also take advantage of layer 2 QoS features if available. The *Internet Engineering*

Task Force (IETF) has developed several technologies that are being deployed to add QoS features to IP networks:

- RSVP, as defined in RFC 2205, is used by a host to request specific QoS from the network for particular application data streams or flows. RSVP is also used by routers to communicate QoS requests to all nodes along the path of the flow and to establish and maintain state. RSVP requests usually result in resources being reserved in each node along the data path.
- *Resource Allocation Protocol* (RAP) is a protocol defined within the IETF. RAP will be used by RSVP-capable routers to communicate with policy servers within the network. Policy servers are used to determine who will be granted network resources and which requests will have priority in cases where there are insufficient network resources to satisfy all requests.
- *Common Open Policy Service* (COPS) is the base protocol used for communicating policy information within the RAP framework. COPS is defined in RFC 2748.
- DiffServ, as defined in RFCs 2474, 2475, 2597, and 2598, uses the *Type of Service* (TOS) field within the IP header to prioritize traffic. DiffServ defines a common understanding about the use and interpretation of this field.
- RTP is used for the transport of real-time data, including audio and video. Using the *User Datagram Protocol* (UDP) for transport, it is used in both media-on-demand and Internet telephony applications. RTP consists of a data and a control part; the latter is called *Real-Time Control Protocol* (RTCP). The data part of RTP is a thin protocol providing timing reconstruction, loss detection, security, and content identification.
- *Real-Time Streaming Protocol* (RTSP), as defined in RFC 2326, is a control extension to RTP. It adds VCR-like functions such as rewind, fast forward, and pause to streaming media.

Given the differences, why should we even consider the use of IP telephony? If the IP telephony world is only going to account for 3 percent of domestic traffic by 2003, is it worth all the hassles? The answer is a mixed bag, but overall the efficiencies of VoIP will outweigh the need to develop better control mechanisms to satisfy the telephony industry.

Applications for VoIP

Voice communications will remain a basic staple for the industry. The PSTNs will not be replaced, or dramatically changed for that matter, in the near term. The immediate need then for VoIP providers is to deliver equal services, similar to the PSTN, at a lower cost of operation and to offer a suitable competitive product as an alternative. This alternative will increase the competition in the industry and force stodgy providers into a new era of meeting the consumer's demands. The first and foremost yardstick that everyone is using is the pricing model. Although switched voice is now down to $.05 to $.10 per minute, just what is the equilibrium price that should be used? When we consider that the IECs' cost per minute is less than $.00125, the $.10 per minute charge they levy on corporate or residential consumers appears excessive. These pricing models are also changing as new technology is implemented in the *Synchronous Optical Network* (SONET) and *dense wavelength division multiplexing* (DWDM) architecture as discussed in previous chapters. In Figure 33-1, we see one of the basic scenarios of how the new providers may challenge the existing infrastructure providers of telephony service. This assumes that a telephone-to-telephone access method is used through the networks. This, by the way, is the preferred method because it uses the tools and techniques to which everyone is accus-

Figure 33-1 Telephone to telephone through the IP network

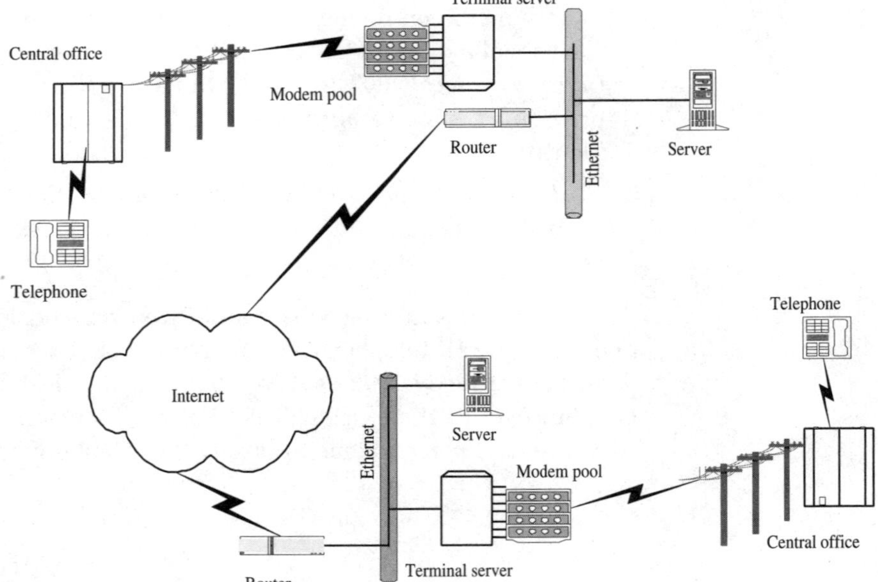

tomed. The basic telephone set has become a staple in the business by having an ergonomically designed device that is easy to use. This is the best way to handle the VoIP for a call. The telephone sets are designed to have the mouthpiece close to the human's mouth and the ear piece that helps to screen out a lot of the ambient noise in the room. Moreover, the sets are not actually doing any of the compression or conversions; these are handled by the devices in the network such as gateways and gatekeepers, specifically designed for this task.

VoIP can be used for just about any requirement for a telephony call. It can be used for fax and for video applications as well. From a single point-to-point call to a multiparty conference call and a web-enabled call center application, the manufacturers have to devise tools and techniques to support the masses.

A second use for VoIP is shown in Figure 33-2, which uses a PC-to-telephone call. This is slightly less desirable because the PC uses an integrated microphone and speakers for the most part. Because they are not designed specifically for this application, they pick up all the noise in the background. Further, to prepare the real-time voice through the PC, the compression and conversion process occurs inside the PC, causing some clipping and chipping, echo, and other forms or latency and distortion. It is not the best solution in the industry, but it does work.

The third way to handle this is to use a PC-to-PC connection, which is the worst of the three scenarios. This one introduces the use of the PC substandard technology on both ends of the connection, compounding the

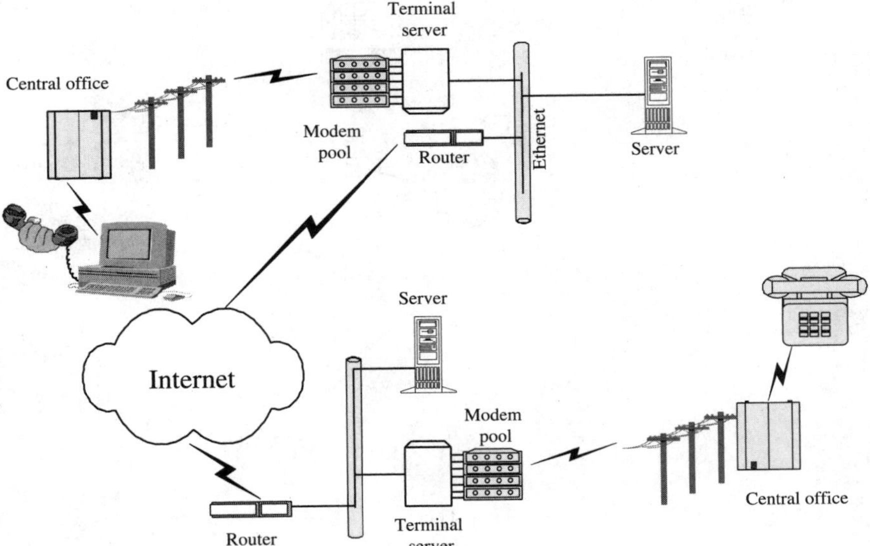

Figure 33-2 PC-to-telephone calls over the IP

problems of echo, delay, clipping, and chipping. This is not the best way to experiment with the technology. Moreover, I have seen people using PC-to-PC communications, whereby the caller places a telephone call over the PSTN to the party desired and tells the recipient to turn on their PC so that they can be called. This is ludicrous: If the IP system does not know whether the called party is active and powered up, then you have to place a telephone call to tell the person to turn on their PC. Why not just have the conversation while you have the called party on the line? This is just an aberration of the technology, but it highlights the fact that VoIP is not yet ready for prime time. There are several tools that IP telephony can bring to the table, enabling the use of VoIP to handle

- Voice-to-voice calls
- Fax-to-fax communication
- Voice-to-call centers
- Voice-to-conference calling arrangements
- PC-to-PC communication

These various paths for handling VoIP are shown in Figure 33-3. The main point here is that there are alternatives to VoIP being used as more than just a voice call. The benefit is enabling technologies to handle myriad options for now and in the future.

Figure 33-3 Various ways of using VoIP

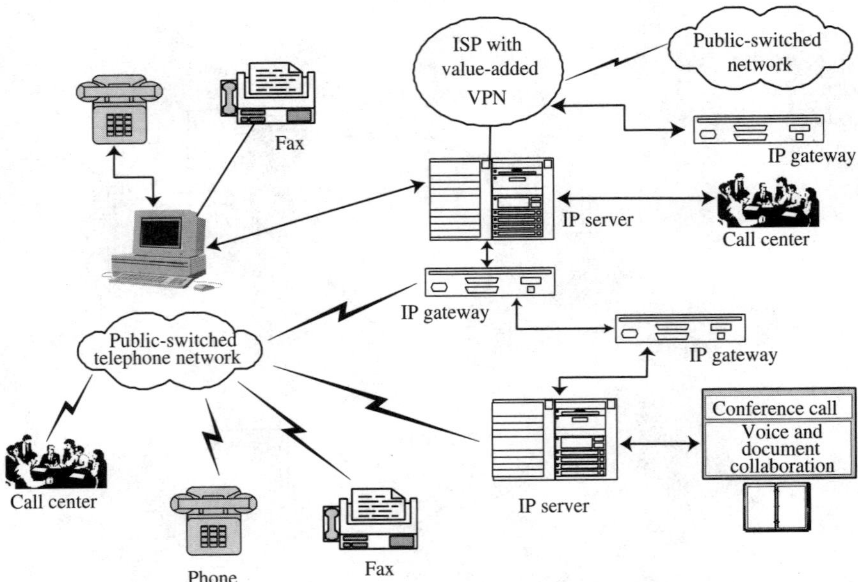

VoIP equipment must have the capability to handle the various options designed around the PSTN and cater to a wide range of configurations. Some of the applications are summarized in Table 33-4 as a means of placing some semblance of order on the VoIP equipment market.

H.323 Protocol Suites

The IP telephony architecture must emulate the PSTN to accommodate the various ways of interconnecting. In Figure 33-4, the *International Telecommunications Union* (ITU) has set the H.323 specification to accommodate the various ways the IP telephony industry must work. Defining gateways and gatekeepers, H.323 is a mechanism for the manufacturers to produce the necessary devices and protocol compatibility for interoperability in the PSTN and the Internet convergence.

Table 33-4

Summary of areas where VoIP equipment must work

Applications	Discussion
Internet telephones	Enhanced telephone sets, both wired and wireless, provide basic telephony services across the Internet instead of the PSTN. These may use a display that will display CLID and possibly some text messaging capabilities.
Remote Access Services (RAS)	Branch office or *Small Office/Home Office* (SOHO) users can gain necessary access to the corporate voice, data, video, and multimedia applications using an intranet. The application here is the home-based worker or telemarketer.
PC-based calls from a traveling person	Using the PC with an Internet connection (dial up from a mobile PC or connected through a *community antenna television* [CATV] modem in a hotel), a travelling user can access all of the corporate services. Excellent for the road warriors in today's generation.
Web-enabled call center services	Call centers can be equipped with access to the Internet. As a user surfs home pages and wants to place an order for a product or if a question arises, the user clicks a button on the PC screen to reach a real-time operator. Customer service applications like this will enhance the overall e-commerce industry and the telephony applications.
PSTN gateways	As shown in Figure 33-3, the gateway function enables the interconnection from the PSTN to the Internet over a gateway. A PC-based telephone can access the PSTN. Another application is for a *universal serial bus* (USB) port telephone connected to a PC that will have unlimited access to customers on the PSTN.

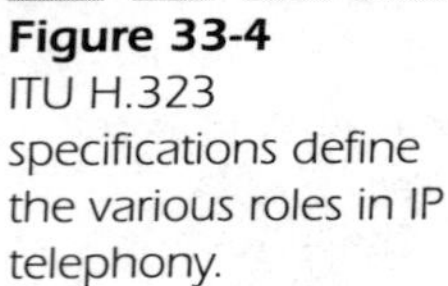

Figure 33-4
ITU H.323 specifications define the various roles in IP telephony.

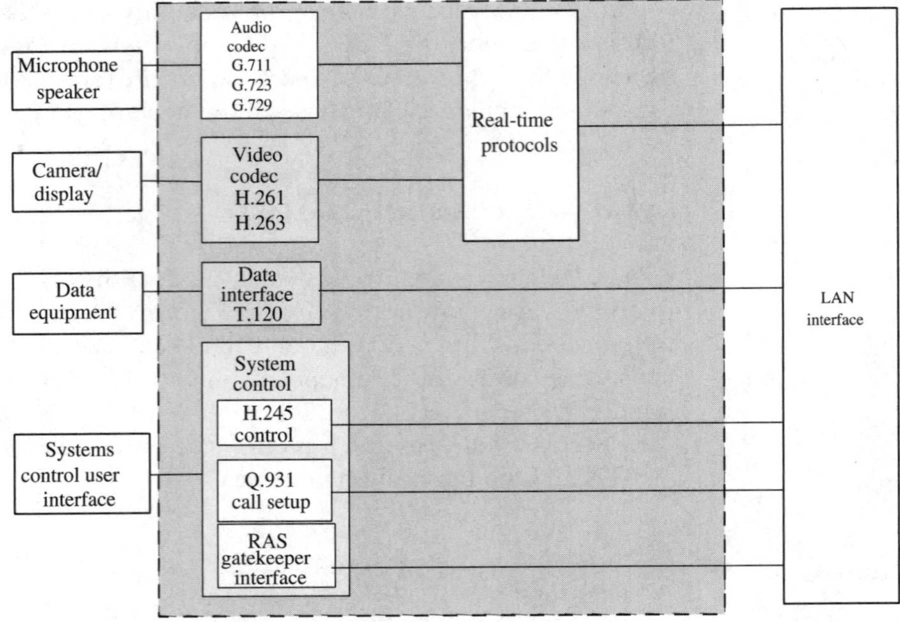

H.323 is an umbrella protocol consisting of a number of different sub-protocols that are used to support audio, video, and data applications. There are three versions of H.323:

- The ITU specified version 1 in 1996.
- Version 2 was specified in 1998.
- Version 3 is the mature signaling protocol.

The H.323 suite was designed to

- Provide video, audio, and data across various network types
- Do point-to-point and multipoint conferences
- Handle call control, multimedia environments, and prioritization
- Function across a *Local Area Network* (LAN) or *Wide Area Network* (WAN).

Its model is similar to *Media Gateway Control Protocol* (MGCP) in that it uses gateways to do the media conversion and gatekeepers (the H.323 term for call agents) to do the signaling. The main difference, however, is that unlike MGCP, H.323 was designed to work primarily with intelligent endpoints (such as IP-based telephones), whereas MGCP was intended to function with *Intelligent Access Devices* (IADs) and simple, dumb, endpoints (such as a normal telephone). H.323 is a complex protocol with a heavy code base. This makes it a very intelligent protocol, but it also causes it to be more expensive to build gateways, and it makes it more difficult to build a simple, reliable gateway. H.323 has a concentrated peer-to-peer architecture that does not enable easy scaling—a key concern to large network providers who need millions of connections. Other shortfalls include the following:

- H.323 is based on TCP/IP, which can cause too much delay to make it practical.
- It will connect over LANs that have no QoS control—an unacceptable solution.
- Neither H.323 nor *Session Initiation Protocol* (SIP) was designed to interface with the PSTN—a drawback.

Figure 33-5 is a different view of the H.323 specifications for the VoIP standards, showing the various other protocol suites needed to support the standard. In this case, the terminal devices running on various telephony protocols, such as ISDN terminals, H.320 terminals, and video conferencing devices, are all accounted for in the H.323 specification. The standard also addresses the various protocols for the gateways and gatekeepers on a

Figure 33-5
H.323 in action

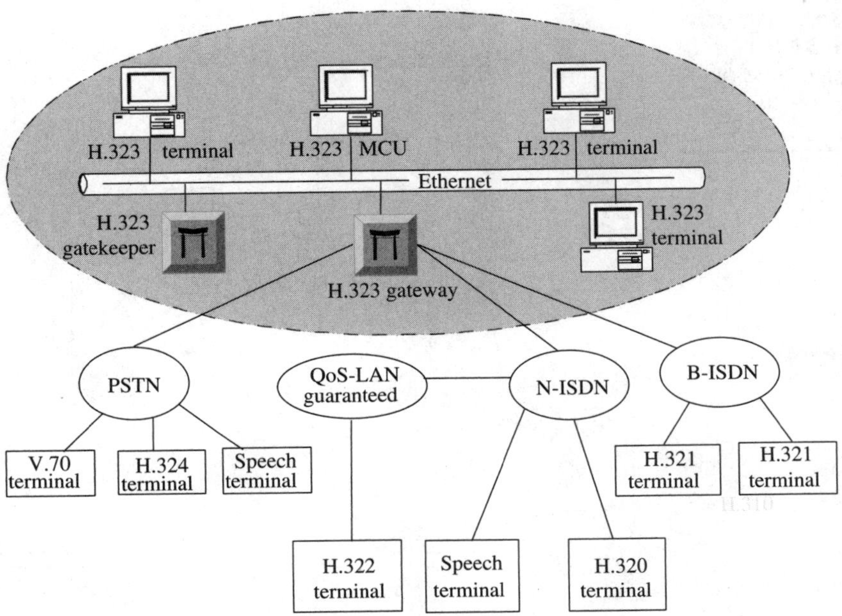

LAN, as well as any compression techniques to manifest the VoIP capabilities in various points along the network.

What also has to occur in the network is to use a layer 2 protocol to carry the IP traffic. Traditionally, a *high-level data link control* (HDLC) frame or a *Point-to-Point Protocol* (PPP) frame is used. However, to assure the QoS, the standards are recommending the use of ATM at the layer 2 switching architecture. In most cases, ATM introduces added overhead by producing the cells to be mapped onto a SONET frame at the layer 1. However, what ATM does assure is the guaranteed bandwidth to deliver the necessary QoS. Only recently has a guaranteed QoS in *Frame Relay* (FR) been introduced. Figure 33-6 is a summary of the bandwidth needs and the QoS issues using the ATM or FR networks to carry the VoIP traffic. It is through this combined use of layer 2 switching and layer 3 network addressing that the systems are able to map and multiplex the traffic on the data networks with any reliability. By the time this book is published, there will probably be three or four new approaches that will emerge, each claiming to be the better way to handle the traffic for VoIP. In most cases, they will be efficient and better than what has been available previously. That doesn't mean a thing because each of the techniques that exist today will be implemented and upgraded or improved over time. No one technology or technique stands to be the winning or killer application today.

Figure 33-6 The use of ATM or FR guarantees the QoS today.

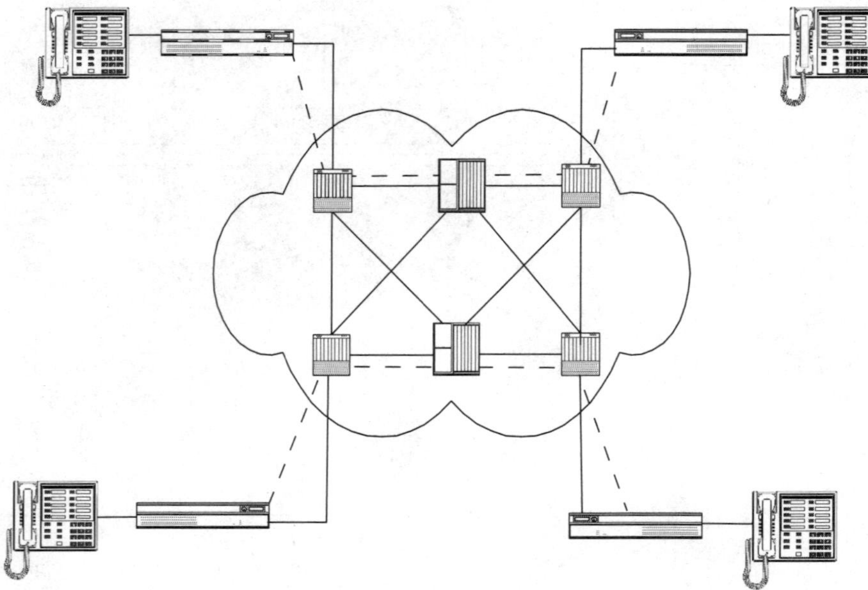

Delay and Jitter on VoIP Networks

Two of the major concerns today are the problems experienced with delay and jitter. The delay factors can be the most disconcerting. In the past, the PSTN has been delivering real-time voice across the network with an average delay of 30 to 50 ms. When we experimented with satellite communications in the early 1980s, the end user community was disgruntled with the potential delays for real-time voice applications. The average delay on a satellite-based network in geosynchronous orbit approximated 250 ms, which was intolerable for the end user. In the VoIP delivery today, the average response and delay factors must be dealt with. On average, if a call is placed across the Internet, the delays are from 800 ms to 2 seconds domestically. This is totally unacceptable for the user community. It certainly does not lend itself to the conducting of business conversations. When placing a call across the Internet overseas, the delay can be as much as 5 seconds! Again, this will never do.

The delays have been the result of all the processing required throughout the network. One problem that results from high end-to-end communications delay is the increase in echo whenever the round trip delay exceeds 50 ms. Because echo is viewed as a quality problem, VoIP systems must accommodate echo control in order to be accepted. Talkers will overlap themselves when the delay increases if one-way transmission exceeds the 200 to 250 ms range. This was a major problem with satellite transmission. Too often, talkers were over talking each other, causing total chaos during the conversations. The budgeted delays for VoIP must be within the 200 to 250 ms range, and that is strictly the worst case. The VoIP systems have to strive for better response and delivery.

Jitter (variable delay) is a variation of the interpacket delivery introduced by the processing of each packet across the network, coupled with transmission delay across the medium. To solve the problem for real-time applications, packets must be collected into a buffer and played out as needed. By holding the packets just long enough to accommodate the longest delay in packet delivery, the added delay can be devastating. This also requires that the VoIP equipment have sufficient buffers to accommodate these interpacket delays on an ongoing basis. The bigger the buffers, the longer the potential delays.

Another major problem that one must contend with is the acceptable level of packet loss. Several studies have been conducted indicating that acceptable packet loss is less than 10 percent. Anything greater than 10 percent packet loss will be intolerable. The network can drop packets when peak loads are being experienced, causing some degree of loss. Moreover, packets can be discarded in case of buffer overrun at the routers and switches across the network. These combined problems may move the packet-loss levels to an unacceptable range.

If we combine the various levels of processing delay and the risk of packet loss, we can see where the networks must accommodate the delay, along with the need for VoIP equipment to address the issue. Figure 33-7 is a representation of the delays that can be cumulatively added to a packet delivery across the network.

Protocol Stack

The VoIP protocol stack uses the benefits of a TCP/IP suite in which to run over the network. Figure 33-8 is a representation of the protocols as they stack on top of each other. The primary areas are as follows:

- In layer 1, we typically use SONET and DWDM for the physical architecture. Although other variations can be used, this is the more common one today.
- At layer 2, we typically use ATM or FR, but the PPP can also be used if some other form of link architecture is employed.

Figure 33-7 Delays across a VoIP network

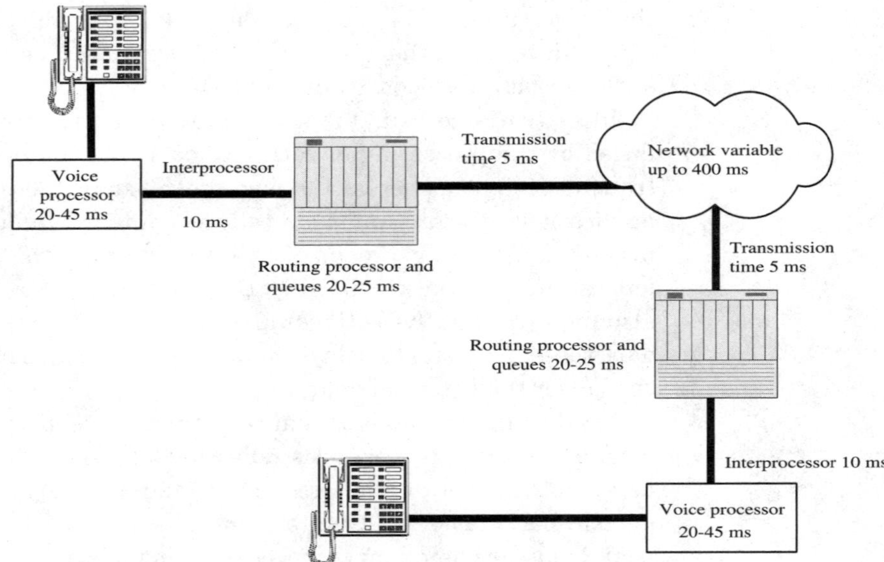

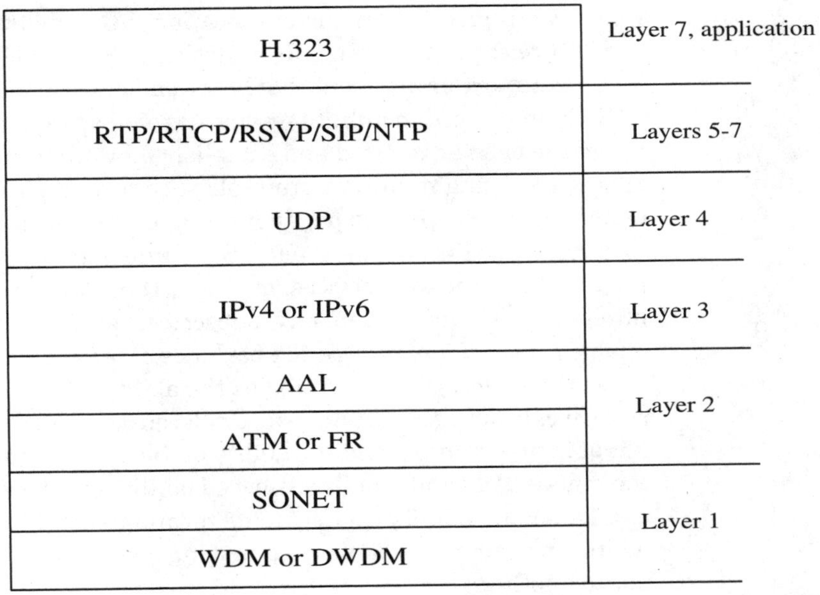

Figure 33-8
The VoIPs stacked up

H.323	Layer 7, application
RTP/RTCP/RSVP/SIP/NTP	Layers 5-7
UDP	Layer 4
IPv4 or IPv6	Layer 3
AAL	Layer 2
ATM or FR	
SONET	Layer 1
WDM or DWDM	

- Layer 3 is where we will find the IP layer (IPv4 or IPv6), using the network layer to handle the datagram protocols.
- At layer 4, we use UDP instead of TCP for the real-time applications. Let's face it, UDP serves as a better set of protocols to use when we can ignore packet loss, or the reliability issues are addressed at a lower level.
- At the upper layers (5 to 7), we see the use of the RTP, the RTCP, and the *Resource Reservation Protocols* (RSVP). Others will include the *Network Time Stamp Protocol* (NTP) and the SIP.
- At the application, we see the H.323 protocols sitting on top of the heap.

SIP SIP is an application-layer control (signaling) protocol for creating, modifying, and terminating sessions with one or more participants. These sessions include Internet multimedia conferences, Internet telephone calls, and multimedia distribution. Members in a session can communicate via multicast or via a mesh of unicast relations, or a combination of these. SIP invitations used to create sessions carry session descriptions, which enable participants to agree on a set of compatible media types. SIP supports user mobility by proxying and redirecting requests to the user's current location.

Users can register their current location. SIP is not tied to any particular conference control protocol. SIP is designed to be independent of the lower-layer transport protocol and can be extended with additional capabilities.

SIP can be used to initiate sessions as well as invite members to sessions that have been advertised and established by other means. Sessions can be advertised using multicast protocols such as *Service Access Points* (SAPs), e-mail, news groups, web pages or directories (*Lightweight Directory Access Protocol* [LDAP]), among others. SIP transparently supports name mapping and redirection services, enabling the implementation of ISDN and intelligent network telephony subscriber services. These facilities also enable *personal mobility.* In the parlance of telecommunications intelligent network services, this is defined as the ability of end users to originate and receive calls and access subscribed telecommunication services on any terminal in any location, and the ability of the network to identify end users as they move. Personal mobility is based on the use of a unique personal identity. Personal mobility complements terminal mobility, that is, the ability to maintain communications when moving a single end system from one subnet to another.

SIP supports five facets of establishing and terminating multimedia communications:

- **User location** Determination of the end system to be used for communication
- **User capabilities** Determination of the media and media parameters to be used for communication
- **User availability** Determination of the willingness of the called party to engage in communications
- **Call setup** Ringing—establishment of call parameters at both the called and calling party
- **Call handling** Transfer and termination of calls

SIP can also initiate multiparty calls using a *multipoint control unit* (MCU) or fully meshed interconnection instead of multicast. Internet telephony gateways that connect PSTN parties can also use SIP to set up calls between them. SIP is designed as part of the overall IETF multimedia data and control architecture currently incorporating protocols such as RSVP for reserving network resources, the RTP for transporting real-time data and providing QoS feedback, the RTSP for controlling delivery of streaming media, the SAP for advertising multimedia sessions via multicast, and the *session description protocol* (SDP) for describing multimedia sessions.

However, the functionality and operation of SIP does not depend on any of these protocols. SIP can also be used in conjunction with other call setup and signaling protocols. In that mode, an end system uses SIP exchanges to determine the appropriate end system address and protocol from a given address that is protocol independent. For example, SIP could be used to determine that the party can be reached via H.323, obtain the H.245 gateway and user address, and then use H.225.0 to establish the call. In another example, SIP might be used to determine that the called party is reachable via the PSTN and indicate the phone number to be called, possibly suggesting an Internet-to-PSTN gateway to be used. SIP does not offer conference control services, such as floor control or voting, and does not prescribe how a conference is to be managed, but SIP can be used to introduce conference control protocols. It does not allocate multicast addresses.

SIP can invite users to sessions with and without resource reservation. It does not reserve resources, but can convey to the invited system the information necessary to do this.

MGCP MGCP, an IETF draft, has been chosen as the initial protocol to provide the call control necessary to enable wide scale telephony over the Internet. MGCP also provides the interoperability between the Internet and legacy telephone networks. MGCP is ASCII based and resides in a UDP packet, relying on a positive acknowledgment mechanism to deal with packet loss. As its basis, MGCP's stateless protocol was designed using a combination of Cisco Systems and Bellcore's *Simple Gateway Control Protocol* (SGCP) and the level 3 Technical Advisory Committee's *Internet Protocol Device Control* (IPDC) specification. Combining these two protocols facilitated a single specification and, in turn, enabled faster deployment. To ensure security, MGCP uses the *IP Security* (IPSec) protocol and has the ability to use encryption on the media streaming at the terminal equipment.

MGCP works with several other protocols to provide the various functions that will be needed in the Internet for VoIP. MGCP is used to communicate between a call agent and a gateway. Together, these help to make up the SoftSwitch architecture. The call agents (also referred to as *media gateway controllers*) are the brains of the operation, and the gateways are the muscle. You need both of them to complete a call over a public network using MGCP. The call agent sends commands to the gateway and receives an acknowledgment for each one. An advantage to call agent/gateway architecture is that new services can easily be introduced without having to swap or upgrade gateways. This model appears as if it is a single gateway, but is actually distributed. This is commonly referred to as the *decomposed gateway architecture*. Putting the call control functions in the call agent and the media streaming functions in the gateway allows for both greater scalability as well as more simple, inexpensive, and reliable local access equipment.

The Gatekeeper The call agents or gatekeepers are scattered around the Internet and act as telephony routers. The call agent is required for signaling within a call, but is not involved in handling the voice traffic. Each call agent, responsible for a number of gateways, accepts event messages from them and tells them how to process calls via a relatively small set of MGCP messages and procedures. The call agents do not use the MGCP to talk to each other. Instead they use SIP and H.323 to intercommunicate. Network providers will normally set up several gateways to provide access for their users or customers. One or more gatekeepers, or at a minimum access to a gatekeeper from another provider, should be set up. These gatekeepers require the ability to route calls to and from the PSTN and also route calls to and from customers of other network providers.

The Gateways A gateway is a device that is designed to join different components. A media gateway joins disparate media (that is, CATV to DSL, and so on). In MGCP, signals are converted between real-time circuits and packets, and between voice and packets. In MGCP, there are a number of different types of gateways performing different tasks. Each type of gateway has a distinguishing name and enables voice to be encapsulated in packets. The gateways are primarily concerned with audio signal conversion and media streaming, while the gatekeepers deal mainly with signaling functions. The gateways may be positioned at the customer premise where they can be directly connected to *Plain Old Telephone Service* (POTS) telephones, a *Private Branch Exchange* (PBX), or a data network. They can also be found at a *point of presence* (POP) or colocation point where they can service larger numbers of users as well as provide other high level tasks. Gateways can be associated with multiple gatekeepers to provide redundancy.

H.248/MEGACO H.248 is the name given to this protocol by the ITU and MEGACO is how it is recognized today by the IETF. This is the first time that these two groups have worked on a standard together and is indicative of the future meshing of telecommunications and the Internet. H.248 has a lot of similarities to MGCP and is typically referred to as the protocol, which will eventually supersede it. H.248 uses the gateway/call agent architecture and also handles the call setup much like MGCP; however, the naming of the parts and functions differs. H.248 also adds enough complexity to enable it to support video traffic. It is still a relatively immature protocol in the world of signaling protocols.

CHAPTER 34

Multiprotocol Label Switching (MPLS)

No one can argue that the growth of the Internet has been exceptional. Depending on what report you read or what research house publishes, the numbers may differ slightly, but the overall result is the same. The Internet is explosive. The events of 2000 and 2001 have caused the growth to slow a bit, but overall we can expect to see continued growth and expansion of the Net. In October 2000, a reported 93 million hosts were attached to the Internet, which represents an increase from 65 million in the same period of 1999. In actuality, there have been over 60 million hosts attached over a three-year period. The graph shown in Figure 34-1 is a representation of the growth and projections for the Internet with some statements estimating over a billion users and 750 million hosts in 2008. Using any calculation method, this is still impressive.

Whether the Internet meets or exceeds the growth projections, one thing is certain—IP networking will be the standard networking protocol for the future, and the growth will place a burden on the ability to process the data packets quickly. The more hosts and users logging onto the network, the exponential growth in routing tables and routers are needed to process the data. Because of the monstrous growth rates, the carriers and vendors alike need to find a more efficient way of dealing with the traffic deluge. Optimizing the network efficiency is a critical goal today and for the future. Inherently, however, IP was not designed to be efficient. Thus, we need to create some better solutions for packet handling. In earlier chapters, we saw that several different techniques were being introduced to handle upper layer protocol traffic. To address lower layer traffic, MPLS became one of the solutions sought after aggressively by the standards committees, vendors, and network operators alike.

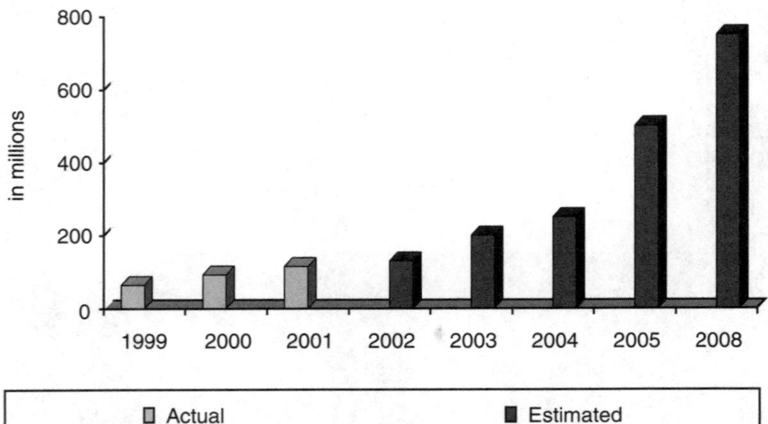

Figure 34-1 Hosts attached to the Internet 1999 to 2008 estimates

Standard IP Networking

Let's start, then with a quick review of the IP addressing scheme. Why do that? Well, you know how big the network is becoming unmanageable! You also know that the way you handle this growth and the demands from the user is to install more routers and bandwidth. With IP, the end user addresses are limited to one of three basic classes. So, in an IP address there are three classes you may have requested. The three addresses include Class A, B, or C. In Figure 34-2, we see the address capacities. First, the IP address is currently based on a 32 bit field that is divided, depending on the class. Using a binary sequence of 2^{32} we can have up to 4,294,967,296 addressable devices. One would think that this is plenty of addresses (and correctly, this is true). However, the numbers were assigned in a problematic way that is now becoming a major addressing issue in the industry.

As shown in the figure, we break the 32 bit sequence into 8 bit bytes. This is called the *decimal dot delimited number sequence*. Using these 8 bits at a time, we can categorize the address as Table 34-1 reflects.

The ranges of addresses by class are shown in Table 34-2 as a means of simplifying the ranges for their expected use.

These classes of addresses (also called *classful addresses*) were based on nice, neat configurations of 8 to 16 or 24 bit sequences for the address and the remaining bits for the hosts that are attached. This also meant that when the addresses were issued, they were somewhat impractical. Make no mistake that the intent was valid, but the wasteful way that the numbers were doled out created longer-term problems.

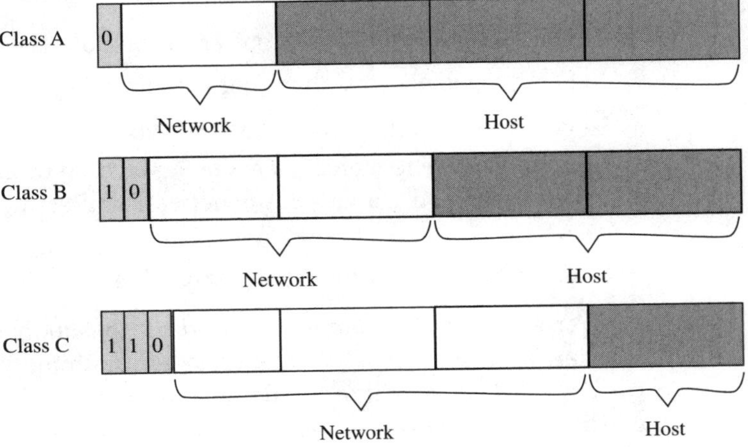

Figure 34-2 The addresses used in IP

Table 34-1 Networks assigned by the decimal dot delimited addresses

Class Address	Bits Allotted for Network Number		Addresses We Actually Get	Bits Allotted for Attached Hosts	Hosts
A	8 (less 1)	7	123	24	16,777,216 (less 2)
B	16 (less 2)	14	16,384	16	65,535 less
C	24 (less 3)	21	2,097,152	8	256 (less 2)

Table 34-2 Ranges for classes of addresses

Address	Network Ranges
A	1.x.x.x through 126.x.x.x
B	128.0.x.x through 191.255.x.x
C	192.0.0.x through 223.255.255.x

The first is that the numbers were assigned to a corporate location. A class "A" address, also called a */8* (pronounced slash 8) address, which denotes that the address prefix (network number), is 8 bits long. This means that the corporation needs to have all 16,777,214 (all 1s and all 0s were eliminated) located at one location. This was impractical. Think of the number of routes that a router will have to maintain and the size of the tables needed to be able to route traffic to these devices. Simply adding memory to the router does not solve the problem, as many people think. Memory is cheap today, but some of the other considerations include

- The demand for more CPU power to compute the routes quickly.
- Routing topologies are changing significantly from the initial networks.
- Changes are happening at an escalating rate.
- Routers use cache memory for the forwarding tables. Caches are affected by the dynamics of the networks, affecting the forwarding cache.
- Sheer volume of information is exploding.

Therefore, we find ourselves needing to subdivide the network into smaller pieces, called *subnets*. A subnet is nothing more than divisions of the larger IP network into smaller pieces where all the hosts in a subnet are on the same physical network (that is, an Ethernet). In 1985, this problem came to the surface and caused the *Internet Engineering Task Force* (IETF) to look at the use of subnetting. This was geared to take on the problems with the two level addressing hierarchies.

There are many reasons we may want to break the network into smaller pieces. These are shown in Figure 34-3. The figure is not broken into priorities or orders of importance but is just a means of listing what the reasons may be for the subnetworking of an address.

This means that we attempt to break down IP addresses depending on the motivations as follows:

- When the sites are geographically dispersed, we use a subnet function so that we do not need different IP addresses in remote locations. Picture a company that has sites spread across the country. If we do not subnet, then each site will need its own IP.
- Because of the departmental uniqueness in certain organizations, we find that they may want to have a separate subnet. Different product lines, for example, in a major conglomerate.

Figure 34-3 Reasons for subnetworking

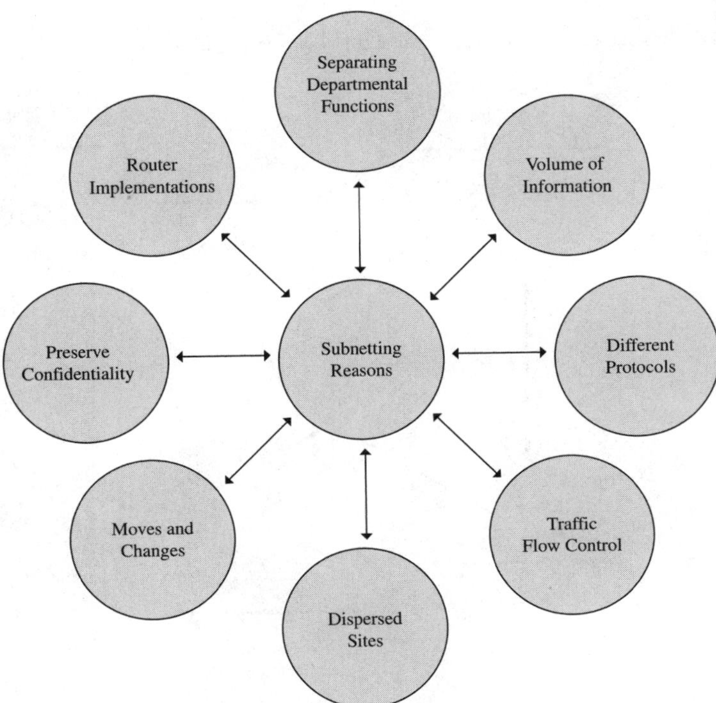

- We may have different protocols running on the network, and we separate the networks into subnets to provide isolation. This is not necessarily a requirement as opposed to a preference. An example may be a financial organization running a mainframe with *Systems Network Architecture* (SNA) traffic on one *local area network* (LAN) and Novell with Front Office word processing on another.
- If traffic is voluminous, we may subnet the IP so that again we can isolate different traffic flows with bridging or routing techniques.
- Some traffic conditions may conflict with others, especially protocols, but also in the amount or type of data that moves between groups.
- Routers used internally between departments or buildings on a campus create a good opportunity to subnet by building or department on a *Campus Area Network* (CAN). This may also be used in a university where the individual colleges (business, engineering, and chemistry) are all on their own subnet. An example of this arrangement in a business is shown in Figure 34-4.
- Confidential traffic moving on a network may create a need to subnet so that we can preserve the confidentiality. An example here is in

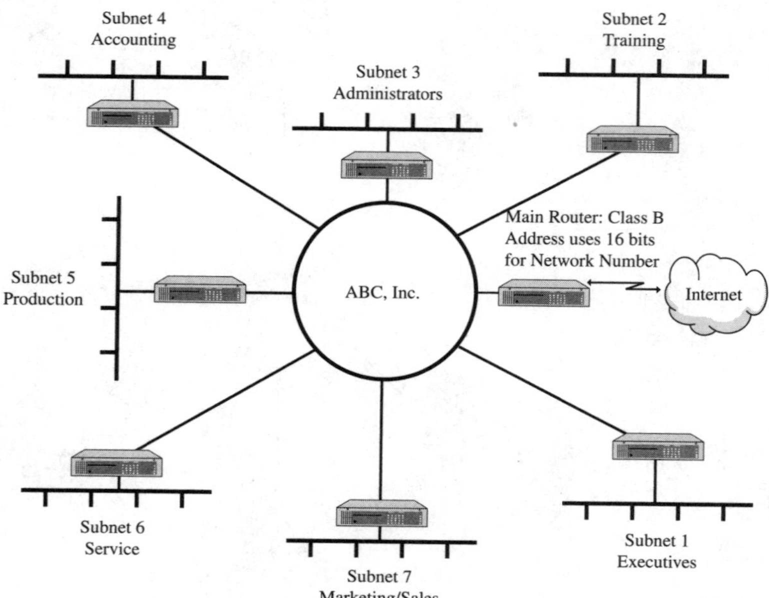

Figure 34-4 An example of subnetting

government networks; we may separate the juvenile information from adult information because of the sensitivity of the data.

- Moves and changes in an organization may cause conflicts in the addressing scheme. Therefore, we may separate the address space using fixed IP addresses for each floor or wing in a building as we hub them back to a centralized closet.

In a class B address, the number of bits allowed for the network address is 16, so we refer to these as /16 networks. By using these IP addressing schemes, we find that the data networking may become more manageable. Many organizations may have subnetted their class C addresses; this is not just a large organization situation. When we refer to a class C address, you may also hear it called a *!24 network*. This just means that 24 of the bits are used for the prefix (network number). When we subnet, we take some of the host address and use these as extended address information for the subnet. An example of this is shown in Figure 34-5 using a class B address and using 8 of the 16 bits allotted for the host address.

Subnetting dealt with the routing problems by making sure that the subnet structure of a network was invisible to the outside world. It is an internal subnetting structure. The route from the Internet to any subnet of an IP address is the same, no matter which subnet the host is on. The outside networks do not care what subnet that the host is on internally; they are primarily concerned with getting the datagram to the entrance of the network, then the internal routing takes place based on the subnet masking.

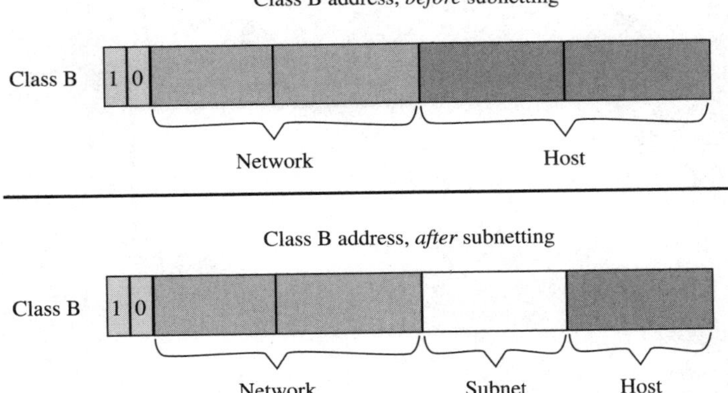

Figure 34-5 A class B address subnetted

Subnet Masking

To be assured that we can separate the address space, each system in the subnet must have the same subnet mask. The subnet mask distinguishes between the network bits, subnet bits, and interface (host) bits. To do this, the mask must have the same number of bits as the IP address and is in the same format as the IP address. Though the mask has the same format as the IP address, its first byte always has a value of decimal 255, which is impossible for any IP addressable interface. A network number 255 is non-addressable. Because the devices process numbers greater than the masked number (so with an 8 bit byte, the numbering range is 0 to 255), the router looks at any number greater than 255, which there are none.

In Figure 34-6, we see a means of dealing with the public and private networking strategies using the subnets in a single router as the external gateway to the outside world. This shows that several logical networks using subnet addressing are used to route the traffic internally. All traffic coming from the outside world comes into the router sitting at the edge of the network. The traffic is addressed to network number 136.8.0.0. The router accepts the traffic, then forwards it to the subnetworks based on the third octet of their address. The addresses have been broken down in blocks of 32 for simplicity.

Natural masks use binary 1s in the network and subnet portion of the IP address while it uses binary 0s to represent the host interface number. Figure 34-7 is an example of the subnet masking and the way it is laid out. Routing protocol standards and definitions describing what happens inside the enterprise often refer to this as the *extended address* or the *extended*

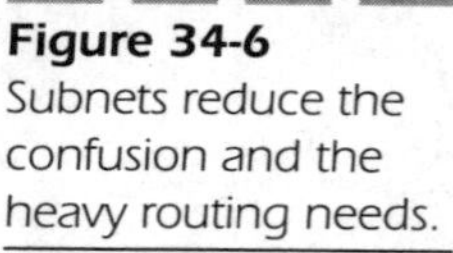

Figure 34-6 Subnets reduce the confusion and the heavy routing needs.

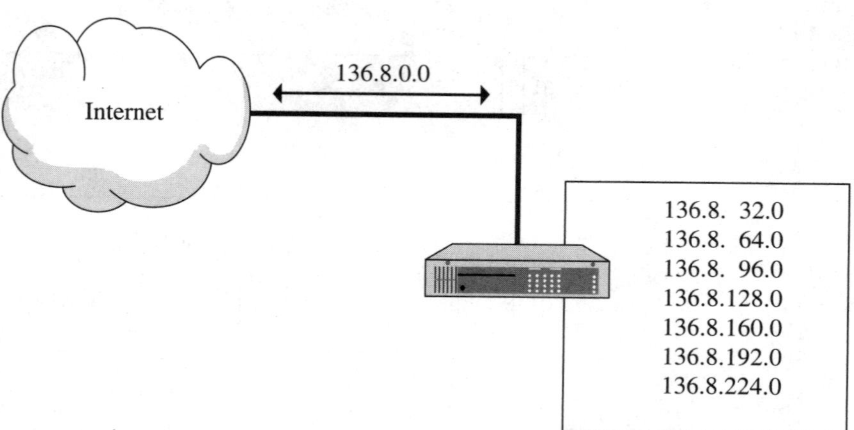

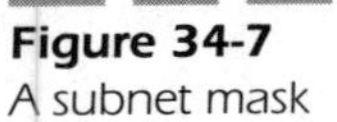

Figure 34-7
A subnet mask

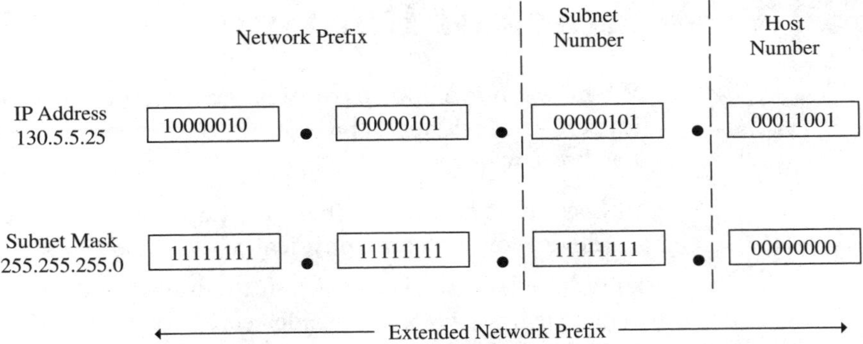

network prefix. Either way, we are looking at a subnetwork mask. The prefix length is equal to the contiguous ones in the traditional mask. Therefore, when we look at address 130.5.5.25 with a subnet mask of 255.255.255.0, we can also call this 130.5.5.25/24. The slash 24 (/24) defines the length of the network prefix (extended) and is easier to understand for many people. We also need to recognize that the routing protocols still must carry the full information of the subnet mask. No standards enable us to define the length of the extended address field, so we must carry the full subnet mask in the routing protocols.

Natural masks were set up in octet boundaries when we first started subnetting. This represents the bits used in a class A, B, or C address. In Table 34-3, we can see the natural masks used by type of address.

Added bits in a mask beyond the natural mask for the network class indicate that subnetworking occurred. The notation of using the slash (/) enables us to define the number of added bit used in the subnets. For example, we may see the notation of /22 or /26. This just indicates the way we have used the subnets and the additional notations to follow.

Table 34-3

The natural mask uses the octet boundaries by class of address

Class	Mask
A	255.0.0.0
B	255.255.0.0
C	255.255.255.0

Rules of Routing

When we move into a routing architecture, we must understand that routers are designed to pass our IP datagrams across the network in one of two basic ways:

- Local routing occurs when the datagram is sent directly to a host (device) on the same physical network as the sender (that is, a LAN).
- Indirect routing occurs when the intended device is not on the same physical network but resides somewhere else. IP must, therefore, forward the datagram to another device (such as a router) to assist in the redirection of the IP datagram to the intended device.

Therefore, IP must decide if the destination device is on a local network by looking at the source and destination addresses. IP does many calculations while determining whether the devices are local to each other or not. Yet, a set of decisions can be made, and forwarding can occur. See Table 34-4 for some of the decisions that IP makes as it evaluate the address of each IP datagram it receives.

This means that the routers look at the address using layer 3 protocols and then use layer 2 to deliver the IP datagrams by encapsulating the datagram in some layer 2 framing format. This is not magic, just the natural evolution of the legacy routing decisions we had to make in the beginning stages of our network evolutions.

The next part is usually one of the more difficult to answer when routing IP datagrams. First, we must ascertain if the source and destination

Table 34-4

The decision matrix for forwarding the IP datagrams

If	Then
Source and destination IP addresses are in different class addresses (that is, an A and B).	Give the datagram to the router for forwarding.
Source and destination IP addresses are in the same class of address (a la both class B), but the networks are different.	Give the datagram to the router for forwarding.
Source and destination address are in the same class and same network but different subnets.	Give the IP datagram to the router for forwarding.
Source and destination IP addresses are in the same network and the same subnet.	Send the datagram directly to the destination.

addresses are on the same network. If they are not on the same network, as seen in Table 34-4, then the datagram is sent to the local router for handling and forwarding. Comparing the subnet field computes if the source and destination address are in the same subnet. If they are in the same subnet, the subnet address field must be the same. This is calculable by looking at the subnet field and seeing if the values are the same.

Variable Length Subnet Masks (VLSM)

In 1987, the IETF approved *Request for Comments* (RFC) 1009, which specifies how subnetted networks can use more than one subnet mask. When there is more than one subnetwork mask assigned, the IP network is said to use variable length subnet masks because the extended-network prefixes have different length.

The *routing information protocol version one* (RIP-1) only enables us to use a fixed length subnet mask across the entire network prefix. This means that RIP-1 only enables a single subnet mask in each network number because it does not provide subnet-masking information as part of its routing table update information. In the absence of this information however, RIP-1 makes very simple assumptions about the subnet mask that is applied to any learned network in its table.

VLSM enables better use of IP address space assigned to an organization. One of the risks we had with the single subnet mask was that we could be locked into a fixed number of fixed-sized subnets. The variable subnets overcome some of the older limitations, but RIP-1 is the problem in this case. VLSM also enables better routing structure as the networks are laid out. It enables the use of route aggregation so that the routing tables can be held to a minimum. As we saw earlier, routing tables are becoming a good portion of the bottleneck in our overall networking strategy. It is not as simple as throwing the memory at the routers to make the systems work better. Many of the combined protocols are working against us. To this end we see that there must be more done to make the networks perform more efficiently. This includes any way we can build the tables and the masking that we can use.

Requirements for the use of VLSM are not that much different from basic subnet masking. These include the following:

- The protocols used to route our traffic must enable and carry extended-network-prefix information in each table advertisement message (*Service Access Points* [SAPs] and solicitations).

- All routers in the network must use a consistent forwarding algorithm for traffic using internal algorithms based on the longest match of address and mask information.
- When route aggregation is used, addresses must have a topological significance in the assignment of the addresses. If not, the routers may have a problem forwarding the traffic.

VLSM is now an accepted means of forwarding the traffic using newer routing protocols such as OSPF, IS-IS, EIS-IS, and now RIP-2. These protocols enable the extended addressing of the network prefix to be variable, depending on the route taken with each advertisement and solicitation message traversing the network.

The Longest Match Syndrome

As already stated, all routers must use a consistent approach in the forwarding algorithms, based on the longest match. The use of VLSM creates an environment where a set of networks with the longer extended-network-prefix may create subnetwork relationships. The routers will gather route advertisement and solicitation information and build their tables. As they capture the information, hundreds or thousands of entries steer the data to the same end-point. The longer address match is considered, the more specific address and forwarding information. The shorter the subnet match, the more generic the route. Routers now use the longest matching extended-network-prefix to be more specific when forwarding traffic.

This, of course, means that the router is doing a lot more work in the assessment of the IP routing and comparisons. Figure 34-8 is an example of the longest match. For the purpose of this graphic, we have kept the route choices limited, whereas in a real-time situation, the address route choice can be in the hundreds or even thousands. This all depends on the addressing and the organization's implementation of their IP addressing. But the real point is that the router must go through many route choices to find the longest match.

Classless Interdomain Routing (CIDR)

Another aspect continued the development of this routing scheme and the ability to route on the various protocols that we use. By 1992, the growth of

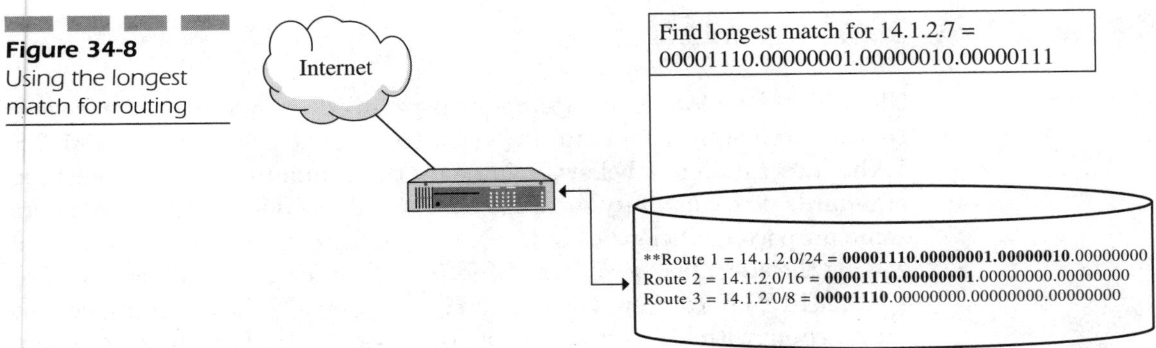

Figure 34-8 Using the longest match for routing

the Internet was getting explosive. Many people were becoming very concerned about the addressing mechanisms and the routing bogging down. Consequently, a study was conducted by the IETF to look at the scaling problems, especially as it relates to address and routing protocols. The concern was that by using the classful addresses, we would deplete the address space available very quickly. Additionally, the growth in the number of routes in the Internet would place a burden on the routers and their ability to deal with the routing tables. No one in his or her wildest imagination ever thought the Internet and the intranets would ever grow the way they were exploding. The protocols, routers, and addressing methods were all coming to a crashing halt. Something had to be done, so the idea of a classless address space became the potential salvation until a new set of addresses could be delivered. Actually, CIDR became an officially supported standard in 1993 in RFCs 1518 to 1520. CIDR supports two major features that were considered beneficial in a global networking strategy. These include

- CIDR eliminates the older version of classful addresses (that is, Class A, B, C, and so on), enabling better utilization of the existing addresses that we had in the IPv4 scheme of doing the addressing using the 32 bit addresses. This also bought us some time until IPv6 could be approved and implemented.
- CIDR supports the idea of route aggregation where a single routing table can be used to deal with the address space of thousands of classful IP address routes. This minimizes the routing table impact and helps to consolidate the information in a routing table entry.

Enter MPLS

The MPLS standard is an attempt by a group of vendors to proliferate and continue the evolution of multilayer switching. The primary goal of MPLS is the integration of label swapping with the standard layer 3 networking standards. The capability to link layer 2 and 3 technologies together offers some purported efficiencies in forwarding data across the networks and in setting the stage for the future of *quality of service* (QoS) functionality. Initially, MPLS efforts were focused on IPv4; however, the basic technology is far reaching with hooks to other network layer protocols such as *Asynchronous Transfer Mode* (ATM), Frame Relay, and so on. The real benefit to the use of MPLS is that it can be used on any media at any layer that can pass data packets. MPLS basic specifications are contained in RFC 3031. With MPLS, conventional network layer routing (IP routing) determines a path through the network. Once the path is selected, data packets are then switched through each node as they traverse the network. Moreover, the use of QoS, traffic engineering, and *Virtual Private Network* (VPNs) are discussions that continually stress the more efficient use of IP networking.

Traffic Engineering

Traffic engineering is the process of selecting network paths so the traffic patterns can be balanced across the various route choices. Routing with the older *Interior Gateway Protocol* (IGP) algorithms may select circuits, resulting in unbalanced utilization. Consequently, some circuits become congested, while others sit idle. Manipulating the IGP metrics can provide the necessary traffic engineering. However, this is difficult to manage when networks are large and have many alternate routes. Figure 34-9 shows an example of the routing protocols used in many networks. *Border Gateway Protocol* (BGP) sits outside the network at the edge, whereas OSPF runs in the core of the network.

Using Figure 34-10, we see the route that will be selected through the network using the OSPF protocols. The traffic will be designed to always travel in the upper segment of the network because this is the shortest path through the net. However, what can happen is the links on this portion of the network will get congested, whereas the lower route will never get used because the path algorithms will not point to the longer route.

MPLS enables us to direct packets over a specific route to traverse the network as shown in Figure 34-11. This explicit form of routing makes sure

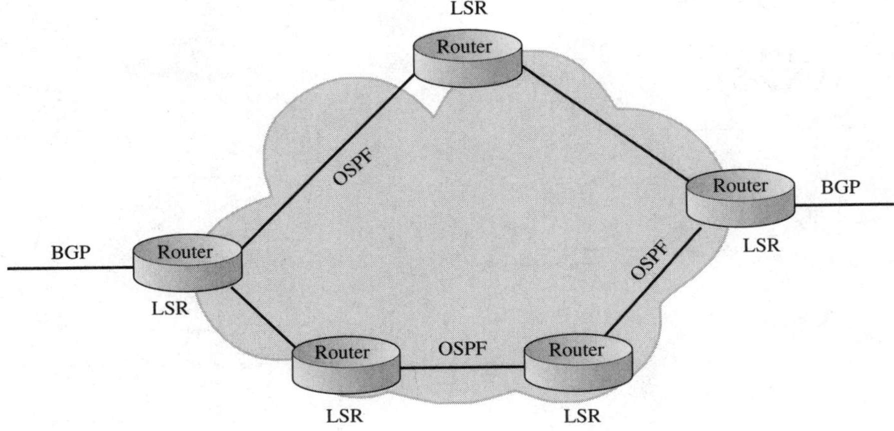

Figure 34-9 Combined protocols used in the network

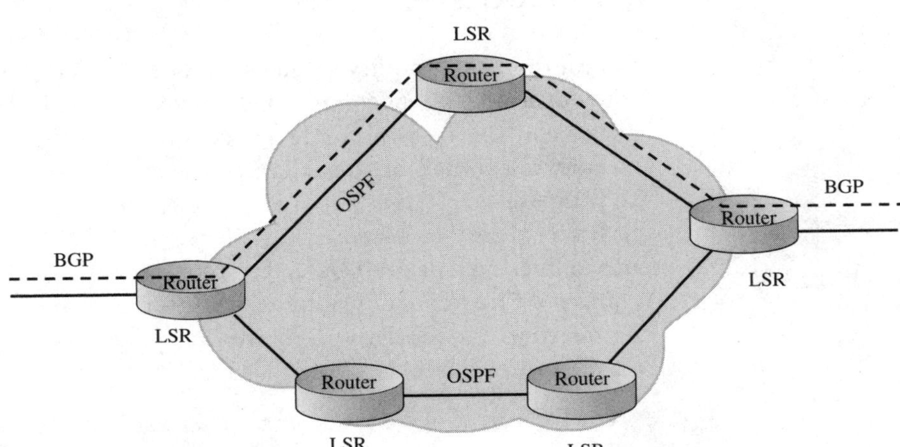

Figure 34-10 The path with conventional routing algorithms

that certain types of traffic follow a prespecified path. Explicit routing is available in the source routing options of traditional IP. Because this is a processor-intensive activity, it was not heavily accepted or implemented. MPLS takes advantage of this explicit routing. MPLS also provides the ability to analyze fields outside the IP packet header when determining the explicit route for a data packet.

Figure 34-11
The chosen route with MPLS specified routing

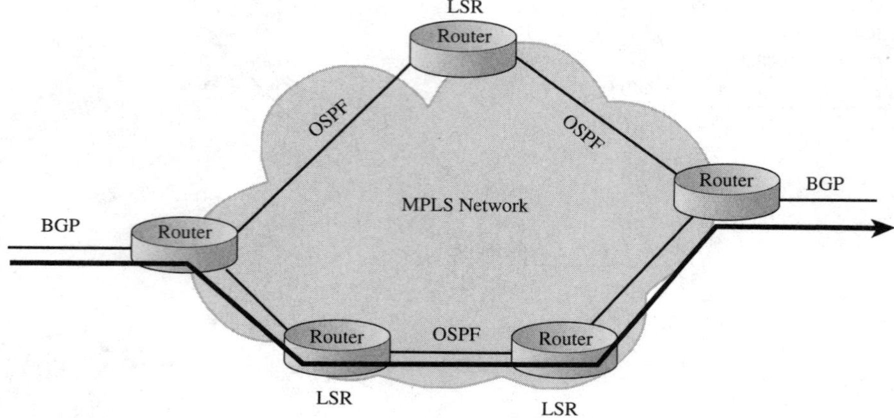

QoS Routing

QoS routing is the ability to choose a route for specific data streams to provide a desired level of service such as for *Voice over IP* (VoIP). The level of service can specify acceptable levels of bandwidth, delay, or packet loss in the network. Much of this information can be seen in the *type of service* (ToS) field in an IP header. This provides the intelligence to deliver different levels of service based on overall network policies. Providing a network path delivering a desired QoS often requires the use of explicit routing. This requires a finer level of granularity than that provided by standard traffic engineering. There are two approaches to providing QoS routing in an MPLS environment:

- The MPLS label contains *class of service* (CoS) information. As traffic flows through the network, this information can be used to selectively prioritize traffic at each network hop.
- The MPLS network can provision multiple paths between entrance and exit points. Each path is engineered to provide a different level of service.

In a traditional connectionless oriented network (such as IP), every router uses a layer-3 routing algorithm (the normal ones we see are *Open Shortest Path First* [OSPF] and BGP). As a packet traverses through the network, each router along the path makes an independent forwarding decision for that packet. Using the packet header addresses, as well as instructions from the routing algorithm, the router chooses the next hop destination for the packet. In an IP network, this involves matching the destination address stored in the IP header in every packet with the most specific route obtained from the IP routing table. This matching and comparing sequence determines the next hop destination for the packet. This is obviously very processor intensive. In a normal connectionless network, this takes place on a hop-by-hop basis to get to the distant end.

MPLS Forwarding Model

MPLS has emerged as the preferred technology to provide QoS, traffic engineering, and VPN services on the Internet. Using MPLS, the optimal path through the network is determined and tagged first. Thereafter, as packets enter the MPLS network, the input routers and switches use the layer 3 header to assign the packets to one of these predetermined paths. A label is attached to the end-to-end path information in the packet. MPLS creates a forwarding hierarchy by using a label stacking process to better handle the traffic given the conditions that exist. The label accompanies the data packet as it traverses the network. All other routers along the path use the label to determine the next hop address (device) instead of the IP address. Because these devices only operate on information in the label, processor-intensive analysis and classification of the layer 3 header occur only at the entrance to the network. This eliminates much of the overhead used in the network and expedites the overall processing of data. Once the MPLS network is established, added enhancements can be used to activate the MPLS services and capabilities. These include other label switching protocols like IP switching, *Resource Reservation Setup Protocol* (RSVP), and others.

MPLS also interacts with other class of service mechanisms and specifically with the IntServ and DiffServ IP mechanisms discussed earlier. MPLS is an advanced form of packet forwarding, which replaces the existing forwarding process based on the best match of destination address with a label-swapping model. Another term that is used is *Forwarding Equivalence Class* (FEC); the MPLS label is the encoded value of the FEC.

In comparison with an ATM core network, non-MPLS ATM switches do exactly that: They switch ATM cells between interfaces, but they know nothing about the payload contained in the cells. MPLS-enabled switches also switch ATM cells (at layer 2) but also use different labels—the VPI/VCI settings in the cells. MPLS brings the performance characteristics of layer 2 network with the connectivity and network services of layer 3 networks.

An ATM label switching router does not reassemble IP packets when switching between two ATM interfaces. The label for any received ATM cell is derived from the VPI/VCI in the cell; so long as the cell is to be retransmitted over another ATM interface, the switch will do so immediately. If the ATM label switching router also has an Ethernet interface, then it will perform frame switching between the ATM and Ethernet interfaces. This means that it will reassemble the ATM cells into a single IP packet but will then switch it onto the Ethernet interface based on the MPLS information derived from the MPLS label and will not need to examine the layer 3 IP information in the frame.

MPLS Components

MPLS networks comprise the following elements and are shown in Figure 34-12:

- *Label Edge Routers* (LERs) Located at the boundaries of the MPLS network. In an ATM network, LERs may well be the routers placed at the edge of the ATM network. LERs apply labels to packets for transmission across the MPLS network; traffic from multiple sources to the same destination can share labels.
- *Label Switch Routers* (LSRs) Switch IP packets or cells based on the labels found in them. LSRs can also support full layer 3 routing functions for unlabeled packets.

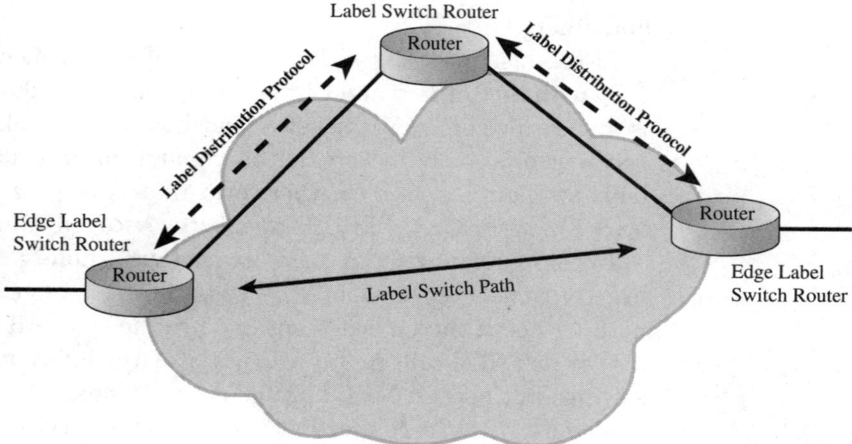

Figure 34-12 The components of MPLS

- *Label Switch Paths* (LSPs) Define the path taken through the MPLS network for a given label. In an ATM network, the LSP is equivalent to an ATM VC.
- *Label Distribution Protocol* (LDP) Used to distribute label information between LSRs and LERs. As well as defining a new protocol, existing protocols, such as BGP and RSVP, have been extended to enable label distribution to be stacked on them.

In ATM networks, the label can actually be the VCI/VPI in the ATM header. In frame relay networks, the *Data Link Control Indicator* (DLCI) in the frame relay header can be used. In LANs, the MPLS label has to be added to the existing MAC frame between the layer 2 header (*dynamic-link library* [DLL] header) and the layer 3 data elements of the frame.

The real issue is that many new and old protocols can work together to meet the demands of the explosive IP networking growth as seen in the Internet. As MPLS rolls out, any layer two functionality will be complemented with the use of the label switching associated with MPLS. A label is a label, so whether we use the Ethernet standards, ATM, Frame Relay, or *Point-to-Point Protocol* (PPP) (VPNs), they all can be improved upon through the use of MPLS.

As the network continues to grow and evolve, other scenarios will likely present themselves to define and improve upon the various protocols that have been discussed throughout this book. No one set of rules ever works for every network scenario. MPLS adds some newer means of providing selective routing through a growing network. This enables us to provide a CoS, QoS, or any other label we want to discuss.

CHAPTER 35

Intranets and Extranets

We begin with an attempt at a definition. Most of the buzzwords around the Internet are easy to define until we get to made-up words like intranet and extranet. These words were invented by journalists to describe Internet technology employed within an organization rather than outside the organization.

Intranet is essentially an Internet with a small *i.* In particular, it implies the web implementation on our internal internet. Now, what do we do about telecommuters of all descriptions? An answer is to give our telecommuters access to our intranet via the Internet, again using web technology.

Extranets are simply our Intranet extended to permit access to our organization by our business partners. That is, a web or a net interface designed specifically to support our business partners and field offices. This also could be a *Virtual Private Network* (VPN) if that level of security were required. You will note here the mixing of field offices and customers. If we were consistent, the field offices would be covered under intranet, while customers were covered under extranet.

Here again, when journalists make up words without clearly defining them, you get this interesting blurring of definitions.

By our definition, if the external employee (telecommuter and so on) has access to our internal network applications, then we are providing an intranet. That is, access isn't limited to our external web site, but extends to our internal web site and perhaps to *File Transfer Protocol* (FTP), telnet, and other applications. If our external business partner, or field office, only has access to our external web site or to an external FTP server, then we are providing an extranet.

Please note that these definitions are not universal. What you call it isn't nearly as important as what it does. The goal of our intranet is to facilitate our business—that is, to help it work better internally and with our correspondent employees.

Whether your organization is geographically diverse or simply growing rapidly, maintaining uniform communications across the organization is the challenge. Distributing information—for example, policy and procedures manuals in paper format—proves to be very ineffective. First, they aren't read because people have work to do that they consider more important. Second, by the time they need the information and read the manual, it is long since out of date. Large companies can keep a full-time employee busy running around putting change pages in all the policy and procedures manuals. Therefore, it is no surprise that the policy and procedures manual is one of the first to be implemented on our intranet. Putting it on web pages with a good index and many cross-references makes it easy to use and find related topics. Keeping it current is also as easy as updating the necessary pages.

Notification of the changes to the manual with a hot link to the affected page may be sent by e-mail to make sure that the latest procedure is followed. This is a lot better than implementing the old change page mentality by using e-mail. That is, instead of sending our change packaged, the changes are sent as e-mail attachments. Although this may seem easy to the sender, it has the effect of killing the e-mail system. These attachments (typically huge files compared to normal e-mail) use up the disk space quickly. If the system administrators are on top of things, they will get more disk space. Most mail systems (actually any system) don't deal well with very full disks, so if they let it slide, the mail system will continue to crash. This reduces the perceived reliability of the whole computing environment.

Managing the Intranet

The group that is selected initially to develop the intranet is the network group. After all, who knows more about networks? And isn't this an intranet? Unfortunately, these folks usually have their plate completely full of moves, adds, changes, and upgrades of all kinds. Their orientation is to get it working and get the heat off.

A better suggestion is to create or give the task to an infrastructure group or tools group whose position in the organization is such that they have an overview of the organization's overall business objectives and direction. To be successful, this group absolutely must have management support. This group will then "design" the intranet. The group will specify what design applications will be served and from which locations each service or capability will be accessible. If this group lacks the technical know-how to do this, it must set forth the overall objectives and hire a consulting firm to figure out a plan and help make cost/performance tradeoffs.

There is also a tendency to make this group, which soon becomes dubbed the webmasters, the bottleneck for all changes to any of the web pages. Their job is to set standards and formats so that the web pages from the various groups have a consistent look and feel. Their job should not be to make and approve all changes to every page. With current web tools, it is possible to design and specify stylesheets, forms, and templates to be used by the individual departmental webmasters. To maintain consistency in our web appearance, we do need to have a coordinated set of webmasters. The department level tends to make the most sense. (Here "department" could also mean "project" in a project-oriented business.) Even in this case, where departments are large, the webmaster tends to become a bottleneck for changes to the pages. Remember that the more dynamic the page (meaning it is frequently updated with new information), the more it will be used.

Another idea is to (within the design guidelines) automate web page creation so that updating the web pages is almost as easy as sending an e-mail. Time spent in getting the proper tools to make web publishing easy will pay off because more people will use this powerful tool. Web publishing should be as easy as drag and drop to encourage folks to communicate via the web pages.

The secondary benefit is that as more information is contained on our web pages, the less need there is to send e-mail attachments.

Web Page Organization

A common complaint is "I can't find anything!"

There are several ways to attack this problem. The first is to calmly sit down and design the organizational structure of the information on the Web. A second, less organized way is to cross-link information everywhere. This cross-linking of pages will cause problems in the future as we outgrow one web server and want to distribute the load across two or more. The problem is particularly acute if each department not only has its own web pages but also runs its own web server. True, the Web was designed to permit access of any machine from anywhere, but when we start to put boundaries around our web pages in the form of firewalls and proxy servers, each machine is essentially treated as a separate entity with separate permissions.

The point for our discussion topic here is that moving pages around can break links to other pages. This just causes a lot of hatred and discontent (not to mention the drudgery of finding and repairing all the broken links), which could easily have been avoided by applying a little design thought ahead of time. Here again, consultants can help. Cross-linking web pages is not the ideal solution, but if properly indexed, it can serve nicely for a relatively small intranet.

The better solution, in general, is to purchase one of the many document management systems. Some of these are also marketed as knowledge management systems. These systems typically provide a nice web-like *graphical user interface* (GUI) on top of a database engine. There are many benefits to using a relational database as the actual storage vehicle for the documentation. It permits referencing that same document from multiple, apparently different locations. The updated document is automatically seen from all of the different locations because there is really only one document in the database. These document management systems permit the storing of all different kinds of documents and drawings in all different formats.

It is often shortsightedly dictated that we will standardize on a given vendor's office automation suite and put everything in that format. What then happens is that we acquire a nearly equal-sized organization that is using a different *operating system* (OS) and different applications. We now have the cross-platform, cross-application problem.

We have actually experienced the following types of problems with e-mail attachments. The Mac people send an attachment to everybody. The PC people who are running the same office automation software can't open the document because it doesn't have the proper file extension, such as .doc. The Mac people never heard of file extensions, so they never remember to put them on in the first place. When the Unix users get the document, they just delete it and never try to read it because it just comes up as garbage. The Unix people have tools that they can use to decode and read the document, but it isn't as easy as point and click. The Unix users have to save the document as a file, launch a compatible program on another server, and then transfer the file before they can read it. It is much easier to throw it away! After all, it is only administrative nonsense!

The document management system does not automatically alleviate the previous problem. Proper design and standards can help this cross-platform problem. Specify the standard for the document management system as some universally readable format, such as HTML or Adobe Acrobat. All platforms can read these document formats directly from the browser. (True, the browser must launch the Acrobat reader, but it can easily do that on all platforms and OSs.)

Again, the document management system makes keeping and organizing different types of documents easier, but it does not do the organizational thinking for you! You must still design the system. There is no substitute for thinking about the organizational structure beforehand. Plan what kinds of data are to be put in each section and who needs to insert, read, and modify each piece of data.

Document management will also permit the storing of any type of document. If desired, each document can be stored in its native word processing format as well as in an Acrobat or HTML page.

It is assumed here (not always a valid assumption) that everyone in a given department has the same desktop environment, so that those working on a document together also have the tools to modify it. Snapshots of the document can be published in Acrobat or HTML format for others to review.

To help with the organization of data with the intranet, several companies have found librarians to be very helpful. Librarians are skilled in the compilation and organization of vast amounts of information, not all of which comes in the forms of books and periodicals. They are also skilled at designing retrieval systems.

Document Security

In this context, security only means access security, such as who has permission to read and write to each document or folder. Document management systems let you set security levels on each folder and each document in each folder. Access can be defined for groups and individuals. (The Unix OS file structure enables you to set up all these same permission structures, but it is a manual process best left to your local Unix whiz. This does mean that you can structure permission on web pages with as much flexibility as with a document management system.)

Collaboration

It is also wise to set up a collaboration workspace within your document management system. The intent here is that this is where work-in-progress documents are kept. Typically, these would be accessible by the folks within the department working on the project. Others within the department could read, but not modify the documents. A separate subarea is set up for leaving comments about each document. Others outside the department perhaps aren't even allowed to see the work-in-progress files. Here again, having this collaborative workspace eliminates the need for e-mail attachments.

Maintaining Interest

We now leave the department workspace-oriented intranet and consider it a mechanism for disseminating information throughout the organization.

Each group (typically department) that is publishing information on the Web needs a webmaster to maintain a consistent image within the department. To the extent we want our web page to be the first resort for folks seeking information about what we are doing, we must maintain an interesting and up-to-date web page. To that end, our webmaster or our page editor must continually provide new articles about what is going on in the department. (A history of these articles serves as a good orientation tool for new employees!)

Jokes

Love them or hate them, but they are a reality! Here again, one can offload a burdened e-mail system by creating a page exclusively for jokes to which anyone can submit a joke. Jokes are then "told" or sent by simply including the hyperlink to the joke on the web page. (You can palpably feel the e-mail system administrator breathing easier.)

There is, of course, the problem of taste in jokes versus company policy. This can be spelled out on the jokes title or index page with a hyperlink to the policy and procedure manual that outlines company policy as to joke content.

Forms

The bane of everyone's existence is having to fill out forms to order everything from pencils to software. Our intranet is a perfect mechanism for putting all of the forms online with required and optional fields. When the form is properly filled out, it is then submitted. Submission can take several forms. In a web-oriented system (as opposed to a document-management-based system), the form is typically e-mailed to the required department. In a document-management-based system, the completed form is simply saved in the proper work-in-progress area. The structure of HTML forms is such that it is easy to have required and optional fields.

These forms can extend to travel request forms, making reservation preferences (if the corporation has an internal travel office), and filling out expense forms.

Transition Intranet Solutions

These applications could be placed in either the intranet or extranet category. Here we are talking about tightly integrated business partners or field offices who resell our product or services and who need real-time access to the status of their order.

Portal Products or Customized Web Pages

Companies that market their products through competing resellers need to support their customers in an evenhanded manner. The reseller will frequently sign onto our web page in the presence of a customer to show that we are an integrated partnership. Typically, this is done by showing our web page with the partner's business logo prominently displayed. Clearly, the provider doesn't want the customer to know that this same product is available from a cross-town competitor. We can do this with customized web pages.

When reseller A signs onto our web site, we show our web page with our and his or her logo along with the product index of our products that he or she is authorized to resell. When reseller B signs onto our web page, we show our web page with a different logo and products.

This can be accomplished by using specialized software (called portal products) designed to present a different face to each customer, your document management system, or your web server software. Each provides the capability to require a login and password to view certain pages.

Along these lines, other passwords may be required of dealers to see their individualized price lists. These can be structured as wholesale and retail sheets so that the sales office can show the customer the retail or wholesale price sheet as required.

These portal products, as summarized previously, don't really do anything you can't do by using your document management system or your enterprise server software. So before you invest, you need to ask the question: "Are they easier to learn to use than using the tools you already have and that your staff is accustomed to using?"

The key to the question lies in the fact that the pages supporting your sales organizations need to have up-to-date information. Old information quickly becomes unused and ignored. When building these web pages, there is no substitute for understanding how information is used within your company.

Building a Community

In large organizations with widely varying territories, we still need a sense of community and the feeling that we are all part of the same team. Sales representatives in different markets need different information because the competitive environment is different in different sections of the country or world. Yet they are all selling the same products. Our web site, therefore, needs to provide the specific market-related information, and it needs to be up-to-date. At the same time, we need to build a sense of community and share, for example, sales strategies that work well.

It is also possible that our direct marketing team and our resellers may be addressing the same customers. (This scenario tends to occur in rapidly growing companies before their marketing plans and structure are fully mature. This condition also occurs with large chain stores that do both wholesale and retail business. The stores may go after the same customer as a corporate account in a common geographic area.)

An intranet can help tremendously with information, if it is used. Therefore, before deploying the intranet, it is wise to contact all departments, agencies, and branches involved to get input about what they would like to see from the intranet. This gives them an opportunity to feel that they are a part of the development and may yield some ideas that have not been thought of by the home office.

Warning: Look out for corporate office arrogance! The people who were given this project are full of themselves and their new high-visibility project. The reality is that no one knows better what to do in order to make sales than the workers in the trenches.

Warning number two: Look out for "too-busy-itus." The field offices have a full plate and, when presented with a request for input, they frequently do the following:

- Don't really understand what corporate is up to
- Don't realize how important their input is
- Won't take the time to understand the issues and create thoughtful input

The predictable result is that when the system is made operational, there comes a hue and cry from these field office folks about what is wrong and how it ought to be changed. Because the system is really for their benefit, as well as the corporate benefit, their involvement is essential from the beginning. Therefore, we recommend forming a working group (ideally, with a working area on our web server or document management system) to work out the design goals and structure of our community-building intranet.

When introducing the intranet to a noncomputer-oriented business, it may require hands-on training. This may have to extend down to how to start up the computer, launch the browser, get online, find data, save or use data, and then sign off. Do not underestimate the amount of training involved! This even goes for high-tech companies. The problem is that current software products are so complex and will do so many things that the operation of the software can't be totally intuitive. Because of the flexibility offered, even setting the initial options requires a systems administrator.

If the system is used for order entry, entering those orders ought to be as easy and as similar to the existing order entry process as possible to minimize training. Building HTML forms from the paper forms minimizes retraining. (If the data needs to be in a different format for the back-end system, let the computer system reformat the data.) Users feel comfortable seeing their old familiar form with a new electronic face. They are not so intimidated, and you will not get as many service or help-desk calls.

For noncomputer-oriented business, outsourcing development of the intranet is the best idea. It will be faster with fewer hassles than developing the skills and technology in house. In any case, it must be treated as a full-blown project with a formal written specification for what the vendor is to do and what constitutes success. The other alternative is a time and materials relationship with your consultants to keep working on the problem until you are satisfied. Understand then that this will take a while and cost more than a predefined contract.

Bulletin Board Service

Although we are touting the benefits of an intranet here, some companies can benefit from a simple bulletin board service where information is posted on the Web to inform people. This is not a document management system, but a simple posting of training seminars—when and where they are held, and what the prerequisites are (if any). It is almost guaranteed that once the intranet is built and information is available, the implementation group will be deluged with suggestions on how it can be further used to improve business operations.

Customer Service

Although we discuss customer service to external customers in the next section, the same issues and attitudes need to be applied to our intranet. The suggestions will come rolling in. To maintain the involvement of everyone using the system, it is imperative that you have a system set up to address and accept these suggestions.

Providing a prompt e-mail response is a good way to acknowledge the receipt of a question, complaint, or suggestion. Having a formal system for posting the arrival of a suggestion, and posting the disposition, keeps everyone involved. Simple suggestions may be handled quickly. The implementation of the change can be posted with a "thank you" for the suggestion note to the originator. Some suggestions may require the formulation of a project team to implement them. That fact and a proposed schedule should also be posted on the system designer's web page.

Thin Clients

Depending on your environment, a thin client intranet is definitely worth considering. Thin client means that each of the client machines (typically a PC) has (in the extreme) no local hard drive and no floppy drive. It is a network-connected device only. When it is turned on in the morning, it receives its download from the server. The server provides it with all the operating software it needs. There are actually two implementations of thin client. One is like a Unix terminal where the terminal device only knows how to run the display. The entire application runs on the server (the application server). The second implementation is where the desktop device is a PC that is intelligent enough to run the application (for example, a browser), but has no local storage and receives the browser and OS software as a download from the file server.

The big advantage is that our intranet is now virus free (more on viruses later). The client systems can't introduce viruses from floppy disks because there aren't any. The main server is managed by the computer services department and backs it up every night. Desktop hard drive crashes are a thing of the past. The downside is (and this is very much a function of your operating environment) that when everyone arrives at 8:01 and turns on the PC, they all request a download at once. This taxes the network and the file server. The bigger the environment (the number of PCs), the greater the load on the system is. A compromise position is to provide the local PC with a hard drive to hold the OS and the local applications.

Using the Unix dumb terminal with an application server can (again depending on your environment) save the expense of a full-time employee who would otherwise be needed to perform system administration on all the desktops.

Extranets

We now take the advantages of intranets and extend them to our business partners and clients. They, like our intranet clients, will access our network over the Internet. Here we describe the applications and don't worry about the implementation. Later, we will cover design of an extranet interface and security measures you may want to implement. Our extranet must project our image to the business community with which we are dealing. It should inspire confidence and project an image of trustworthiness. We should indicate what security steps are taken when handling credit card transactions, for example.

Inventory Management

Wholesale

The first example of an extranet is a wholesaler who must maintain inventory to handle transient demand as well as his contract quantities. The normal cost of distribution is about 18 percent of the cost of goods. This can be reduced to about 5 percent with the application of an extranet and electronic commerce. The wholesalers can now do just-in-time inventory control. Depending on the goods involved, the wholesaler may be able to get overnight shipment from providers, thus reducing the need for large inventories. The goods can be drop-shipped directly to the client's location. With a good e-commerce extranet, the wholesaler also has visibility into the source's inventories to immediately determine what delivery promises to make to clients.

Secondary Markets

The next revolution in e-commerce is in secondary markets. Here the source (manufacturer or grower) has made a product, satisfied major contracts, and has remnants left over from the production run or parts that don't quite meet the original specifications. These could be textiles with flaws or paint off color for the major buyer.

Integrated circuits (chips) don't perform at 600 MHz, but do perform at 300 MHz. Historically, these circuits were either thrown out (which effectively raised the cost of production on the remaining "good" parts), or a wholesaler (speculator) would buy the stuff at deeply discounted prices in hopes that they could be sold.

The extranet offers the capability to put the inventory online and accept bids. This can be done by the *Original Equipment Manufacturer* (OEM) or by a wholesale dealer who markets the parts on the Web without having to transport them from the OEM location. When the product is sold, it is drop-shipped by the OEM.

Privacy Issues

One could say that privacy is at odds with security. The better we can verify who the calling party is, the more secure our Web (extranet). The trick is to verify identity without compromising or appearing to compromise the privacy of our customers. This is a double-edged sword. You need the tools to maintain info about your visitors to verify their identity, but these tools can also be abused.

In an attempt to improve security on the Web and to facilitate tracking down hackers, Intel put a serial number in each chip that could be read by browsers (any software really). Thus, we could track the users of the machine as they visited various web sites if the browser put that number in each transaction. Although much excitement was created over tracking individuals, remember that some machines have multiple users. Therefore, we really only know which computer visited the site. That is okay for tracking hackers (how many times does a burglar sneak into the house to use the Internet connection?), but it is insufficient for tracking purchasing profiles.

The privacy advocates Primary Information Center and Junk Busters lead the attack against this invasion of privacy. The *World Wide Web Consortium* (W3C) has developed a platform for *Primary Preferences Project* (P3P). The intent is to build an industry-wide mechanism for controlling privacy information. The goal is that the web site publicly publishes its privacy control mechanisms to every browser that visits the site.

The site's privacy practices profile might define what data they collect and what their rules are for passing (selling?) that data to other marketing entities. (The obvious question is what is the enforcement mechanism for this policy? What assurances do we, the customer, have that it is being followed?) Here is where we as a site provider must engender confidence and trust.

When configuring your browser, you specify or set up the preferences for privacy practices that you expect from the sites you visit. You also fill in a user profile that stores some or all of your personal data. These data are then sent to web sites whose privacy practices match your privacy practices profile that you set up. These data could include e-mail addresses, name, billing information, and personal preferences in clothing, music, food, hobbies, and so on. Our privacy profile might include the conditions under which we permit our data to be shared, sold, or used. We could specify the other types of sites with whom we would permit our data to be shared.

P3P is really a plan to create a common protocol between the browser and the web site (server) to collect and exchange privacy data. It defines the structure and words that define privacy practices. It also defines a protocol used for exchanging these data between browser and server. Thus, the privacy data is only transferred if the privacy profile set by the user matches the practices published by the site. Both Netscape and Microsoft intend to support the P3P standard.

This is a sensitive subject and therefore getting agreement between those who want the information and want to traffic in it and those who want to protect the information totally will be in perpetual conflict. Our observation is that the probability that the browser profile and the site profile will indeed match up goes down geometrically with the number of elements or rules in the profile.

(Just for a fun example: If the site has four rules, A, B, C, and D, each of which can be true (1) or false (0), and the browser has the matching four permissions, then the probability of all four matching up is $0.5 \times 0.5 \times 0.5 \times 0.5 = 0.0625$ or about 6 percent. If we add one more rule E, then it halves the probability to 3 percent.)

Our prognostication is that acceptance of P3P will be slow because of these conflicting goals. More on this can be found at www.w3c.org.

Perishable Goods Application

Most producers of perishable goods (for example, fruit, vegetables, and flowers) have long-term contracts with transportation and distribution companies who market their goods and provide a guaranteed income for a certain quantity of goods. What becomes of surplus production? It typically goes on the spot market. That is, the goods are transported to a physical intermediate market/warehouse where they are then purchased by wholesalers.

This can mean that the goods are sold at deeply discounted prices, or that they are put on a consignment spot market. The producer has no idea if the product will sell or not and absolutely has no price guarantee. The consignee may sell the entire product and tell the producer that only half of it was sold. The ultimate buyer has no idea of how old (fresh) the product really is and has no real recourse if the product turns out to be substandard.

Several companies are using extranets to provide a direct link between the source producer and wholesalers. The trick is to contact a reliable transport company that provides timely, reliable delivery. This is another case of seeking to eliminate the physical transportation to an intermediate staging point. This takes time and diminishes the freshness of the product.

Offering the goods for sale on our extranet lets the wholesalers buy directly, thereby improving the product freshness at the retailer by days to a whole week. Working financing into the deal allows the producers to get paid upon shipment while giving the wholesaler 30 to 60 days to pay. The product is fresher and therefore worth enough extra to cover the financing costs.

Here we have a product clearinghouse that greatly cuts down on time to market and at the same time matches up buyer and seller. When a financial clearinghouse backs this business, it instills confidence in the producer and buyer that the transactions will be handled properly in a timely manner and everyone will get paid.

Purchasing Cooperatives

The capability to form purchasing cooperatives that can negotiate volume pricing with suppliers is another growing extranet application. Smaller businesses from parts stores to building contractors can pool their requirements and approach the supplier with volume deals.

The second major advantage is getting the orders correct. This almost can't be emphasized enough. It can take a full-time employee to verify orders, process the paperwork, send back the incorrect stuff, and get credit for it while reordering the proper parts.

Finding out immediately if the product is in stock and when the delivery time will be cuts down on local inventory requirements. (Many parts and even drugstores practice just-in-time inventory where they will order the part for next-day delivery.) The ability to see the supplier's inventory online permits shopping at alternate sources should your normal supplier be out of stock.

This system can work across the Web or, in the case of banks and brokerage houses, across the extranet with special software. The client portion runs on the user's desktop workstation and the server software portion runs on the server or mainframe.

Brokerage houses, for example, supply their good clients with desktop software. In this manner, they standardize the software and guarantee interoperability. The same model works for construction projects. If the customer (for example, a construction company) uses the standardized project management software, the supplier can then read the bill of material required and bid on supplying the components. The suppliers then become business partners with their customers. This also helps the suppliers plan their production schedules and inventory levels. (Ford has had such a system in place for over 10 years. The suppliers see the production schedule and supply the parts by contract just in time. The reduction of on-site inventory has more than paid for the implementation of the system. The original system used the X.25 network and custom software.) The objective is eventually to be able to run paperless projects.

Boeing's 777 was the first paperless airplane. They were able to maintain schedule and, more importantly, the parts fit better, minimizing production line rework.

Outsourcing

Whether the problem is to manage a large project with just-in-time delivery of component parts or subcontracted services or a just-in-time maintenance service, all can be done over the Net.

The requirements are the following:

- A common database
- A common mechanism (interface) for accessing that database
- The ability to set up different views of the data

Each contributor to the project needs only to see that set or subset of data relative to his or her portion of the project.

A good example of this might be an airline that outsources some or all of its services. For example, it has a schedule for the following:

- Flight crew
- Cabin crew
- Airframe maintenance
- Engine maintenance
- Reservation agents
- Gate agents
- Ticket counter agents
- Flights themselves
- Seat reservations
- Meals (catering)
- Wheelchairs
- Cabin cleaning
- Refueling

The list is almost endless. Most airlines outsource catering and fueling services, but they could outsource everything. Although the crews typically work for the airline, they are treated almost as outsource subcontractors. The flight and cabin crew members individually bid on each flight. Their bid is accepted or rejected, not on price, but on seniority. Maintenance could be handled the same way and for small startup airlines it typically is. (Small startup airlines, to keep initial investment costs down, outsource many more services than the major carriers.)

The flight hours of the plane are published on the Extranet page. Then the maintenance company prepares a bid and crew to handle the scheduled maintenance stop. Note that the maintenance company doesn't need to see the catering schedule, nor almost any of the other schedules. Each contributor to the overall project has its own view of the common project database that contains all of the schedules. The capability to provide these different views, as discussed, can be implemented by using the inherent system file permissions, a "portal" software product, or a data management system.

Watch for these outsourcing trends using extranets not only for airlines, but also for design and construction projects both large and small.

Computer Hardware Vendor

Typically, the delivery of a complex hardware server system (especially complex customized configurations containing multiple product lines) requires months of delivery time. When the customer is dealing with a local representative who may be halfway around the world from the productions source, getting the order right and delivered on time is nearly impossible.

The field office creates the order for the system, doing the best job it can in specifying the configuration. Unfortunately, they typically have no idea of what the production schedules or inventory levels are. This can result in long delays in the fulfillment of the order.

Here is a typical example: The customer orders a system with 2, 9GB hard drives. What the configurator in the field office couldn't know is that the factory has stopped ordering 9GB drives and has gone to 18GB drives as the standard drive. The order waits on the shelf until some 9GB drives can be ordered and delivered. Note that the order-processing folks can't take it upon themselves to substitute an 18GB drive. Two 9GB drives were ordered, and that is what must be delivered!

Now let's examine the same delivery problem with an extranet, where the production schedules and inventory levels are available to the field offices by signing onto our home office web site.

The sales representative, or technical support representative, can log into the site with the customer right there. The parts and price list are right there. They can use the online system configuration software to select hardware components, including redundant power supplies and processors, if required. When they get to specifying the 2 9GB drives, they find that inventory is nonexistent, and there is a 60-day wait if they are backordered. They find that the 18GB drives are in stock. The customer, on the spot, opts for the 18GB drives because it means a difference in delivery schedule from 2 weeks to 2 months. (The customer planned on mirroring the data on the drives and therefore needed 2 physical drives instead of the single 18GB drive.)

The customer is ecstatic with our ability to be responsive to his or her needs and to deliver the equipment so quickly.

Automating Customer Service

The two issues to be addressed here are the following:

- Improving the responsiveness of our customer service organization or function
- Reducing the cost of providing the service

(Some companies have found the ideal solution. They charge their customers to provide service that a well-designed product wouldn't require.)

The basic problem is that having technical folks respond to e-mail customer service queries is boring for the individual. They soon develop a bad attitude and quit. Having nontechnical personnel respond has the danger of responses not being technically correct (or even close in some cases). Therefore, the need to create a low-cost, accurate, automated e-mail response system is absolutely necessary.

The answers to simple and standard e-mail queries can be handled by a completely automated system. An example of this is that a customer can order parts by e-mail. If the customer is willing to include the part number and catalog number, it greatly minimizes the probability of error. Although this is an idealized example of a query/response from an automated system point of view, it illustrates one end of the spectrum of capabilities.

In real life, customers will mix multiple unrelated topics in an e-mail. (We know we do it all the time.) The trick then for the automated system is to accurately parse the content of the e-mail for subjects, verbs, and objects. The analysis of the query then eliminates the chaff and identifies the important content. The success or failure of the system depends upon the complexity of the topics and issues. Take a test drive before you buy!

The two goals of improving service while lowering costs are often in conflict. Our opinion is that good responsive customer service should outweigh the cost consideration. Who is the company to the customers? Well, at first it is the salesperson who sold the product (unless it is bought off the shelf in a store) and after that it is the customer service representative/system.

Future sales (via word of mouth) rest upon the responsiveness of our customer service.

Ever more, our corporate image is our web site backed up by the customer service system. As technology improves, more and more automation can be applied to the customer service problem. Our preferred design approach for an automated system is to automatically scan and parse the e-mail looking for issues. It's important to fabricate an immediate, automatic response outlining the issues detected. This response always includes a hyperlink to the specific web page that addresses those issues. This would include a pointer to frequently asked questions. For those subjects not recognized or identified, that fact is relayed to the customer. "The following issues have been sent to our research department."

If more information is known to be required, then the automated response can make that query. The goal is to keep the customer in the automated system while still being responsive. For some number of queries (ideally, a small number), a human will have to respond. This implies a hierarchy of technical ability, each of which in turn analyzes the query, asks for more information if needed, and passes the request to the next level if it can't resolve the problem at that level.

Within a short while, you will learn what standard customer queries are and be able to automate the responses to these. The trick is to detect the "out of the normal" queries and quickly pass them to the human hierarchy of knowledge response team.

This system structure provides the customer with an immediate response that is correct (or nearly so) or a response indicating that this problem requires some research. What is most important is that the customer feels that he has been heard and his problems are being handled. Ideally, the customer is also told approximately how long it will take for the answer to his or her problem to be researched and formulated. Our experience is that customers are willing to wait a couple of days if they know that the response they will receive will be accurate and to the point.

Our interaction with vendors who have a poorly designed human interface on their product is that they have many more requests for help than if the product were more intuitive. Their product is fairly technical, their customer service is overloaded, and their business is handled by people in a hierarchical system. The problem is that because there is no continuity in a given customer's request, a sequence of queries about the same subject receives answers that are mostly tangential to the query at hand. This does not instill confidence in the customer base. Needless to say, this customer is losing market share.

Implementing Extranets

In order to provide some sort of consistency, some definitions are required. These are not universal (and will be violated frequently in the trade press), but will help with the following discussions.

Intranet

An internal network is comprised of one or more LANs, interconnected with one or more routers running the *Internet Protocol* (IP). (So far, this is only an internet with a small *i*.)

The bulk of our traffic on this Net is web-based. That is, we have one or more web servers, and all of the clients are running a browser. Our primary means of communicating is using the Web. (Technically, mail is a different set of application protocols, but most browsers also provide a mail capability.) Perhaps we should say that the primary interface that our users use for their data is via their browser. This network may be geographically far flung where dedicated lines (or Frame Relay) are used to connect our routers. There is no connection to the Internet. Now, here is where things get touchy. If we were to use VPN technology between our corporate locations, one could argue that it is still an intranet.

Extranet

We are connected to the Internet and specifically permit access to our network (or portions of our network) by our customers.

We are now at risk from all sorts of hackers, crackers, and freaks. The goal then is to keep them out (or at least slow them down and make them work for it) while giving our friends and customers "free," that is, unfettered, access.

The first step then is to install a firewall. A firewall is simply a machine whose job it is to examine each packet arriving and verify whether it is permitted, based on a set of rules. This set of rules is referred to as the rule base. Firewalls essentially operate at the IP and *Transmission Control Protocol* (TCP) levels of the protocol stack. This means that the firewall examines each packet to see if it is coming from and going to the proper IP address, and that it contains the proper or permitted TCP port number. You can set up firewalls to be a one-way valve. If you do this, your employees can surf the Net at will and send out any kind of packet, but your firewall will not permit any unsolicited packets. (This means that mail won't work; for example, the firewall won't let in any packets, including mail.) The other problem with this approach is that it will permit your employees to hack on the Net. If they are caught, the corporation may be liable, so think about this issue. This is not to say that firewalls are not useful, but to point out that simplistic solutions will not achieve your objectives.

Performing basic packet filtering on source and destination IP addresses is a useful feature when combined with additional firewall capabilities. It was initially believed that a firewall was the solution to all our Internet security problems. The limitations of a firewall-only implementation are now apparent. If we only need to communicate between specific networks or specific devices with fixed IP addresses, a firewall is a fine solution.

TCP Filtering

The next level of checking that the firewall provides is to check inside the TCP packet for a port number. For example, the standard TCP port for *Hypertext Transfer Protocol* (HTTP) is port 80. (Any port could be used in this example. Port 8080 is another port that is frequently used by proxies.)

The firewall now checks for IP addresses (which might say any IP address) and the port number in the TCP packet. This means that unless the port is open through the firewall, you can't get that service (for example, HTTP or FTP). This provides for a more general audience. Now instead of filtering an IP address, one can say, "Let anybody from any IP address come through the firewall, but only if he is doing HTTP on port 80 to a specific server." This means that we could have other web pages on other servers on port 80 internally, but they would not be available because the firewall limits access to a specific server. Clearly, we can allow access to as many servers as desired.

Our presentation is a little simplistic since it makes the assumption that the OS on which the firewall software is running is itself secure and unhackable. Not only is this not generally true, but it is impossible to secure the OS perfectly. (For example, there are known attacks to a Unix-based system that is running send mail. Therefore, the send mail application should not be present on your firewall machine. A properly maintained OS will have this door closed—see the following explanation.) So, it isn't just enough to run a firewall; that firewall must be run on a "locked down" OS.

An aside on attacks: There are two general kinds of attacks against systems. One is called the *denial-of-service* attack. In this case, the hacker sends one or thousands of packets to the server, causing it to spend all its processing power processing these garbage packets so that it has no time left over to do its real job, such as serve up web pages. The second kind of attack is more invasive. In this case, the attacker tries to gain access to the machine, preferably as the super user. If the attacker achieves this, the system is literally wide open. As a super user, the attacker can set up accounts on the machine, read any file, wipe out any file, and so on. Very clever hackers will therefore edit the log files so that they show no trace of the hackers' presence. There are a couple of ways to keep log files. One is to frequently copy the log file to an obscure directory and change its name so that its name does not contain "log" (for example, call it bob). The second is to encrypt the file. The latter is only convenient if the firewall application provides this option. The former only works if the log file is copied often enough to capture the hacker's presence before the log file gets cleaned.

Constant vigilance *must* be maintained to ensure that the OS has the latest patches that close off discovered holes. Because the Unix community is open, as each attack is discovered, patches are written and posted to newsgroups where your system administrator can download and install them.

Despite the fact that Unix isn't perfect, the rapid availability of fixes for bugs and holes nearly as quickly as they are discovered means those diligent system administrators can maintain a very secure OS. Our security experts generally agree that the preferred OS is Unix for firewalls and proxy servers. There are frequently reports of break-ins to web servers, and the more noteworthy are those belonging to the Department of Defense—because they should know better. In practically every case, the hole (or weakness) exploited by the hacker is a known hole, the fix for which has been available from the newsgroup for six to nine months before the attack took place. The fault lies then with sloppy or nonexistent system administration. An equal number of holes exist in proprietary OSs; however, one must wait for the vendor to release a patch. This isn't always a speedy process.

Stand-Alone System

It is highly recommended that the firewall run on a dedicated machine. Figure 35-1 is a classic firewall drawing with the external network to the left and our protected network to the right. To save on hardware costs, it is tempting to run the web server as another task on the same machine. This is not wise because, if the hacker manages to crack the OS, they have access to the whole machine and any applications running on it. As a stand-alone machine, they can crack into the firewall, but can't go any farther. This assumes that the other machines on your network won't allow access (for example, they don't allow telnet) from the now defunct firewall.

To further protect against this problem, we introduce the DMZ (see Figure 35-2). It is highly recommended that you not implement the network in Figure 35-1. The DMZ, as we shall see, is a protected area into which all external traffic goes. That means, ideally, that no external traffic flows through the firewall to our internal network. It all goes to servers in the DMZ.

Figure 35-3 shows the relationship of the DMZ to the logical firewall. Although we think and talk about firewalls as a single entity, when we are using a DMZ, it is logically two firewalls.

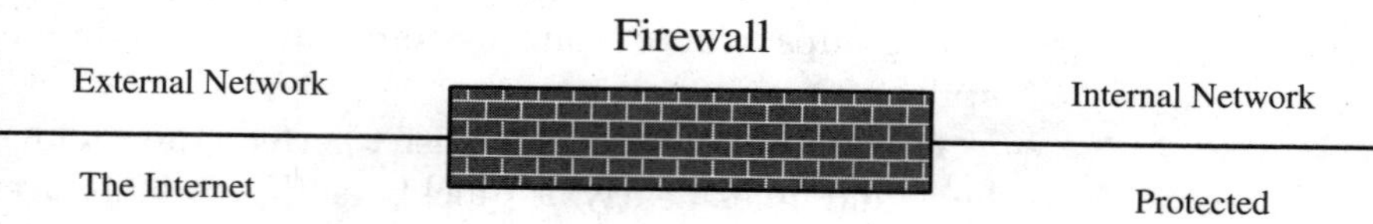

Figure 35-1
A classic firewall implementation

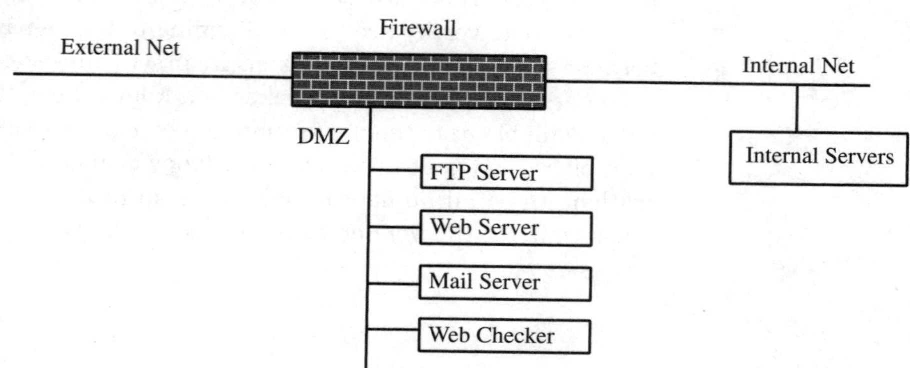

Figure 35-2
The DMZ provides a protected area where all traffic flows.

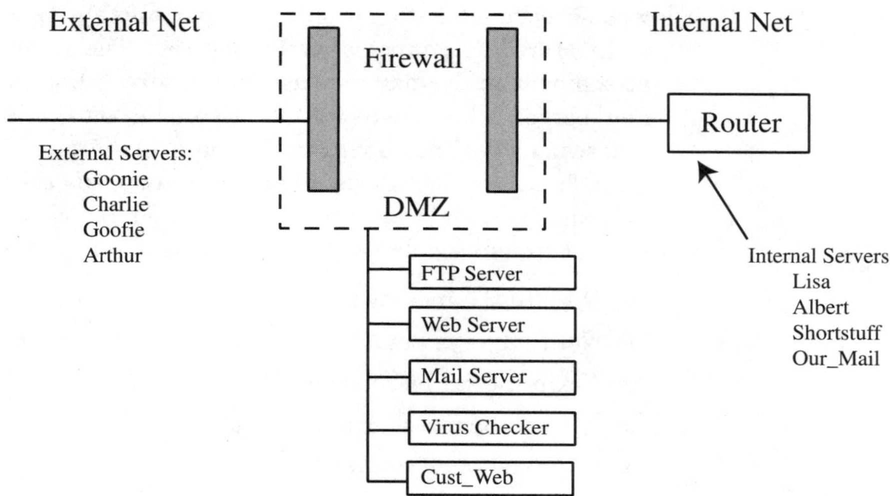

Figure 35-3 The relationship of the DMZ to the firewall actually creates two firewalls.

Note that the DMZ is another communications port on the firewall (typically, an Ethernet port with its own subnet address). The firewall may now have up to six sets of rules for traffic from the following:

- Outside to inside
- Inside to outside
- Outside to DMZ
- DMZ to outside
- DMZ to inside
- Inside to DMZ

In general then, we will put all services accessible by external folks in the DMZ and allow none onto our internal network. (We all recognize that there will be exceptions to this rule. In all cases, make sure that you design the firewall rules carefully so as not to get any unintended consequences.) Figure 35-3 shows the FTP, mail, and web (HTTP) servers in the DMZ.

Virus Checking

With the constant threat of viruses arriving in mail attachments and via other mechanisms, some firewall vendors also offer virus checking. Because the processing load on the firewall can be enough for one machine, the virus-checking function is offloaded onto another machine that also resides in the DMZ. The process works like this. The firewall recognizes a mail attachment on the mail message. Before forwarding the mail message to the mail server, it first sends it to the virus checker that opens the attachment and scans it for known viruses. It would do the same for a file forwarded to the FTP server. If there is no problem, the virus checker tells the firewall that all is OK, and the message is delivered. If a virus is detected, there are several options available, depending on the vendor's implementation:

- Discard the message.
- Notify the system administrator by e-mail or pager.
- Clean the file by removing the virus and pass the message on.
- Return the message to the sender.
- Send an e-mail to the sender.
- Send an e-mail to the recipient.

In some cases, these options can be combined. Note also that not all vendors support all these options.

The virus-checking function is no better than the latest information it has about viruses. Therefore, if you choose to implement virus checking, it places another task on the system administrators to periodically log into the virus vendor's web site and download the latest virus profiles. Although this sounds like an easier solution to manage than desktop virus checking, according to the vendors themselves, it is not a substitute for desktop virus checking. Again, the desktop virus checking is only effective if the virus profiles are kept up-to-date. If you are getting the feeling that maintaining network security is a perpetual struggle and (depending on the size of your shop) requires a dedicated system administrator, you are getting the right feeling.

Firewall Rules Bases

Most firewalls operate (or should operate) from the position that nothing is permitted except that which is specifically allowed by a rule in the rule base. Some early firewall products required that the administrator learn the syntax of the rule base. Most firewalls available today offer an easy-to-use GUI interface. Table 35-1 shows a typical rule base as one might see it from a GUI.

Table 35-1

A typical rule base viewed through a GUI

source	destination	protocol
any	mail server	http
any	mail server	smtp
client1	cust_Web	http
client2	cust_web	http
goofie	ftp_server	ftp
charlie	ftp_server	ftp
goonie	ftp_server	ftp
mail_server	our_mail	smtp
goofie	shortstuff	telnet
lisa	ftp_server	ftp
albert	ftp_server	ftp
shortstuff	ftp_server	ftp

The basic rule here is to put the most frequently used rule first. You don't want the firewall searching through and applying a whole bunch of rules that rarely apply before it gets to the one that is needed. All this searching takes precious processing power. Therefore, the permission of any source, external or internal, can contact the Web_server machine, but it may only use HTTP. (Since the port isn't specified, it is assumed to be the default port 80.) Note that any source can send mail to mail_server. We have also installed another web server (cust_Web) especially for our customers. This one is only accessible by devices attached to specific network addresses belonging to our customers (for example, custnet1 and custnet2). There is a list of network addresses in the firewall's rule base—one for each customer. There are normally other rules in the table that tend to be vendor-specific.

For example, the tracking column lets you specify if you want a short or long log entry every time this rule is invoked.

Next, the FTP users are at machines called goonie, charlie, and goofie (these are really aliases for specific IP addresses). They are the only devices allowed to FTP to FTP_server. In the last rule of our example, we are also restricting FTP to internal machines called lisa, albert, and shortstuff.

The rule allowing mail_server to talk to our_mail server indicates that mail_server is only a mail forwarder. It isn't the real mail server. It is only allowed to talk to our real mail server with our_mail. If somehow, mail_server is hacked, it won't stop our internal mail server. In addition, you don't want to put the real mail server or your internal web server in the DMZ because then all your internal traffic has to go through the firewall, increasing its burden (see "Firewall Performance (Again)").

The next to last rule allows goofie to telnet to shortstuff. Although this seems innocent enough, the question is, Where can the user go from here? This gets into password administration, which is beyond our scope. The issue is that normally a single username/password file is used system-wide so that we only have to administer access permissions in one place. If we allow telnet from goofie, then the user at goofie has a login account and could telnet to other servers in the internal network. The solution is to have (and maintain) separate accounts on the server lisa.

Note that there are no rules permitting anyone to talk to virus_checker. There are also no rules permitting icmp-echo (ping), so that would-be hackers cannot use this tool to discover servers in the DMZ or on our network.

Firewall Performance (Again)

Although we mentioned it previously, firewall performance can be a bottleneck. The more traffic we run through the firewall, the greater the potential for a bottleneck. Again, it is essential to have an understanding of the traffic flow. How many packets per second must the firewall handle? In our example, we would add up mail, web, FTP, and telnet traffic. Remember that mail is handled twice by the firewall and three times if the virus checker is employed. After you have determined the traffic volumes, check with the hardware and software vendors of your firewall package to determine if you have sized the system correctly.

Proxies

Simplistically, you may think of a proxy as an application-level firewall. In the case of web proxies, it can serve to direct, redirect, or limit access through the proxy to servers and specific URLs.

Although the proxy may be placed in parallel with the firewall, our preference is to put it behind the firewall. Remember, we may use several firewalls in parallel to handle the traffic load. Extreme care must be taken to make sure that you have a consistent rule set across all firewalls.

Forward Proxy

There are two kinds of proxies. Your company uses a forward proxy to limit where your employees are permitted to surf. ISPs use proxies to limit the options that their members have. They don't limit destination web sites (although they could exclude known porn sites), but they limit their members to HTTP and FTP. This limits some of their hacking options. Also, the queries to the remote web sites appear to be coming from the proxy instead of from the individuals user's terminal because the proxy masks the user address by replacing it with its own.

Reverse Proxy

A reverse proxy is used as a frontend for a web server. (A large server farm would implement an array of proxies.) Requests to www.anywhere.com are routed to our proxy that forwards the request to server content_server (the keeper of the real web page). The surfer thinks it is talking to server www for all requests. This makes it nice for the webmaster, who is in the process of replacing content_server with bigger_server. All of our user's links remain the same; we just change where the proxy points. As previously, we can limit which URLs the proxy permits. So we could have two web servers, but only the content of one is visible to the outside world.

The proxy (both forward and reverse) also helps reduce network traffic and content_server load. The proxy caches every requested web page. The next request to that same server, whether it is from the same or a different source, is satisfied out of cache without having to go back to the content_server. If the ISP has a proxy array, it will eventually contain all the popular web pages, thus minimizing the number of requests sent out across the Internet. The reverse proxies at the content provider's location also cache popular pages. The rules as to how much to cache and when to go back to the content_server for the latest version of the page may be set up in the proxy server. In fact, at a large site with a large array of servers, the popular pages are downloaded into the proxies at low traffic time, and they serve the web requesters. This improves site reliability and flexibility.

Figure 35-4 shows a typical proxy array behind a firewall. Each of the proxies directly connected to the firewall provides an interface to one or more servers, usually handling a given subject. The proxy in front of the other proxies has servers of related subjects, and the front-end proxy steers the request to the appropriate proxy that is caching that subject. This array then acts to share the load and provides for backup in case of a proxy failure.

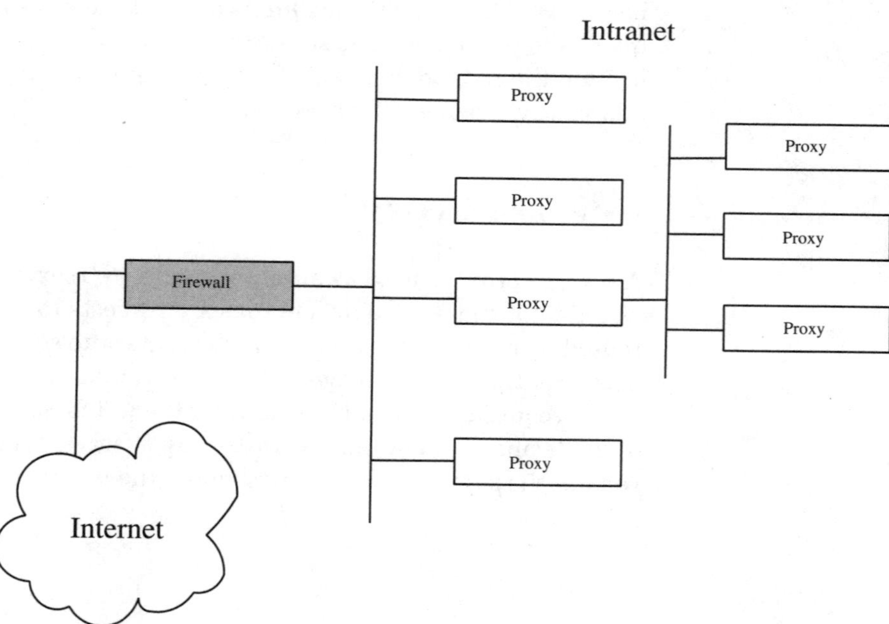

Figure 35-4 A typical array of proxy devices sitting behind the firewall

Proxy Security

Figure 35-5 shows a hypothetical configuration. Note that there are two proxy functions. One is a proxy for the internal web server bigger_server, and the other is a proxy for cust_Web. At this point, our options begin to expand (read "get complex"). Whereas the proxy could simply map request from the surfer to the server, it could also require that the surfer has an account on the proxy and log in to see a particular server. We can take this to whatever extreme satisfies our paranoia. For example, we could issue account IDs and passwords to our customers. First, they must log into the proxy. Then they must log into cust_Web to see their specific information. (If we were paranoid, we could make them log into the firewall first.) We would only let our remote offices see the web pages on bigger_server and they must log into the proxy first. This is another safeguard against hackers mucking with our web page.

The LDAP server is the keeper of the user database of username and password. These users can be segregated into groups such as sales, customer1, customer2, marketing, engineering, and customer support. The proxies then could restrict access, based on group membership. Note that

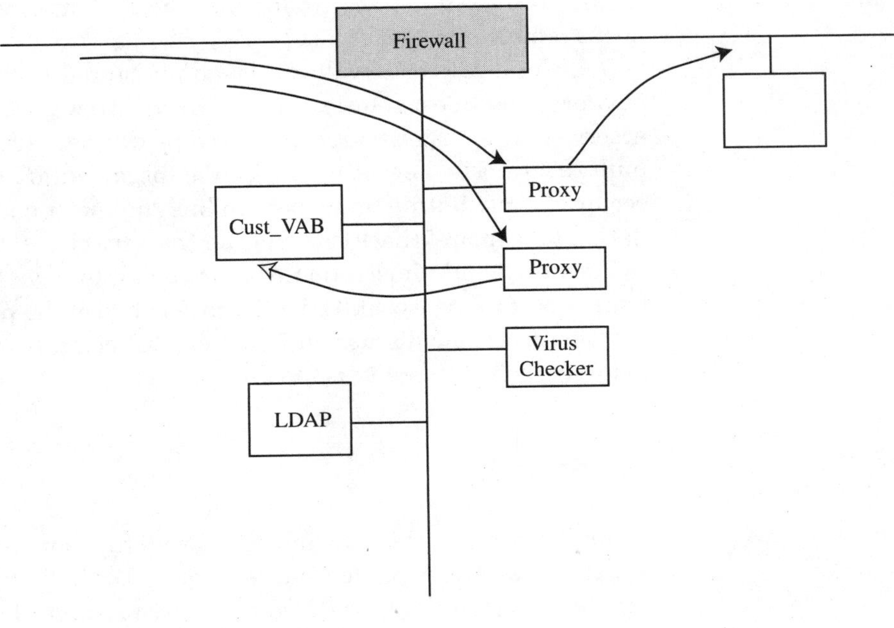

Figure 35-5 A hypothetical flow through the firewall to the proxy and then on to the applicable server in the DMZ

the LDAP function may be a single point of failure. Therefore, consideration should be given to providing access to a backup LDAP that has been synchronized to the main one. Setting up (designing the database structure) and maintaining the LDAP server is a significant effort depending on the experience of your staff. Maintenance of the LDAP database will also take administrative and support staff time.

Although the servers in the DMZ here have been drawn as separate boxes for clarity, some of the functions could be combined into one machine. This appears to be cost effective, but remember, if it is that machine that is being hacked, you could lose all services it provides.

Proxies can also do redirection. When a server is taken down and another takes its place at a completely different address, the proxy can simply redirect the user to the new URL.

We discussed securing the OS on the firewall. The same should be done for the proxy. It too should be secured with all of the latest OS patches installed. Applications such as send mail that are not needed should not be running.

Administration

For both firewalls and proxies, the safest way to administer them is to require the administrator to log into the system console. That said, what about remote administration?

Your local proxy/firewall can be administered from inside your secured network. The safest solution here is for the firewall to limit administration access to a few specific machines. Here we mean desktop machines, not multi-user hosts. Few of us work in an organization where everyone is so occupied contributing to the bottom line that he or she has no time or inclination to "explore" (hack) systems on the network.

For remote administration, you are going to have to open the administration port to the world. The difference between the proxy and the firewall function from our discussion is not crystal clear. In addition, in products today, their functionality is melding.

Firewalls

In our discussions, we have limited firewall functionality to IP and TCP layers. It is the gatekeeper for our network or DMZ. Its functionality and configuration should be considered *company private* information. The extent that a hacker knows what kind of firewall we are running and its configuration gives the hacker a better clue as to where to start when attacking our network.

Proxy

We have used the proxy as an application-level filter that is able to filter (or force) access to specific servers. We use it to hide the true identity of our servers and limit the pages that our visitors are allowed to see and the functions (HTTP, FTP) that they are allowed to perform at each server. When designing a web site, all references within that web site should be relative references (for example, relative to the existing server). Web pages containing absolute references will be passed directly back to the user who will now have the name (or IP address) of the server. The proxy won't help because, in this case, only diligence on the part of the webmaster will fix this problem. It is also a good idea to avoid using IP addresses when identifying servers for two reasons. First, you don't want the external world to know your IP addresses. Second, if you need to upgrade a server from a smaller machine to a bigger one, you have to ferret out every reference to the old IP address and change it—a maintenance nightmare. The solution is to use aliases via the *Domain Name System* (DNS).

Domain Name System (DNS)

The DNS is simply a listing of the device IP and the human-readable name (alias). A site or company usually runs two DNS systems—one for the internal (protected) network and one for the external network. When a query for www.company.com arrives, the DNS returns the IP of the server (or proxy) that is designated as www. The external DNS then will only point to proxies, while the internal DNS will point to the actual servers.

Internal web pages that are available or viewable by external users should only reference other web pages by their external DNS alias. You can have as many aliases as you want for an IP address. If you fail to follow this rule, here is what happens. A page is delivered to a user that references another page in the internal system by its IP address. When the user selects that link, the firewall says permission denied.

If you were to use the internal alias for the server in the delivered web page, when the user selects that reference, he gets a *no DNS entry* error. Therefore, the rule is when you deliver a page to an external user, that user must have a mechanism to resolve the actual address *and* have permission via the proxy/firewall to access the designated server.

If the pages are viewable by both internal and external folks, then the internal and external DNS entries have to be the same so that all of the links work properly.

Fungible Services

Clearly, the firewall can take on proxy services and the proxy can take on firewall services. We recommend using two separate boxes to improve security. Investigate your firewall vendor thoroughly. Research as much information as you can before making a decision. A beautiful, easy-to-use GUI is wonderful, but what you really want is the best security on the market. Remember, setting up the firewall, proxy, and web servers is not a "spare-time" operation. You will have to dedicate individuals to these tasks. The maintenance of these servers must also be budgeted.

We also recommend designing an internal policy with a review board to review changes to the firewall and proxy configurations. The goal here is to put a damper on statements like "Oh, we'll just open up this port for that function." Indeed, that may be a good idea, but it may also have other security implications.

Since the firewall is the keeper of the keys to the kingdom, its configuration should be treated as company private. The more a hacker knows about your firewall, the easier his job is. The proxy configuration on the other hand is only going to permit access to selected protocols on selected servers so it is easier to manage and maintain. Unfortunately, as new servers and proxies for them are added, firewall adjustments must also be made to permit those connections.

CHAPTER 36

Network Management SNMP

Simple Network Management Protocol (SNMP) is the Internet standard for monitoring and managing devices connected to the Internet or your own intranet. It defines a data set structure of information that each device may provide called a *Management Information Database* (MIB). It also makes provisions for individual vendor-based, customized MIBs. A vendor then can provide the standard set of parameters and also extend that set with "custom" information, such as "vendor-specific" information. These data are collected, kept, and reported by an agent that runs on the managed device. We call this agent the managed function.

A management function resides on a workstation or host, and interacts with the managed function or agent. The management function must have a copy of the MIB profile loaded for each managed function, plus any vendor-specific profile.

In this manner, a standard management function can interact with the standard MIB, while a vendor-specific management function may be used to glean more specific information. These custom MIBs are also registered with the *Internet Advisory Board* (IAB).

Thus, you find standards-based systems, such as HP's OpenView, and vendor-specific systems, such as Fore's ForeView and Cisco's CiscoView, that are compatible and work with OpenView.

Network Management Goals

The ultimate goal of network management is to improve system uptime. The CEO of an organization may see it as another overhead expense. The CIO should see it as a tool for maintaining a near 100 percent uptime.

The higher the system availability, the less equipment is needed, the lower the cost of running the network, and the more efficient the organization will be that utilizes its resources.

Everyone in the corporate management structure needs a clear understanding of the goals of network management, and the amount of time and effort required to reach those goals. Network management is a great tool for proactively managing the network and collecting alarms and alerts at threshold levels that are below the "Houston, we've got a problem" level. It has often been said that if you do have a failure, then the network management didn't work! Unfortunately, many network management systems are viewed as disaster recovery mechanisms rather than disaster avoidance and prevention mechanisms.

One must also consider the point of diminishing returns. Whereas you may be able to achieve 99.9999 percent uptime, the cost in additional redundant equipment and failover systems may not be justified for your enterprise. Here is a good time to plug for disaster recovery plans. Every enterprise must have a disaster recovery plan. In the context of network management systems, what is the plan of action if multiple (unlikely) failures occur? The likely events should already be handled by the network design. (In this context, the building burning down or an earthquake rendering it uninhabitable is an unlikely event.) At one extreme, the plan may state, "We close up shop and go out of business" or "We wait until the flood waters recede, clean up, buy new equipment, and start over." Although these aren't recommended plans, they are plans and indicate that management has done some thinking about the problem. At the other extreme is a company, for example, a stock exchange or a hospital, which must keep going at all costs. In this case, alternate facilities are identified that contain computing equipment that can take over if their own equipment fails. A set of network connections is in place to provide connectivity to all of the customers/branch offices, including a backup network management. The downtime is minimal. Namely, the downtime would be the time it takes to load the backup configuration on backup machines, plus the time it takes to switch over the network connections and—we're up!

History

In the days of centralized systems, the host or mainframe was about the only smart device in the system. The rest of the devices in the network were relatively dumb and could be reset when necessary by the host.

Figure 36-1 shows a typical centralized network configuration. Most of the remote devices in the network were connected to the host by dedicated lines. At the larger remote locations, a remote concentrator, or controller, is used to multiplex multiple terminals on the single data line. At smaller locations, a single terminal may exist. Dial-up lines may serve some low-traffic locations. But, in any case, there were not multiple paths in the network. It was strictly hub and spoke. Network management consisted of monitoring the status of all phone lines and modems. Dial backup for the dedicated lines was often used when high availability was required. This technique is still used today to back up our Frame Relay connection. Today, an ISDN line can provide 128 Kbps dial backup.

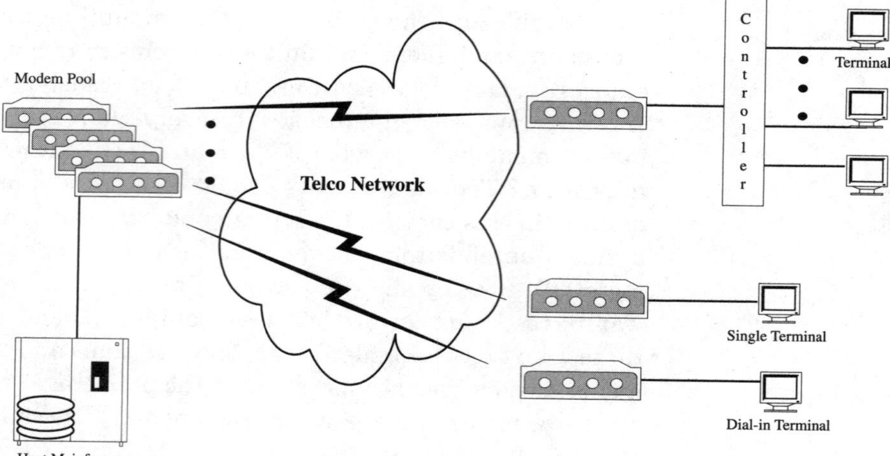

Figure 36-1 Typical centralized network configuration

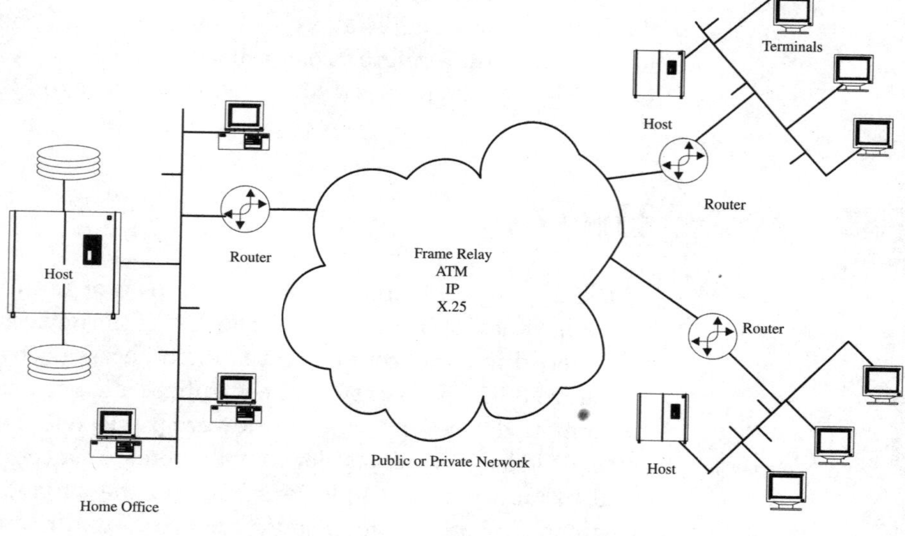

Figure 36-2 Typical distributed processing system

As we have migrated from the centralized model to the distributed computing model, every element in the network is now intelligent.

Figure 36-2 depicts a typical distributed processing system. The *Wide Area Network* (WAN) technology isn't important per se in that it may be leased as a service from a network provider who will generally not allow our management system visibility into his network components. What is important is that the WAN connection must be available in all of our locations and that it provides an inexpensive high-performance interconnection.

If the WAN is of our own construction, then we consider it to be part of our overall network. That is, we lease lines between locations and provide our own networking equipment, whether it is the Frame Relay switches or routers.

In either case, the distributed system has many more elements to manage and, most importantly, that can be controlled managed. The failure of any one of these elements could cause a significant number of our users to be cut off.

In Figure 36-2, the "router" may be taken to represent a router specifically, but in a general diagram like this it could be representative of other network elements such as a *Frame Relay Access Device* (FRAD), an *Asynchronous Transfer Mode* (ATM) switch, a *Local Area Network* (LAN) bridge, or even a lowly hub.

Network Management Function Interaction

Philosophically, there is a management function and a managed function. Their physical location in the network is not important. Managed elements may be physically collocated with the management function, or may be at the far edge of the network. The management function is typically located at the headquarters' location. However, it does not have to be that way. As a case in point, the major network providers, such as AT&T, Sprint, and WorldCom, have distributed *Network Operations Centers* (NOC).

The management function is the user's interface to the network management system, and it is typically a GUI, which provides status, reports, and statistics gathered from the managed function.

Figure 36-3 depicts the logical relationship between the managed and management functions. The managed element, or object, is the actual device, whether it is a hub, bridge, or router. The agent function is the program, or process, that runs on the managed system and provides the interface to the management function. Associated with the managed element or device is a set of attributes. These attributes may include memory size and utilization, interface speed, traffic load, and so on, depending on the type of device.

As one might suspect, there is a protocol defined between the management function and the agent that defines the format and content of each of

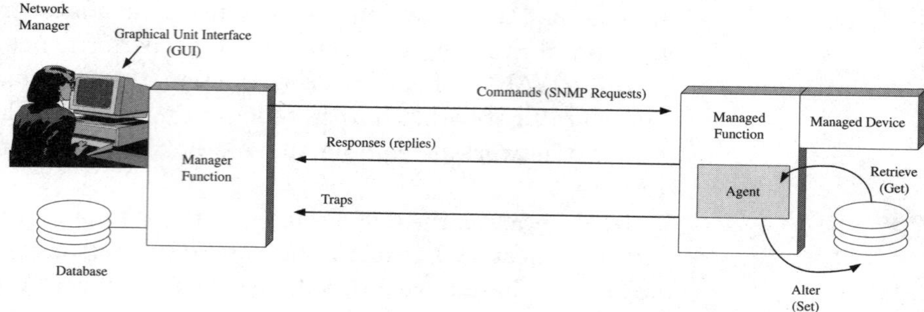

Figure 36-3 Logical relationship between the managed and manager functions

the commands, responses, and traps. RFC 1905 governs these protocols. RFC 1907, in turn, governs the content of the SNMP messages. SNMPv2, also known as MIB-II or mib-2, is the dominant definition set today and has superseded the SNMPv1 mib-1. Version 2 greatly expands the detail available from the managed agent.

The agent is the interface to the management function and may accept commands to set parameters (such as timeouts, thresholds, and, in some cases, system operational parameters, such as group membership addresses or routing tables) as well as provide status upon request. The agent can also send unsolicited messages (called traps) indicating alarm conditions. The name "trap" comes from the capability of the manager function to set a threshold (trap), and should this condition be met, the agent sends off an unsolicited message. All other messages sent by the managed function agent are a result of GetRequest or GetNextRequest queries.

It should be clear that the data reported by a device like a hub is a lot less complex than the data from something like a router. The hub needs to report Ethernet collisions on the LAN, the number of frames handled, the number of errored frames, and so on. The router, on the other hand, has routing tables, queue sizes, memory utilization, *central processing unit* (CPU) utilization, and so on.

To keep this all straight, the management function must know about the content of each MIB on each device. In other words, it must have a complete database identical to the one on the managed device so that when it receives the information, it can place it in the appropriate location. When a new device is installed in the network, its MIB content and format must be loaded into the machine (typically, a host or a workstation) providing the management function. Likewise, the agent software must be installed in the managed device.

This emphasizes an important point. Setting up and running a network management system is more than a casual undertaking. It requires a commitment on the part of management to provide the resources and purchase the appropriate tools.

Database Structure

The MIB content is defined by the RFCs listed in Table 36-1.

Table 36-1 is a listing of the relevant standards that apply the network management in general and SNMP and RMON in particular. The complete listing is available from http://rfc.fh-koeln.de/rfc/html_gz/rfc.html.gz.

It is worth noting that the contents of the MIBs are defined by using *Abstract Syntax Notation 1* (ANS.1). This is a language that allows a succinct definition of the content in terms of numeric or alpha characters.

It takes a little practice to be able to read the descriptions because ANS.1 is very much like a programming language. If it becomes necessary to delve into the contents of the MIBs, we recommend one of the books in the bibliography.

The content is organized according to a logical structure, called a *Structure of Management Information* (SMI). The actual SMI structure is defined internationally by the RFC as a tree, using branches of the tree for various organizations. Figure 36-4 is a partially filled-in example of this tree. Each data element is unique because its path from the root through the various branches and twigs to the leaf is unique. Vendors may choose to customize the content of their MIBs under their internationally assigned vendor number.

Those familiar with computer file system structures will recognize this tree. Whether you call the nodes or levels "folders" or "directories," one contained within another allows different leaf notes to have the same name yet be unique because the path to that data is different. E-mail presents a simple example of this. JohnSmith6BigCompany.com and JohnSmith6Big University.edu are different people with a common name. They are distinguished globally because the path to each is different. (First, you must go to the .edu as opposed to the .com domain and then you go to the BigCompany or the BigUniversity.)

Figure 36-4 shows the internationally agreed-to structure and the numbers to each level. Like a file system, this sequence is read from the highest-level (tree root) to the lowest-level leaf, where the actual data element resides. For example, you can find out what kind of a system you are managing by going to 1.3.6.1.2.1.1.1. This is the path: iso.identified-organization.dod.internet.management.mib-2.system.SysDescription.

Table 36-1

Relevant standards that apply the network management

RFC	Definition
1155	Structure and identification of management information for TCP/IP-based Internets
1156	Management Information Base for network management of TCP/IP-based Internets
1157	Simple Network Management Protocol (SNMP)
1158	Management Information Base for network management of TCP/IP-based Internets: MIB-II
1159	Message Send Protocol
1160	Internet Activities Board
1161	SNMP over OSI
1270	SNMP communications services
1271	Remote network monitoring Management Information Base
1900	Renumbering needs work
1901	Introduction to community-based SNMPv2
1902	Structure of Management Information for Version 2 of the Simple Network Management Protocol (SNMPv2)
1903	Textual Conventions for Version 2 of the Simple Network Management Protocol (SNMPv2)
1904	Conformance Statements for Version 2 of the Simple Network Management Protocol (SNMPv2)
1905	Protocol Operations for Version 2 of the Simple Network Management Protocol (SNMPv2)
1906	Transport Mappings for Version 2 of the Simple Network Management Protocol (SNMPv2)
1907	Management Information Base for Version 2 of the Simple Network Management Protocol (SNMPv2)
1908	Coexistence between Version 1 and Version 2 of the Internet-standard Network Management Framework
1909	An Administrative Infrastructure for SNMPv2

The SNMP data are contained in 1.3.6.1.2.1.11. Vendor-specific MIBs are under 1.3.6.1.4.1; that is iso.identifiedorganization.dod.internet.private.enterprises.

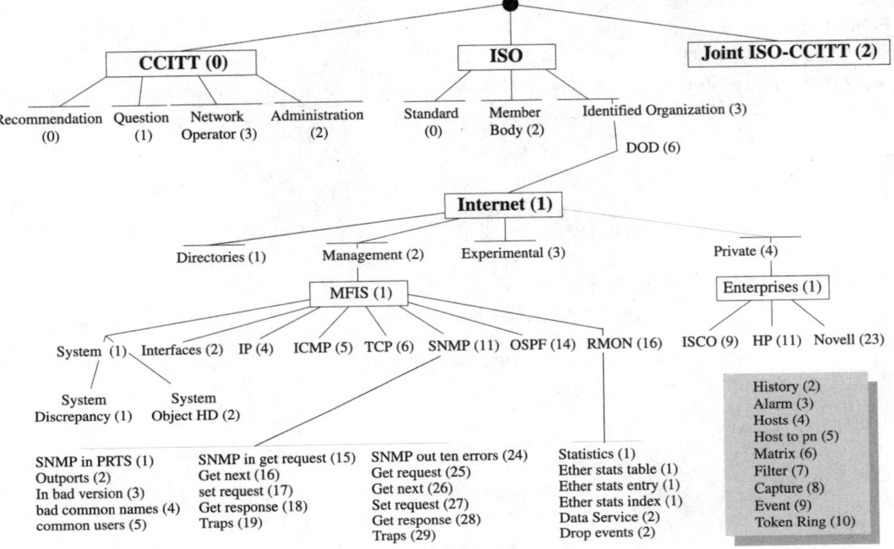

Figure 36-4 SMI structure defined as a tree

Architecture

Commands are sent from the network management function to the agent either to set parameters or to collect status information. When status is requested, the agent replies with a Response message. The agent is also capable of sending an unsolicited status message if some previously defined threshold is exceeded.

Figure 36-5 shows the architecture of the SNMP function and how the functions are related to the ISO's *Open Systems Interconnect* (OSI) model.

As shown, SNMP runs on top of a standard Internet protocol stack where *User Datagram Protocol* (UDP) is used at the transport. UDP is a connectionless protocol and, like IP, does not set up a connection between communicating entities. In a query/response protocol like SNMP, a connection is not necessary. If the target device fails to respond to a query, simply send the query again. If it were necessary to set up a connection for short queries, it would actually make the system slower.

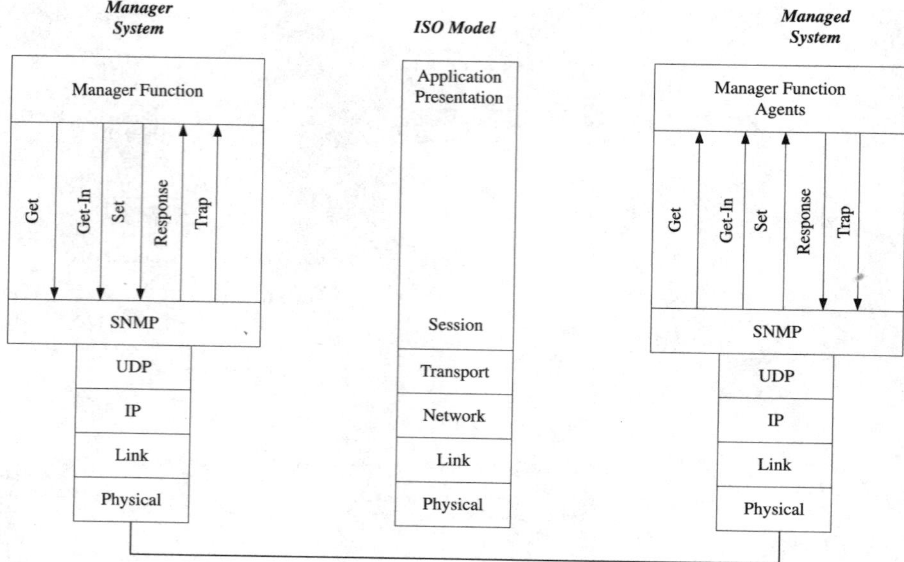

Figure 36-5 SNMP functions and how they are related to the ISO's OSI model

The underlying physical network is irrelevant because it can be Ethernet or a Frame Relay network. From an OSI point of view, SNMP and its interaction with the manager and managed functions encompass the upper three layers of the protocol stack. A discussion as to which functions belong to which layer usually winds up in a religious argument, so we will consider that SNMP and the management functions are an undivided upper-layer set of protocols.

The commands and responses are indicated in Figure 36-5. The actual names (and assigned numbers) for them are delineated in Figure 36-4.

Figure 36-6 shows how the SNMP message is carried nested within the protocol layers. Since our discussion must remain general, we take it that the link layer protocol typically has some leader and header information and some trailer bits. The link layer frame carries the IP packet across the link layer connection to the next router. The IP datagram header has the routing information that lets the routers direct this packet to its final destination. The UDP header, frankly, doesn't do much. The SNMP message is contained in the UDP datagram. The version field insures that we are conversing with another agent of the same version. The community field is important because it is the security function. Our SNMP function can therefore only collect information from members of our own community.

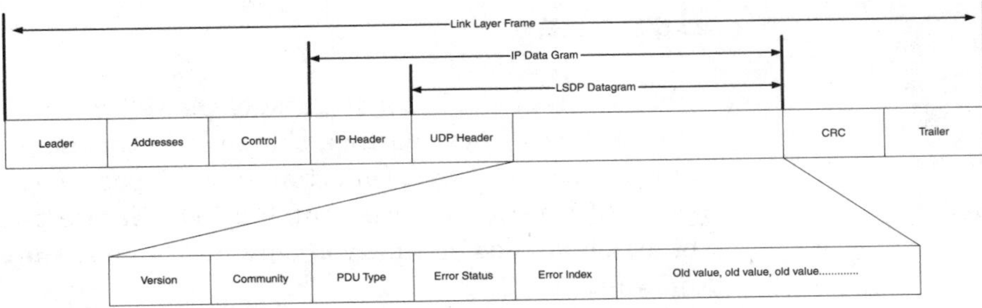

Figure 36-6
The SNMP message carried within the protocol layers

This prevents your competitor from picking the brains of your equipment attached to the Internet. The *Protocol Data Unit* (PDU) type specifies whether there is a GetRequest, GetNextRequest, SetRequest, Response, or a Trap. The error fields are used to identify SNMP errors, such as tooBig, noAccess, or badValue. The error pointer points to the location of the offending data field. Each *Object Identifier* (OID) then is included, followed by its value (see Figure 36-4).

It is clear that this is a Simple Network Management System. The messages are simple and straightforward, and all of the items in the MIB are defined either by the standards bodies or by the vendors themselves. (If it is so simple, why are the books on the subject two inches thick?)

Remote Monitoring (RMON), although not nearly as widely used as SNMP, is simply another mechanism for keeping a database on the performance of network elements. Figure 36-4 breaks out a small portion of the Ethernet statistics, part of the statistics category under RMON. These tables exist for Token Ring as well.

Network Management System Issues

If one had to ascertain the performance of one's network by evaluating the detailed content of each of these message types, it could drive one to drink. Fortunately, there are network monitoring systems (the network management function) with GUI interfaces. Unfortunately, these products are not fully mature and easy to use. In the following paragraphs, we provide some insight as to what to look for when purchasing one of these systems.

Bundling

It has always been in the interest of the seller to bundle features into feature packages. They can then charge more for the package because it contains so many features. Unfortunately, the package usually contains only a few of the features that we want. It is in the interest of the user then to have things unbundled so that you only need to buy the specific features that you want.

The first question that needs to be asked of the vendor is whether his feature sets are bundled. What is included in the basic package? What features are extra? Is installation included? Is implementation support included? (As indicated, the installation and configuration of a network management system is not easy for the uninitiated. You can anticipate the need for plenty of vendor help.)

The GUI

The fact that the network management has a *graphical user interface* (GUI) does not mean that it is intuitive or even easy to use. Each vendor has a different approach. The people who use the tool to watch the network don't necessarily understand the interrelationships of all of the protocols and elements that are being managed. The GUI should permit high-level management to be easily accomplished while providing the capability to see deeper into the system as the operators become more sophisticated.

Network Size

How large a network will the network management system handle? Most network management systems will handle the average to small network with no problem. It should not be assumed that the network management system is infinitely extensible. This issue goes to the core of the network management system's internal database design. (Ideally, the purchaser/user/operator shouldn't have to know or worry about this, but limitations in the database design will affect the number of nodes that can be handled.) We point out here merely that the design of the database affects the efficiency with which data can be extracted. It, therefore, has a direct impact on system performance. Performance can be improved by keeping the entire database in memory. Unfortunately, the size of the memory either limits the size of the database or creates a performance impact when the database can no longer be contained within the system memory. The other alternative is to grow the system memory to match the size of the network. In the worst case, the network management system application may become unstable and crash as the database size is exceeded. (This behavior has been noted in published tests of the leading network management system products.)

The user of the GUI, therefore, should not have to know how the database objects are structured in order to retrieve information. For instance, it is frequently desirable to be able to compare the performance statistics of one device with another similar, but not identical, device. This may be accomplished with varying degrees of success with different products. Although routers and bridges perform different functions, it is frequently useful to be able to compare the total number of frames, or packets, handled by each one as the load on the network varies. Another useful metric is the number of errored frames, or packets, per period of different network loads from these different devices. Unfortunately, not all network management systems permit such comparisons. In some cases, it is impossible; in others, it is merely difficult.

Another important question is: what percent of your network bandwidth is used by the network management system? Ideally, it should be 1 percent or less. This is significant since most SNMP data are collected by polling (sending GetRequests). Trap data only arrives if the thresholds that have been set in the devices have been exceeded. Therefore, it is important that the network management system have a user-programmable polling priority system so that important devices, like routers and firewalls, are sampled more frequently than end devices, such as workstations. This maintains the current status of the important devices and eventually collects information on the end stations while minimizing the impact on network traffic by the network management system.

Web-Enabled GUI

Being able to view reports via a browser eliminated the need to collect, duplicate, and distribute reports. With this feature, anyone who wants the report (and has the proper permission) may collect the report from the network management system. This worthwhile feature is far from universal.

Alarm History

Keeping a history of the exception condition is so useful that we count it as a necessity. It is not only necessary to be able to identify problem areas and equipment, but also to keep a history of similar problems on that and similar equipment. Keeping history alone is not sufficient. The network management system must permit searching the data and displaying historical trends. Trend analysis is the key here. Remember that the goal of a network management system is to maximize uptime. This cannot be achieved by reacting to problems. It can only be accomplished by anticipating problems. Therefore, in addition to keeping a historical record that lets us find out when and where this problem last appeared, the network management system can, with the proper software, do a linear regression on the historical data and predict when (but probably not where) the problem might next appear. If, for example, the problem were a hardware failure associated with a particular component, the frequency of failure might tell us when it might appear again. If the problem were traffic related, such as buffer overflow, and the consequence was a system slowdown or crash, then the historical traffic data could allow us to identify potential trouble spots that exhibit similar traffic profiles to the failed node. Then we could go back and look at traffic through the failed node from last week/month/year and use that data as a threshold, or trap value, to predict failures of similar nodes under similar traffic conditions.

Alarm Presentation

Alarms must be programmable and presented in a logical fashion so that they convey the relative importance of the message. Programmable filters are a "must" to permit concentration on the most important alarms first.

Statistics

It should be noted that the more statistics the network management system keeps, the more data it must collect from all components (agents) in the network, and therefore the more traffic will be added to the network by the network management system. Again, the goal is to try to keep the bandwidth requirements of the network management system down to 1 percent of network traffic. In small networks, this is easy to achieve. In networks of thousands of nodes, a thoughtful use of alarms and a prioritized polling sequence must be employed to achieve this goal.

Free Trials

There is no better way to learn about a network management system than to live with it for a few months before you buy it. This provides the opportunity to find out how user-friendly the network management system really is. The vendor's demonstration always looks good. They showcase their strengths and minimize their weaknesses. Their presenter is thoroughly familiar with the product and can really make it sing. (How long did it take her to become that proficient?)

In reality, it is often difficult to take advantage of a prepurchase trial offer. Although the vendor is willing to let you try the software for a few months, it takes management commitment to dedicate resources within the organization to this trial. A network management system is not something that can be evaluated in someone's (nonexistent) spare time. Remember, it isn't the initial outlay that is important, but the ongoing lifecycle costs.

Network Mapping

Much is made of this "gee-whiz" feature. And, to be sure, it is a convenience to have a network management system that will create a network map. (One could argue that we should have been keeping the network map up-to-date as we designed and grew the network. Unfortunately, most networks and especially LANs were never really designed; they just grew.) How the mapping function works and what information is available from the map is more important than the map itself. First, the map should be readable. Second, it should permit zooming in and out. In the "show-me-the-whole-network" mode, the map of a large network will be too dense to read. It should provide the capability to zoom in until the subsystem of interest is displayed.

The mapping function from different vendors works with varying degrees of success. There are two basic ways of implementing the mapping function. The network management system can use SNMP information from each agent, or it can observe the traffic on the network. The SNMP approach yields a map more quickly, but can only display managed devices. Tracking traffic flow can theoretically find every interface or port on the network, but could take months of operation to finally determine a map. (In some cases, it is impossible to create a map.)

Thus, two points must be made. First, accumulating the information to draw the map is not a 10-minute job. Depending on the size of the network, it could take hours to days to months. Second, the accuracy of the map is dependent on how extensively each node on the network supports SNMP.

For example, several prominent vendors don't support standard MIBs. As noted, there are standard MIBs and vendor-specific MIBs. In order for your network management system to utilize the vendor-specific MIBs, they must be loaded into the manager function. For normal (here read "simple") networks, the mapping function works well and fairly quickly. That means it works for Ethernets connected with routers and bridges, where each interface on the router is its own subnet. Today, however, we tend to use fully switched Ethernets where each device is on its own switched port from a LAN switch. The advantage is that each port is buffered so that there are no collisions on the network. A second advantage is that one can create *virtual LANs* (VLAN) where devices on different ports of the switch can be made members of the same broadcast domain or subnet. This permits you to design traffic flows on the various LAN segments in order to minimize response times and keep broadcast traffic confined to those devices that are members of the same logical net. The disadvantage from a network management system point of view is that it is difficult or impossible for the mapping function to create a map because there are managed devices hiding behind, or within, other managed devices. This is especially true if the LAN switch doesn't support transparent bridging or standard SNMP MIBs. The next fly in the ointment is the development of *Emulated LANs* (ELANs). These are similar to VLANs, where we are emulating LANs by using an ATM switch and edge devices that are LAN switches. Here again, the ATM switch and edge device may not provide the MIB support necessary to build an accurate map of your network.

We tend to expect too much from the mapping function. Consider the difference between the logical and physical network. In the first case (see Figure 36-7), the logical and physical networks are the same. We would expect the mapping function to find or draw this picture.

In Figure 36-8, we have a common configuration for today's networks. All the devices (1 to 9) are on the same IP subnet. However, for traffic reasons there are only two VLANs. A and B segments comprise one, while the other is made up of segments B and D. For example, devices 1, 2, 3, and 5 are in one broadcast domain, while devices 4, 5, 6, 7, 8, and 9 are in another. This appears logically as shown in Figure 36-9. Different vendors' network management systems will yield different network maps in this case. The best (for example, as close as you are going to get) will be from a network management system that watches network traffic.

An exercise left to the reader is to make up a traffic scenario and then see what you can learn about this from the other side of the router. If the MIB in the switch has good data, you are in fairly good shape. If it doesn't support the standard MIB, it is hopeless. Although most of the industry has

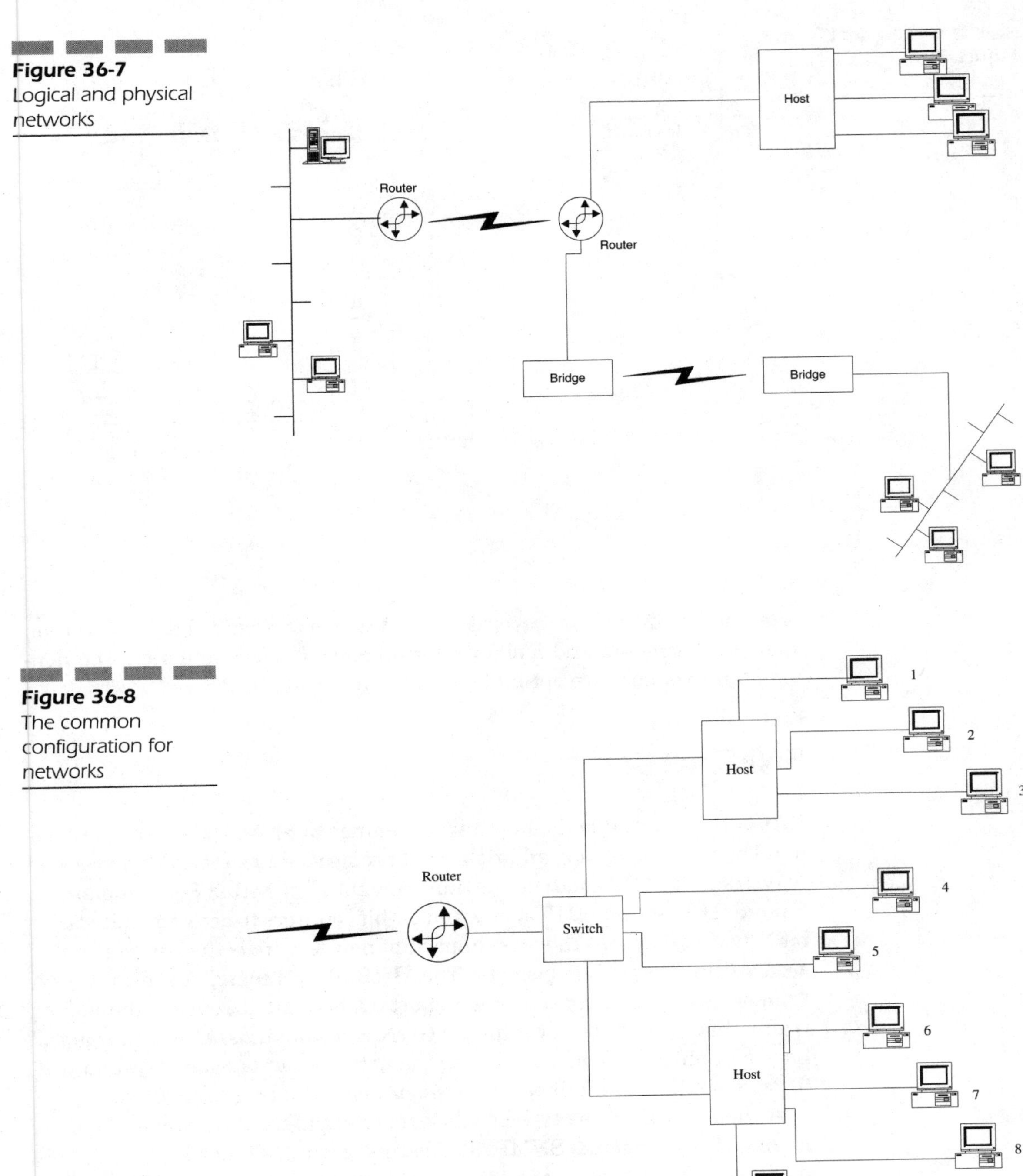

Figure 36-7
Logical and physical networks

Figure 36-8
The common configuration for networks

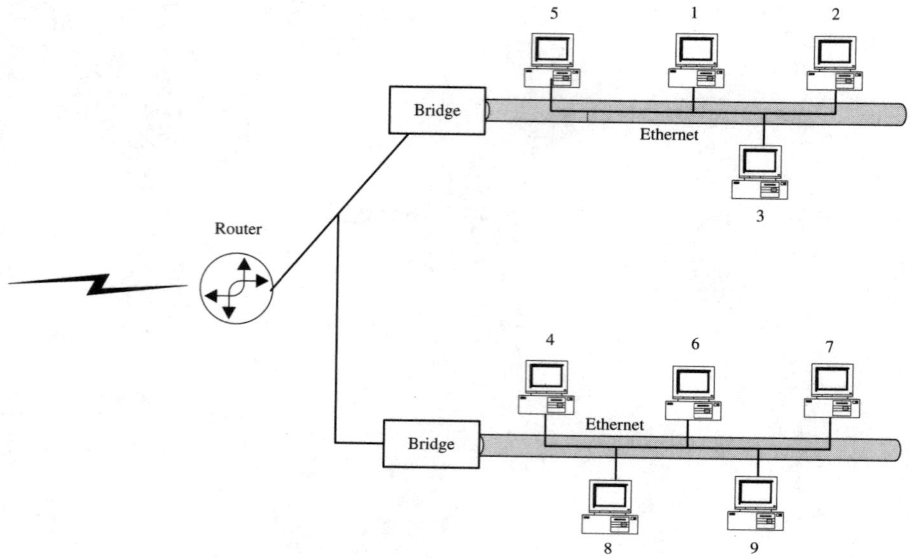

Figure 36-9
Two VLANs

supported SNMPv2 for several years, there has always been contention among the vendors and a significant amount of noncompliance. The noncompliant vendors are opting for their own custom MIB.

SNMPv3

Part of this contention is due to shortcomings of SNMPv2 that require fixing. There are also some additions that are needed to support higher-speed networks. SNMPv3 adds a GetBulk command, a better Set command, a unique ID for each SNMP agent, and 64-bit counters to accommodate Gigabit Ethernet. Perhaps the most important new feature is the addition of real security in the SNMP packets. The GetBulk command essentially says "gimme the whole MIB." This is important because it reduces the polling traffic that was created by multiple GetNext commands. Although this will have a minimal impact on reducing the network management system traffic on a small network, it will be a major benefit to large networks.

Although MIB data can be fetched remotely, there is no way to remotely manage SNMP agents. SNMPv3 adds this feature, along with the capability to describe agents within agents (the problem identified in Figures 36-8 and 36-9). Namely, the problem is that switches and hubs behind other switches and routers are difficult (to impossible) to find. SNMPv3 provides the mechanism to do this *if* (and it is a big if) SNMPv3 is supported by all the vendors involved. The same problem exists here as with the previous versions of SNMP—standards compliance is not universal. Many vendors have opted for vendor-specific MIBs (shown in Figure 36-4) under 1.3.6.1.4.1. This means that the network management system must have a current complete copy of the MIB in order to be able to manage the device (agent). The *Desktop Management Task Force* (DTMF) is trying to standardize the various data types in a more useful form via the *Common Information Model* (CIM).

Although the forgoing seems critical of SNMPv1 and v2, one must remember the era in which they were born. The functionality and memory in managed devices (for example, hubs) were extremely limited. Today, with cheap memory and processing power in every device, many more capabilities can be added at little or no cost.

Security

The initial design of SNMP included a modicum of security by using the community field (see Figure 36-6) to identify devices belonging to our management community. This was satisfactory before the days of inexpensive and capable network sniffers. With today's technology, it is not difficult to collect SNMP packet header information and either steal the information or substitute the content. Stealing content allows the hacker to build a map of our network for later exploitation. Substituting the content of our Set commands could be disastrous. One hopes that there are no hackers on our internal network, but in today's world, one is never sure. Our paranoia turns to prudence if we are running interconnections of our network over the public Internet without the benefit of a *Virtual Private Network* (VPN). Before becoming too paranoid, one must remember that the hacker must have physical access to the network backbone, a sniffer, the time and will to hack, and the desire to do mischief. The probability of the confluence of all these attributes is relatively low in absolute or real terms.

Given our paranoia, if the hacker is able to collect frames and packets, he can collect the community information and therefore control any of our devices that are preconfigured for remote management.

SNMPv3 solves this problem by encrypting the packet content and by establishing an authentication mechanism. All this is possible today because memory and processing power are readily available in the managed devices.

The new RFCs 2274 and 2275 provide for a *User-based Security Model* (USM) and a *Views-based Access Control Model* (VACM). These RFCs set forth a scheme for today and provide for future expansion to include possibly public key authentication and directory integration.

The USM uses a distributed security control mechanism with a user name and password disseminated from a centralized system. The user list, or access list, is distributed (securely using encryption) to each managed element. The remote agent does enforcement. Thus, the network management system user must log into each managed agent. Access to each MIB element can be put under password control. The USM specifies authentication and encryption functions, while the VACM specifies the access control rules. Each managed device can therefore keep a log of accesses and by whom (just as a firewall or proxy server logs each access). Although perhaps not important for hubs, it is very important for sensitive devices such as routers, switches, and firewalls. Previous versions of SNMP could not create this important audit trail.

In addition to encrypting the packet contents (for example, using *Data Encryption Standard* [DES]), each packet contains a time stamp to synchronize the network management system with the managed agent. This prevents the man-in-the-middle from recording the queries and commands, analyzing them, substituting his content, and playing them back at a later time. The USM also specifies its own MIB so that the passwords and user names can be remotely and securely maintained. Thus, although enforcement of access is done by each MIB agent, the user name and password can be centrally maintained and distributed to each managed agent by using the specified encryption technique. This means that a network authentication server must exist just for SNMP.

In the final analysis, it is your choice whether or not to implement the security features. There are definite benefits and costs associated with installing and maintaining the authentication server and its database.

Java

Sun Microsystems is pushing Java as a mechanism for enhancing the flexibility of SNMP. As we have seen, SNMP defines MIB-1, MIB-2, and MIB-3. Even though we can capture data on higher-speed elements and control network components securely, if changes are needed to the SNMP agents, software must be loaded (typically, manually) in each managed device.

The Java concept is that each agent has memory and CPU cycles to spare. Why not have it as a Java virtual machine? The network management system would then download the new functionality to each or all devices.

SNMP was designed to manage elements like routers, switches, and hubs. Managing remote servers, desktop machines, or applications' processes was never considered. The introduction of a Java-based system permits the management of higher-layer functions while maintaining compatibility and coexistence with SNMP.

There are several problems to overcome before Java can become a popular network management system tool. The first is performance. As an interpretive language, it is processor intensive and slow. Solutions for this problem involve compiling the Java code on the fly as it is downloaded to the target machine. The code then runs in native mode. Sun's Hot Spot compiler and Microsoft's *Just in Time* (JIT) compiler are designed to do this.

The good part is that the network manager can easily maintain the same revision level of the agents in all devices throughout the network. Unfortunately, this threatens the market for vendor-specific network management system solutions. The best advice is to stay tuned as vendors and standards organizations try to solve the problems of complex networks of today and the even more complex networks of tomorrow. History has shown that during the early stages of development and deployment, there are multiple solutions from a variety of vendors—all incompatible with one another. As the technology (and markets) mature, the industry hones in on the essential features and functions that are accepted and supported industry-wide. These then become the core functionality on top of which each vendor builds his proprietary extensions.

INDEX

Symbols

A

G

H

I

J

K

L

N

Q

R

S

T

U